USING
AutoCAD® 2000

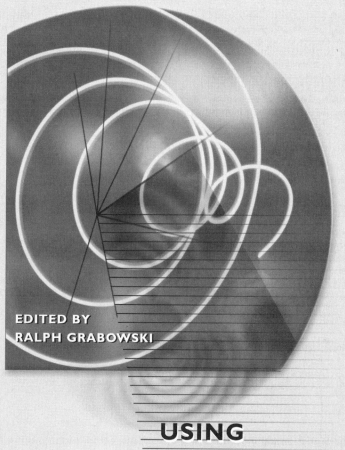

**EDITED BY
RALPH GRABOWSKI**

USING
AutoCAD® 2000

Autodesk.

Press

Thomson Learning™

Africa • Australia • Canada • Denmark • Japan • Mexico • New Zealand • Philipines • Puerto Rico •
Singapore • Spain • United Kingdom • United States

NOTICE TO THE READER

Publisher does not warrant or guarantee any of the products described herein or perform any independent analysis in connection with any of the product information contained herein. Publisher does not assume, and expressly disclaims, any obligation to obtain and include information other than that provided to it by the manufacturer.

The reader is expressly warned to consider and adopt all safety precautions that might be indicated by the activities herein and to avoid all potential hazards. By following the instructions contained herein, the reader willingly assumes all risks in connection with such instructions.

The publisher makes no representation or warranties of any kind, including but not limited to, the warranties of fitness for particular purpose or merchantability, nor are any such representations implied with respect to the material set forth herein, and the publisher takes no responsibility with respect to such material. The publisher shall not be liable for any special, consequential, or exemplary damages resulting, in whole or part, from the readers' use of, or reliance upon, this material. Autodesk does not guarantee the performance of the software and Autodesk assumes no responsibility or liability for the performance of the software or for errors in this manual.

Trademarks

AutoCAD ® and the AutoCAD logo are registered trademarks of Autodesk, Inc. Windows is a trademark of the Microsoft Corporation. All other product names are acknowledged as trademarks of their respective owners.

Autodesk Press Staff

Executive Director: Alar Elken
Executive Editor: Sandy Clark

Developmental Editor: John Fisher
Editorial Assistant: Allyson Powell
Executive Marketing Manager: Maura Theriault
Executive Production Manager: Mary Ellen Black
Production Coordinator: Jennifer Gaines
Art and Design Coordinator: Mary Beth Vought
Marketing Coordinator: Paula Collins
Technology Project Manager: Tom Smith

Cover illustration by Brucie Rosch

COPYRIGHT © 2000 Thomson Learning™.

Printed in the United States of America
    3 4 5 6 7 8 9 10 XXX 04 03 02 01 00

For more information, contact
Autodesk Press
3 Columbia Circle, Box 15-015
Albany, New York USA 12212-15015;
or find us on the World Wide Web at http://www.autodeskpress.com

Library of Congress Cataloging-in-Publication Data
Using AutoCAD 2000 / edited by Ralph Grabowski.
    p.   cm.
  ISBN 0-7668-1236-7 (pbk.)
  1. Computer graphics.   2. AutoCAD.   I. Grabowski, Ralph.
T385.U855   2000
604.2'0285'5369—dc21

99-32667
                                                    CIP

BRIEF CONTENTS

CONTENTS

CHAPTER 7—CREATING THE DRAWING

CHAPTER 11—WORKING WITH LAYERS

CHAPTER 12—INTRODUCTION TO DIMENSIONING

CHAPTER 16—CONSTRUCTING SECTIONAL AND PATTERNED DRAWINGS

CHAPTER 17—USING INQUIRY COMMANDS

CHAPTER 18—LEARNING INTERMEDIATE DRAW COMMANDS

CHAPTER 19—APPLYING INTERMEDIATE EDIT COMMANDS

INTRODUCTION

With more than 2,200,000 users around the world, AutoCAD offers engineers, architects, drafters, interior designers, and many others a fast, accurate, extremely versatile drawing tool. Welcome to the first book about how to use AutoCAD and make it the productivity tool for you!

Now in its 10th edition, Using AutoCAD 2000 makes using AutoCAD a snap by presenting you with easy-to-master, step-by-step tutorials through all the commands of AutoCAD. Using AutoCAD is designed to lead the novice through the basics of AutoCAD to more advanced features like customizing, 3D, and AutoLISP.

WHAT'S NEW IN THIS EDITION

- Reordered chapters and topics for easier access to key topics and exercises.

- New illustrations! Clearly display what you will see on the screen as you perform each command.

- Fully updated to Release 2000 for Windows 95/98 and NT. New sections added include:

Realtime Object Snap and new object snap modes

Polar Mode

Add-A-Plotter Wizard

MDE (multiple drawing environment)

AutoCAD DesignCenter

Properties Window

Hyperlinking

Summary Info

Layouts

QDim

QSelect

Lineweights

3dOrbit

Solids Editing

OLE Object Properties

Non-rectangular Viewports

CONVENTIONS

Command lines are indented: User response is boldfaced (Author instructions are italicized within parentheses.)

Key are displayed in small caps when instructed to use:

ENTER	BACKSPACE	F1
CONTROL	ESCAPE	F2

FEATURES

PULL-DOWN MENUS

Pull-down menus throughout the text offer an easy way for users to follow their progress from the menu bar.

NOTES

 Note: These notes highlight programming and user hints on working effectively with the AutoCAD commands.

TUTORIALS

Most chapters contain a tutorial that is outlined by a box. These tutorials serve to rein force groups of topics learned and range in complexity from brief to chapter-long.

An icon indicates if you are to use a file found on the accompanying CD-ROM.

2000 ICON

This icon indicates features that are new in AutoCAD 2000.

EXERCISES AND CHAPTER REVIEW

All exercises are pulled to the back of each chapter for easy access after you have read the concepts of the chapter. Review questions test key chapter concepts.

ONLINE COMPANION™

The Online Companion™ is your link to AutoCAD on the Internet. We've compiled supporting resources with links to a variety of sites. Not only can you find out about training and education, industry sites, and the online community, we also point to valuable archives compiled for AutoCAD users from various Web sites. In addition, there is information of special interest to users of Using AutoCAD. These include updates, information about the author, and a page where you can send us your comments. You can find the Online Companion at:

http://www.autodeskpress.com/onlinecompanion.html

When you reach the Online Companion page, click on the title Using AutoCAD.

E.RESOURCE™

This is an educational resource that creates a truly electronic classroom. It is a CD-ROM containing tools and instructional resources that enrich your classroom and make your preparation time shorter. The elements of e.resource link directly to the text and tie together to provide a unified instructional system. With *e.resource* you can spend your time teaching, not preparing to teach.

Features contained in *e.resource* include:

- **Syllabus:** Lesson plans created by chapter. You have the option of using these lesson plans with your own course information.

- **Chapter Hints:** Objectives and teaching hints that provide the basis for a lecture outline that helps you to present concepts and material.

- **Answers to Review Questions:** These solutions enable you to grade and evaluate end-of-chapter tests.

- **PowerPoint® Presentation:** These slides provide the basis for a lecture outline that helps you to present concepts and material. Key points and concepts can be graphically highlighted for student retention.

- **World Class Test Computerized Test Bank:** Over 800 questions of varying levels of difficulty are provided in true/false and multiple-choice formats so you can assess student comprehension.

- **AVI Files:** AVI files, listed by topic, allow you to view a quick video illustrating and explaining key concepts.

- **DWG Files:** This is a list of .dwg files that match many of the figures in the textbook. These files can be used to stylize the PowerPoint® Presentations.

WE WANT TO HEAR FROM YOU!

Many of the changes to the look and feel of this new edition came by way of requests from and reviews done by users of our previous editions. We'd like to hear from you as well! If you have any questions or comments, please contact:

> The CADD Team, c/o Autodesk Press
> 3 Columbia Circle, PO Box 15015, Albany, NY 12212

> or visit our web site at **http://www.autodeskpress.com**

ABOUT THE EDITOR

Ralph Grabowski, is a contributing editor to *CADENCE* magazine, and former Senior Editor at *CADalyst* magazine. Mr. Grabowski has written about AutoCAD since 1985, and is the author of over three dozen books on computer-aided design. He now publishes the *upFront.eZine* newsletter, the weekly email newsletter on CAD and the internet. His email address is ralphg@xyzpress.com.

ACKNOWLEDGMENTS

We would like to thank and acknowledge the many professionals who reviewed the manuscript to help us publish this AutoCAD text. A special acknowledgment is due the following instructors, who reviewed the chapters in detail:

> Adrian Dwarka, Red River College, Winnipeg, Manitoba
> Lynn Gurnett, York County Technical College, York, ME
> Jerry Hayes, Texas State Technical College, Abilene, TX
> Dennis Jorgensen, Bakersfield College, Bakersfield, CA
> Robert McMicken, Dixie College, St. George, UT
> James Overton, Southeast College of Technology, Memphis, TN

> Technical Editor: John Sprung, Kwantlen College, Surrey BC.

CHAPTER 1

A Quick Start in AutoCAD

As a new user, you need an opportunity to experience the "feel" of the AutoCAD program. The concept and operation of graphic design software can be unique to a first-time user. This "quick start" chapter gets your feet wet and introduces you to some of the features discussed in detail in other chapters. After completing this chapter, you will be able to:

- Start and exit AutoCAD

- Create a new drawing

- Place objects in the drawing

- Experience the basic 3D capabilities of AutoCAD

INTRODUCTION TO USING AUTOCAD

Welcome to *Using AutoCAD*! This chapter is specifically designed to acquaint you with the AutoCAD drawing program from the start. Subsequent chapters cover the subject more thoroughly. Let's get started!

STARTING AUTOCAD

Before you start AutoCAD 2000 for the first time, you must have Microsoft® Windows® 95, Windows 98, or Windows NT™ loaded and running on your computer.

1. Double-click (press the left mouse button twice) on the icon labeled **AutoCAD 2000**. Alternatively, click the **Start** button on the window task bar. Choose **Programs**, choose **AutoCAD 2000**. Notice that your computer now loads the AutoCAD program. After an opening screen is displayed (called the *splash screen*), you see AutoCAD and the **Startup** dialog box.

2. For now, choose the button labeled **Start From Scratch**.

AutoCAD 2000

Figure 1.1 *Launching AutoCAD from the Desktop*   Figure 1.2 *Launching AutoCAD from the Start Button*

3. Choose **OK**, and the dialog box disappears. You now see the drawing window. This window should look similar to Figure 1–4. AutoCAD has two other windows that you will see from time to time. One is called the **Text** window because it displays only text; the second is called the **Render** window because it displays 3D renderings.

DRAWING IN AUTOCAD

AutoCAD draws or edits objects by using *commands*. Commands are words such as line, circle, and erase that describe the object to be drawn or the operation you wish to perform.

At the bottom of the drawing editor is an area known as the *command prompt* area (refer to Figure 1.3). This area lists the commands you have entered. You should see the word **Command**: on that line now. You can specify commands from menus, toolbar icons, or by keying (typing) them from the keyboard.

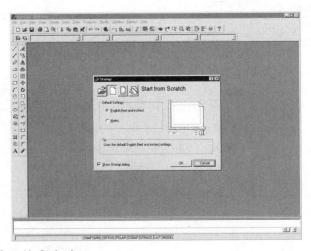

Figure 1.3 *The Start Up Dialog box*

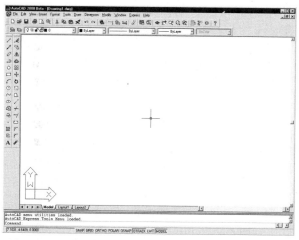

Figure 1.4 *AutoCAD Drawing Editor*

If you wish to key in any command from the keyboard, you must first have a "clear" command line. The current line must display the **Command**: prompt, without anything after it. If there is other text, just clear it by pressing ESC from the keyboard before typing in the new command.

In addition to commands, you use a pointing device to control AutoCAD. This device is usually a mouse or a *digitizing tablet*. From this point forward, we refer to the pointing device as the mouse, since that is the most commonly used input device.

1. Move the mouse and notice how the crosshairs cursor moves around the screen. The intersection of the crosshairs is used to specify points on your drawing.

USING DROP-DOWN MENUS

Like all other Windows software, a common method for specifying commands via the *drop-down menu*. Drop-down menus contain commands that can be selected with the mouse. To see how it works, let's enter a command from the drop-down menu. We'll choose the **Line** command from the **Draw** menu.

2. Move the crosshairs cursor above the top of the screen on the menu bar that extends the width of the drawing area.

3. As you move the arrow across each word in the menu bar, it is *highlighted*.

4. Choose the word **Draw** (click the left mouse button when **Draw** is highlighted). Notice that a drop-down menu extends downward into the drawing area (Figure 1.5).

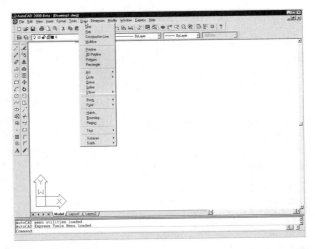

Figure 1.5 *Accessing Drop-Down Menus*

5. Move the pointer down to highlight the word **Line**, then click the left mouse button. Notice that the words **_line Specify first point:** appear on the command line at the bottom of the screen. This is called a *prompt*. On the command line, AutoCAD always tells you what it expects from you. Here, AutoCAD is asking you for the point from which the line starts (the first point).

6. Move the crosshairs cursor into the screen area and enter a point (click the left mouse button) approximately at the location shown in Figure 1.7.

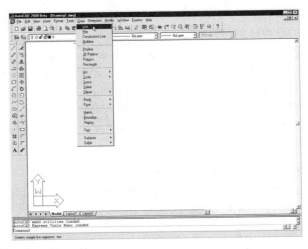

Figure 1.6 *The Draw Drop-Down Menu*

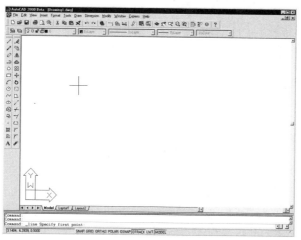

Figure 1.7 *Drawing a Line*

7. Continue to move the crosshairs cursor around the screen. Notice how the line "sticks" to the intersection of the crosshairs; the line stretches and follows your movement. This is called *rubber banding*.

8. You are now prompted, **Specify next point or [Undo]:**. Move the crosshairs and enter a point at approximately the point shown in Figure 1.8. (The **Undo** option "undraws" the previous line segment; it is useful when you make a mistake.)

9. Continue to enter lines as shown in Figure 1.9. The prompt changes to **Specify next point or [Close/Undo]:**. The **Close** option creates a polygon by drawing a line segment from the end of the last segment to the start of the first segment.

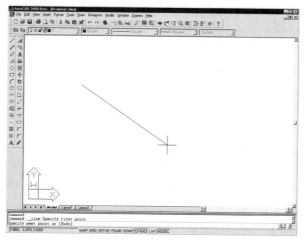

Figure 1.8 *Designating the Endpoint of the Line Segment*

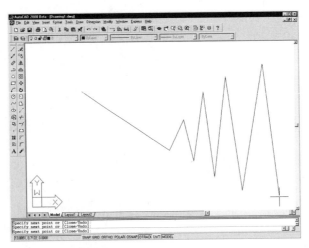

Figure 1.9 *Drawing Multiple Line Segments*

10. The **Line** command remains active so you can draw as many line segments as you desire without having to reselect the command repeatedly. After you enter the last point, click the right mouse button and choose **Cancel** to end the **Line** command. Alternately you can press ESC from the keyboard to stop any command. Some commands terminate automatically, while others must be deliberately terminated when you are finished.

BLIP MARKS

If you look closely at the endpoints of the lines, you might see some small crosses. These are called *blip marks*. These temporary blips are not part of your drawing. They are displayed for reference purposes at points you have entered.

Let's remove the blips from your drawing. In the menu bar, pull down the **View** menu. Choose the **Redraw** command. AutoCAD redraws the screen and removes any blip marks.

USING TOOLBARS AND ICONS

AutoCAD places a strong emphasis on toolbars with icon buttons. AutoCAD has more than two dozen toolbars. In a default installation, AutoCAD normally displays just four toolbars as follows:

- The **Standard** toolbar (top of the window) contains many of the same commands as found in other Windows applications.

- The **Object Properties** toolbar (just below the Standard toolbar) allows you to change the properties of objects.

- The **Draw** toolbar (upright, at extreme left) is used to draw objects, such as lines, arcs, and hatch patterns.

- The **Modify** toolbar (just to the right) is used to change object characteristics such as size.

These four hold most of the everyday commands you use with AutoCAD. Each toolbar can be floating or docked. If you were to open them all, AutoCAD's 27 toolbars would obscure much of the AutoCAD window!

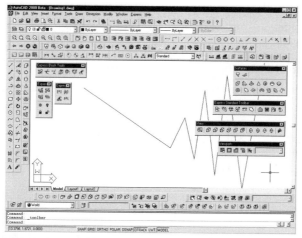

Figure 1.10 *The Result of Opening Every Toolbar*

Using a Toolbar Button

Every button on a toolbar holds a command. For example, the first button of the **Draw** toolbar is the **Line** command for drawing line segments. The symbol of the line with two dots at either end is called an icon for the **Line** command.

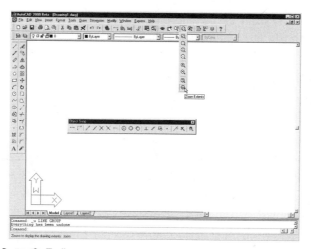

Figure 1.11 *The Parts of a Toolbar*

When you cannot remember the purpose of a button, move the cursor over the button and wait for a second or two. AutoCAD displays a *tooltip*, a one- or two-word description of the button's purpose. On the status line (at the bottom of the window), AutoCAD displays a one-sentence description of the command's purpose. In the case of the Line command, the status line reads, **Creates straight line segments: line**.

To draw a line with the icon button

1. Position the cursor over the **Line** button, and click the left mouse button. At the command line, you should see AutoCAD prompting you, as follows:

   Command: _line Specify first point:

2. You may now draw some lines or press ESC to exit the **Line** command.

Accessing Flyouts

Some buttons represent a set of options rather than a single command. You recognize these by the small black triangle in the corner of the button. The triangle indicates the presence of a *flyout*. The flyout is a group of buttons that "flies out" from a single button.

To access a flyout

1. Position the cursor over any button that has the tiny triangle on it, such as the **Zoom** button.

2. Hold down the left mouse button. AutoCAD displays the flyout buttons.

3. Without letting go of the mouse button, move the cursor over the buttons on the flyout. As the cursor passes over a button, it changes slightly to give the illusion of being depressed.

4. When you reach the button you want to use, release the mouse button.

5. AutoCAD starts the command associated with the button. In addition, the button you accessed moves to the top and appears on the toolbar where it is quick and easy to access if you need that same command again next.

Manipulating the Toolbar

The toolbar has many controls hidden in it. You can move, resize, dock, and dismiss toolbar. Here's how:

1. To move a docked toolbar, position the cursor over the two "bars" at the left end of a horizontal toolbar (or at the top of vertical toolbars). Hold down the mouse button, and *drag* the toolbar away from the edge of the window.

2. To move a floating toolbar, position the cursor over the title bar. Hold down the mouse button and drag the toolbar to another location.

3. To *dock* the toolbar means to move it to the side or top of the drawing area, such as the two docked toolbars at the top of the window in Figure 1.10. To

dock the toolbar, move it all the way against one of the four sides of the drawing area. To *float* the toolbar, grab it by the edge and drag it away from the edge of the drawing area.

4. To resize (or stretch) the toolbar, position the cursor over one of the four edges of the toolbar. Press the left mouse button, then drag the toolbox to a new shape.

5. To *dismiss* (get rid of) the floating toolbar, click the tiny X in the upper corner. The toolbar disappears.

6. To get back the toolbar, type the **Toolbar** command and select the toolbar's name from the list. Alternatively, right-click any toolbar button and select a toolbar from the list.

EXITING A DRAWING

How you exit your drawing depends on the command you use. The following four sections outline the possibilities. Choose the one you want and follow the instructions.

Discard the Drawing and Quit AutoCAD

If you don't want to keep your drawing, and you wish to stop work now, enter the **Quit** or **Exit** command (you can do this from the keyboard) or choose **Exit** from the **File** menu. AutoCAD displays a dialog box that asks you to confirm your choice (see Figure 1.12). Move the cursor to and choose the **No** box to exit without saving your work. AutoCAD will not record your work to disk and will return you to the Windows desktop.

Save the Drawing and Quit AutoCAD

If you want to save your work to disk and quit AutoCAD, enter the same **Quit** command. This time choose **Yes** to save your work, and AutoCAD displays the **Save Drawing As** dialog box. Type "MYWORK" for the file name and choose **Save**. AutoCAD saves your work under the name "MYWORK" and returns you to the Windows desktop.

Save Your Work and Remain in AutoCAD

If you want to keep your work and remain in AutoCAD, enter the **SaveAs** command. The same **Save Drawing As** dialog box is displayed on the screen. Name the file, choose **Save**, and you will be returned to your drawing in AutoCAD.

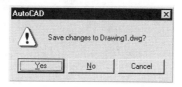

Figure 1.12 *Drawing Modification Dialog Box*

Figure 1.13 *Save Drawing As Dialog Box*

Remain in AutoCAD and Start a New Drawing

If you would like to start a new AutoCAD drawing, use the **New** command. AutoCAD displays the **Create New Drawing** dialog box, which looks exactly like the **Startup** dialog box of Figure 1.13.

Now let's practice a little bit using AutoCAD to get a "feel" for the program.

TUTORIAL

Figure 1.14 shows a rendering of the drawing you will construct. It consists of a base with four holes and a shaft with rounded edges.

Getting Started

The following is a listing of items to be typed and entered to complete the drawing. The items you type appear at the bottom of the screen on the command line. The word **Command:** appears at this location now. Your response is shown by **boldface** in this tutorial. You can type the items in either uppercase or lowercase. Be sure to press ENTER after each response. Items enclosed in parentheses ( ) are instructions and are not typed. If ENTER is shown, press ENTER on the keyboard.

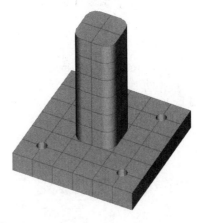

Figure 1.14 *Finished Drawing*

If you mess up, just type **U** and then press ENTER. This undoes the previous step. You may use it several times to undo each step in reverse order.

To enter a new command, the prompt on the command line must say **Command:** If it does not, press ESC. This cancels the current command and places AutoCAD ready for the next command.

1. From the **File menu**, choose **New**. AutoCAD displays the **Create New Drawing** dialog box.

Figure 1.15 *Create New Drawing Dialog Box*

2. Choose the **Use a Wizard** button.

3. From the **Select a Wizard** list, select **Quick Setup**, and then choose the **OK** button. AutoCAD displays the **Quick Setup** dialog box.

4. **Quick Setup** presents a series of setup screens. In the first screen you specify the units. Accept the default of **Decimal** units by choosing the **Next** button. AutoCAD displays the **Area** screen.

5. In the **Area** screen, define the width of the area by clicking the **Length** box. Clear the current value using the DEL or BACKSPACE key. Then type **10** and press TAB to see the effect in the preview image.

6. Choose **Finish**. AutoCAD dismisses the dialog box.

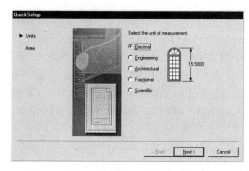

Figure 1.16 *Quick Setup Dialog Box*

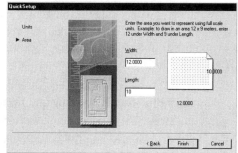

Figure 1.17 *The Area Setting of the Quick Setup Dialog Box*

Let's turn on some drawing aids, such as the snap and grid.

- The *grid* is a visual guide that helps to show you distances. Think of graph paper without the lines, just dots at the intersection of lines. The grid is displayed and not printed. For example, when the grid is set to 1 inch, the drawing is covered with dots spaced 1 inch apart.

- The *snap* is like setting a drawing resolution. When on, snap causes the cursor to move in increments. For example, when the snap is set to 1 inch, the cursor moves in 1-inch increments. The grid and snap can have the same or a different spacing.

1. From the menu bar, choose **Tools | Drafting Settings**. That notation is shorthand for two steps: (1) choose **Tools** from the menu bar and then (2) choose **Drafting Settings** from the **Tools** menu. AutoCAD displays the **Drafting Settings** dialog box.

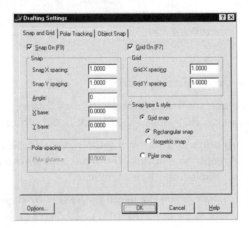

Figure 1.18 *Drafting Settings Dialog Box*

2. Turn on snap by clicking the check box at the **Snap On** option. A check mark appears in the box to indicate snap mode has been turned on. The **F9** next to **Snap On** is a reminder that you can press the **F9** function key to turn snap on and off. Alternatively, click the word **SNAP** near the bottom of the screen. The word changes from an "out" button to an "in" button, indicating whether snap is turned off (out position) or on (pushed in).

3. If the **Snap X spacing** and **Snap Y spacing** are not **1.0000**, change them to that value now.

4. Do the same for **Grid**: turn it on and set the grid spacing to 1.0000.

5. Choose **OK** to dismiss the dialog box. Notice that a grid of fine dots appears in the drawing.

6. Move the mouse to see the crosshairs cursor move in increments of 1.

Drawing the Baseplate

The drawing contains a baseplate. You draw it first. To start, you are looking down in plan and in 2D view; you change into a 3D view later. Start the **Box** command:

>   Command: **Box** ENTER
>
>   Specify corner of box or [CEnter] <0,0,0>: **3,2**
>
>   Specify corner or [Cube/Length]: **9,8**
>
>   Specify height: **1**

Your drawing should look similar to Figure 1.19.

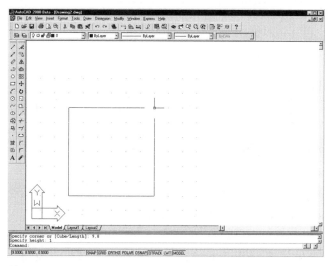

Figure 1.19 *Baseplate*

Drawing the Shaft

Now you draw the shaft by using the **Box** command again. The shortcut to repeating a command (without typing the command name a second time) is to press the spacebar.

>   Command: *Press* SPACEBAR
>
>   BOX Specify corner of box or [CEnter] <0,0,0>: **5,4,1**
>
>   Specify corner or [Cube/Length]: **7,6,1**
>
>   Specify height: **6**

Your drawing should look similar to Figure 1.20.

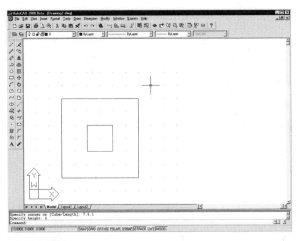

Figure 1.20 *Baseplate and Shaft*

Creating the Holes

You now draw the holes. A hole is created in two steps: drawing a cylinder, then subtracting the cylinder from the box.

> Command: **Cylinder**
>
> Current wire frame density: **ISOLINES=4**
>
> Specify center point for base of cylinder or [Elliptical] <0,0,0>: **4,3**
>
> Specify radius for base of cylinder or [Diameter]: **.25**
>
> Specify height of cylinder or [Center of other end]: **I**

You copy the hole to the other parts of the baseplate. Make use of an option of the **Copy** command called **Multiple**. This allows you to make multiple copies more easily. AutoCAD shows a command's options in square brackets, such as **[Center of other end]**, as seen above.

Copy the last-drawn object using the L (short for last) option:

> Command: **Copy**
>
> Select objects: **L**
>
> I found
>
> Select objects: *Press* ENTER
>
> Specify base point or displacement, or [Multiple]: **M**
>
> Specify base point: **4,3**

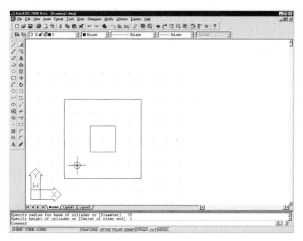

Figure 1.21 *Drawing the first Circle*

Specify second point of displacement or <use first point as displacement>: **8,3**

Specify second point of displacement or <use first point as displacement>: **8,7**

Specify second point of displacement or <use first point as displacement>: **4,7**

Specify second point of displacement or <use first point as displacement>: *Press* ENTER

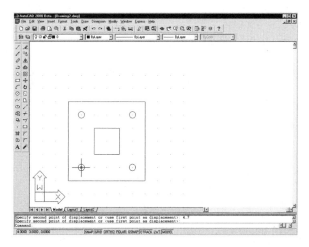

Figure 1.22 *Copying the Circles*

You have drawn four cylinders that represent the holes, but they are solid cylinders. You need to remove the cylinders from the baseplate so that they become holes. You will now use the **Subtract** command to remove the cylinders from the baseplate.

> Command: **Subtract**
>
> Select solids and regions to subtract from ..
>
> Select objects: *(select the baseplate)*
>
> 1found objects: *Press* ENTER
>
> Select solids and regions to subtract...
>
> Select objects: **fence**
>
> First fence point: *(place cursor and click at center of circle)*
>
> Specify endpoint of line or [Undo]: *(click at center of second circle)*
>
> Specify endpoint of line or [Undo]: *(click at center of next circle)*
>
> Specify endpoint of line or [Undo]: *(click at center of last circle)*
>
> Specify endpoint of line or [Undo]: *Press* ENTER
>
> 4 found
>
> Select objects: *Press* ENTER

The baseplate won't look any different after the cylinders are subtracted from the base plate until you remove the hidden lines at the end of this tutorial.

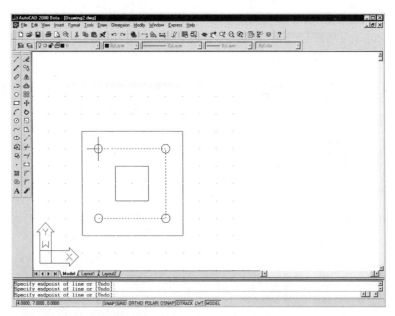

Figure 1.23 *Subtracting Cylinders from Baseplate*

Viewing the Drawing in 3D

Let's have some fun and view the drawing in 3D. (You learn more about these capabilities in Chapter 30.)

> Command: **Vpoint**
>
> Current view direction: **VIEWDIR=0.0000,0.0000,1.0000**
>
> Specify a view point or [Rotate] <display compass and tripod>: **.5,-1,1**
>
> Regenerating model.

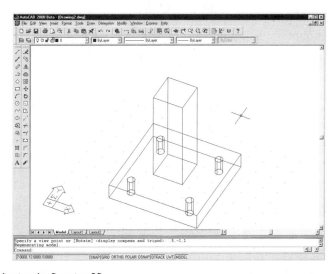

Figure 1.24 *Viewing the Drawing 3D*

Filleting the Corners

To round the corners of the shaft, use the **Fillet** command.

> Command: **Fillet**
>
> Current settings: **Mode = TRIM, Radius = 0.5000**
>
> Select first object or [Polyline/Radius/Trim]: *(select one vertical edge of the shaft)*
>
> Enter radius <0.5000>: *Press* ENTER
>
> Select an edge or [Chain/Radius]: *(select another vertical edge of the shaft)*
>
> Select an edge or [Chain/Radius]: *(select another vertical edge)*
>
> Select an edge or [Chain/Radius]: *(select last vertical edge)*

Select an edge or [Chain/Radius]: *Press* ENTER

4 edge(s) selected for fillet.

After a few seconds, AutoCAD fillets the four corners of the shaft.

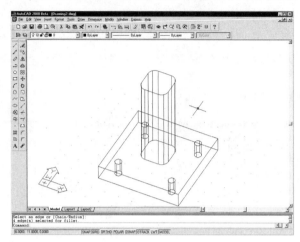

Figure 1.25 *Filleting the Shaft*

Removing Hidden Lines

You are seeing the 3D object in wireframe view. Remove the hidden lines. Removing the hidden lines takes a small amount of time.

Command: **Hide**

Regenerating model.

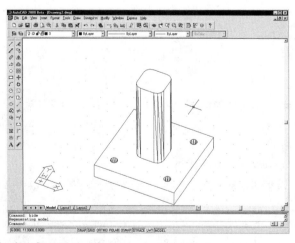

Figure 1.26 *Hidden Line Drawing*

To return to the plan view, type the **Plan** command and press ENTER twice. Save or discard your drawing in the manner you learned earlier in this chapter.

SUMMARY

- AutoCAD drawings are constructed with drawing and editing commands.

- Commands are entered from the keyboard, drop-down menus, toolbars, and from a digitizer template (you learn about this later in another chapter).

- The command line shows your command activity and displays prompts that tell you what input AutoCAD expects from you. You can use either uppercase or lowercase when entering a command from the keyboard.

- The command line must be "clear" before typing a new command from the keyboard.

- The **Line** command is used to draw lines. Use the **Redraw** command to clear the temporary blip marks from your drawing.

- Cancel a command by press the ESC key.

USING AUTOCAD

So, that was a quick start to using AutoCAD. Welcome to the rest of *Using AutoCAD*. For you to start in the right direction while working through this text, it's best to know what lies ahead.

- Chapter 2 gives you an overview of the computer hardware used by typical CAD systems.

- Chapter 3 introduces you to the Windows operating system.

- Chapter 4 begins to explain in more detail some of the concepts you learned in this chapter.

From there, the remaining chapters take you step by step through the program. Before long you'll be using AutoCAD like a pro—enjoy!

 NOTE: You can move chapter by chapter, use the table of contents, or use the index to find a specific topic. Each chapter is page-numbered separately.

CHAPTER 2

Hardware of an AutoCAD System

To operate your AutoCAD program best, it is helpful to have at least a basic understanding of your computer and its operating system. This chapter provides an overview of computer operation. The next chapter discusses the Windows operating system, covering the basic operations you need to compute effectively. After completing this chapter, you will be able to

- Identify the basic parts of the personal computer

- Understand the purpose of peripheral equipment

- Demonstrate the proper care and operation of storage media

THE COMPUTER

The computer is the central part of the CAD system. The *peripheral* equipment is connected to the computer, as discussed later in this chapter.

Let's look at the categories of computers. Computer systems are divided into five main categories. These are

- Palm computers

- Notebook computers

- Desktop computers

- Workstations

- Mainframe computers

Palm Computers. The smallest of computers, a palm computer fits in the palm of your hand. Typically, you input data by writing with a stylus on the computer's screen; handwriting recognition software translates your writing into text and graphics. Palm computers, such as the 3Com Palm, are not (yet) used for CAD; they are, however, used to collect data that can be used by CAD software, such as building dimensions and GPS (global positioning system) data for creating maps or base plans.

Figure 2.1 *A Palm Computer with Stylus*

Notebook Computers. The notebook computer is the smallest practical computer for running CAD. This computer is about the size of a notebook (hence the name), which makes it easily transportable. Notebook computers tend to be more expensive and somewhat less powerful than the fastest desktop computer; any of today's notebook computers, however, is powerful enough to run AutoCAD.

Desktop Computers. The desktop computer is the type you most commonly see on a desktop. This versatile machine is sometimes referred to as a "personal" computer because it is mostly designed for use by one person at a time. Today's personal computers are often connected to a network of many computers, which allows files and some programs to be used by more than one operator at a time.

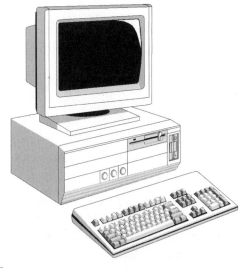

Figure 2.2 *A Personal Computer*

Personal computers usually consist of a case that contains the central processing unit, memory, one or more disk drives, and additional adapters, such as for sound and video. The display device, keyboard, and other input devices (such as mouse) are external to the case.

Workstations. Workstations are larger, faster, and more expensive than personal computers. They typically contain one or more faster processors, much more memory, and higher-resolution graphics. This class of computer is meant for running larger and more sophisticated programs than personal computers typically handle. For example, the special effects of Hollywood movies are created on workstations.

Mainframe Computers. This is the largest type of computer. Mainframes are capable of processing a large amount of data. Mainframes are used by government and companies who handle large amounts of data, such as credit card and other financial transactions.

COMPONENTS OF A COMPUTER

A computer is made up of several parts that are essential to its operation. In addition, components can be added that speed up and enhance the operation of the computer. Let's look at some of the components that make up a computer.

System Board

The system board (sometimes referred to as the *motherboard*) is an electronics board that holds most of the computer's chips and adapter boards. The CPU (central processing unit), memory chips, ROM (read-only memory), and other parts are mounted on this board.

In addition, the board contains expansion slots. These are the slots in which you mount add-on circuit boards such as video capture cards and drive adapters. There are four types of expansion slots:

- **ISA** (industry standard architecture): the oldest format of expansion slot. Most computers have some of these slots for compatibility with older adapter cards.

- **PCI** (peripheral connect interface): the most common slot in today's computers; it is found in PCs and Macintosh computers.

- **AGP** (advanced graphics port): this slot is designed for high-speed data transfer between the CPU and a graphics board.

- **PCMCIA** (personal computer memory card interface adapter) or PC Card: designed for notebook computers. The adapter card looks like a thick credit card, and might contain a modem, network interface, or even a small hard disk drive.

Figure 2.3 *An adapter card. This card fits an ISA expansion slot.*

Figure 2.4 *A PCMCIA (or PC Card) adapter card. This is a network card.*

Central Processing Unit

The central processing unit (or CPU) is the center of activity of the computer. It is here that the software is processed. After processing, instructions are sent out to the display, printer, plotter, or other peripheral. With few exceptions, all information passes through the central processing unit.

Physically, the central processor is a computer chip mounted on the system board. Today's high-speed CPUs are often packaged with *cache* memory, which helps speed up calculations. Because of the heat generated by the CPU, it is common to have a *heat sink* (looks like a series of black fins) and a small fan mounted on the chip.

There are several types of processing chips. Your computer's graphics board probably has its own CPU for processing graphical images. Your printer and plotter have

Figure 2.5 *A Pentium CPU*

their own CPUs for processing print data. Intel, AMD, Cyrix, IBM, and others manufacture CPUs. Today's Pentium line of processors is the most commonly used for AutoCAD.

Memory

Computer memory can be divided into two categories: ROM (read-only memory) and RAM (random access memory). ROM memory is contained on preprogrammed chips on the system and is used to store basic sets of command for the computer.

Random access memory (RAM) is the memory that is mostly referred to when computer memory is being discussed. RAM is used to temporarily store information in your computer. It is temporary storage because all data in RAM is lost when the computer is turned off.

Software programs (such as AutoCAD) have a minimum requirement for the amount of RAM necessary to run the program. This amount is given in kilobytes. A common reference might be "16 MB", meaning 16 megabytes. A megabyte is 1,024 kilobytes; a kilobyte is 1,024 bytes. You might see these numbers are rounded, such as 1,000 bytes in a kilobyte.

Disk Drives

Disk drives are used to store large amounts of data, such as the Windows operating system, applications programs, drawings, and other documents. A disk drive is either non-removable (or fixed drives) or removable.

In almost all cases, a fixed drive is a hard drive. Today's hard drives hold 8 GB (gigabyte) or more of data. A gigabyte is one thousand megabytes.

Removable drives include diskettes, CD-ROMs, large capacity "diskettes" such as Iomega's Zip and Jaz drives, and even tape drives.

Figure 2.6 *3 1/2" Diskette*

Diskettes are used to hold small amounts of data since their capacity is typically limited to 1.44 MB. Some computers support diskettes that hold 2.88 MB of data.

CD-ROM drives (short for compact disc, read-only memory) are the most common method for distributing software today. Your copy of AutoCAD probably arrived on a CD-ROM. This disc can be read, but not written to. That makes the data secure from accidental erasure and attack from computer viruses. More recently, CD-ROM discs have become available that can be written to. These allow you to store up to 650 MB of data.

The Zip drive is a popular way to store and exchange large files, such as AutoCAD drawings and desktop publishing documents. The Zip disk looks like a fat diskette and stores either 100 MB or 250 MB of data. The Jaz drive is a removable hard drive, storing 1 GB or 2 GB of data on a cartridge. Jaz drives are often used for data backup.

Tape drives were the original backup medium, because they were relatively cheap. They suffered, however, from slow speed. Today's tape drives can store 4GB or more data on a single tape similar to that used by 8mm video cameras.

Disk Care

All of your work is recorded to disk, so caring for the disk is very important. If a disk is damaged, you might lose your work! Frequent back-ups (copying files to a second disk, tape, or recordable CD-ROM) and proper handling of disks minimizes the possibility of file loss.

Hard disks are installed inside the computer and are not handled. This eliminates the danger of improper handling, but does not prevent damage. A hard drive can be dam-

Figure 2.7 *A 2GB Jaz disk (at left) and a 100MB Zip disk (at right).*

aged by shock. If you move the computer, be sure the power is off and move it gently. Today's hard drives have *self-parking heads* that, upon shutdown, move the head to a sector that does not have data stored on it. When the hard drive is on, do not move or tilt the computer. This is particularly important with notebook computers.

Floppy disks are subject to damage from rough handling. The 3 1/2" diskettes and Zip disks are protected by a sliding door. The purpose of this door is defeated if you open it to look or touch the magnetic material. Touching the disk surface through the opening leaves oil from the skin on the disk, possibly making it unreadable.

Dust and smoke can also leave particles on the disk surface that can prevent the drive from reading the disk properly. Heat and cold can cause the disk material to expand or contract, causing problems. Magnets scramble the data on the disk's tracks. Spilling a liquid onto a disk leaves a residue; whether the liquid is hot or cold, it can cause temperature damage.

CD-ROMs are relatively sturdy forms of data storage. They too must be handled with care that protects the surface, from which the laser reads the data. Avoid touching the surface of the CD-ROM; handle it by its center and edges.

Write-Protecting Data

Normally, a disk can be read by the computer or written to (new data can be saved on it). If you do not want to be able to write to a disk to help preserve what is already

Switch in Write-enable position

Switch in Write-protect position

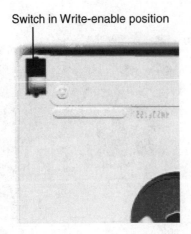

Figure 2.8 *Write-Protecting a 3 1/2" Diskette*

stored on the disk, you can write-protect the disk. This allows the disk to be read by the computer, but not written to.

The standard high density 3 1/2" disk has a small slide switch in one corner of the disk. Sliding the switch alternates between write-protecting and write-enabling.

PERIPHERAL HARDWARE

Peripheral hardware consists of add-on devices that perform specific functions. A properly equipped CAD station consists of several peripheral devices. Among these are

Plotter	Displays
Printers	Input devices
Scanners	

PLOTTERS

Plotters are used to produce a "hard copy" of your work. The three main types of plotters used most often are inkjet plotters and laser printers. Some offices might still have older pen plotters.

Inkjet Plotter

The most common plotter today is the inkjet plotter. The plotter works by sending an electrical current to the print head. The electricity heats up the ink, causing it to squirt out the end of a nozzle and hit the paper.

In a typical print head, there are 48 nozzles lined up vertically. Large-format inkjet plotters can have as many as 512 nozzles. For color plots, there are four heads: one each for cyan (light blue), magenta (pink), yellow, and black. By combining those four colors using a process called *dithering*, an inkjet plotter can produce color plots containing 256 or more colors. For higher quality output, some inkjet plotters have as many as seven color cartridges.

Although a typical resolution is about 360 dots per inch, some inkjet plotters are capable of 1440 dots per inch. Inkjet plotters can use normal paper, but specially coated paper results in brighter colors.

Laser Printer

Laser printers create prints similar to a photocopy. They use a laser-copy process that produces sharp, clear prints. Although some laser printers have excellent graphics capabilities, they are usually restricted to a maximum of B paper size (11" x 17"). Resolution is typically 600 dpi, but can be 1200 dpi or higher.

Pen Plotter

Pen plotters are the oldest way of producing a plot. The pen plotter uses technical ink pens to draw on vellum, Mylar, or other suitable surfaces. Some pen plotters can use other types of pens such as ballpoint, pencil, and marker-type pens.

The plotter is controlled by signals sent by the computer to the plotter. The pen movement results from the response to the signals by servo or stepper motors. These motors control both the pen movement and the up and down motion of the pen. The accuracy (resolution) of the plotter is determined by the increment by which the servomotors are capable of moving.

DISPLAY SYSTEMS

Displays are used to view the work while in progress. A display device is often referred to as a *monitor* or *CRT* (cathode ray tube). The quality of the image on the display is determined by its resolution. The resolution is controlled by the number of dots (pixels) contained on the screen. These pixels make up the image. A typical 1024 x 768-resolution display contains 1,024 pixels horizontal and 768 pixels vertical on the screen. Many professional CAD users prefer higher resolution displays such as 1280 x 1024 or higher.

To display higher resolution, the device driver, the operating system, and the graphics board contained in the computer must be capable of displaying the higher resolution. The monitor must be matched to the graphics board.

INPUT DEVICES

Just as a word processing program requires a keyboard to input the individual letters, numbers, and symbols, a CAD program requires an input device to create and manipulate drawing elements. Although many programs allow input from the keyboard arrow keys, an input device speeds up the drawing process. The most common input devices used for CAD drawing are digitizer, mouse, and scanner.

Digitizer

A digitizer is an electronic input device that transmits the absolute X and Y location of a cursor that rests on a sensitized pad. A digitizer may be used as a pointing device to move a point around the screen, or as a tracing device for copying drawings into the computer in scale and proper proportion. The points are located on the pad by means of a stylus (similar in appearance to a pencil), or a puck (alternately referred to as a cursor). Digitizers use a fine grid of wires sandwiched between glass layers. The cursor is then moved across the pad and the relative location is read and transmitted to the computer.

In "tablet" mode, the digitizing pad is calibrated to the actual absolute coordinates of the drawing. When used as a pointing device, the tablet is not calibrated.

Digitizers are available in several sizes. You can use small pads to digitize drawings larger than the pad surface by moving the drawing on the pad and recalibrating. However, this can be very annoying if you frequently work with large-scale drawings to be digitized.

Mouse

A mouse is an input device that is used for pointing and positioning the cursor. The name comes from its appearance. A mouse can only be used for command input; it cannot digitize. There are primarily two types: a mechanical mouse and an optical mouse.

A mechanical mouse has a ball under the housing. The rolling ball transmits its relative movement to the computer through a wire connecting the mouse to the computer.

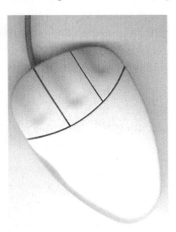

Figure 2.9 *A Three-Button Mouse*

An optical mouse uses optical technology to read lines from a special pad to sense relative movement. Although an optical mouse uses a pad, it is not capable of digitizing.

Today's mouse devices come with one button (Macintosh), two buttons (Microsoft), three or four buttons (Logitech). More buttons is better since you can use them for additional functions. For example, you can make the middle button of a three-button mouse act like a double-click, reducing the number of times you click the mouse button. Some mouse devices come with a roller wheel, which allows you to quickly zoom and pan with programs, like AutoCAD 2000.

Scanner

The scanner is another type of input device. It is used to "read" paper drawings and maps into the computer. The scanner works by shining a bright light at the paper. A head made of many CCDs (charge couple devices, the same technology used in a video camera) measures the amount of reflected light. A bright area is the paper, while a dark area is part of a line.

The scanner sends data to the computer, where it can be displayed as a raster image by AutoCAD or converted to a vector file by specialized software. Scanners are available in a variety of sizes, ranging from very small units that read business cards to E-size scanners that read large engineering drawings.

CHAPTER REVIEW

1. What are the five main categories of computer?
2. Name some types of peripheral hardware.
3. What does the term *resolution* mean?
4. How does a mouse differ from a digitizing pad as an input device?
5. What is the center of activity of the computer?
6. Explain the difference between RAM and ROM memory.
7. Compare and contrast laser and ink jet printers.
8. How do an optical mouse and a mechanical mouse differ?
9. Would a complete CAD system with necessary peripherals be appropriate for architectural as well as manufacturing applications? Why?
10. What space capacities are floppy disks usually available in?
11. What might cause the loss of a file?
12. What does the term *write-protect* mean? How is this accomplished?
13. What might happen to the information on a disk if it passes through a magnetic field?

CHAPTER 3

The Windows Operating System

The Windows operating system is the most basic level of computing. You must achieve a basic level of understanding of Windows to perform fundamental computing operations. After completing this chapter, you will be able to

- Understand Windows Explorer

- Demonstrate file management techniques using Windows commands

- Change directory paths

- Use the AutoCAD file dialog boxes

THE OPERATING SYSTEM

To use AutoCAD and other Windows applications software, your computer must have a program that translates program instructions so that the electronics in the computer can understand and act on them. That program is called the *operating system*, or OS for short. The operating system is originally supplied on floppy disk or CD-ROM and is often included in the purchase price of your computer. If the Windows operating system is installed on your hard drive, it loads automatically when you start your computer.

Windows can be thought of as an umbrella program under which all other programs are run. Windows must be loaded and operating first before you begin other programs. When you start your computer, it looks for certain files. If they are present, they are loaded. You might notice disk operating system messages and version numbers when your computer starts.

Windows names the disk drives with a letter and colon, such as A: and C:. In all Windows computers, drives A: and B: are disk drives for removable media, such as floppy disks. Drives C and above are usually for hard drives, network drives, and additional removable drives. For example, E: is often, but not always, the CD-ROM drive.

Network drives are the disk drives of other computers that you can access over a network. Your computer must be connected to the other computers through a network cable. By being networked, you and your co-workers can share drawings and other files. The networked version of AutoCAD allows a single copy to be used by two or more operators, depending on the terms of the network license.

Removable drives are any disk drives in which the medium can be removed. This includes, for example, floppy diskettes, CD-ROM discs, Zip disks, and tape cartridges. When reading or writing using a removable disk, you must be careful that the correct disk has been inserted.

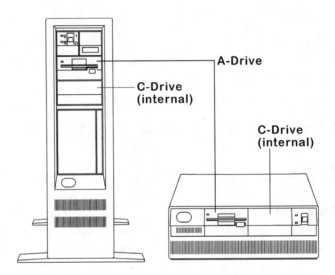

Figure 3.1 *Cabinet Designs*

The position of the drives in the computer can differ according to the computer cabinet design. Figure 3.1 shows a typical setup for different cabinet designs.

FILE NAMES AND EXTENSIONS

File names can have up to 255 letters and characters (some characters are exempted). Older versions of operating systems, such as Windows v3.1 and DOS, are limited to file names with eight characters. When a file name of more than eight characters is copied to a computer running the older OS, the file name is truncated (chopped off) to six characters and ~1 is added. For example, **houseplan.dwg** becomes **housep~1.dwg**.

A dot (.) followed by a three-letter code is usually added to the end of the file name; this is called the *file extension*. The file extension denotes the type of file. For example, an AutoCAD drawing file has a .DWG (or .dwg) extension. The correct notation is the file name, followed by a period, then the file extension. For example, an AutoCAD drawing named **widget** is stored by the OS as **widget.dwg**. Some programs add their own file extension code, while you must specify the extension in other programs.

Wild-Card Characters

When using file dialog boxes, some Windows functions, and certain AutoCAD commands, you can use "wild-card" characters to specify files. The two most common wild-card characters are the question mark (?) and the asterisk (*).

The question mark fills in for any single character. By using two or more question marks, you represent two or more characters. When you enter

CAR?.DWG

you would refer to files such as:

CAR1.DWG

CAR2.DWG

CART.DWG

CARD.DWG

The asterisk represents *all* characters. For example, when you enter

*.DWG

you refer to all the files with the .dwg file extension. Alternately, when you type

FLPLAN.*

you refer to all the drawing files named FLPLAN, regardless of their file extension (or even if no file extension exists).

DISPLAYING FILES

There are four ways to display and manipulate files in Windows 95/98 and NT.

1. **Windows Explorer.** This is the "official" method for viewing files in Windows 95/98 and NT v4.x. Explorer lets you view files in all subdirectories on all drives, including computers networked to yours.

2. **File Manager.** This is the old method of viewing files from Windows v3.x and NT v3.x. The File Manager has one drawback and one advantage over Explorer. The drawback is that File Manager is limited to displaying eight-character file names; file names longer than eight characters are truncated to six characters and given a ~1 suffix. The advantage is that File Manager can sort files by file extension, something Explorer cannot do; the equivalent in Explorer is to sort files by Type, which is not as useful for the power user.

3. **File-Related Dialog Boxes.** Almost all of AutoCAD 2000's file-related dialog boxes—such as **Save** and **Import**—let you view and manipulate files. This is a handy shortcut that obviates the need to switch to Explorer or File Manager.

4. **DOS Session.** The oldest method is to start a DOS session, and then type in commands—such as **Dir**, **Tree**, **Copy**, **Move**, and **Delete**—together with parameters, such as **/s**, ***.***, and **a:** . This is of use to power users, since neither Windows Explorer nor File Manager can perform some file-related tasks, such as renaming all files in a subdirectory.

WINDOWS EXPLORER

The Windows Explorer (not to be confused with Internet Explorer) lets you view and manipulate files on your computer's drives and all networked drives that you have permission to access. You can change the way Explorer looks, but the most useful configuration has two windows, as shown in Figure 3.2.

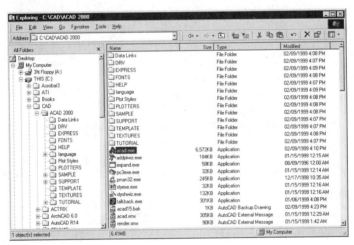

Figure 3.2 *Windows Explorer*

Viewing Files

In Windows Explorer, the left window pane displays the complete tree of drives (including networked drives) and *folders* (the new name for *subdirectories*). This is called the **Folder** view. To see more of the tree, use the scroll bar.

A + (plus) sign beside a folder means that the folder has subfolders. Click the + sign to "open" the folder. The + changes to a − (minus) sign; click the − sign to close the folder.

In the right window pane, Explorer displays the files and subfolders contained in the folder currently selected in the left window pane. This is called the **Contents** view. You change the sorting order of the Name, Size, Type, and Modified by clicking on the sort bar. (To see the sort bar, set the View to display file Details.) Click a second time and Explorer displays the list in reverse order. That is useful for displaying files sorted by alphabetical order or by size, either largest first or smallest first.

Explorer lends itself to displaying files and folders in different views, each giving you more detail. The View item on the menu bar lets you switch between large icons, small icons, list, and detail. These four views are shown in Figure 3.3, front to rear.

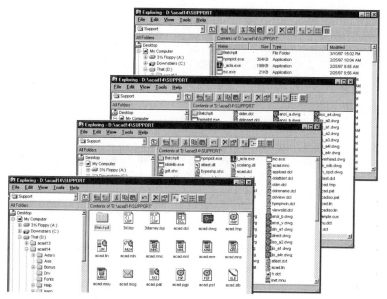

Figure 3.3 *Changing Explorer's View*

Manipulating Files

You can right-click on any folder or file to change its name, etc. (Alternatively, you can use the **File** and **Edit** items on the menu bar.) When you right-click, Explorer displays a pop-up menu with the following options. The options differ, depending if you right-click a folder or file:

> **Open (Folder):** Brings up another Explorer window displaying the contents of that folder.

> **Open (File):** Opens the file with the associated program. For example, a LSP (AutoLISP source code) file is opened by Notepad. An EXE file, such as Acad.exe, starts to run.

> **Explore (Folder only):** Displays the contents of the folder in the window pane at right.

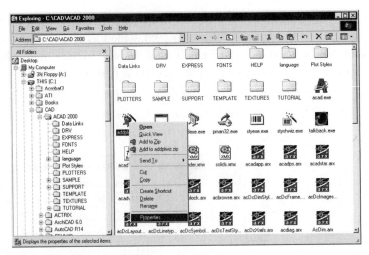

Figure 3.4 *Right-Click Menu for Folders*

Find (Folder only): Displays the **Find: All Files** dialog box. Similar to AutoCAD's Find File, this dialog box lets you find a file anywhere on the computer and (if connected) the network.

Sharing (Folder only): Brings up a dialog box that allows you to give others on a network access to this folder. You can set three types of permission: (1) full read-write (2) read only and (3) access with password.

Send to (File and Folder): This option lets you send the folder or file to one of several destinations, such as the floppy drive, the Briefcase folder, and the printer. You cannot send folders to the printer; to send a file to the printer, Explorer first opens the associated application.

Create Shortcut (File and Folder): A *shortcut* in Windows is a little file that points to the original file's location. This lets you have icons on the Desktop as shortcuts for accessing files without having to wander through the subdirectories to locate them each time.

Delete (File and Folder): Erases the folder or file.

Rename (File and Folder): Changes the name of the folder or file. As an alternative, click twice (slowly, do not double-click) on the name and Explorer lets you change the name directly.

Properties (File and Folder): Lets you set the attributes for the folder or file, including archive, read-only, hidden, and system.

Copying And Moving Files

Copying or moving a file using Explorer takes three steps:

1. In the Folders pane (on the left side), make sure you can see the name of the folder you will be copying the file to. If you cannot see it, scroll the list to see more folders or open the appropriate closed folder by clicking the + sign.

2. In the Contents pane (on the right side), *drag* the file to the folder. (Drag means to click on the file and move the cursor without letting go of the mouse button until you reach its destination.)

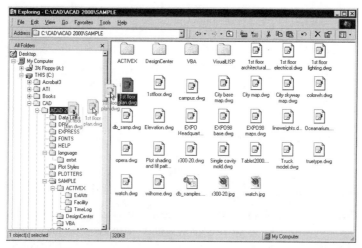

Figure 3.5 *Moving a File*

3. To copy the file, hold down the CTRL key while dragging the file. As a reminder that you are copying the file, Explorer displays a small + sign near the dragged icon.

If you drag the file on top of an application, the application will attempt to open the file. For example, dragging a DWG file onto the **Acad.exe** icon causes AutoCAD to launch with that drawing.

AUTOCAD FILE DIALOG BOXES

Although Explorer is the official way to view and manipulate files in Windows, you can perform some of those functions from within AutoCAD's file-related dialog boxes. That makes sense, because it is when you are saving and opening files that you probably want to work with them. Those dialog boxes are displayed when you use the **Open, Save, Import, Export,** and other file-related commands (see Figure 3.6).

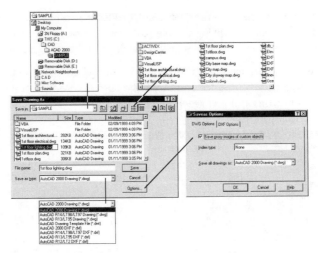

Figure 3.6 *Typical AutoCAD File Dialog Box*

AutoCAD's **File** dialog box provides the following functions:

Switch Views. Choose the **List view** icon to see a compact listing of files in a folder. Choose the **Detail view** icon to also see the size, type, and modified date-time for each file. Click the column headers to sort in alphabetical order; click a second time to sort in reverse order.

Select a Different Folder. You can select a different folder (subdirectory) in two different ways: (1) the list box lets you move directly to another folder; or (2) single-click the **Up One Level** icon to move up the folder tree one folder at a time; double-click the folder to move down into it.

Create a New Folder. Click the **Create New Folder** icon to create a new folder in the current folder.

Delete a File. Select the file name and press the DEL key. Windows displays a dialog box that asks, "Are you sure you want to send 'drawing.dwg' to the Recycle Bin?" When you respond **Yes**, the file is not erased but moved to the Recycling Bin. If deleting the file was an accident, you can go to the Recycling Bin and retrieve the file.

Rename a File. Click twice (slowly) on the file name, then type a new name.

Move or Copy a File. Drag the file to another folder. Hold down the CTRL key to copy the file.

Context-Sensitive Menu. Right-click a file or folder to bring up the same context-sensitive menu as seen in Explorer.

Load the File. Double-click a file name to load it without having to choose **OK**.

Save the File. Double-click a file name to write over an existing file.

CHAPTER REVIEW

1. What function does Windows perform?

2. What is the purpose of these two wild-card characters?

    ?

    *

3. How many different ways can you display a list of files?

4. Using Windows Explorer, how can you copy a file instead of moving it?

5. What is the **Open** command for?

6. Which functions does AutoCAD's **File** dialog box provide? (True or False)

    a. Rename a file: T / F

    b. Delete a file: T / F

    c. Create a new folder: T / F

    d. Open or save a file: T / F

    e. Display the size of a file: T / F

CHAPTER 4

General AutoCAD Principles

The operation of a computer-aided design (CAD) software program is unique among software programs. Primarily, CAD is used to do many design jobs more effectively. This chapter reviews the benefits and applications of CAD. It also discusses the fundamentals of working with AutoCAD. After completing this chapter you will be able to

- Explain the benefits of using computer-aided design

- Identify some standard applications of computer-aided design and the benefits of using CAD for these disciplines

- Understand the concepts of a computer-aided design program

- Understand the AutoCAD screen layouts

- Understand the terms used in this text and in AutoCAD

- Understand the functions of special keyboard keys

TRADITIONAL DRAFTING TECHNIQUES

The need for drawings as a means of communication has been present throughout history. While other types of work have benefited from technology, traditional drafting techniques have remained relatively unchanged until the mid to late 1980s.

The introduction of CAD has had an impact on the industry that is greater than all the previous changes combined. The acceptance of the graphics world of CAD and its capabilities has been phenomenal. CAD presents advantages that are undeniably superior to traditional techniques.

BENEFITS OF COMPUTER-AIDED DRAFTING

CAD is a more efficient and versatile drafting method than traditional techniques. Some of the advantages are

Accuracy. Computer-generated drawings are drawn and plotted to an accuracy of up to fourteen decimal places of the units used. The numerical entry of critical dimensions and tolerances is more reliable than the traditional methods of manual scaling.

Speed. The ability of the CAD operator to copy, to array items, and to edit the work on the screen speeds up the drawing process. When the operator customizes the system to a specific task, the speed of work is greatly increased.

Neatness and Legibility. The ability of the plotter to produce exact and legible drawings is an obvious advantage over the traditional methods of hand-drawn work. The uniformity of CAD drawings, which produce lines of constant thickness, print quality lettering, and no smudges or other editing marks, is preferred.

Consistency. Because the system is constant in its methodology, the problem of individual style is eliminated. A company can have a number of drafters working on the same project and produce a consistent set of graphics.

Efficiency. The CAD operator must approach a drafting task in a different manner than when using traditional techniques. The CAD program is capable of performing much of the work for the operator, so the job should be pre-planned to use all the benefits of the system.

APPLICATIONS OF CAD

Applications of CAD are now present in many industries. The flexibility of the many programs now available has had a major impact on how various tasks are performed. Some applications of CAD now include architectural, engineering, interior design, manufacturing, mapping, piping design, yacht design, and entertainment industries. A brief discussion of each of these applications follows.

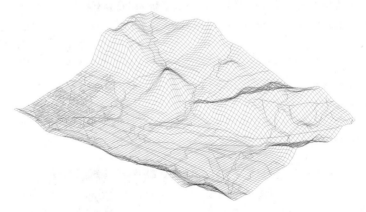

Figure 4.1 *AutoCAD Drawing (Digital Terrain Modeling.) (Courtesy Autodesk Inc.)*

ARCHITECTURAL

Architects have found CAD to be one of the most useful tools that has ever been available to them. Designs can be formulated for presentation to a client in a shorter period of time than is possible by traditional techniques. The work is neater and uniform. The designer can use 3D modeling capabilities to help the client better visualize the finished design. Changes can be performed and resubmitted in a very short time.

The architect can assemble construction drawings using stored detail. Database capabilities are used to extract information from the drawings, perform cost estimates, and prepare bills of materials.

Many third-party applications are available to help the architect customize AutoCAD for his/her discipline. Software is available for doing quick 3D conceptual designs, providing building details, generating automatic stairs, and more.

ENGINEERING

Engineers use CAD in many ways. They also use programs that interact with CAD and perform calculations that would take more time using traditional techniques.

Among the many engineering uses are

Electronics engineering	Mechanical engineering
Chemical engineering	Automotive engineering
Civil engineering	Aerospace engineering

Autodesk has a CAD product called the Mechanical Desktop, which customizes AutoCAD for mechanical design. Third-party developers have add-on software for roadway design, digital terrain modeling, linkage evaluation, stress analysis, and more.

INTERIOR DESIGN

AutoCAD is a valuable tool for interior designers. The 3D capabilities can be used to model interiors for clients. Floor plan layouts can be drawn and modified very quickly. Many third-party programs are available that make 3D layouts of areas such as kitchens and baths relatively simple.

MANUFACTURING

Manufacturing uses for CAD are many. One of the main advantages is in integrating the program with a database for record keeping and tracking purposes. The ability to maintain information in a central database simplifies much of the work required in the manufacturing process.

Technical drawings used in manufacturing can be constructed quickly and legibly.

MAPPING

Mapmakers use CAD to construct maps that store much data. For example, a municipal map can contain information about all buildings, transportation systems, and underground services (such as storm sewer and water pipes). By using search criterion, the CAD system returns a map showing, for example, all water pipes older than 20 years. This map then shows where municipal engineers should replace old piping.

GPS (global positioning system) is used to create maps. GPS consists of 20 satellites that always circle the earth, broadcasting data about their location. A handheld GPS receiver provides a fix on your location by analyzing data from at least three satellites.

PIPING AND INSTRUMENTATION DIAGRAMS (P&ID)

The piping in industrial plants is designed with CAD software that has been adapted to created 3D piping and instrumentation diagrams (P&ID). The 3D P&ID drawings show where two pipes might interfere with each other, allowing the problem to be corrected before the plant is constructed.

YACHT DESIGN

Yacht and ship designers use AutoCAD to create the complex drawings for their unique designs.

ENTERTAINMENT

The entertainment industry is largely based in the electronic media, so the use of CAD graphics is a perfect match. Television stations and networks use electronic graphics in place of the traditional artwork that must be photographed and converted. The weather graphics you can view on television and in the newspaper are a form of CAD graphics. Movie and theater set designs are often done in CAD. Moviemakers are turning to forms of electronic graphics for manipulations and additions to their filmed and animated work. The introduction of AutoCAD compatible programs, such as Discrete and 3D Studio Max, has brought life-like realism and animation to many entertainment applications.

OTHER BENEFITS

Using a computer to draft and design also makes available a new world of information technology. Many systems now incorporate electronic mail software networks (email) and access to the World Wide Web with a modem.

Commercial online services, such as America Online®, as well as the worldwide network of the Internet, can be accessed through a computer to gather information on numerous topics, putting you in touch with others using AutoCAD. There are Web sites specific to AutoCAD, as well as discussion groups and email newsletters.

AUTOCAD OVERVIEW

AutoCAD is a powerful CAD software package. A CAD system is to drafting what a word processor is to writing.

Your drawings are displayed on a graphics monitor screen (see Figure 4.2). This monitor takes the place of the paper in traditional drafting techniques. All the additions and changes to your work are shown on the monitor screen as you perform them.

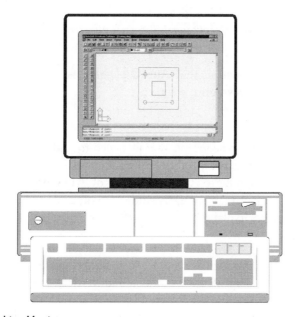

Figure 4.2 *Graphics Monitor*

A drawing is made up of separate elements consisting of lines, arcs, circles, strings of text, and others supplied for your use. These elements are called *objects* or *entities*.

The commands are selected from menus, icons, or typed on the keyboard. A menu is a list of items from which you can choose what you want. AutoCAD provides for menu selections to be placed on the screen and on a digitizing pad. Icons are small pictorial representations of commands.

Objects are placed in the drawing by means of *commands*. A command is started in one of several ways: you can type the command's name at the keyboard; you can select the command from a menu; or you can choose the command from a toolbar button. Some commands start when you select an object in the drawing. You are then asked to identify the parameters of the command. These parameters are called *options*. After

Figure 4.3 *AutoCAD Drawing (Architectural) (Courtesy Autodesk Inc.)*

you identify all the information that AutoCAD requests, the new objects, or any changes to objects, are shown on the screen.

TERMINOLOGY

This book contains terms and concepts that you need to understand to use AutoCAD properly. Some of the terms you need to know are explained briefly in this chapter. If you need additional help, refer to Appendix B "AutoCAD Command Summary" and Appendix H "Glossary." In addition, the index contains all the terms used and the pages on which they are explained.

COORDINATES

The Cartesian *coordinate* system is used in AutoCAD. Figure 4.4 illustrates the system. The horizontal line represents the X axis. The vertical line represents the Y axis. Any point on the graph can be represented by an X and Y value shown in the form of X,Y. For example, 2,10 represents a point 2 units in the X direction and 10 units in the Y direction.

The intersection of the X and Y axis is the 0,0 point. This point, called the *origin*, is normally the lower left corner of the screen. (You can, however, specify a different point for the lower left corner.)

AutoCAD also works with 3D (three-dimensional) drawings. The third axis is called the Z axis, which normally points out of the screen at you.

DISPLAY

In this book, the term *display* refers to the part of the drawing that is currently visible on the monitor screen.

DRAWING FILES

The *drawing file* is the file that contains the information used to describe the graphic image you created. A drawing file automatically has a file extension of .dwg added to it.

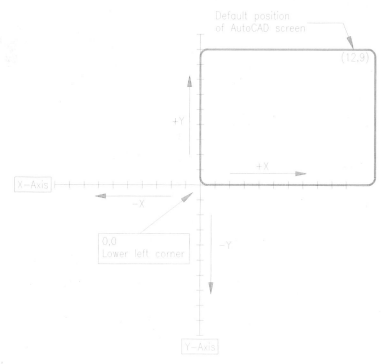

Figure 4.4 *Coordinate System*

LIMITS

When using AutoCAD, you draw in a rectangular area. The borders of this area can be defined by *limits*. You may draw anywhere inside and outside these limits. You can set the limits to whatever size you wish. You can also change them at any time. The limits are described by *X,Y* coordinates for the lower left corner and the upper right corner. Your drawing limits can be thought of as the "sheet size" that you are drawing on.

UNITS

The distance between two points is described in *units*. The units for AutoCAD can be set to any of the following:

1. Scientific
2. Decimal (Metric)
3. Engineering
4. Architectural
5. Fractional

One unit is equal to whatever form of measurement (for example, feet or meters) you wish.

Figure 4.5 *AutoCAD Drawing (Site Planning) (Courtesy Autodesk Inc.)*

ZOOMING AND PANNING

You often want to enlarge a portion of the drawing to see the work in greater detail, or to reduce it to see the entire drawing. The **Zoom** command facilitates this. AutoCAD's zoom ratio is many trillions to one!

The **Pan** command allows you to move around the drawing while at the same zoom level. You can pan the view up, down, and side to side, for example, to see parts of the drawing that might be off the screen.

USE OF THE MANUAL

Each chapter in this book is designed to build on the previous chapters. Many sections contain tutorials that you should follow. The problems found at the end of a chapter are specifically designed to use the commands covered in the manual to that point.

This method allows your learning to be self-paced. Remember, not everyone will grasp each command in the same amount of time.

KEYS

Several references are made to different keyboard keys in this manual. The following keys are referred to in this manual. Note that the keys might be found in a different location on some keyboards.

Control Key

Some commands are executed by holding down one key, while pressing a second key. The Control key is labeled **Ctrl** and is always used in conjunction with another key.

Flip Screen

The text and graphics screens can be alternately displayed by using the Flip Screen key. On most keyboards, the **F2** key is used for this.

Alt Key

To access menu commands from the keyboard, hold down the **Alt** key while pressing the underlined letter.

COMMAND NOMENCLATURE

When a command sequence is shown, the following notations are used:

> **Boldface.** When text is in **boldface**, it designates user input. This is what you enter. Entry can be made from either the screen menu or the keyboard.

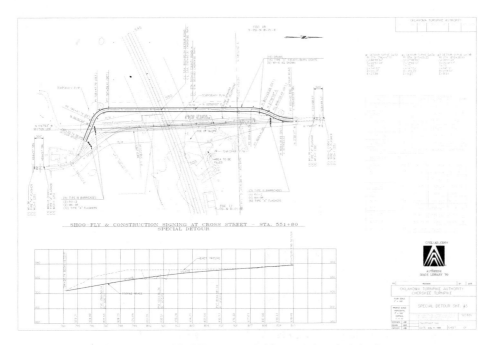

Figure 4.6 *AutoCAD Drawing (Civil Engineering) (Courtesy Autodesk Inc.)*

<Default>. Entries enclosed in brackets are the default values for the current command. The default value is executed when you press ENTER.

Enter Point. When prompted to "enter point," you should enter a point on the screen at the designated place. (To enter the point, you can either click on the point, or type the *X,Y,Z* coordinates.) The program usually shows you a point on a drawing. The points are designated, such as "point A."

*. Means to press the ENTER key after the entry. Do this after each input.

Choose or Select. Make the desired choice. You can use either uppercase or lowercase in response to any command inquiry.

When a response option has one letter capitalized, entering the capital letter for that response is all that is required. For example, if the option is **eXit**, simply entering **X** is sufficient for choosing that option.

The menus shown in tutorials serve to guide you through the complex menu structure. The boxed part of the menu represents the correct choice to make from the screen menu.

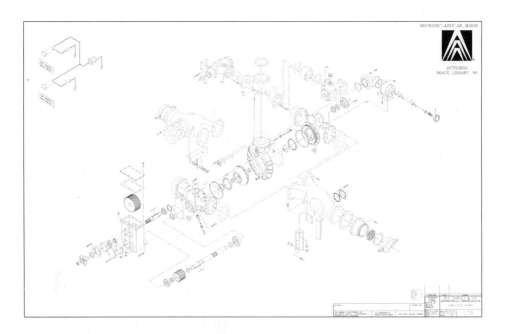

Figure 4.7 *AutoCAD Drawing (Parts Assembly) (Courtesy Autodesk Inc.)*

AUTOCAD FILE EXTENSIONS

AutoCAD uses many different types of files to support your drafting work. The contents of a file are often described by the file name extension. For example, linetypes are stored in files that end with .lin, while drawings are stored in files with an extension of .dwg. Many of the file extensions used by AutoCAD 2000 are listed here.

DRAWING FILES

Extension	Description
.$ac	Temporary file created by AutoCAD
.bak	Backup drawing file
.dwf	Drawing Web format file
.dwg	AutoCAD drawing file
.dwt	Drawing template file
.dxb	AutoCAD binary drawing interchange file
.dxf	AutoCAD drawing interchange file

AUTOCAD PROGRAM FILES

Extension	Description
.arx	ObjectARx (AutoCAD Runtime eXtension) program file
.dll	Dynamic link library
.exe	Executable files, such as AutoCAD itself
.lsp	AutoLISP program files

SUPPORT FILES

Extension	Description
.ahp	AutoCAD-format help file
.cfg	Configuration file
.cus	Custom dictionary file
.dct	Dictionary file
.dxt	Translator file used by DXFIX utility program
.err	Error log file
.fmp	Font mapping file

.hlp	Windows-format help file
.hdx	Help index files
.lay	Layer Manager layer settings (Express tools)
.lin	Linetype definition files
.lli	AutoVision Landscape library
.log	Log file created by the Logfileon command
.mli	Rendering Material Library
.mln	Multiline library file
.mnc	Compile menu file
.mnd	Uncompiled menu file containing macros
.mnl	AutoLISP routines used by AutoCAD menus
.mns	AutoCAD-generated menu source files
.mnu	Menu source files
.msg	Message file
.pat	Hatch pattern definition files
.pfb	PostScript font file
.pgp	Program parameters file for external commands and command aliases
.ps	Font Map file
.psf	PostScript support file
.scr	Script files
.shp	Shape and font definition files
.shx	Compiled shape and font files
.xmx	External message file

PLOTTING SUPPORT FILES

Extension	Description
.ccp	CalComp color palette files used by CalComp printers and plotters
.pcp, .pc2, .pc3	Plot configuration parameters files
.plt	Plot file

.pm	Océ plotter support
.rpf	Raster-pattern fill definition file for Hewlett-Packard plotters

IMPORT-EXPORT FILES

Extension	Description
.3ds	3D Studio file (import and export)
.bmp	Windows raster file (device-independent bitmap)
.cdf	Comma delimited file (created by AttExt)
.dxx	DXF file created by AttExt
.eps	Encapsulated PostScript file
.pcx	Raster format
.sat	ACIS solid object file (Save As Text)
.sdf	Space delimited file (created by AttExt)
.slb	Slide library file
.sld	Slide file
.stl	Solid object stereo-lithography file (solids modeling)
.tif	Raster format (Tagged image file format)
.tga	Raster format (Targa)
.wmf	Windows metafile format

AUTOLISP, ADS, AND OBJECTARX PROGRAMMING FILES

Extension	Description
.ase	Names and locations of ASE (AutoCAD SQL Extension) database drivers
.cpp	ADS and ObjectARX source code
.dcl	DCL (Dialog Control Language) descriptions of dialog boxes
.def	ADS and ObjectARX definitions
.fas	AutoLISP fast load program
.h	Definitions for ADS and ObjectARX functions
.hdi	Heidi support file
.lib	ADS and ObjectARX function libraries

.mak	ADS and ObjectARX make files
.pif	Program information file; used by Windows for DOS applications
.rx	List of ObjectARX applications that load automatically
.tlb	ActiveX Automation type library
.unt	Unit definition file

REMOVING TEMPORARY FILES

While it is working, AutoCAD creates "temporary" files. When you exit AutoCAD, those temporary files are erased. However, if your computer crashes, those temporary files might be left on your computer's hard disk. You can erase those files, but only when AutoCAD is *not* running. Never erase files with these extensions while AutoCAD is still running:

.$a

.ac$

.dwk

.dwl

.ef$

.sv$

.swr

Figure 4.8 *AutoCAD Drawing (3D Drawing and Rendering) (Courtesy Autodesk Inc.)*

These files can be erased to free up disk space when AutoCAD is *not* running.

USING NOTE BOXES

Note boxes are provided throughout the text. These note sheets contain helpful hints for using AutoCAD.

 Note: Be sure to read the note boxes in each chapter for helpful hints for using AutoCAD.

CHAPTER REVIEW

1. Name some applications of CAD.

2. Why is CAD a more flexible method of drafting than traditional drafting techniques?

3. Why would an office using several CAD stations produce more consistent work than a traditional office?

4. Why is CAD more accurate than traditional methods of scaling?

5. What does CAD do for mapping?

6. How can programs that interact with CAD be useful for engineering purposes?

7. What benefits does CAD provide for the user over traditional drafting techniques?

8. Does the construction or editing capabilities of CAD provide the greater increase in production speed?

9. The area on the screen in which you draw is surrounded by borders called _____.

10. Some commands prompt for additional information in order to continue execution of the command, and in some instances a default setting can be chosen. If a default setting is present, how can it be identified and what must be done to select it?

11. What are the drawing elements used in computer-aided drafting called? Give three examples of drawing elements.

12. For each of the following options, list the required response you would enter at the prompt:

    Close:

    eXit:

    circle:

    LAyer:

    Edit vertex:

13. From where in AutoCAD can you select the commands used to draw and edit?

14. Using the figure below, label each portion of the axis indicated. Next, draw a rectangle on the figure to represent the drawing screen position relative to the coordinate axis.

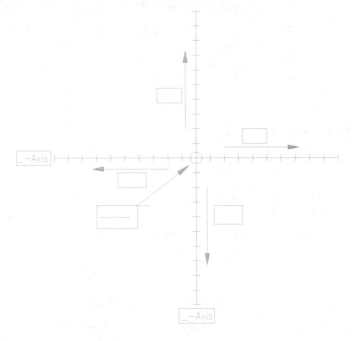

Label Each Portion of the Axis Indicated.

15. To enlarge or reduce all or a portion of your drawing on the screen, what command would you use?

16. What term is used to refer to the portion of the drawing that is currently visible on the screen?

17. What is the increment of length in which a drawing is constructed?

18. When using this book, where would you look for extra help or information about commands listed?

19. What happens when you press the CTRL key?

20. If you have zoomed in to a portion of the drawing for close examination, how can you move to another part of the drawing while staying in the same zoom?

21. What procedure/command is used to place objects in a drawing?

22. What is a temporary file?

23. When can you erase a *.ac$ file?

CHAPTER 5

Getting In and Around AutoCAD

To use the AutoCAD program, you must learn how to start the program and know your way around the AutoCAD user interface. After completing this chapter, you will be able to

- Start a new drawing

- Open an existing drawing

- Search for drawings

- Save the drawing

- Recognize all elements of the AutoCAD user interface

- Enter commands through the keyboard, drop-down menus, and toolbars

- Use keyboard shortcuts

- Understand the elements of a dialog box

BEGINNING A DRAWING SESSION

Your drawing session begins by starting the AutoCAD program. As described in Chapter 1 "A Quick Start in AutoCAD," double clicking the AutoCAD icon starts the program.

If there is no icon on the Windows desktop, click the Windows **Start** button, then choose **Programs**. Choose the **AutoCAD 2000** folder, and choose **AutoCAD 2000**.

When AutoCAD is loaded, the **Start Up** dialog box appears. For now, choose **OK** to dismiss the dialog box. You see the drawing area. The drawing area is the window where your drawing activities take place.

THE DRAWING AREA

After AutoCAD starts, you are presented with a drawing area in which you perform your work. Notice that the top and bottom of the window display information (see Figure 5.1). Let's look at these areas, starting at the top of the AutoCAD and moving down the window.

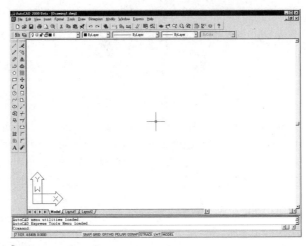

Figure 5.1 *AutoCAD's User Interface*

TITLE BAR

At the very top of the AutoCAD window is the title bar. The title bar tells you the name of the software (AutoCAD 2000) and the name of the drawing, such as Drawing1.dwg.

Figure 5.2 *The Title Bar*

TOOLBAR INDICATORS

Near the top of the window, just above the drawing area, AutoCAD displays indicators on the toolbar. This toolbar is called the "Object Properties" toolbar. The indicators include the following:

Figure 5.3 *The Toolbar Indicators*

Layer Buttons. The first button (at the far left) is called **Make Object Layer Current**. Choose this button, then select an object in the drawing. AutoCAD makes that object's layer the current layer. Choose the second button to display the **Layer Property Manager** dialog box.

Layer Status, Color, and Name. Lists the status, color, and name of the current layer. The four little icons prefixing the layer name indicate the layer status as follows:

- Lightbulb: layer is on or off.
- Shining sun: layer is thawed.
- Snowflake: layer is frozen.
- Padlock: layer is locked or unlocked.
- Printer: layer is printable or non-printable.

The small square prefixing the layer name indicates the current color of the layer. Every drawing contains at least one layer called "0" (zero). You learn all about layers in Chapter 11 "Working with Layers."

Color. Lists the current working color. In most cases, the color is called BYLAYER, which means that the layer dictates the color. However, this can overridden on an entity-by-entity basis. AutoCAD can work with as many as 255 colors. You learn more about colors in Chapter 20 "Working with Intermediate Operations."

Linetype and name. Lists the look and name of the current linetype (line pattern). In most cases, the linetype is called BYLAYER, which means the layer determines the linetype. As with colors, this can be overridden for each entity. Before you can assign a linetype, its definition must be loaded into the drawing. AutoCAD has two styles of linetypes: simple dot-dash patterns, and lines with text and 2D shapes. Note that linetypes are sensitive to the scale of the drawing. You learn more about linetypes in Chapter 15 "Constructing Multiview Drawings."

Lineweight. Lists the weight (width) of lines. As with colors and linetypes, most lines have a weight of BYLAYER, which means the layer decides the width of the line. This can be overridden with a specified width. You learn more about lineweights in Chapter 16 "Constructing Sectional and Patterned Drawings."

Plot styles. Lists current plot styles, if any, in the drawing. A plot style collects together settings, such as lineweight, end caps, line fill, and screening, that affect the appearance of objects in two ways: (1) as they are drawn and (2) when they are plotted. Plot styles are discussed in greater detail in Chapter 9 "Plotting Your Work."

THE DRAWING AREA

Even the drawing area contains information. Figure 5.1 shows three pieces of information. The crosshairs cursor shows you where you are in the drawing. When you move the cursor out of the drawing area, it changes to an arrow cursor, which you are probably familiar with from other Windows software.

The small box at the center of the cursor indicates that you can select (or "pick") an object in the drawing. The box is called the *pickbox* and is three pixels wide. The size of the box can be made larger (or smaller) with a system variable called **PickBox**.

In the lower-left corner is a double-ended arrow. This is called the *UCS icon*; UCS is short for user coordinate system. The UCS icon is primarily used when you draw in three dimensions; it helps you orient yourself in three-dimensional space. Sometimes, the UCS icon will change to other shapes. You learn how to use the UCS icon in Chapter 32 "The User Coordinate System."

Figure 5.4 *Layout Tabs*

 At the bottom of the drawing area are several tabs, labeled **Model, Layout1**, etc. These tabs allow you to switch between model space and layout mode. You learn more about layout mode in Chapter 23 "Working with Viewports and Layouts."

THE COMMAND PROMPT AREA

Below the drawing area is the *command prompt area*. This area is where your commands are listed and the resulting prompts are displayed. If you become confused as to what AutoCAD expects of you, the prompt line is the place to look for the answer. You can see the history of your commands by pressing function key **F2**. AutoCAD displays a text window.

Figure 5.5 *The Command Prompt Area*

If you like, you can drag the command prompt area away from the AutoCAD window. With the mouse, drag the lower right corner (the square area between the scroll bars) away from the AutoCAD window. Release the mouse button.

X, Y, Z COORDINATES

At the bottom left of the window, on the *status line*, are three numbers, which represent the *X, Y, Z* coordinates of the crosshairs cursor. The format of the numerical readout is controlled by the **Units** command, as you learn in Chapter 8 "Setting Up a New Drawing."

[3.7669, 0.1497, 0.0000]   SNAP GRID ORTHO POLAR OSNAP OTRACK LWT MODEL

Figure 5.6 *The Status Line*

The coordinates constantly update, as you move the cursor. You switch the display between absolute coordinate display, polar coordinate display, and off by pressing CTRL+D or CTRL+4. If the coordinates do not update as you move the cursor, then they are toggled off; double-click the coordinate display to turn them on. *Double-click* means to rapidly click the mouse button twice on the entry you desire.

MODE INDICATORS

The bottom of the screen, next to the coordinates, is a display of the *mode* indicators (see Figure 5.6). These modes are sometimes called *toggles* since they are either on or off, just like a light toggle. When on, the mode's button looks as if it is pressed in. As shown in Figure 5.6, **OTRACK** and **MODEL** are on; the other modes are turned off.

> **Tip:** In AutoCAD Release 13 and 14, you had to double-click these modes to turn them on and off. In AutoCAD 2000, you simply click the mode buttons once.
>
> AutoCAD 2000 no longer displays the **TILE** mode indicator. It does, however, display three new mode indicators: **POLAR**, **OTRACK**, and **LWT**.

The modes include the following:

SNAP: Indicates whether the snap mode is on or off.

GRID: Indicates whether the grid is on or off.

ORTHO: Indicates whether the orthogonal mode is on or off.

POLAR: Indicates whether polar mode is on.

OSNAP: Indicates whether any object snap mode is on.

OTRACK: Indicates whether object tracking mode is on.

LWT: Indicates whether line weights are displayed or not.

Model or Paper: Indicates whether AutoCAD is in model space or paper space (layout mode).

SCREEN POINTING

You enter points, distances, and angles simply by "showing" AutoCAD the information on the screen. Entering two points could indicate the distance and angle requested.

On some commands (such as **Line**) the absolute coordinate display converts to a relative distance and angle display. Using this as a method of measurement is quite suitable in most circumstances, although actual numerical entry using coordinates is more accurate.

SHOWING POINTS BY WINDOW CORNERS

Some commands require input of both a horizontal and a vertical displacement. Requesting a "window" and using the X and Y distances from the lower left corner and the upper right corner as the displacement may show both points at one time.

For example, the window in Figure 5.7 displaces a value of (5,3) from the lower left corner of the window.

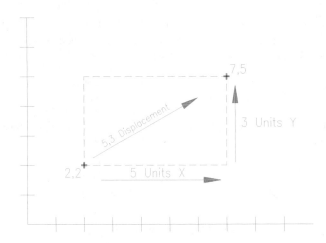

Figure 5.7 *Using a Window to Show Displacement*

ENTERING COMMANDS

Commands can be entered in AutoCAD in several different ways. Let's look at the methods you can use to enter commands in AutoCAD.

KEYBOARD

The most basic method of input is the keyboard. You can type the command directly from the keyboard. Your input is displayed on the prompt line. If you make a mistake, simply backspace to correct. (Other editing keystrokes are listed below.) After typing, you must press ENTER or the Spacebar to activate the command. AutoCAD doesn't care if you use uppercase, lowercase, or a combination of each.

To type a command from the keyboard, the command line must be "clear." That is, another command must not be in progress, unless it is a transparent command. If the command line is not clear, you can clear it by pressing ESC.

You can edit the text on the command line using these keys:

Keystroke	Meaning
LEFT ARROW	Moves the cursor one character to the left
RIGHT ARROW	Moves cursor to the right
HOME	Moves the cursor to the beginning of the line of command text
END	Moves the cursor to the end of the line
DEL	Deletes the character to the right of the cursor
BACKSPACE	Deletes the character to the left of the cursor
INS	Switches between insert and typeover modes
UP ARROW	Displays the previous line in the command history
DOWN ARROW	Displays the next line in the command history
PGUP	Displays the previous screeenful of command text
PGDN	Displays the next screeenful of command text
CTRL+V	Pastes text from the Clipboard into the command line
ESC	Cancels and clears the current command and returns the command line. You can press ESC at any time. If entered while an operation is in progress, it will terminate the command at its present point.

Keyboard Shortcuts

In addition accepting the entry of the full name of commands, AutoCAD allows you to type shortcuts. There are several types of shortcuts, such as aliases, function keys, and control keys. An *alias* is an abbreviation of the full command name. For example, the alias of the **Line** command is **L**. To execute the **Line** command more quickly, simply press **L**, followed by ENTER. Examples of other aliases are:

Alias	Meaning
a	Arc
adc	ADCenter
aa	Area

b	Block
-b	-Block
bh	BHatch
co	Copy
d	DimStyle
dal	DimAligned
la	Layer
-la	-Layer
le	QLeader
xl	XLine
xr	XRef
z	Zoom

There are many other command aliases, which are defined in the Acad.Pgp file found in the \acad2000\support folder. You can learn more about aliases in Chapter 39 "Tailoring AutoCAD."

Function Keys

A *function key* is one of the keys on your keyboard prefixed with an F. AutoCAD assigns the following meaning to these function keys:

F1	Calls up the help window
F2	Toggles between the graphics and text windows
F3	Toggles object snap on and off
F4	Toggles tablet mode on and off; you must first calibrate the pad before you can toggle the tablet mode on
F5	Toggles to the next isometric plane when in isomode; the planes are activated in a rotating fashion (left, top, right, and then repeated)
F6	Toggles the screen coordinate display on and off
F7	Toggles grid on and off
F8	Toggles ortho mode on and off
F9	Toggles snap mode on and off
F10	Toggles polar tracking on and off
F11	Toggles object snap tracking on and off
ALT+F4	Exits AutoCAD

ALT+F8　　Displays the Macros dialog box

ALT+F11　　Starts Visual Basic for Applications editor

Control Keys

You use a *Ctrl-Key* shortcut by holding down the CTRL and then pressing another key. You may already be familiar with some, such as CTRL+C to copy to Clipboard and CTRL+Z to undo. AutoCAD displays its Ctrl-key shortcuts in the drop-down menus, next to the related command.

Ctrl-key	Meaning
CTRL+1	Displays Properties dialog box
CTRL+2	Opens AutoCAD Design Center
CTRL+6	Launches **dbConnect**
CTRL+A	Toggles group mode on or off
CTRL+B	Turns snap mode on or off
CTRL+C	Copies to Clipboard
CTRL+D	Changes coordinate display modes
CTRL+E	Switches to the next isoplane
CTRL+F	Toggles object snap on and off; display **Object Snap** dialog box when no object snaps are set
CTRL+G	Turns the grid on and off
CTRL+H	Backspaces on command line
CTRL+K	Creates a hyperlink
CTRL+L	Turns ortho mode on and off
CTRL+N	Starts new drawing
CTRL+O	Opens drawing
CTRL+P	Prints or plots drawing
CTRL+Q	Records command text to a log file
CTRL+R	Switches to the next viewport
CTRL+S	Saves drawing
CTRL+T	Toggles tablet mode
CTRL+V	Pastes from Clipboard
CTRL+X	Cuts to Clipboard
CTRL+Y	Performs a Redo
CTRL+Z	Performs an Undo

Repeating Commands

To repeat the previous command given to AutoCAD, press ENTER (or Spacebar) at the **'Command:'** prompt. The last command entered will repeat. As an alternative, right-click to display the cursor menu and then select the **Repeat** option.

Some commands automatically repeat until you press ESC. To force a command to automatically repeat, first type **Multiple**, as in

> Command: **multiple circle**

This forces the **Circle** command to repeat until you press ESC.

MOUSE BUTTONS

AutoCAD takes control over the mouse buttons anytime the cursor is within the AutoCAD window. Thus, AutoCAD might not react as you might expect when you press a mouse button. The most significant difference is the effect of pressing the right mouse button. In most cases, the right mouse button displays a *cursor menu* when inside the AutoCAD window.

A cursor menu is a small menu that appears at the cursor, as shown in Figure 5.8. Context-sensitive means that the content of the menu changes, depending on where you right-click.

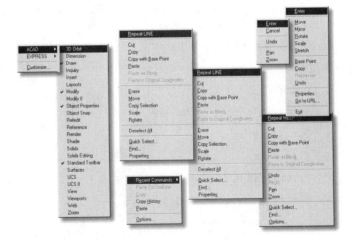

Figure 5.8 *Some of AutoCAD's Many Cursor Menus*

AutoCAD recognizes as many as 16 buttons on a mouse or digitizer puck. While most mouse devices are limited to two or three buttons, a digitizer puck commonly has 4, 12 or 16 buttons. By default, AutoCAD defines the first 10 buttons found on an input device as follows (it is possible to change the meaning of buttons 2 through 16):

Mouse Button	Tablet Button	Meaning
Wheel	...	Zoom or Pan
Left	1	Select object or item
Right	2	Context-sensitive menu
Middle	3	Object snap menu
...	4	Cancel command
...	5	Turn snap on and off
...	6	Turn ortho on and off
...	7	Turn grid on and off
...	8	Change coordinate display
...	9	Switch isoplane
...	10	Toggle tablet mode
...	11 - 16	User-definable

Table 5.1

It is possible to change the meaning of all buttons except the first button (which is always the Select button). You learn more about button customization in Chapter 40 "Customizing Menus and Icons."

The IntelliMouse Wheel

The wheel on a mouse compatible with Microsoft's IntelliMouse can be used to zoom and pan the drawing, without invoking the **Zoom** and **Pan** commands.

To zoom in or out, roll the wheel forward (zoom in) and backward (zoom out). To zoom the drawing to its extents, double-click the wheel button.

To pan about the drawing, hold down the wheel and drag the mouse. An alternative pan method is to hold down the CTRL while also holding down the wheel, then drag the mouse.

In some cases, you might prefer that the wheel act as a button. You can change the action of the wheel from performing zooms and pans by setting the value of the **MButtonPan** system variable to 0. Now when you click the wheel, it acts like the middle button of a three-button mouse.

Note that some brands of IntelliMouse come with three or four buttons. This allows you to use them like a three-button mouse, yet have all the benefits of rolling and pressing the wheel.

TOOLBARS

AutoCAD provides toolbars that conform to the standard set by Microsoft for its Office line of software. The use of toolbars is described in detail in Chapter 1 "A Quick Start in AutoCAD."

All toolbars and all icons can be changed, except for portions of the **Standard** and **Object Properties** toolbars. To learn more about toolbar customization, read Chapter 41 "Writing Toolbar Macros."

TABLET MENUS

AutoCAD includes a tablet menu (Figure 5.9). This is a printed template that is placed on a digitizing tablet. The user can customize the tablet menu for a particular application. Items chosen from a tablet menu respond in the same manner as those chosen from the screen menu.

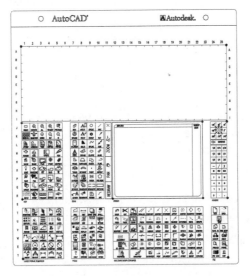

Figure 5.9 *The Digitizing Tablet Overlay Provided with AutoCAD 2000*

DROP-DOWN MENUS

The drop-down menu (see Figure 5.10) is displayed under the title bar. It categorizes commands in groups. For example, all file-related commands are listed under **File**, while all drawing commands are listed under **Draw**.

Selecting items from a drop-down menu has the same effect as typing them at the keyboard, except you don't press ENTER.

Using Drop-down Menus

Clicking on one of the listings will cause a drop-down menu to appear on the screen. You can choose a command or function from the drop-down menu. If you pull down a menu and do not want to select an item, you can exit the menu by clicking the menu name a second time.

You can access items on the drop-down menu from the keyboard. Press the ALT , then press the underlined letter. For example, to start the **Line** command, which is found in the **Draw** menu:

Press the ALT. Notice that the **Draw** menu has the D underlined.

Press D. Notice that the **Line** command has the L underlined.

Press L.

The menu uses punctuation marks as shorthand to indicate special meaning. Look for these marks in Figure 5.10:

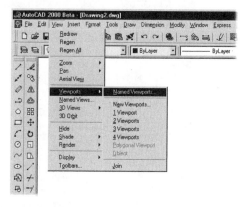

Figure 5.10 *Drop-down Menu*

View: The underlined letter on the menu bar means you can access it from the keyboard. Hold down the ALT key and press **V** to access the **View** menu.

Tiled: The underlined letter in the menu can also be accessed from the keyboard, but you don't hold down the ALT key: just press the **D** key.

>: The arrowhead means that the menu item displays a submenu. One menu item can have one, two, or more submenus. In Figure 5.10, the **Tiled**

Viewports menu item displays a submenu of nine more options.

✓: The check mark means a menu item is turned on.

...: The ellipses means that the menu item displays a dialog box. Think of the ... as "more to come." In Figure 5.10, **Named Views...** would display the **View Control** dialog box.

Gray Text: When a menu item is displayed in gray text, it means it is not available at the current time. In Figure 5.10, **Polygonal Viewports** is not available.

Drop-down menus are arranged in a logical manner. Figure 5.11 shows the sequence used to pull down and choose an item from a drop-down menu.

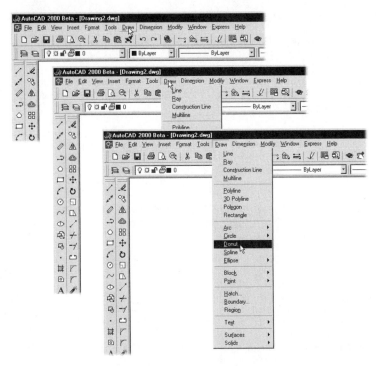

Figure 5.11 *1. Select Menu Category*
2. Click To Pull Down The Menu
3. Choose Command From Menu

DIALOG BOXES

Many commands allow you to enter or select information from a *dialog box*. The benefit of a dialog box is that it displays all your options at the same time.

In AutoCAD 2000, more commands display a dialog box than ever before. A number of commands give you the option to use the dialog box or the command line. If a command does not use a dialog box, then it displays a prompt at the command line. Sometimes you might not want a dialog box. There are a couple of reasons for this. One reason might be because you prefer using the command line since it can be faster than a dialog box in some cases. Another reason is because script files and AutoLISP routines cannot control dialog boxes.

There are two ways to suppress the display of a dialog box:

1. Prefix the command with a hyphen to suppress the dialog box and display the command line instead. For example, the **BHatch** command normally displays the **Boundary Hatch** dialog box. Prefixing the hyphen displays the command line options, as follows:

   Command: **-bhatch**

   Current hatch pattern: **ANSI31**

   Specify internal point or [Properties/Select/Remove islands/Advanced]:

2. When the **FileDia** system variable is set to 0, AutoCAD suppresses the display of file-related dialog boxes, such as the **SaveAs** and **VSlide** commands. If you want to force the display of the dialog box, type the ~ (tilde) character when AutoCAD prompts for 'File name:'.

Using a Dialog Box

When a dialog box is displayed, an arrow pointer replaces the crosshairs. You use the arrow pointer to choose items in the dialog box. When done, choose the **OK** button to accept the changes and dismiss the dialog box. If you just want to dismiss the dialog box, choose **Cancel.**

Most dialog boxes are *modal*. That means you must dismiss them before continuing to work in AutoCAD. A few dialog boxes, however, are *non-modal*. That means they continue to hover on your Windows desktop as you continue working in AutoCAD. An example of a non-modal dialog box is the **Properties** dialog box. To dismiss a non-modal dialog box, click the small x in the upper right corner of the non-modal dialog box.

Some dialog boxes contain *sub-dialog* boxes. If a sub-dialog box is displayed, you must reply to the prompt or choose **Cancel** to continue. Other dialog boxes contain *tabs*. The tabs allow a single dialog box to display a lot of information, separated into similar-looking boxes. For example, the **Tools | Options** command displays a dialog box with nine tabs!

Many dialog boxes contain a **Help** button. Selecting the Help button displays a help box that explains the purpose and use of the dialog box.

You move a dialog box by dragging its title bar. That means that you move the cursor over the title bar and holding down the left mouse button while moving the box. Moving dialog boxes is useful if you need to see a screen object that is under the box.

Let's look at the parts of a dialog box. Figure 5.12 shows a typical dialog box, with many of its parts labeled.

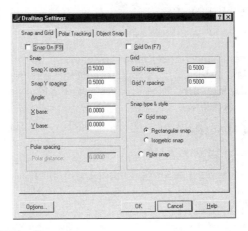

Figure 5.12 *Typical Dialog Box*

Buttons in Dialog Boxes

Buttons are used to select items. Select buttons with the pointer or by pressing keys on the keyboard.

- A button with a heavy border is the default. Pressing **Enter** is like choosing the button with the mouse.

- Three periods (...) after a button opens a sub-dialog box.

- Buttons with a single arrow (<) refer to an action that is required in the graphics screen, such as selecting an object. An example is the **Select Objects** button in the **Attribute Extract** dialog box.

- Buttons with two arrow pointers (<<) expand the size of the dialog box. Alternatively, the dialog box may have a **Show Details** button, such as in the **Layer Properties Manager** dialog box.

- If the text on the button is in gray, the button's function is not available at this time.

The following is a summary of the buttons used in dialog boxes.

Check Box. Boxes following items that show a check mark if selected. These are mostly used to "turn on" a function. If a ✓ is present, the function is "on."

Radio Button. Two or more round *radio buttons* allow you to select one from series of items. Only one option can be active at a time. The black dot in the button indicates which is currently selected.

List Box. A *list box* contains a list of items that you can choose. List boxes contain scroll bars that allow you to scroll among the available choices.

Drop-Down List Box: A *drop-down list box* is a type of list box, except that it "drops down" if an item with an arrow follows it. An example of a drop-down list box is shown in Figure 5.13.

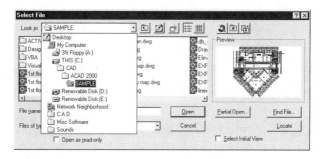

Figure 5.13 *List Box and Scroll Bars*

Scroll Bars. Some list boxes (Figure 5.13) contain more entries than can be displayed at a time. *Scroll bars* exist on some list boxes to bring additional items or options into view by scrolling up or down. To display another entry below the current filed of view, click on the down arrow of the scroll bar.

You can also move through the scroll list by clicking on the bar and, while continuing to hold the mouse button down, sliding the bar up or down with the cursor. When you click again, the entries are redisplayed at the new location. Note that the position of the scroll bar is relative to the position of the displayed items. Thus, if there are many items in the list, a relatively small movement of the scroll bar will scroll several items. If you click the scroll bar and then wish to cancel before clicking the second time, entering ESC will cancel, returning the list to the original location.

Edit Box. An *edit box* contains a single line of text (or, in some cases, a paragraph of text) that can be edited. To edit the text in the box, click in the box. A cursor bar will appear at the text. You can move the cursor bar with the arrow keys on the keyboard.

Image tile. Some dialog boxes contain an *image tile*. This is a small window that displays a selected item such as a drawing file or line type.

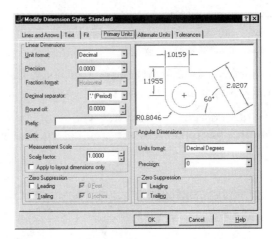

Figure 5.14 *Dialog Box with Image Tile and Tabs*

Tabs. Some dialog boxes contain *tabs*. These allow a single dialog box to display much more information in addition dialog boxes.

ICON MENUS

Icon menus are displayed on the page as graphic images instead of words. Figure 5.15 shows a sample icon menu. The icon menu is sometimes known as a *palette*. Icon menus are rarely used in AutoCAD anymore. The sole remaining icon menu is displayed when you choose **Draw | Surfaces | 3D Surfaces** from the menu bar.

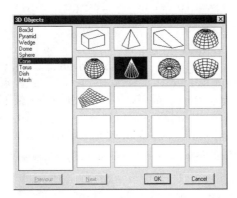

Figure 5.15 *Icon Menu*

When a palette appears, the crosshairs change to an arrow pointer. Each selection has a box next to it. If you move the pointer to the box, an outline appears around the selection. Double clicking on the selection box chooses the item. You can also use the cursor arrow keys to choose an item. A palette is cleared from the screen when you choose an item or press **Esc**.

STARTING A NEW OR EXISTING DRAWING

When you start AutoCAD, you either begin a new drawing or call up an existing drawing you saved previously. Let's look at how to perform these functions.

STARTING A NEW DRAWING

When AutoCAD first starts, it displays a new, blank drawing. At the same time, it displays the **Startup** dialog box.

After AutoCAD is started, the **New** command is used to start another new drawing. Type **New** at the keyboard and press **Enter**. AutoCAD displays the **Create New Drawing** dialog box (see Figure 5.16), which is almost identical to the **Startup** dialog box.

Figure 5.16 *Creating a New Drawing*

(As of AutoCAD Release 14, you give the drawing its file name when you first save it. This is unlike earlier versions of AutoCAD, in which you decide on the file name with the **New** command.)

Both dialog boxes give you three options for starting a new drawing:

1. **Start from Scratch.** AutoCAD sets up the drawing only in metric or English units.

2. **Use a Template.** AutoCAD 2000 includes 54 DWT template files that have preset sizes, standards, and plot styles.

3. **Use a Wizard.** There are two wizards—**Quick** and **Advanced**—which lead you through the steps of setting up a new drawing.

START FROM SCRATCH

The Start from Scratch option creates a new drawing. This is the fastest way to start a new drawing. Select either the English or Metric units options.

> **English:** The new drawing uses feet and inches for its measurement system; sets the drawing limits to 12" x 9"; and sets dimension styles for inches. AutoCAD reads the settings stored in the Acad.Dwt template file.

> **Metric:** The new drawing uses the metric system for its measurement system; sets the drawing limits to 429 mm x 297 mm; and sets the dimension styles for metric. AutoCAD reads the settings stored in the AcadIso.Dwt template file.

In addition, AutoCAD stores your selection in the **MeasureInit** system variable, which can have a value of 0 or 1, as follows:

> **0:** (English) AutoCAD reads the hatch patterns and linetypes defined by the **ANSIHatch** and **ANSILinetype** registry settings.

> **1:** (Metric) AutoCAD reads the hatch patterns and linetypes defined by the **ISOHatch** and **ISOLinetype** registry settings.

TEMPLATE FILES

When you begin a new drawing, AutoCAD uses a standard list of settings for the drawing. These settings are taken from a template drawing with the file extension of .dwt. A *template drawing* is a separate drawing containing settings and, optionally, drawing elements that your new drawing can use as a template. Your new drawing will be exactly equal to the template drawing. (The template file is the equivalent to the *prototype* file in earlier versions of AutoCAD.)

Although the default template drawing is Acad.dwt, you can use *any* drawing as a template drawing. For example, you might want to create a drawing that contains all the settings you normally use and has a title block already drawn. You then use this template drawing to create a starting point for all your new drawings. Each new drawing that used this template has the same settings and the title block at the moment it was created.

It is not necessary, however, to use a named template drawing. If you do not select a template drawing, AutoCAD assumes the standard settings. Note that you can change any of the settings, regardless of the template drawing.

Choosing a Template Drawing

In the **Create New Drawing** dialog box, choose **Use A Template**. AutoCAD displays the list of the template drawings. As you select a template name from the list, AutoCAD displays a small preview image.

Figure 5.17 *Selecting the Template Drawing*

AutoCAD includes many template drawings for a variety of sizes and standards. Choose the **Browse** button to find other template files.

WIZARDS

As mentioned earlier, AutoCAD includes two "wizards" that step you through the stages needed to set up some of the many parameters that define a drawing. The two are called **Quick** and **Advanced**.

The Quick Wizard takes you through just two steps—units and area—in creating a new drawing. You worked through the Quick Wizard in Chapter 1 "A Quick Start in AutoCAD." We leave discussion of the Advanced wizard for the next chapter, "Setting Up a New Drawing."

EDITING AN EXISTING DRAWING

Many times, a previously saved drawing must be edited. To edit an existing drawing, you must first "open" it. To open an existing drawing, use the **Open** command.

When you enter **Open** from the keyboard (or choose **Open** from the **File** menu), the **Select File** dialog box is displayed. Figure 5.18 shows the **Select File** dialog box.

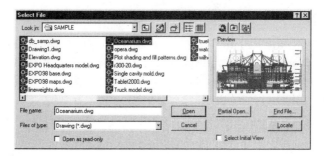

Figure 5.18 *Select File Dialog Box*

To open an existing drawing, choose a drawing name from the files listing. When you select the drawing, it is displayed in the **Preview** window.

 In AutoCAD 2000, you can select more than one drawing to be opened at the same time: Hold down the CTRL while selecting file names.

Choose the **Open** button to open the drawing(s).

If you wish to choose a different directory, double-click the desired directory (yellow folders) in the directories listing.

FINDING FILES

Sometimes you cannot remember the name of a drawing file, but you can remember what it looks like. While the **Select File** dialog box's **Preview** lets you see one file at a time, it is faster to see previews of all drawings. To help you preview any drawing file, AutoCAD includes the **Find File** button on its file-related dialog boxes.

When you choose the **Find File** button, Autodesk displays a tabbed dialog box: **Browse** and **Search** (see Figure 5.19).

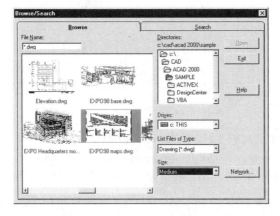

Figure 5.19 *Browse/Search Dialog Box*

Choose the subdirectory where you think the file is kept. AutoCAD displays a small preview image of every AutoCAD LT, R13, R14, and 2000 drawing in the subdirectory. The **Size** option lets you display the images in small, medium, or large size. Select the drawing you want, and choose **Open**.

SEARCHING FOR FILES

Sometimes you might not remember the location of the drawing file you need. It might be located in another subdirectory, another drive, or on another computer altogether. To help you find any file, the **Find File** feature is able to look at every subdirectory on every drive in your computer. Drives include the hard drives, floppy drives, CD-ROM drives, connected network drives, and other storage devices. In addition, if your computer is connected to a network of other computers, the **Find File** feature is able to search every subdirectory of every networked drive your computer has permission to read.

To find a specific file, click the **Search** tab of the **Browse/Search** dialog box (Figure 5.20), and fill in as much information as you can:

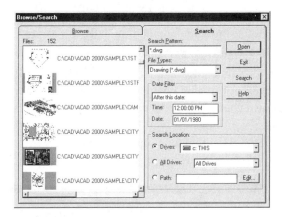

Figure 5.20 *Result of a Search*

Search Pattern: Allows you to specify the parts of the name that you may remember. Type in as much of the file's name as you can remember. Using wild-card characters lets you specify just part of a name. Recall that ? is a place holder for any character, while * searches for all files that match the rest of the criteria you supply. For example, Door????.* searches for all files that start with Door and have eight-character filenames, while Door????.dwg searches for all AutoCAD DWG drawing files that start with Door and have eight-character filenames.

File Types: A shortcut for selecting DWG drawing, interchange DXF, and template DWT files.

Date Filter: Useful for when you know roughly the date when the drawings were created. For example, if the drawing was created last month, specify "Before this date" and enter this month.

Search Location: Allows you narrow or broaden the search to specific drives, subdirectories, and paths.

After you have entered the search parameters, choose **Start Search**. AutoCAD spends a bit of time rummaging through your computer's drives and produces a list of all matching files. (You can at any time choose **Stop Search** to bring the process to a premature halt.) If the drawing was created with AutoCAD Release 13, 14, 2000, or LT, then the dialog box displays preview images.

SAVING AND DISCARDING DRAWINGS

During a drawing session or when you complete a session, you want to either save or discard the work you just completed. If you save your work, it is recorded to the hard drive on your computer. A saved file is recorded with the name you specify the first time you save it.

SAVING YOUR DRAWING AUTOMATICALLY

If you wish, you can set a time interval for AutoCAD to automatically save a temporary backup of your work. You can also save your work and remain in the drawing file. It is good practice to save your work to disk periodically. If you experience a power outage or your computer hangs up, you will lose the work completed since the file was last saved to disk. AutoCAD 2000 is set up to automatically save your work every 120 minutes to a temporary file, but you can set it up to back up more often.

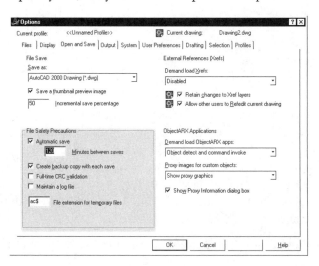

Figure 5.21 *Changing the Automatic Save Time*

To change the time between automatic backups, choose **Tools | Options | Open and Save** from the menu bar. In the **File Safety Precautions** area, change the value to 15 minutes. (It is not recommended that you select a time shorter than this.) Choose **OK**.

As with any software program, it is wise to save your work periodically, even if AutoCAD automatically saves to a temp file. See the next section for more information on saving drawings.

If you choose to discard your work, only the part of the drawing performed since the last file save is discarded. If you have not saved any of the work since the drawing was created, discarding the work deletes all instances of the drawing.

SAVING DRAWINGS

There are four commands you can use to save drawings. The following explains how each method works.

QSave

Use the **QSave** command when you wish to save your work and remain in AutoCAD. This is useful if you want to save your work at periodic intervals and continue to work.

When the drawing has not yet been named, the first time you use **QSave**, the **Save Drawing As** dialog box is displayed. (The default name for the first new drawing is Drawing1.Dwg.) You type the file name by which to save the drawing.

Save and SaveAs

The **Save** and **SaveAs** commands operate identically in AutoCAD. Both display the **Save Drawing As** dialog box. This command allows you to save the drawing in older formats that can be read by AutoCAD Release 14, 13, 12, and AutoCAD LT.

If AutoCAD is set to suppress dialog boxes (using the **Filedia** system variable), this prompt appears on the command line:

> *Save Drawing as <default>:*

If you enter the name of a drawing that already exists, AutoCAD displays the message box shown in Figure 5.22.

Figure 5.22 *Message Box*

If you want to replace the existing drawing with the new one, choose **Yes**. If you do not, choose the **No** button.

Save As R14, R13, and R12

The drawing files created by AutoCAD 2000 cannot be read by earlier versions of AutoCAD until they are translated. (AutoCAD 2000 can read drawings created by earlier AutoCADs.) For this reason, the **SaveAs** command includes options to translate the drawing to earlier formats.

To save the drawing in an earlier version, choose the **Save As Type** list box. The list box includes the following file format options:

AutoCAD R14 Drawing (*.dwg): Saves the drawing in Release 14 format, which can also be read by AutoCAD LT 97 and 98.

AutoCAD R13/LT 95 (*.dwg): Saves the drawing in a format that can be read by Release 13 and AutoCAD LT for Windows 95.

AutoCAD R12/LT R2 (*.dwg): Saves the drawing in a format read by Release 12, Release 11, and AutoCAD LT Release 2 and 1. Caution: some objects are lost in the translation.

OBJECTS LOST IN TRANSLATION

When you save an AutoCAD 2000 drawing in a format for an earlier version of AutoCAD, some objects are lost or modified because they are new to 2000. These objects cannot be represented in earlier versions of AutoCAD, so the translation must erase them or convert them to next-best objects.

Drawing Template File (*.dwt)

To save a drawing as a template file, choose the **Save As Type** list box. Then choose the template DWT file format. You can save a file as a template when you know you will want to use the exact same elements and settings in future files.

DISCARDING YOUR WORK—QUIT

Occasionally you will create a drawing that you do not wish to keep. If you want to exit the current drawing and discard the changes, use the **Quit** command. AutoCAD displays a message box that the drawing has been changed and asks you what you want to do.

If you want to exit the drawing, discard the changes, and exit the AutoCAD program, use the **Quit** (or Exit) command. A message box is displayed asking you whether you wish to save the changes, discard the changes, or cancel the **Quit** command and re-enter the command. The message box is shown in Figure 5.23.

Figure 5.23 *Message Box*

EXERCISES

1. Move the cursor around with your mouse or digitizer. Do the coordinates at the lower left corner of the screen move? Press F6. Move the cursor again and watch the coordinates.

2. Press F8. Watch the status line to see the display show the ortho mode turn on and off. Press F9 for snap mode.

3. Move the crosshairs to the top of the drawing area and into the menu bar.

4. Move the arrow from left to right over the entries. Position the arrow over the name **Draw** and click. A drop-down menu extends down into the drawing area. Move the cursor back to the name **Draw** and click. The drop-down menu disappears.

5. Choose the **Draw** drop-down menu again, and choose **Line**. Draw some more lines.

6. From the menu bar, choose **Format**, then **Layers**. Do you see the **Layer Properties Manager** dialog box?

7. Let's create a new layer. Move the cursor. Notice the arrow pointer. Choose the **New** button. Type **MYLAYER**.

8. Move the arrow to a blank area and click. Notice how the new layer name is now listed in the layer names listing.

9. Make the layer current by clicking on the layer name, then in the **Current** box.

10. To remove the dialog box from the screen, choose **OK** button at the bottom of the dialog box.

11. Look at the layer name in the **Layer** toolbar near the top of the screen. Is the layer "MYLAYER" current?

12. Press F2. Do you see the text screen? Press F2 again to return to the drawing screen.

13. Choose **Line** from the menu bar again. Draw some lines and use ESC to cancel.

14. Press the ENTER key. Did the **Line** command repeat?

15. Clear the command line with ESC again. Press the Spacebar. The **Line** command repeats again. Either ENTER or the Spacebar can be used to repeat a command.

16. Clear the command line again.

17. Press F1 from the keyboard and press ENTER. You should see the **Help** dialog box. To get more detailed information on a topic or to access a different section of the Help menu, position the cursor over any text that is green and click with the left mouse button. Familiarize yourself with using the **Help** dialog box.

CHAPTER REVIEW

1. Is it necessary to add a .dwg extension to the drawing name when you create a drawing file from the main menu?

2. What would be some of the benefits of using template drawings?

3. What are mode indicators? Where are they located when turned on?

4. How does a menu indicate the presence of a submenu? A dialog box?

5. Must you exit AutoCAD to copy or delete a file?

6. When you use the F2 key to use the flip screen function, in what way will the display be altered and what information is presented by using this option?

7. List the control and/or function key associated with the following definitions:

    Turns Ortho mode on and off _____

    Removes characters from the prompt line one at a time _____

    Allows you to toggle between graphic and text screens _____

    Toggles Tablet mode on and off _____

    Cancels all the characters on the command line _____

    Toggles to the next isometric plane in iso mode _____

    Used the same as the backspace key _____

    Cancels the current command and returns to the command line _____

    Toggles the screen coordinate display modes _____

    Toggles the grid on and off _____

    Toggles snap on and off _____

8. After a command has executed and you have returned to the command prompt, what happens when you press ENTER?

9. How is the Help facility invoked?

10. What is the purpose of the prompt area?

11. What is the danger of translating a AutoCAD 2000 drawing to an earlier version?

CHAPTER 6

Setting Up a New Drawing

A new drawing in AutoCAD is almost like a blank sheet of paper. Before starting work on a new drawing, you should first set up standards. Drawing standards ensure that all drawings come out looking the same. The firm you work for, whether as an employee or as a contractor, usually has established drawing standards; some drafting disciplines, such as architecture, also have established drafting standards. Drawing standards include the type of measurement units to be used and the scale factors. After completing this chapter, you will be able to

- Perform drawing setup

- Set the type of measurement used in your drawing

- Set the size of the workspace in the current units

- Set units and limits that will match the size of plot media and scale

- Display the status of the current drawing

- Work through the Advanced Setup wizard

SETTING UP A DRAWING

When you begin a drawing on the traditional drawing board, you start by determining the scale and size of the drawing. You would not draw the first line before choosing the paper size and selecting the scale at which you will construct the drawing.

AutoCAD uses settings that somewhat parallel these decisions. You could, as you read in Chapter 1, just start drawing with AutoCAD's default settings. The problem is that those settings are not suitable for all drawings. Consider, for example, the difference between architectural (fractional inches) and engineering (decimal inches) scales; the difference between European drawings (metric units) and North American (imperial units); and the differences between drawing on small 8 1/2" x 11" and large 36" x 48" media.

To use AutoCAD properly, you must learn to perform some basic settings. In this chapter, you learn the following concepts and commands:

Units: Sets up AutoCAD to draw in specified units, such as architectural, engineering, decimal, and scientific

Limits: Sets up the "real world" drawing area

Scaling: Sets up units and limits to match the desired plot scale

Status: Displays the settings in the drawing

Summary Info: Records information about the drawing

SETTING THE DRAWING UNITS

Since CAD programs are used in many types of work, the units by which distances are measured can be in a variety of formats. The engineer, scientist, and architect require different notations for distances, coordinates, and angles. AutoCAD provides the capability for each through the **Units** command. Setting the units is the first step in preparing a new drawing.

To set the units format for your drawing, select **Units** from the **Format** drop-down menu. AutoCAD displays a dialog box that you use to set the type of units. Figure 6.1 shows the dialog box.

The first section of the dialog box is titled **Length**. Let's look at the available choices.

There are five selections for the **Type** of units:

Architectural: Architectural drawings use the feet, inches, and fractions style of showing measurements. For example, 1' 9 3/4" is an example of architectural units.

Note: The Architectural and Engineering formats designate that one unit equals 1 inch.

Decimal: Decimal units are commonly used for metric units; decimal units are the *default* units (the units used in a new AutoCAD drawing, if no other units are selected). An example of decimal units is 21.75. Note the lack of unit notation, such as m (meter); for this reason, decimal units are sometimes called "unitless."

Figure 6.1 *Drawing Units Dialog Box*

In the metric system, there are 100 cm (centimeters) in a meter; 10 mm (millimeters) in a centimeter.

Engineering: Engineering units are used in applications such as civil engineering and highway construction. In engineering units, 1' 9 3/4" are shown as 1'-9.75". There are 12 inches in a foot; 3 feet in a yard.

Fractional: AutoCAD can also be set to display fractional units. An example of fractional units is 1 9 3/4. Note the lack of unit notation, such as ' (feet) or m (meters).

Scientific: AutoCAD provides scientific units for use in projects that deal with very large dimensions. An example of scientific units represents 21.75 units as 2.1750E+01. Note the lack of unit notation, such as ' (feet) or m (meters).

To set the desired units, simply choose the type from the drop-down list.

Note that the **Units** command affects only the *display* of distances, coordinates, and angles by AutoCAD. You are free to input data in any format and any precision. AutoCAD stores the data internally in unitless (decimal) format to an accuracy of 14 decimal places.

When displaying units, for example, on the status line, AutoCAD converts to the unit type and precision you select by applying the **Units** command. You are free to change the units display at any time.

Figure 6.2 shows the user inputting data in decimal units (at the 'Command:' prompt), such as 7,5.25. On the status line, AutoCAD is displaying the coordinates in architectural format: 0'7",0'5 1/4".

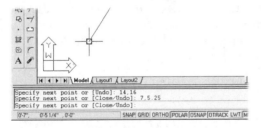

Figure 6.2 *Displaying Architectural Units on the Status Line*

Setting the Precision for the Units

After you have selected the format, you should set the precision for the coordinates and distances. This is done in the **Precision** section. For example, if you choose **architectural** format, you need to determine the smallest fractional denominator. To do this, use the **Precision** drop-down list box to change the number in the precision box. Precision ranges from 0" to 1/256" for factional units, and 0 to 0.00 000 000 for decimal units.

SETTING THE ANGLE MEASUREMENT

After selecting the format and the precision for the units, set the method of angle measurement. The angle measurement setting determines the coordinate display. The type of angle measurement is set in the right half of the **Drawing Units** dialog box (see Figure 6.1).

AutoCAD can display five types of angle measurement. The following is a list of the different types of measurement and an example of each.

> **Decimal Degrees:** This option instructs AutoCAD to display whole degrees as whole numbers, and partial degrees as decimals, such as 45.1234. Decimal degrees are the default units. Note the lack of units notation, such as ° (degrees).

> **Deg/Min/Sec:** Degrees are represented as whole degrees; partial degrees are represented as minutes; partial minutes as seconds. There are 360 degrees in a circle; one degree contains 60 minutes; 1 minute contains 60 seconds. AutoCAD uses **d** to represent degrees, ' to designate minutes, and " to represent seconds. For example, 22 degrees, 14 minutes, 45.23 seconds is listed as 22d14'45.23".

> **Grads:** AutoCAD can specify angles in grads. An example is 35.0000g. There are 400 grads in a circle.

> **Radians:** For programming work, you use radians to calculate angles. An example is 0.678r. There are 2π (approximately 6.282) radians in a circle.

> **Surveyor's Units:** Surveyor's units are used to designate property lines. It is a form of the degrees/minutes/seconds type of format, except that they are relative to compass points. An example is N22d15'32"W, with N and W representing north and west, respectively.

To set the angle measurement for your drawing, select the angle type from the drop-down list.

Setting the Precision for the Angle Measurement

After you have specified the angle format, set the precision of the angle format. You can select up to eight decimal places for decimal degrees, and up to four decimal places for D. M'S" degrees. The angle precision is set at the angle precision box.

Setting the Angle Orientation

After you have selected the unit types, you can designate whether AutoCAD measures angles in a clockwise or counterclockwise direction. As you become familiar with the operation of some commands, you will better understand the effect of this setting. The default setting is counterclockwise. Check the **Clockwise** box to measure angles clockwise.

Setting the Zero-Angle Direction

Set the zero-angle direction by first selecting the **Direction** button. The resulting dialog box appears as follows.

Figure 6.3 *Direction Control Dialog Box*

AutoCAD by default sets the angle zero to the right (east). This setting makes all angles that have a direction of zero read to the right. You can set the zero angle to display in any direction you wish. This will not affect text in the drawing, since you set the text angle relative to this setting.

For example, you might want the zero angle to be straight up on the screen. Selecting **North 90.0** achieves this. To set the desired zero angle direction, select the appropriate radio button.

 Note: You can, if you wish, "show" AutoCAD the angle for "0 degrees." To show AutoCAD the zero direction, choose the **Other** radio button, then choose the **Pick an Angle** button, then select any point on the screen. Next, move the cursor in the direction of the desired angle and enter a second point.

This can be useful if you must work in reference to a certain angle.

Once you have set the type of unit measurement and the method of angle definition, the settings will remain in effect throughout the remainder of the drawing. If you save your drawing and return, the settings will remain intact. It is only necessary to set the units when you are beginning a new drawing.

Alternately, you can use a template drawing to preset the drawing units.

SETTING THE DRAWING LIMITS

Setting the limits of your drawing allows you to determine the size of working area you need to draw in. This can be helpful because the area needed to construct different types of drawings can be dramatically different. For example, a grid map showing an area of a city and the detail section of a watch each require different limits.

To better understand how drawing limits work, you should also read the section following on scaling your drawings.

 Note: Limits are set in the current drawing units. For this reason, you should set the units before specifying the limits.

To set the limits, choose **Drawing Limits** from the **Format** drop-down menu. The **Limits** command does not have a dialog box; the sequence is shown at the command line:

> Command: '_limits
>
> Reset Model space limits:
>
> Specify lower left corner or [ON/OFF] <0'-0",0'-0">: *(Pick lower left corner.)*
>
> Specify upper right corner <1'-0",0'-9">: *(Pick upper right corner.)*

Let's look at each of the command line options.

> **On:** Turns on the limits. When system variable **LimCheck** is on, you can only draw within the area of the limits.
>
> **Off:** Turns off the limits. When system variable **LimCheck** is off, you can draw either inside or outside the area defined by the limits.
>
> **Lower Left Corner:** The default allows setting of the lower left corner (by coordinates) for the drawing area. If you enter a coordinate for the lower left corner, you are prompted for the coordinate of the upper right corner.
>
> **Enter:** Retains the current (default) value for the lower left corner and prompts you for the coordinates for the upper right corner.

The initial setting of the drawing limits is determined by the template drawing. You can, however, change the drawing limits at any time using the **Limits** command.

If you wish to prevent yourself from drawing outside the limits, turn on the **LimCheck** system variable, as follows:

> Command: **limcheck**
>
> Enter new value for LIMCHECK <0>: **1**

Among system variables, 0 (zero) is shorthand for off, while 1 (one) means on. With **LimCheck** turned on (set to 1), AutoCAD warns you "**Outside limits" when you attempt to draw outside the drawing limits; AutoCAD does not draw anything.

Notes About Limits:

The screen does not display the new limits after they are set. Make a habit of executing the **Zoom All** command after setting new limits. This displays the actual drawing page size on the screen.

To see the size of your drawing page, turn on the grid (press F7). This allows you to visualize the edges of your "paper."

Limits have an effect on two of AutoCAD's commands. The **Grid** command displays a grid of dots; the extent of this grid is constrained by the setting of the **Limits** command (see Figure 6.4). Similarly, the Zoom All command zooms to either the: (1) drawing limits; or (2) current extents—whichever is larger.

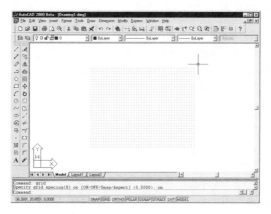

Figure 6.4 *Grid Extents are Defined by Limits*

SCALING THE DRAWING

One of the hardest concepts for a new CAD user to grasp is that of scale. Think of when you sketch a picture of your house on a piece of paper. You draw the house small enough to fit the paper (there isn't a piece of paper big enough to for you to draw the house full size!). This is called *scaling* the house to fit the paper.

A typical house is 50 feet long. To fit on your 10"-wide paper, you sketch the house 60 times smaller. We say that the scale is 10" = 50' or 1:60. The 1:60 is called the *scale factor*. One way to think of scale factor is to think to yourself, "One inch on my drawing is 60 inches on the house."

The math works out like this: 50 feet x 12 inches/foot = 600 inches. Divide 600 inches by 10 inches = 60. Another way of looking at it is:

$$\frac{50 \text{ feet} \times 12 \text{ inches/ft}}{10 \text{ inches}} = \frac{600}{10} = 60$$

Here are some more examples of scale factors:

- The truck is 20 feet long. The paper is 10 inches wide. The scale factor works out to be 1:24.

- The sailboat is 30 m long. The paper is 1m wide. The scale factor is 1:30 (it's much easier in metric, isn't it?).

- The gear is 1 inch in diameter. The paper is 8 inches wide. The scale factor is 8:1 (notice that the scale factor numbers are "reversed" when small objects are drawn at a larger scale factor).

SCALING A CAD DRAWING

In CAD, scaling isn't done in the same way. Instead of creating a drawing to scale, you create the CAD drawing at full size. That means the 50-foot long house is drawn 50 feet long; the one-inch gear is drawn one inch in diameter. There is no limit to the size of drawing that AutoCAD can create; you could draw the entire solar system—full size—if you want to. Full size is shown as 1:1.

Scale comes into the scene when it comes time to plot your drawing. That huge drawing, whether of your house or your solar system, must be made small enough to fit the paper it is plotted on. That's where scale factor comes into play. You tell AutoCAD's **Plot** command to make the house 60 times smaller to fit the paper.

There is a catch, however. Some of the items in your drawing cannot become 60 times smaller. These include text, hatch patterns, linetype patterns, and dimensions. These are called *scale-dependent*. If you made the text 60 times smaller, you would not be able to read it. The solution is simple: draw the text 60 times larger. When it is plotted, the text is the correct size.

You may be puzzled. How big should text be when plotted on a drawing? The standard in drafting is 3/8" for normal-sized text. When you place text in your house drawing, specify a height of 22.5" tall (= 3/8" x 60). That might seem much too high for you (nearly two feet high!) but trust me: when it is plotted, it will look exactly right.

In the same way, you must apply the scale factor to other scale-dependent objects, including dimensions, linetypes, and hatch patterns. For the house drawing example, all these items are drawn 60 times larger in order that they look right when plotted.

This is one of the most important things to learn about CAD: *scale-dependent items must be drawn at the inverse of the scale at which you plan to plot the drawing.*

SCALED DRAWING LIMITS

You can use the drawing limits to help you "see" the size of the paper on which you will be plotting the drawing. For example, before you begin drawing, you decide the scale is 1"=10 units. You have determined that one AutoCAD unit equals a foot. Then, the scale of the plot will be 1"=10'.

Let's suppose that you are planning to plot on paper that measures 36" x 24" (also known as "D-size media"). How would you set limits that are in true relation to the page size?

You multiply the dimensions of the paper by the units per inch. This determines the number of units in the *X* and *Y* dimensions of the paper:

$$10 \text{ units/inch} \times 36" = 360 \ (X \text{ limit})$$

$$10 \text{ units/inch} \times 24" = 240 \ (Y \text{ limit})$$

If you now set limits of 0,0 (lower left) and 360,240 (upper right), the limits will match the page size.

You can, of course, change the plotting scale later. If you do, you should change the limits if you want to continue to match the limits to the paper size.

Knowing the methodology of calculating the sheet size to the intended plotting scale allows you to set up your drawing to see the available drawing page size.

Table 6.1 shows the relationship between drawing scale and plot sheet size. You can use this table to set up the limits for your drawings.

CHECKING THE DRAWING STATUS

AutoCAD has a multitude of modes, defaults, limits, and other parameters that you will occasionally need to know. The **Status** command displays the current state or value of each of these. The **Status** command may be used any time you are in AutoCAD. To view it, choose **Tools** from the menu bar, then **Inquiry**, and **Status**. You are presented with the Text window as shown in Figure 6.5.

The third line, **Model space limits**, refers to the the limits you set using the **Limits** command. The fourth line, **Model space uses**, describes the *X,Y* coordinates of a rectangle containing the drawing. ****Over** means part of the drawing is outside the limits.

Press F2 to return to the AutoCAD drawing window.

DRAWING SUMMARY

You can store information about the drawing, as well as view additional data, with the **Drawing Properties** dialog box. From the **File** menu, choose **Drawing Properties**. The **Drawing Properties** dialog box has four tabs.

	SHEET SIZE				
FINAL PLOT SCALE	A 11 x 8½	B 17 x 11	C 24 x 18	D 36 x 24	E 48 x 36
1/16	176', 136'	272', 176'	384', 288'	576', 384'	768', 576'
3/32	132', 102'	204', 132'	288', 216'	432', 288'	576', 432'
1/8	88', 68'	136', 88'	192', 144'	288', 192'	384', 288'
3/16	66', 51'	102', 66'	144', 108'	216', 144'	288', 216'
1/4	44', 34'	68', 44'	96', 72'	144', 96'	192', 144'
3/8	29'-4", 22'-8"	45'-4", 29'-4"	64', 48'	96', 64'	128', 96'
1/2	22', 17'	34', 22'	48', 36'	72', 48'	96', 72'
3/4	14'-8", 11'-4"	22'-8", 14'-8"	32', 24'	48', 32'	64', 48'
1	11', 8'-6"	17', 11'	24', 18'	36', 24'	48', 36'
1½	7'-4", 5'-8"	11'-4", 7'-4"	16', 12'	24', 16'	32', 24'
3	3'-8", 2'-10"	5'-8", 3'-8"	8', 6'	12', 8'	16,' 12'

	SHEET SIZE				
FINAL PLOT SCALE	A 11 x 8½	B 17 x 11	C 24 x 18	D 36 x 24	E 48 x 36
10	110, 85	170, 110	240, 180	360, 240	480, 360
20	220, 170	340, 220	480, 360	720, 480	960, 720
30	330, 255	510, 330	720, 540	1080, 720	1440, 1080
40	440, 340	680, 440	960, 720	1440, 960	1920, 1440
50	550, 425	850, 550	1200, 900	1800, 1200	2400, 1800
60	660, 510	1020, 660	1440, 1080	2160, 1440	2880, 2160
100	1100, 850	1700, 1100	2400, 1800	3600, 2400	4800, 3600
Full Size	11, 8.5	17, 11	24, 18	36, 24	48, 36

Table 6.1

```
AutoCAD Text Window - Drawing1.dwg                              _ □ X
Edit

Command: status
46 objects in Drawing1.dwg
Model space limits are X:   14.5262   Y:    4.7351
                        X:   35.8408   Y:   20.3172
Model space uses        X:   10.9685   Y:    7.4550 **Over
                        X:   12.7531   Y:   15.5073
Display shows           X:    2.9380   Y:    0.1709
                        X:   47.4290   Y:   24.8813
Insertion base is       X:    0.0000   Y:    0.0000   Z:    0.0000'
Snap resolution is      X:    0.5000   Y:    0.5000
Grid spacing is         X:    0.5000   Y:    0.5000

Current space:          Model space
Current layout:         Model
Current layer:          "0"
Current color:          BYLAYER -- 7 (white)
Current linetype:       BYLAYER -- "Continuous"
Current lineweight:     BYLAYER
Current plot style:     ByLayer
Current elevation:      0.0000  thickness:      0.0000
Fill on  Grid on  Ortho off  Qtext off  Snap off  Tablet off
Object snap modes:      Center, Endpoint, Intersection, Extension
Free dwg disk (C:) space: 2047.7 MBytes
Free temp disk (C:) space: 2047.7 MBytes
Free physical memory: 8.7 Mbytes (out of 191.3M).
Free swap file space: 1746.9 Mbytes (out of 1856.7M).

Command:
```

Figure 6.5 *Drawing Status Information*

Figure 6.6 *Drawing Properties Dialog Box*

The **General** tab tells you where the drawing file is located, its size, and its attributes (such as whether it is read-only or hidden). This tab reports also when the drawing was created, last modified (edited), and last accessed (opened).

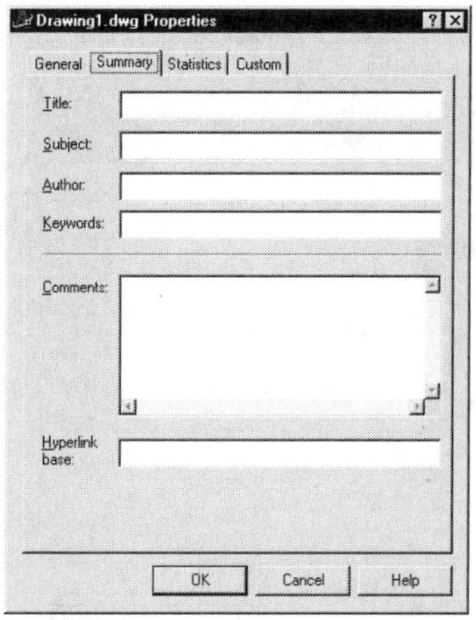

Figure 6.7 *Summary Tab of the Drawing Properties Dialog Box*

The **Summary** tab allows you to enter predefined properties such as **Author** (the drafter), a **Title** for the drawing (such as the project name), and **Subject**. The **Keywords** field makes it easier to use search software to find the drawing.

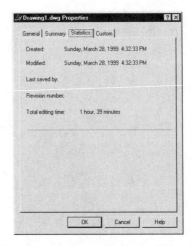

Figure 6.8 *Statistics Tab of the Drawing Properties Dialog Box*

The **Statistics** tab reports the time (in hours and minutes) that the drawing has been open. Note that **Total Editing Time** is *not* the actual time spent editing; rather, it is the time that the drawing has been open in AutoCAD.

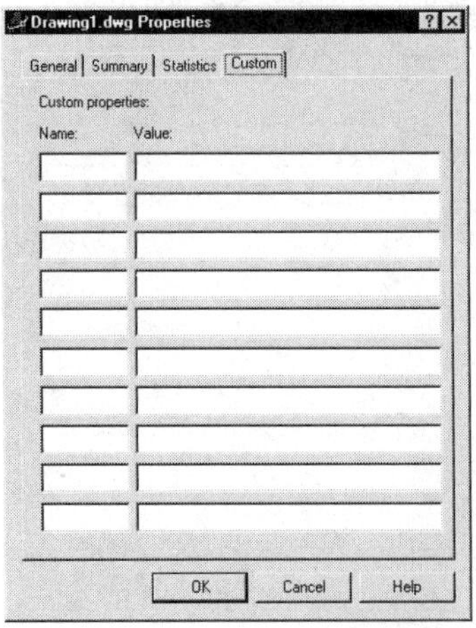

Figure 6.9 *Custom Tab of the Drawing Properties Dialog Box*

The **Custom** tab allows you to include any kind of information you want. There are ten fields available for your use.

Choose **OK** to dismiss the dialog box.

EXERCISES

1. Let's suppose that you would like to set limits that have 180.00 units horizontal (*X* value) and 120.00 units vertical (*Y* value). First, verify that the units are set to decimal units. Next, from the menu bar, choose the **Format** menu and choose **Drawing Limits**.

   Lower left corner = 0,0

   Upper right corner = 180,120

   This sets the limits with a lower left corner (the origin) at 0,0 and the upper right corner at the coordinates of 180,120. You do not have to retain 0,0 for the lower left corner; however, this should be maintained unless you have special circumstances that require a change.

2. Calculate drawing limits for a drawing that will be plotted on a 24" x 18" sheet of paper at 1_4" = 1'-0" scale.

3. Calculate drawing limits for a drawing that will be plotted on a 36" x 24" sheet of paper at 1" = 50.00'.

4. Use the **Status** command to check the limits of your current drawing.

TUTORIAL

ADVANCED SETUP WIZARD

To help step you through all stages of setting up a drawing, AutoCAD includes the Advanced Setup wizard. In this tutorial, you set up a drawing that needs to be 180 feet by 120 feet in size. You'll work in architectural units and will want to display the accuracy to the nearest 1/2-inch.

1. Choose the **New** command from the **File** menu.

2. When the **Create New Drawing** dialog box appears, choose **Use A Wizard** (the fourth button).

Figure 6.10 *Create New Drawing Dialog Box.*

3. Choose **Advanced Setup**, and then choose **OK**. AutoCAD displays **The Advanced Setup** dialog box. Along the left edge, notice the five steps: Units, Angle, Angle Measure, Angle Direction, and Area.

AutoCAD uses the units setting to know how to display units on the status line, in dimensions, and in dialog boxes.

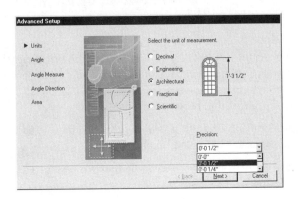

Figure 6.11 *Advanced Wizard Dialog Box*

4. Choose the **Architectural** radio button.

5. In the **Precision** list box, choose 0'-0-1/2".

6. Choose **Next**. AutoCAD uses the angle setting for formatting the display of angles in dimensions, on the status line, and in dialog boxes.

7. Click the radio button next to **Deg/Min/Sec**.

8. From the **Precision** list box, choose 0d00'00". Notice how the preview image updates to show your selection.

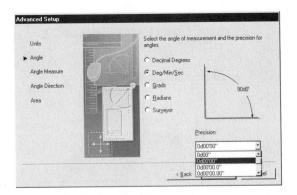

Figure 6.12 *Angle*

9. Choose **Next**. By default, AutoCAD measures angles from west. However, you can change the setting by picking one of the four directions, or typing any angle you like. This then becomes the 0-degree point from which AutoCAD starts measuring angles.

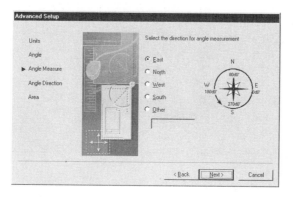

Figure 6.13 *Angle Measure*

10. Accept the default (west=0 degrees) by choosing **Next**. By default, AutoCAD measures its angles in counterclockwise direction. If your drawing needs it, here is where you tell AutoCAD to measure angle in the clockwise direction.

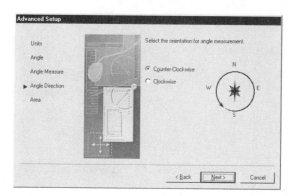

Figure 6.14 *Angle Direction*

11. Accept the default (angles measured counterclockwise) by choosing **Next**.

12. The default size of 1 foot by 9 inches is too small for your drawing. Click the **Width** text box and type **180'**. The apostrophe (') suffix is important, otherwise AutoCAD interprets the number as inches.

13. In the **Length** box, type **120'**. Notice how AutoCAD updates the **Sample Area** preview to show the new limits of 180' by 120'.

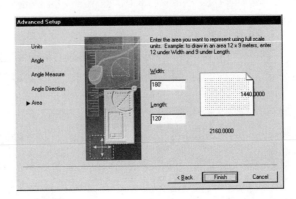

Figure 6.15 *Area*

14. Choose **Finish**. AutoCAD spends a few seconds setting up the new drawing. Note that AutoCAD does not automatically perform a **Zoom All**, so you might want to execute that command. The drawing is now ready, as shown in Figure 6.16.

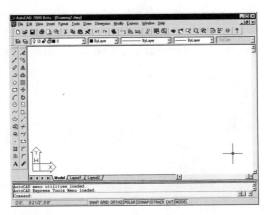

Figure 6.16 *Completed Drawing*

CHAPTER REVIEW

1. How might you obtain a listing of drawing parameters?

2. How are angles measured in surveyor's units?

3. What choice of unit systems is available in AutoCAD?

4. What can you do to visualize the drawing limits you have set?

5. Can the zero angle position be altered? How?

6. Why should the units of a drawing be set before the limits?

7. What restrictions apply to your drawing when the limits are turned on?

8. Why would you want to change the zero angle setting from the default position?

9. What are the only exceptions to drawing in generic units?

10. At what point does AutoCAD need to know the units you have used to construct your drawing?

11. At what scale factor do you create your drawing in AutoCAD?

12. How large should text be in a drawing that will be plotted at a scale factor of 1:50?

CHAPTER 7

Creating the Drawing

Most CAD drawing activities involve the most basic drawing commands. After completing this chapter, you will be able to

- Demonstrate the most basic drawing commands
- Use display commands to magnify and maneuver around the drawing
- Understand the commands that help you draw more accurately
- Reverse the undesirable effects of a command

GETTING STARTED

Now that we know how to set up a drawing, let's jump in and start drawing right now. We will be using some basic drawing commands, modes, and assistance. They are

LINE: Draws line segments.

POINT: Constructs a point object.

CIRCLE: Draws a circle by several different methods.

ARC: Constructs an arc by eleven methods.

OSNAP: Assists in connect precisely to geometric features of an object.

AUTOSNAP: Automatically displays object snap modes.

POLAR: Automatically snaps to specific angles.

REDRAW: Clears the drawing screen of clutter.

REDRAWALL: Redraws all viewports.

REGEN: Regenerates the drawing display from the database.

ZOOM: Enlarges and reduces the view of the drawing.

PAN: Moves the screen around the drawing.

AERIAL VIEW: Displays the entire drawing in a small window.

GRID: Places a "grid paper" of dots on the screen.

SNAP: Sets a drawing increment.

ORTHO: Forces the cursor movement to be either perfectly horizontal or vertical.

DIRECT DISTANCE ENTRY: Draws relative distances.

TRACKING: Moves relative distances.

U (single undo): Undoes a single drawing operation.

REDO: Negates a single undo operation.

DRAWING LINES

Drawing lines is the most basic part of a CAD program. In AutoCAD you can draw different types of lines and apply various line options, but all lines begin with the **Line** command.

DRAWING A LINE

To draw a line, you must first use the **Line** command. The **Line** command can be accessed from either the menu bar, a toolbar button, or in the command window. From the menu bar, choose **Draw**, and then in the **Draw** menu, choose **Line**. (From the **Draw** toolbar, choose the **Line** button.)

> Command: **line**
>
> Specify first point:
>
> Specify next point or [Undo]:

The command line prompts to "Specify first point:". Before we proceed to draw, let's first talk about "first points" and "next points."

You must first show AutoCAD the point from which you wish to start drawing. Think of this as the place where you first put your pencil to the paper. This is the "first point." Now you must determine the point at which the line ends. This is the "next point."

Many drawings contain lines that connect (such as a box consisting of four consecutive and connecting lines), so the "Specify next point:" prompt is repeated until you terminate the command by pressing ESC, or pressing ENTER a second time. This allows you to enter a series of lines without the inconvenience of invoking the **Line** command for each line segment.

Note: AutoCAD can draw more than one type of line:

- Polyline: Lines are drawn in the same manner as with the **Line** command but they can include arcs, splines, and variable width—all as a single object.

- Multiline: This feature allows you to draw up to 16 parallel lines at one time.

- Sketch: Sketching is "free hand" drawing.

- Construction Line and Ray: This feature draws infinite construction lines (xlines) and semi-infinite construction lines (rays) that display and are plotted.

LINE OPTIONS

Some drawing aids are associated with the **Line** command. These aids provide assistance that is unique to a CAD drawing system.

They are

1. Continue

2. Close

3. Undo

4. @

Let's look at each option.

Continue. When you terminate the **Line** command, you are free to begin a new line elsewhere. However, if you wish to go back and begin at the last endpoint, you can "reconnect" by using the **Continue** option. This option is hidden in AutoCAD: the next time you start the **Line** command, press ENTER at the "Specify first point:" prompt.

Close. When you construct a rectangle, the last line you draw is connected to the beginning point. Aligning the end of the line with the beginning of the first line can be a tedious process. The **Close** option automatically performs this for you.

To see how this works, let's construct the rectangle now. You will connect the lines at the final intersection with the **Close** option.

> Command: **line**
>
> Specify first point: *(Enter Point 1.)*
>
> Specify next point or [Undo]: *(Enter Point 2.)*
>
> Specify next point or [Undo]: *(Enter Point 3.)*
>
> Specify next point or [Close/Undo]: *(Enter Point 4.)*
>
> Specify next point or [Close/Undo]: *Close*

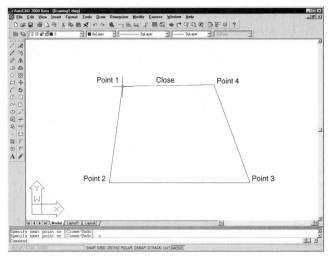

Figure 7.1 *Applying the Close Option*

The **Close** option is used to return to the start (the first point) of any consecutive string of lines. Using **Close** creates a closed polygon. Notice that the **Close** option is not available until you have drawn at least three line segments, the minimum number required to construct a polygon (a triangle, in this case).

Remember that you need only type the capitalized portion of an option. In the example above, you could have simply typed C, as follows:

> Specify next point or [Close/Undo]: **c**

Undo. Sometimes you draw a line segment that you want to erase. Instead of erasing and then reconnecting with the **Continue** option, you can "step back" through the series of line segments. This is accomplished with the **Undo** option. Type **U** at the "Specify next point" prompt to go back one line segment at a time. When you have backed up to the desired point, simply continue the line segments.

Undo is used while you are still active in the **Line** command. Notice that the **Undo** option in the **Line** command is similar to the **U** command, except that it is used on a *single* line segment while the **Line** command is active. Using the **U** command after the **Line** command is ended undoes *all* lines drawn by that **Line** command.

@. The **@** ("at") symbol is the prefix for entering relative coordinates, which you learn about later in this chapter.

Canceling the **Line** Command

You cancel the **Line** command by pressing ENTER, by using ESC, or by choosing another command from the menu. As of AutoCAD 2000, you can also right-click and choose **Cancel** from the cursor menu.

DRAWING LINES BY USING COORDINATES

Accurate drawing construction requires that lines and other objects in a drawing be placed in a precise manner.

One method of placing lines precisely is through the use of coordinates. There are three types of 2D coordinates: Cartesian (or absolute), relative, and polar. Let's look at each type and learn how to use it.

Cartesian Coordinates

Cartesian coordinates specify a point on an *X, Y* grid, with the 0,0 point as a reference. Let's look at how the *X,Y* coordinate grid works. Figure 7.2 shows a coordinate grid. The 0,0 point is at the center of the grid.

Figure 7.2 *The Cartesian Coordinate System*

Notice that the grid has two lines, known as axes. The *X* axis runs horizontally, and the *Y* axis runs vertically. The intersection of the *X* and *Y* axes is located at the 0,0 point.

The increments increase in positive numbers as you move to the right on the *X* axis and upward on the *Y* axis. Likewise, they decrease to the left in the *X* axis and down along the *Y* axis.

The AutoCAD drawing screen can be overlaid on the grid as shown in Figure 7.3. Notice that the 0,0 point is in the lower left corner of the screen. This is the "normal" work screen in AutoCAD. If you remember, when we learned how to set the drawing limits in Chapter 6, we set the lower left drawing corner to 0,0. This places the AutoCAD screen in the upper right quadrant of the grid where all *X* and *Y* values are positive.

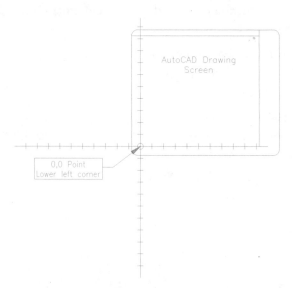

Figure 7.3 *AutoCAD Drawing Area Superimposed on the Coordinate System*

A single point on the grid can be identified by a coordinate. This coordinate is called an absolute coordinate and is specified by the numeric value of the location where it lines up with the *X* and *Y* axes, with respect to the 0,0 point. That designation is listed in an "*X, Y*" format. Let's look at an example.

Figure 7.4 shows a coordinate grid with the AutoCAD screen placed in the normal location. The point 6,4 is shown on the grid. Notice that it is 6 units to the right along the *X* axis and 4 units up on the *Y* axis.

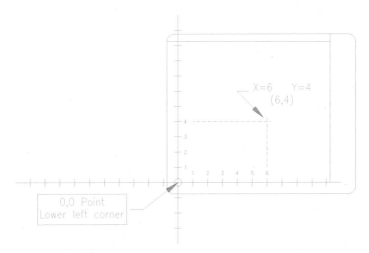

Figure 7.4 X,Y *Coordinate of 6,4*

You use absolute coordinates to specify points for our commands. Consider the example of a **Line** command. When you choose the **Line** command, you are first prompted for the starting point, then the endpoint of the line segment. If you enter 2,2 for the beginning point and 5,4 for the endpoint, AutoCAD draws a line as shown in Figure 7.5.

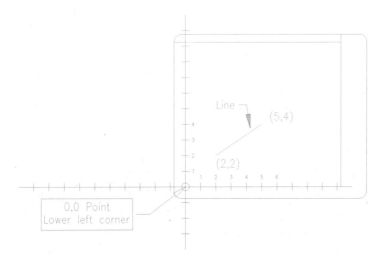

Figure 7.5 *Drawing a Line with* X,Y *Coordinates*

Relative Coordinates

Relative coordinates are not specified from the 0,0 point, but from any given point. Relative coordinates are typically specified from the last point entered. Relative coordinates can be used as the first point to offset from the last point.

Relative coordinates are specified in the same *X,Y* format as absolute coordinates. The difference is that relative coordinates describe the distance from the last point entered. (Recall that absolute coordinates describe the *X,Y* distance from the 0,0 origin.) Let's look at an example.

You can draw the same line we constructed in Figure 7.5 with relative coordinates. Refer to Figure 7.6.

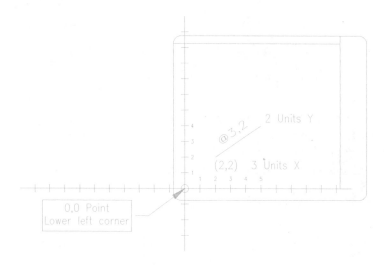

Figure 7.6 *Drawing a Line with Relative Coordinates*

When you start the **Line** command, you specify the first point with the absolute coordinate of 2,2. When prompted for the endpoint of the line segment, you can enter

@3,2

You can think of this as "at the last point, go three units in the *X* direction and two units in the *Y* direction." Notice the format used—the @ symbol is used to specify a relative coordinate. The "@ *X,Y*" format is always used in this manner.

Your line is drawn three units *X* and two units *Y* from the first point (2,2) as shown in Figure 7.6. Note also that you could have entered the first endpoint of the line by simply moving the cursor on the screen and entering the beginning point wherever you wished.

Polar Coordinates

Polar coordinates are a type of absolute and relative coordinate. Polar coordinates, however, specify a point by defining a distance and *angle* from the last point. The format is

@distance<angle

Before you can properly specify an angle, you need to look at the default AutoCAD angle specification. Figure 7.7 shows the angle specifications used by AutoCAD. The default direction for angle zero is to the right (east). The default angle rotation is counterclockwise. (The direction for angle zero and the angle rotation direction can be changed with the **Units** command.)

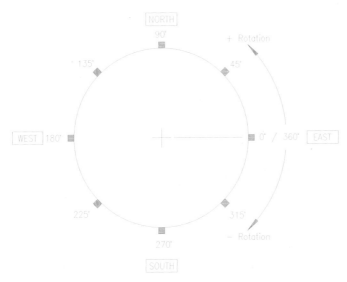

Figure 7.7 *AutoCAD's Default Angle Specification*

Now that you know AutoCAD's angle specification, let's look at an example using polar coordinates. You could specify the endpoints of a line by entering the first endpoint, then enter

@10<30

for the endpoint of the segment. This would draw a line segment that is 10 units and 30 degrees from the first point, as shown on the graph in Figure 7.8.

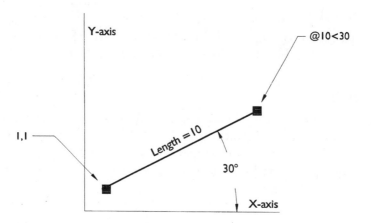

Figure 7.8 *Drawing a Line with Polar Coordinates*

As you proceed through each of AutoCAD's commands, you will acquire a perception of the best method of point entry to be used for the task.

DRAWING POINTS

A point is a dot (like a single touch of a pen point) drawn in the drawing. To place the point, choose **Draw | Point | Single Point** from the menu bar. (Note that **Draw | Point | Single Point** uses a vertical bar to separate menu choices. In this case, you choose the **Draw** menu, followed by the **Point** sub-menu, followed by the **Single Point** item.)

Command: **Point**

Current point modes: **PDMODE=0 PDSIZE=0.0000**

Specify a point: *(Pick)*

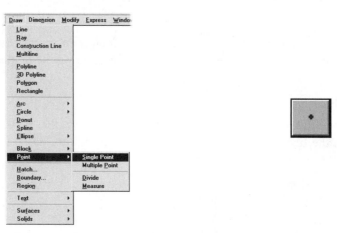

To draw more than one point, choose **Draw | Point | Multiple Point**.

POINT AND SIZES

Points can differ in design. You specify the type and size of point you want by using the **Pdmode** and **Pdsize** system variables. Above, the **Point** command indicated that both variables were set to 0.

Pdmode designates the type of point drawn, and **Pdsize** controls the size of the point entry. It is however, much easier to use the **DdPType** command to display a dialog box (see Figure 7.9), or choose **Point Style** from the **Format** menu.

Figure 7.9 *Point Types*

You can have the point drawn as a circle, square, line, or cross—or a combination of these. If you use **Pdmode** to specify a point type, Figure 7.10 shows the possible combinations and their corresponding point objects:

Figure 7.10 *Point Types by Reference Number*

Pdsize controls the size of the point objects. (**Pdmode** values of 0 and 1 are not affected by **Pdsize**.) If a positive number is entered, an absolute size for the point is specified. If the number is negative, AutoCAD uses the number as a percentage of the screen size. Thus, if you zoom in or out, the size is approximately the same in relation to screen size.

 Note: Pdmode and Pdsize are retroactive. Setting new Pdmode or Pdsize values will change all existing points to reflect the new size and style following the next Regen command.

DRAWING CIRCLES

AutoCAD provides six methods of constructing circles. These are

Center and radius	Three-point circles
Center and diameter	Tangent circles
Two-point circles	Tangent three-point circles

Let's look at each type of circle construction. It is helpful to read each brief section, then follow the short exercise that follows.

DRAWING A CIRCLE BY SPECIFYING CENTER AND RADIUS

AutoCAD allows you to enter the center point of the circle and specify the radius. Designate the radius either by entering a numeric value or by showing AutoCAD the distance on the screen by entering a point on the circumference of the circle.

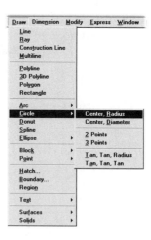

Command: **Circle**

Specify center point for circle or [3P/2P/Ttr (tan tan radius)]: *(Select center.)*

Specify radius of circle or [Diameter]: *(Select radius.)*

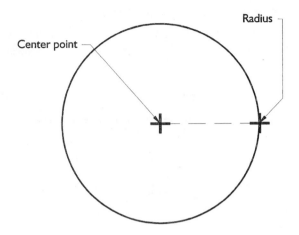

Figure 7.11 *Drawing a Circle by Center and Radius*

DRAWING A CIRCLE BY SPECIFYING CENTER AND DIAMETER

You can construct a circle by stipulating the center point and a diameter. Either choose **Center, Diameter** from the **Circle** menu, or enter **D** when prompted for the diameter or radius. You will then be prompted for the diameter.

> Command: **Circle**
>
> Specify center point for circle or [3P/2P/Ttr (tan tan radius)]: *(Pick center.)*
>
> Specify radius of circle or [Diameter]: **D**
>
> Specify diameter of circle <1.0000>: *(Pick diameter.)*

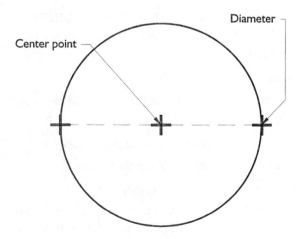

Figure 7.12 *Drawing a Circle by Center and Diameter*

DRAWING A CIRCLE BY SPECIFYING TWO POINTS

Responding to the **Circle** command's initial prompt with **2P** allows you to construct the circle by showing AutoCAD two points on the circumference.

Command: **Circle**

Specify center point for circle or [3P/2P/Ttr (tan tan radius)]: **2P**

Specify first end point of circle's diameter: *(Select first point.)*

Specify second end point of circle's diameter: *(Select second point.)*

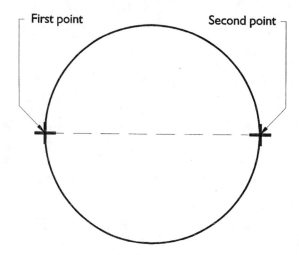

Figure 7.13 *Drawing a Circle by Two Points on the Diameter*

DRAWING A CIRCLE BY SPECIFYING THREE POINTS

Simply entering three points on the circumference also draws a circle. You do this by specifying the **3P** option.

Command: **circle**

Specify center point for circle or [3P/2P/Ttr (tan tan radius)]: **3P**

Specify first point on circle: *(Select first point.)*

Specify second point on circle: *(Select second point.)*

Specify third point on circle: *(Select third point.)*

DRAWING CIRCLES TANGENT TO OBJECTS

AutoCAD allows you to construct circles tangent to lines or circles. You can do this by identifying two lines, two circles, or one line and one circle to construct a tangent

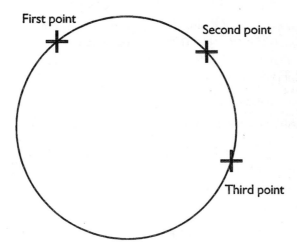

Figure 7.14 *Drawing a Circle by Three Points on the Circumference*

circle to, and the tangent circle's radius. To construct tangent circles, choose the **Tan, Tan, Radius** option from the **Circle** submenu.

> Command: **Circle**
>
> Specify center point for circle or [3P/2P/Ttr (tan tan radius)]: **TTR**
>
> Specify point on object for first tangent of circle: *(Select first circle/line-point "A".)*
>
> Specify point on object for second tangent of circle: *(Select second line/circle-point "B".)*
>
> Specify radius of circle <1.0000>: *(Enter value.)*

The following illustrations show the different effects of the **Circle** command's **TTR** option in constructing circles tangent to lines and/or circles.

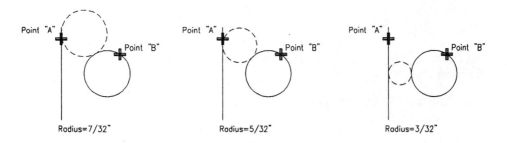

Figure 7.15 *Drawing a Circle by Two Tangents and a Radius*

DRAWING TANGENT THREE-POINT CIRCLES

You can construct three-point circles tangent to any combination of lines and circles by specifying each point using the **TANgent snap mode** option. Let's look at an example of this. Construct a square box of two units on each side as shown in the following illustration. From the **Draw | Circle** menu, choose **3 Points**.

Command: **circle**

Specify center point for circle or [3P/2P/Ttr (tan tan radius)]:   **3P**

Specify point on object for first tangent of circle: **tan**

to *(Choose point "A".)*

Specify point on object for second tangent of circle: **tan**

to *(Choose point "B".)*

Specify third point on circle: **tan**

to *(Choose point "C".)*

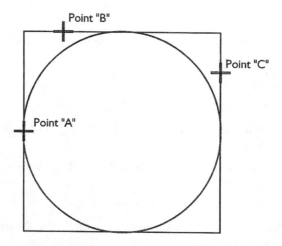

Figure 7.16 *Drawing a Circle by Three Tangent Points*

Set up the **TANgent** option in continuous mode from **Running Object Snap** in the **Options** menu.

The following illustrations show three-point circles constructed using the **TANgent object snap** method.

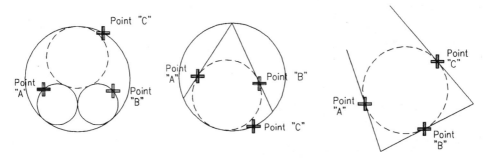

Figure 7.17 *Different Ways to Draw a Circle Using Tangent Object Snap*

DRAWING ARCS

Arcs are segments of a circle. Like a circle, an arc has a center point and a radius (or diameter). But an arc also has a starting point and endpoint—like a line. In some disciplines, the "chord" length defines the arc. For these reasons, there are 11 methods of constructing an arc in AutoCAD. Note that an arc must have an angle of less than 360 degrees.

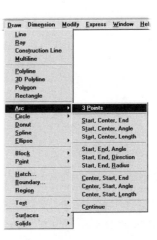

To use the **Arc** command properly, you must have a thorough understanding of the different ways in which an arc can be constructed. You can specify an arc in the following ways:

Start, end, and one point on the arc (3 points)

Start, center, and end point

Start, center, and included angle

Start, center, and length of chord

Start, end, and included angle

Start, end, and radius

Start, end, and tangent direction

Center, start, and end point

Center, start, and included angle

Center, start, and length of chord

Continuation of a previous line or arc

References to "center" mean the center point of the arc. If you do not specifically choose any of the above combinations of arc construction, method one (three points on an arc) is the default.

Although 11 methods might seem like a lot to learn, you really only have to understand a few principles to use them all.

By tradition, AutoCAD uses letters to designate methods of arc construction. The command options are sometimes referred to by these abbreviations (though not necessarily with the **Arc** command):

A—included Angle

C—Center

D—Direction

E—End point

L—Length of chord

R—Radius

S—Start point

Let's review each of the methods of arc construction and look at some examples.

DRAWING THREE-POINT ARCS (3-POINT OR S,S,E)

The "three points on an arc" is the default method of constructing an arc. The first and third points are the endpoints of the arc, while the second point is any point on the arc that occurs between the beginning and end points.

Command: **Arc**

Specify start point of arc or [CEnter]: *(Enter point "1".)*

Specify second point of arc or [CEnter/ENd]: *(Enter point "2".)*

Specify end point of arc: *(Enter point "3".)*

Three-point arcs can be constructed from either direction (clockwise or counter-clockwise). The arc is drawn from the first point toward the second and third points.

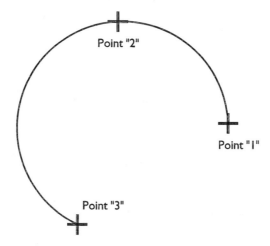

Figure 7.18 *Drawing an Arc by Three Points*

START, CENTER, END (S,C,E)

This method constructs an arc counterclockwise from the start to the specified end point. The arc is constructed from a radius using the specified center point. The radius is equal to the actual distance from the center point to the start point. For this reason, the arc does not pass through the specified end point if it is not the same distance from the center.

Relative coordinates and specified angles from the center point can be used if desired.

> Command: **Arc**
>
> Specify start point of arc or [CEnter]: *(Pick starting point.)*
>
> Specify second point of arc or [CEnter/ENd]: **CE**
>
> Specify center point of arc: *(Pick center point.)*
>
> Specify end point of arc or [Angle/chord Length]: *(Pick end point.)*

Some examples of angles constructed with this method are shown in Figure 7.19.

START, CENTER, INCLUDED ANGLE (S,C,A)

This method draws an arc with a specified start and center point of an indicated angle. The arc is drawn in a counterclockwise direction if the indicated angle is positive and clockwise if the indicated angle is negative. Examples of arcs constructed by this method are shown in Figure 7.20.

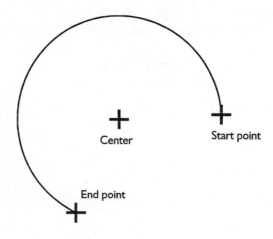

Figure 7.19 *Drawing an Arc with a Center Point*

Command: **Arc**

Specify start point of arc or [CEnter]: *(Pick starting point.)*

Specify second point of arc or [CEnter/ENd]: **CE**

Specify center point of arc: *(Pick center point.)*

Specify end point of arc or [Angle/chord Length]: **A**

Specify included angle: *(Specify on angle.)*

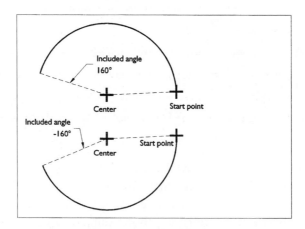

Figure 7.20 *Two Ways of Drawing an Arc with an Included Angle*

START, CENTER, LENGTH OF CHORD (S,C,L)

A *chord* is a straight line connecting an arc's start and end point (see Figure 7.21). In some applications, such as the design of curves in a railroad track, an arc of a specified chord length is required. AutoCAD allows construction of such an arc and allows the user to specify the chord length.

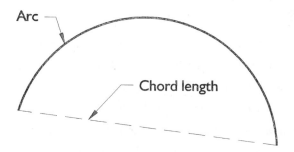

Figure 7.21 *The Chord Length of an Arc*

For construction of this type of arc, the chord length determines the ending angle. The arc is always drawn in a counterclockwise direction. When chord length is positive, a "minor" arc is drawn. When the chord length is negative, AutoCAD draws the "major" arc. (A "minor" arc has a shorter length than a "major" arc.) An example of an arc with a specified chord length is shown in Figure 7.22.

Command: **Arc**

Specify start point of arc or [CEnter]: *(Pick starting point.)*

Specify second point of arc or [CEnter/ENd]: **CE**

Specify center point of arc: *(Pick center point.)*

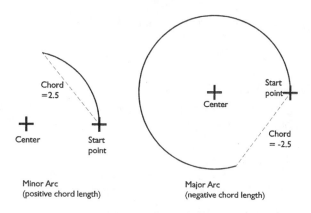

Figure 7.22 *Two Ways of Drawing an Arc with a Chord*

Specify end point of arc or [Angle/chord Length]: **L**

Specify length of chord: *(Type a number or pick the length.)*

START, END, INCLUDED ANGLE (S,E,A)

This type of arc is drawn counterclockwise if the specified angle is positive and clockwise if the specified angle is negative.

Command: **Arc**

Specify start point of arc or [CEnter]: *(Pick starting point.)*

Specify second point of arc or [CEnter/ENd]: **EN**

Specify end point of arc: *(Pick endpoint.)*

Specify center point of arc or [Angle/Direction/Radius]: **A**

Specify included angle: *(Type a number or pick an angle.)*

Two examples, one using a positive angle and one using a negative angle, are shown in Figure 7.23.

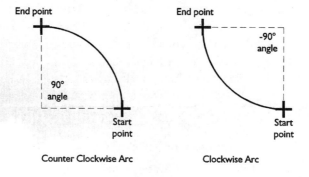

Figure 7.23 *Two Ways of Drawing an Arc with an Included Angle*

START, END, RADIUS (S,E,R)

This type of arc is always drawn counterclockwise from the start point and normally will draw the minor arc. However a negative value for the radius will cause the major arc to be drawn.

Command: **Arc**

Specify start point of arc or [CEnter]: *(Pick starting point.)*

Specify second point of arc or [CEnter/ENd]: **EN**

Specify end point of arc: *(Pick endpoint.)*

Specify center point of arc or [Angle/Direction/Radius]: **R**

Specify radius of arc: *(Type a number or pick a radius.)*

Two examples of this type of arc are shown in Figure 7.24.

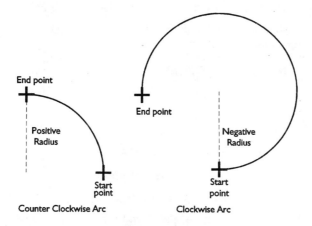

Figure 7.24 *Two Ways of Drawing an Arc with a Radius*

START, END, DIRECTION (S,E,D)

This method allows you to draw an arc tangent to a specified direction. It creates an arc in any direction, clockwise or counterclockwise, major or minor. The type of arc depends strictly on the direction specified from the starting point. You specify the direction by using a single point.

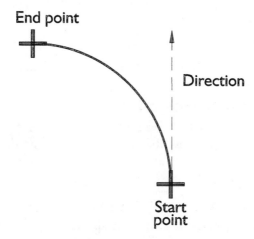

Figure 7.25 *Drawing an Arc with a Tangent Direction*

> Command: **Arc**
>
> Specify start point of arc or [CEnter]: *(Pick starting point.)*
>
> Specify second point of arc or [CEnter/ENd]: **EN**
>
> Specify end point of arc: *(Pick endpoint.)*
>
> Specify center point of arc or [Angle/Direction/Radius]: **D**
>
> Specify tangent direction for the start point of arc: *(Pick a point to indicate the direction.)*

CENTER, START, END POINT (C,S,E)

This construction method is similar to the SCE option, except that you specify the center point first.

> Command: **Arc**
>
> Specify start point of arc or [CEnter]: **CE**
>
> Specify center point of arc: *(Pick the center point.)*
>
> Specify start point of arc: *(Pick the starting point.)*
>
> Specify end point of arc or [Angle/chord Length]: *(Pick the ending point.)*

CENTER, START, INCLUDED ANGLE (C,S,A)

The construction method is similar to the SCA method, except for the order of entry. This method constructs an arc from a specified center point, using a start point and an included angle that describes the length of the arc.

> Command: **Arc**
>
> Specify start point of arc or [CEnter]: **CE**
>
> Specify center point of arc: *(Pick the center point.)*
>
> Specify start point of arc: *(Pick the starting point.)*
>
> Specify end point of arc or [Angle/chord Length]: **A**
>
> Specify included angle: *(Type the angle or pick two points showing the angle.)*

CENTER, START, LENGTH OF CHORD (C,S,L)

This is similar to the SCL method, except for the order of entry. This method of construction draws an arc that is described from a center point, then a point that designates the start of one end of the arc. The length of a chord is from the arc's first endpoint to the second endpoint.

> Command: **Arc**
>
> Specify start point of arc or [CEnter]: **CE**

Specify center point of arc: *(Pick the center point.)*

Specify start point of arc: *(Pick the starting point.)*

Specify end point of arc or [Angle/chord Length]: **L**

Specify length of chord: *(Type the length or pick two points showing the length.)*

CONTINUE ARC

This "hidden" option allows you to precisely attach an arc to a previously drawn line or arc. To invoke this method, simply respond with SPACEBAR or ENTER when prompted for the first point. The start point and direction are taken from the endpoint and starting direction of the previous line or arc drawn.

Command: **Arc**

Specify start point of arc or [CEnter]: *(Press* ENTER.*)*

Specify end point of arc: *(Pick the ending point.)*

This method is very useful for smooth connection of arcs to lines and other arcs. An example of this method is shown in Figure 7.26.

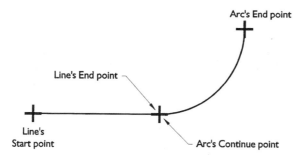

Figure 7.26 *Continuing an Arc from a Line*

USING OBJECT SNAP

You might notice that you spend a great deal of time lining up points on the screen. Connecting to desired points on lines, circles, arcs, and other objects can be a tedious procedure. AutoCAD has provided a drawing aid that makes this process much easier. It is called "object snap."

Object snap provides a "window" that is attached to the intersection of the crosshairs. This window is called the *aperture*. Object snap causes the cursor to grab onto geometric features (such as the endpoint of a line or the center of a circle) when they are within the aperture (sort of like target practice). The *X,Y* coordinate of the geometric feature is then treated as though you typed the coordinates precisely at the keyboard.

Object snap can be used in two ways: (1) in the middle of a command any time the command requests a point, such as at the "Specify next point:" prompt, or (2) turned on, so that object snap mode is always available.

For example, if you are using the **Line** command and wish to place the endpoint of the currently drawn line segment in the middle of a circle, you would choose object snap mode **CENter** before placing the point. We refer to this use of object snap as *temporary mode* because the object snap is effective for only one operation. (See additional discussion of aperture and temporary and running modes in the later section called "Methods of Using Object Snap.")

The following command sequence illustrates how you would connect a line to the center point of a circle as shown in Figure 7.27.

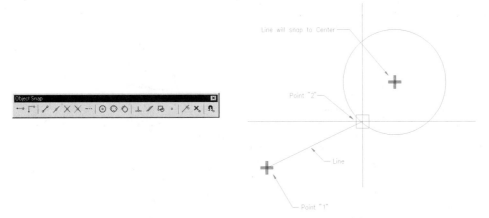

Figure 7.27 *Object Snap to the Center of a Circle*

Command: **Line**

Specify first point: *(Point 1.)*

Specify next point or [Undo]: **CEN**

of *(Point 2.)*

To point:

When you type **CEN**, you are typing the abbreviation for Center object snap mode. This mode snaps to the precise center of a circle, arc, or polyline arc.

OBJECT SNAP MODES

The following modes describe the different object snap capture methods, with an illustration of each application.

CENter: Captures the center point (radius point) of a circle, arc ellipse or elliptical arc. The aperture must contain a part of the circle or arc to identify the object.

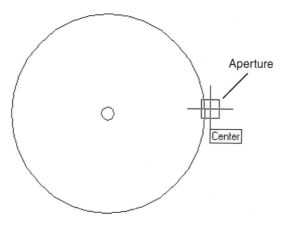

Figure 7.28 *Object Snap: Center*

ENDpoint: Causes the nearest endpoint of a line, arc, elliptical arc, mline, polyline, or ray to be captured. **ENDpoint object snap** also snaps to the nearest corner of a trace, solid, and 3D face.

INSertion: Captures the insert point of a block, attribute, text, or shape.

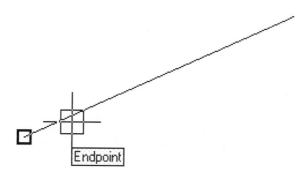

Figure 7.29 *Object Snap: Endpoint*

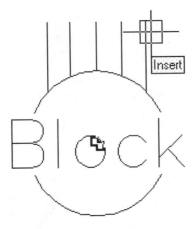

Figure 7.30 *Object Snap: Insertion Point*

INTersection: Captures the intersection of two lines, of a line and either a circle or an arc, or the intersection of two circles and/or arcs. **INT** also snaps to the intersection of ellipse, elliptical arc, mline, polyline, ray, spline, and xlines.

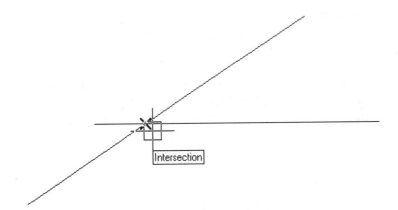

Figure 7.31 *Object Snap: Intersection*

APParent Intersection: Captures the apparent or projection intersection of two objects that do not physically cross. APP mode should not be turned on at the same time as INT mode, since the two modes can conflict with each other.

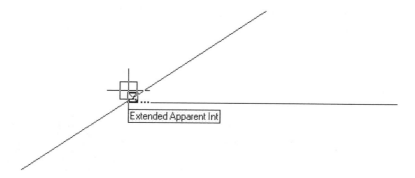

Figure 7.32 *Object Snap: Apparent Intersection*

MIDpoint: Captures the midpoint of a line, arc, elliptical arc, ellipse, mline, polyline, 2D solid, spline, and xline.

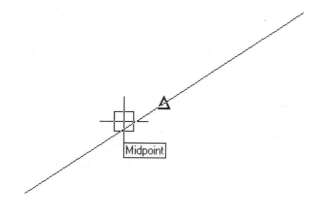

Figure 7.33 *Object Snap: Midpoint*

NEArest: Captures the nearest point on a line, circle, arc, elliptical arc, ellipse, mline, point, polyline, spline, and xline.

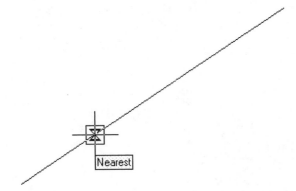

Figure 7.34 *Object Snap: Nearest*

NODe: Snaps to a point.

Figure 7.35 *Object Snap: Node*

QUAdrant: Snaps to the nearest quadrant point of a circle, elliptical arc, ellipse, or arc. The quadrants are the parts of a circle or arc that occur at 0, 90, 180, and 270 degrees (even when snap **Rotate** is set to an angle other than 0). If the circle or arc is part of a rotated block, the quadrant points are rotated with it. Only the parts of an arc that are visible will be captured.

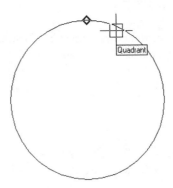

Figure 7.36 *Object Snap: Quadrant*

PERpendicular: Snaps to a point on the object that forms a perpendicular from the last point or to the next point. Eligible objects include: line, mline, polyline, ray, xline, spline, circle, arc, ellipse, elliptical arc, and 2D solid. (Polylines include polygen, donut, etc.)

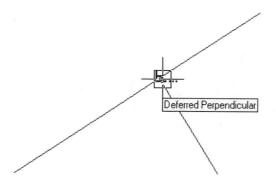

Figure 7.37 *Object Snap: Perpendicular*

TANgent: Snaps to a point on a circle, arc, ellipse, or elliptical arc that constructs a tangent to the last or next point entered.

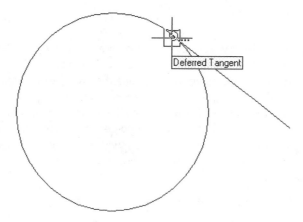

Figure 7.38 *Object Snap: Tangent*

EXTension: Displays a temporary extension line from the endpoint of an object. The tooltip displays the distance and angle (in polar coordinates) from that endpoint.

Figure 7.39 *Object Snap: Extension*

PARallel: Snaps to a point in parallel with an object. The cursor snaps to the parallel alignment path of the object's endpoint.

Note: the parallel alignment path is calculated from the "First point."

Figure 7.40 *Object Snap: Parallel*

QUIck: Quick is a modifier that is entered before you choose each snap function. Choosing **Quick** causes the object snap routine to choose the first point that it finds instead of choosing the one that is closest to the intersection of the crosshairs. This might cause problems if there are several objects in the aperture. **Quick** is used to save computer search time when there are several objects to choose from. If you encounter problems using **Quick,** cancel the point captured and try again without using **Quick. Quick** is a modifier that you should use only after you are proficient with the use of the object snap feature.

NONe: Turns off running object snaps for one point.

You can see a cursor menu of object snap modes by pressing CTRL along with the right button on a two-button mouse (on a three-button mouse, press the middle button). In AutoCAD 2000, right-click the word **OSNAP** on the status line. When the cursor menu appears, choose **Settings**.

Figure 7.41 *Object Snap Cursor Menu*

METHODS OF USING OBJECT SNAP

Object snap can be used in two different ways. The first is what we call *running mode*. When used this way, AutoCAD captures points in the *aperture* whenever a command prompts for a point. The aperture is a small, square cursor that shows the area AutoCAD searches for geometry. At these times, the aperture appears on the screen, then disappears when it is not required. (The aperture does not affect normal drawing activities.)

The second way to use object snap is what we refer to as *temporary mode*. In temporary mode, object snap is used for a single operation only.

Let's look at how to use each mode.

Running Object Snap Mode

To set the object snap modes, choose **Drafting Settings** from the **Tools** drop-down menu. AutoCAD displays the **Object Snap** tab of the **Drafting** Settings dialog box for setting running object snaps.

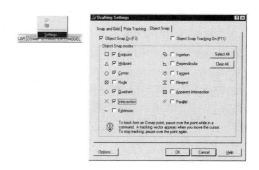

Figure 7.42 *Object Snap Tab*

Select the check box of each object snap mode that you want in effect. The selected modes are highlighted and contain a check in the check box next to them. You can choose more than one; choose the **Select All** button to select all modes.

☐ Endpoint
△ Midpoint
○ Center
⊠ Node
◇ Quadrant
✕ Intersection
— Extension
⠃ Insertion
┗ Perpendicular
ᵒ Tangent
⊼ Nearest
⊠ Apparent intersection
∥ Parallel

Selecting the same mode a second time deselects it; choose the **Clear All** button to turn off all modes.

Note: Autodesk recommends that **INTersection** and **APParent** snap not be on at the same time.

Temporary Mode Object Snap

Object snap can also be invoked in what we previously referred to as temporary mode. This is for those special times when you need an object snap for just one capture.

Note: The temporary mode is designed to be used in the middle of other commands.

You access the object snap cursor menu by pressing CTRL and the right mouse button.

You can override the running object snap functions by invoking temporary mode. For example, if you are using running mode object snap and want to enter a point without using object snap, access the object snap cursor menu, and choose **None**. After you enter the point, the running mode returns automatically.

The **From** Command

The object snap **From** command is used together with other object snap modes to create a place holder, a temporary reference point. It is useful for object snapping "around a corner." When entering coordinates during a **From** object snap, use relative coordinates, as follows:

Command: **Line**

Specify first point: **from**

Base point: <Offset>: **end**

of: **@6,4**

Specify next point or [Undo]: *(Pick a point)*

While you can use absolute coordinates, they cancel the **From** object snap mode.

OBJECT SNAP OVERRIDE

Sometimes, you may find that you want to override the current object snap settings. There are several ways to do this.

> **OSNAP Status Button.** One situation is when you want to turn off all object snap modes but not lose the settings. For example, you have set **MID**, **INT**, and **END** object snap modes. Then, you need to turn them off for a couple of commands.
>
> The old way is to use the **Osnap** command to turn off the three modes, then later return to the **Osnap** command to turn on **MID**, **INT**, and **END** again.
>
> Look at the **OSNAP** button on the status line. When the word **OSNAP** looks pushed out, object snap is off; when **OSNAP** looks pushed in, one or more object snap modes are turned on, as shown in Figure 7.43. To temporarily turn off object snap modes, choose the **OSNAP** button. To turn on the modes again, choose a second time.

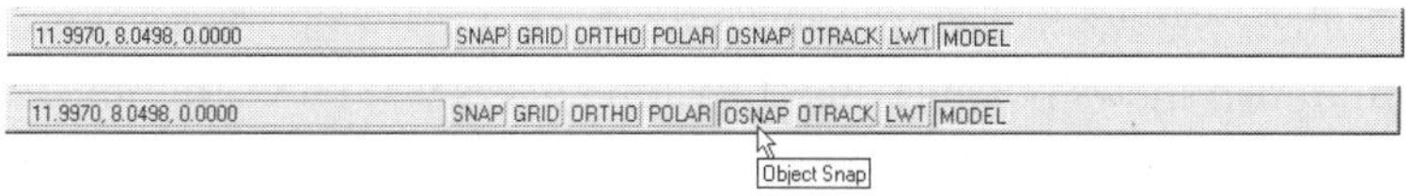

Figure 7.43 *Object Snap Button on Status Bar: Off (upper) and On (lower)*

> **OsnapCoord.** Here's another situation. You enter coordinates from the keyboard when object snap is turned on. Which should win out: your typed coordinates or the object snap? The **OsnapCoord** system variable lets you pick from three different situations:

- 2: The default setting of **OsnapCoord** is 2, which lets your keyboard entry override all object snap settings (except in scripts).

- 0: The opposite setting is 0, which means that the object snap settings override your keyboard coordinates.

- 1: The other setting is 1, which means keyboard entry and scripts override the object snap settings. (We discuss scripts later in this book.)

To change the value of **OsnapCoord**, type a number—either 0, 1, or 2—matching the settings listed above:

Command: **osnapcoord**

New value for OSNAPCOORD <2>: **0**

AUTOSNAP

Object snap is excellent for aligning drawing elements to geometric features. However, you sometimes work "blind" with object snap, not really sure what AutoCAD will snap to. There are several solutions to this problem:

- Turn off all object snaps except the one you want to use. For example, you may want to start drawing a line from the end point of an arc. In this case, turn on only **ENDpoint** object snap.

- In a crowded drawing, there may be many endpoints that AutoCAD could snap to. Use the transparent **Zoom Window** feature (discussed later in this chapter) to get a close-up view of the exact endpoint (or other geometric feature) to snap to.

- To see object snap at work, turn on **AutoSnap**, where AutoCAD shows a small symbol over the geometric feature it will snap to. (The drawback to AutoSnap is that it slows down processing and can get confusing in cluttered drawings when you have many object snap modes turned on.)

When you use any object snap settings, AutoSnap displays a visual cue (or marker) and a snaptip when you move the aperture box over a snap point. After you start a command, AutoSnap shows possible object snap points as you move the cursor around the drawing.

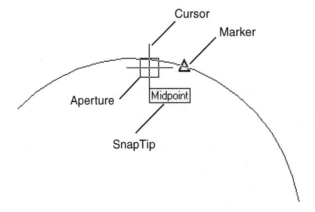

Figure 7.44 *AutoSnap Features*

AutoSnap has these features:

Cycling. By pressing the TAB key, you cycle through all possible object snaps available within the aperture. For example, when you press TAB while the aperture box is over an arc, AutoSnap shows you the object snap possibilities (assuming all object snaps are turned on): NEArest, QUAdrant, ENDpoint, MIDpoint, TANgent, another QUAdrant, CENter, the other ENDpoint, INTersection, and then repeats the list.

Marker. AutoSnap displays an icon representing the geometric feature's location. For example, the icon for CENter object snap looks like a small circle. (The Quick object snap has no marker.)

Snaptip. A small text box that looks like a tooltip. It describes with a word or two the object snap mode. For example, the snaptip for PERpendicular reads "Deferred Perpendicular" because the perpendicular line will be drawn when you select the second point.

Magnet. When the cursor is close enough, AutoCAD locks the cursor onto the current object snap point.

To change the AutoSnap settings, from the **Tools** menu choose **Options**, then choose the **Drafting** tab, which displays the AutoSnap settings.

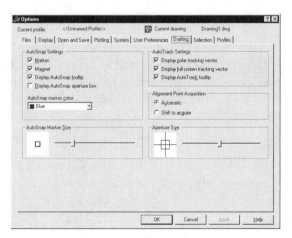

Figure 7.45 *Drafting Tab Contains AutoSnap Settings*

To turn the marker, magnet, snaptip, and aperture box on or off, select the check box next to each option. (A check mark means the option is turned on.) To increase the size of the marker, move the **Marker Size** slider to the right; moving the slider to the left decreases the size of the marker icon. For the color of the marker, you can choose from seven colors. The default color is blue; when the monochrome vectors option is on, the marker icon is always black.

OBJECT SNAP TRACKING

Object snap tracking allows the cursor to track along an *alignment path* based on object snaps. The alignment path is a fine dotted line, as shown in Figure 7.46. In addition, a tooltip displays the alignment, its distance, and angle. It is possible for two or more alignment paths to be displayed at one time, depending on the geometry of the drawing and the number of object snaps turned on.

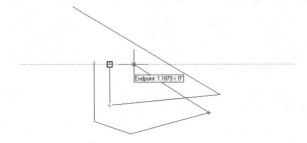

Figure 7.46 *Object Snap Tracking*

To turn on Osnap tracking, choose the **OTRACK** button on the status line (or press F11 or CTRL+W). Note that for Osnap tracking to work, at least one object snap must also be turned on. That means that both the **OSNAP** and **OTRACK** buttons must be depressed on the status bar.

POLAR TRACKING

Polar tracking displays an alignment path similar to Osnap tracking. The alignment path, however, is displayed whenever the direction angle is an increment of 90 degrees.

Choose the **POLAR** button on the status bar (or press F10) to turn on polar tracking. When polar tracking is on, alignment paths are displayed whenever the drawing angle is 0, 90, 180, and 270 degrees. In addition, a tooltip displays the word Polar, the distance from the last point, and the angle, such as Polar: 3.8550<180°, as shown by Figure 7.47.

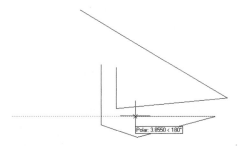

Figure 7.47 *Polar Tracking*

You can have AutoCAD 2000 display the alignment path at other angles, as well. For example, it is convenient to have polar tracking set for every 15 degrees. To do so, follow these steps:

1. Right-click **POLAR** on the status line. From the cursor menu choose **Settings**. Notice that AutoCAD displays the **Drafting Settings** with the **Polar Tracking** tab.

Figure 7.48 *Polar Tracking Tab*

2. In the **Polar Angle Settings** section, select the **Increment Angle** list box.

3. Choose **15**, and then choose **OK**.

4. Test the polar tracking by starting the **Line** command, selecting a point, then moving the cursor around the screen at the "Specify next point" prompt. Notice that the polar tracking appears at increments of 15 degrees, such as 15, 30, and 45 degees. Figure 7.49 shows a composite figure.

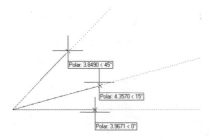

Figure 7.49 *Polar Tracking at 15-degree Increments*

CLEANING THE SCREEN (REDRAW)

After you have been drawing for a while, you begin to build up a number of "markers." While these markers, called blips, are often helpful in locating points on a drawing, too many of them can be distracting. AutoCAD provides a command for cleaning up a drawing called **Redraw**, which clears the screen and redraws the objects.

Redraw is also useful when the **Erase** command causes a partial loss of some object that is to remain. This is common when a previously erased object overlays an object that remains.

Command: **Redraw**

Some commands, such as **Zoom**, automatically execute a redraw, so it is not necessary to perform the redraw before executing these commands.

REDRAWING TRANSPARENTLY

A redraw can be executed while another command is active. This is called a transparent redraw. To perform a transparent redraw, enter an apostrophe (') before the command. For example, to perform a transparent redraw during the **Line** command, enter

'Redraw at the 'To point:' prompt. When the transparent redraw is completed, the previously active command will be resumed. The **Redraw** command contained in the screen menu will always redraw transparently. This means you can choose **Redraw** from the menu while you are currently in a command without exiting that command.

> Command: **Line**
>
> Specify first point: *(Pick.)*
>
> Specify next point or [Undo]: *(Pick.)*
>
> Specify next point or [Undo]: **'redraw**
>
> Resuming LINE command.
>
> Specify next point or [Undo]: *(Pick.)*
>
> Specify next point or [Undo]: *(Press* ENTER.*)*

REDRAWING ALL VIEWPORTS

AutoCAD lets you create more than one view of the drawing, as we shall see later in this book. The **Redraw** command cleans up the current viewport. To simultaneously clean up all viewports, use the **Redraw** command (the **Redraw** command shown on the **View** menu is actually the **RedrawAll** command). Like the **'Redraw** command, the **'RedrawAll** command is transparent.

REGENERATING THE DRAWING

The **Regen** command causes AutoCAD to regenerate the drawing display from the database. This differs from the **Redraw** command, which just redraws the screen, in that AutoCAD actually recalculates line endpoints, hatched areas, etc., as well as cleaning up erased objects by removing them from the database. To invoke the **Regen** command, enter:

> Command: **Regen**

Several commands cause an automatic regeneration. Among these are sometimes **Zoom**, **Pan**, and **View restore**. The regeneration occurs if the new display contains areas that are not within the currently generated area. The **RegenAuto** command controls the automatic regeneration in AutoCAD. See Chapter 39 "Tailoring AutoCAD" for an explanation of **RegenAuto**.

You cancel the regeneration by pressing ESC. If **Regen** is terminated, however, some of the drawing might not be redisplayed. To redisplay the drawing, you must reissue the **Regen** command. Normally, a **Redraw** is used when you want to simply clean up the screen, because regeneration takes longer to perform.

Just as **RedrawAll** refreshes all viewports, so does **RegenAll** regenerate all viewports. Both **Regen** and **RegenAll** are found in the **View** menu.

Note: Regen a drawing after extensive editing to see the "actual" state of the modified drawing.

ZOOMING THE DRAWING

The **Zoom** command allows you to enlarge or reduce the view of your drawing. You can think of zoom as a magnifier and a shrinker. Most drawings are too large and too detailed to work with on the computer monitor's small 15"—or even 20"—drawing screen. CAD operators routinely zoom in to small areas to show detail.

Let's consider an analogy to zooming. Imagine that your drawing is the size of a wall in a room. The closer you walk to the wall, the more detail you see. But as you get closer, you only see a portion of the wall. If you move a great distance away from the wall, you see the entire drawing, but not much detail. Note that the wall does not change size; you just change your viewing distance.

The **Zoom** command works in much the same way. You can enlarge or reduce the drawing size on the screen, but the drawing does not actually change size. If you zoom in closer, you see more detail, but not all of the drawing. If you zoom out (further away), you see more of the drawing, but less detail.

USING THE ZOOM COMMAND

A zoom is performed with the **Zoom** command:

Command: **Zoom**

Specify corner of window, enter a scale factor (nX or nXP), or
    [All/Center/Dynamic/Extents/Previous/Scale/Window] <real
    time>:

The prompt indicates that the **Zoom** command has no less than three defaults: (1) click a point on the screen and AutoCAD goes into the **Window** option; (2) type a number and AutoCAD goes into the **Scale** option; and (3) when you press ENTER, AutoCAD goes into real-time zoom mode.

While the prompt indicates seven options, there are two missing: **Left** and **Vmax**. Let's look at *all* the options for the **Zoom** command:

- **Specify corner of window:** Same as the **Window** option, below.

- **Enter a scale factor (nX or nXP):** Same as the **Scale** option, below.

- **All:** Zooms out to show everything in the drawing, or to the limits of the drawing, whichever is larger.

- **Center:** Zooms about a center point.

- **Dynamic:** Displays a zoom/pan box.

- **Extents:** Zooms out to show everything in the drawing.

- **Previous:** Shows the previous view, whether a zoom or pan.

- **Scale:** Absolute and relative zoom factors in model space and paper space.

- **Window:** Zooms in to a rectangular area specified by two points.

- **<real time>:** Zooms in real time along with the mouse movement.

- **Left:** Zooms relative to a lower left point (a "hidden" option).

- **Vmax:** Zooms out to the maximum that is possible without a **Regen** being required (a "hidden" option).

It may seem confusing to have so many options. I find that I use only a few of the options most often, specifically the **Extents, Window, Previous,** and **<real time>**options.

Zoom All

A **Zoom All** causes the entire drawing to be displayed on the screen. This typically results in the entire area of the limits to be shown. If, however, the drawing extends outside the current limits, the **All** option shows the extents of the drawing, including the area of the drawing outside the limits.

Note this difference between the **Zoom** command's **All** and **Extents** options: **Extents** displays the extents of the drawing, while the **All** option displays *either* the drawing extents *or* the drawing limits, depending on which is larger.

Command: **Zoom**

Specify corner of window, enter a scale factor (nX or nXP), or
    [All/Center/Dynamic/Extents/Previous/Scale/Window] <real
    time>: **A**

Occasionally, **Zoom All** has to regenerate the drawing twice. If this is necessary, AutoCAD displays the following message on the prompt line:

* * Second regeneration caused by change in drawing extents.

When the limits are changed, the entire drawing area is not shown until a **Zoom All** is performed.

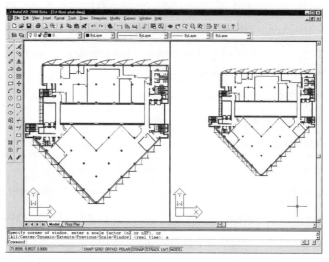

Figure 7.50 *Zoom All (in righthand viewport)*

Zoom Center

The **Zoom** command's **Center** allows you to choose a center point for the zoom. You then specify the magnification, or a height for the new view in units.

Command: **Zoom**

Specify corner of window, enter a scale factor (nX or nXP), or
    [All/Center/Dynamic/Extents/Previous/Scale/Window] <real
    time>: **C**

Specify center point: *(Select.)*

Enter magnification or height <45.0000>: *(Enter a number.)*

If an "X" follows the magnification value, the zoom factor is relative to the current display.

Zoom Dynamic

The **Dynamic** option allows dynamic zoom placement. Note that this option is an alternative to **Aerial View** and **Real Time Zoom and Pan**.

Command: **Zoom**

Specify corner of window, enter a scale factor (nX or nXP), or
    [All/Center/Dynamic/Extents/Previous/Scale/Window] <real
    time>: **D**

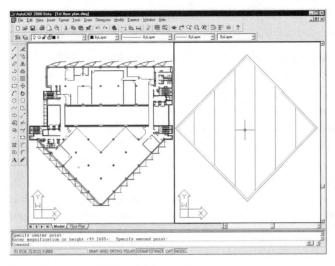

Figure 7.51 *Zoom Center (in righthand viewport)*

When you choose **Dynamic Zoom,** you are presented with a special view selection screen containing information about current and possible view screen selections. The screen shown in Figure 7.52 is representative of a typical display for **Dynamic Zoom**.

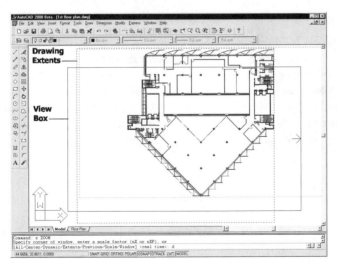

Figure 7.52 *Zoom Dynamic Display*

Each of the view boxes is labeled and noted as to its colors. (Monochrome displays do not, of course, show these colors.) Let's look at the meaning of each of these boxes.

Drawing Extents: The drawing extents box is the blue, dotted-line box. The drawing extents can be thought of as the "sheet of paper" on which the drawing resides.

View Box: The green dotted-line box indicates the view when you began the **Zoom Dynamic** command. The black solid-line view box defines the size and location of the desired view. You can manipulate this box to achieve the view you want. The view box is initially the same size as the current view box. It operates in two modes: pan mode and zoom mode.

Pan Mode: The view box can be moved to the desired location. A large X is initially placed in the center of the box. This denotes panning mode. When the X is present, moving the cursor causes the box to move around the screen.

Figure 7.53 *View Box in Pan Mode*

Zoom Mode: Pressing the pick button causes the X to change to an arrow at the right side of the box. The arrow denotes the zoom mode. Moving the cursor right or left increases or decreases the size of the view box. The view box increases and decreases in proportion to your screen dimensions, resulting in a "what you see is what you get" definition of the zoomed area. This differs from the standard window zoom, which works from a stationary window corner and may show more of the screen, depending on the proportions of the defined zoom window.

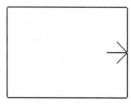

Figure 7.54 *View Box in Zoom Mode*

You can toggle between zoom and pan modes as many times as you wish to set the size and location of the view box. When you have windowed the desired area, press ENTER and **Zoom** is performed. The area defined by the zoom box is now the current screen view.

Using Dynamic Zoom Without a Pointing Device: Although recommended, it is not necessary to have a pointing device installed to use **Dynamic Zoom**. Simply use the arrow keys to move the view box. If you have used the arrow keys, pressing ENTER will toggle between pan mode and zoom mode. If you have not used the arrow keys since last toggling with the [Enter] key, pressing it will perform the zoom at the current location of the view box.

The easy way to perform this is to manipulate the view box to the desired location and size and press ENTER twice.

Zoom Extents

Zoom Extents displays the drawing at its maximum size on the display screen. This results in the largest possible display, while showing the entire drawing.

Command: **Zoom**

Specify corner of window, enter a scale factor (nX or nXP), or
    [All/Center/Dynamic/Extents/Previous/Scale/Window] <real
    time>: **E**

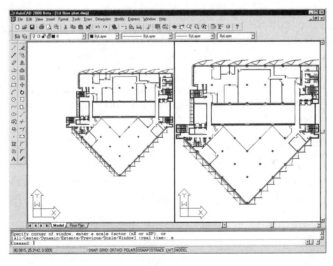

Figure 7.55 *Zoom Extents (in righthand viewport)*

Zoom Previous

The **Previous** option allows you to return to the last zoom or pan you viewed. This option is useful if you need to interact between two areas frequently. AutoCAD remembers the last 10 view changes; simply use the **Zoom P** command several times in a row. Since the view coordinates are stored automatically, you are not required to use any special procedure to use them.

> Command: **Zoom**
>
> Specify corner of window, enter a scale factor (nX or nXP), or
> [All/Center/Dynamic/Extents/Previous/Scale/Window] <real
> time>: **P**

Zoom Scale. The **Zoom** command's **Scale** option allows you to enlarge or reduce the entire drawing (original size) by a numerical factor. For example, entering **5** will result in a zoom that shows the drawing five times its normal size—a zoom in. (The entire drawing, of course, cannot be displayed on the screen in this instance.) The zoom is centered on the previous screen center point.

If an **X** follows the zoom factor, the zoom is computed relative to the current display. For example, entering **5x** makes the drawing five times larger than its current zoom factor.

Only positive values can be used in zooms. If you want to zoom out (display the drawing smaller than current), use a decimal value. For example, **.5** results in a view of the drawing that is one-half its normal size. CAD operators often follow the **Zoom Extents** command with the **Zoom 0.9x** command. This lets them see the entire drawing, with a bit of "breathing room" around the edges.

> Command: **Zoom**
>
> Specify corner of window, enter a scale factor (nX or nXP), or
> [All/Center/Dynamic/Extents/Previous/Scale/Window] <real
> time>: **S**
>
> Enter a scale factor (nX or nXP): **5**

Zoom Window. The **Zoom Window** command allows you to determine the area you wish to see zoomed in. The **Window** option uses a rectangular window to show the area, which you specify with two screen picks.

> Command: **Zoom**
>
> Specify corner of window, enter a scale factor (nX or nXP), or
> [All/Center/Dynamic/Extents/Previous/Scale/Window] <real
> time>: **W**
>
> Specify first corner: *(Select P1.)*
>
> Specify other corner: *(Select P2.)*

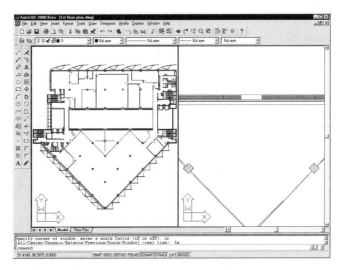

Figure 7.56 *Zoom Scale (in righthand viewport)*

A box is displayed around the area to be zoomed.

Zoom Real Time. With Pentium computers and fast display drivers, AutoCAD has enough horsepower available to zoom in real time. *Real time* means that the zoom changes continuously as you move the mouse. To enter real-time zoom mode, choose the **View | Zoom | Realtime** from the menu or enter the following at the 'Command:' prompt:

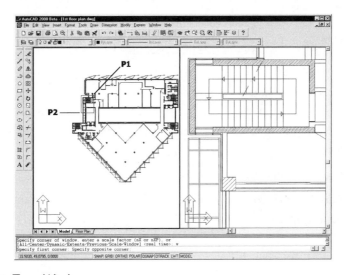

Figure 7.57 *Zoom Window*

Command: **Zoom**

Specify corner of window, enter a scale factor (nX or nXP), or [All/Center/Dynamic/Extents/Previous/Scale/Window] <real time>: *(Press* ENTER*)*

Press Esc or Enter to exit, or right-click to activate pop-up menu.

Once in real-time zoom mode, press the left mouse button over the drawing. The cursor changes to a magnifying glass. Move the mouse up (toward the monitor) and down (away from the monitor). As you move the mouse up, the drawing becomes larger; as you move the mouse down, the drawing becomes smaller.

When the drawing is the size you want it to be, you have two options:

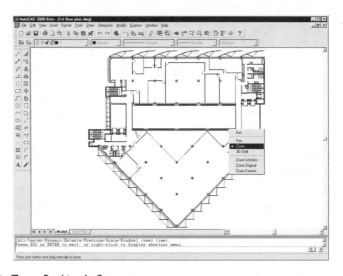

Figure 7.58 *Zoom Realtime's Cursor menu*

- Press ENTER or ESC to exit real-time zoom mode and the **Zoom** command.

- Or, right-click the mouse. AutoCAD displays the cursor menu shown in Figure 7.58. The menu gives you these options:

Exit. Exits the **Zoom** command.

Pan. Switches to realtime pan mode, discussed later in this chapter.

Zoom. If in pan mode, switches to real-time zoom mode.

3Dorbit. Allows real-time rotation of 3D objects.

Zoom Window. Displays a cursor with small rectangle. Select two points and AutoCAD displays that as the new zoomed view. Remains in real-time zoom mode.

Zoom Previous. Displays the previous zoom view. Remains in real-time zoom mode.

Zoom Extents. Displays the drawing extents. Remains in real-time zoom mode.

To dismiss the **Zoom** menu without choosing one of its options, simply click anywhere outside the menu.

TRANSPARENT ZOOMS

A transparent zoom can be executed while another command is active. Enter **'Zoom** at any non-text prompt; notice the apostrophe prefix. The following notes apply to using transparent zooms:

- The fast zoom mode must be on, which it is, by default. This is set with the **ViewRes** command.

- You cannot perform a transparent zoom if a regeneration is required (zoom outside the generated area—see **Zoom Dynamic**).

 Note: The drop-down menus and toolbar buttons automatically use transparent zooms.

For example, here is the **Zoom** command used transparently in the **Line** command:

Command: **Line**

Select first point: *(Pick.)*

Select next point: **'zoom**

\>\>Specify corner of window, enter a scale factor (nX or nXP), or [All/Center/Dynamic/Extents/Previous/Scale/Window] <real time>: **E**

Resuming LINE command. Select next point: *(Pick.)*

- Transparent zooms cannot be performed when certain commands are in progress. These include the **Vpoint**, **Pan**, **View**, and another **Zoom** command.

PANNING AROUND YOUR DRAWING

Many times, you are zoomed in to an area of the drawing to see more detail. You might want to "move" the screen a short distance, to continue work while still at the same zoom magnification. This sideways movement is called "panning." You can think of a pan as similar to placing your eyes at a certain distance from a paper drawing, then moving your head about the drawing. This would allow you to see all parts of the drawing at the same distance from your eyes.

Panning is performed with the **Pan** command. To use the **Pan** command, enter:

Command: **Pan**

Press ESC or ENTER to exit, or right-click to activate pop-up menu

The Pan command automatically enters real-time pan mode. This mode has the same effect of dragging the drawing at the specified distance and angle. This method of real time panning is the same as real time zoom: move mouse while holding down the left button. AutoCAD displays a "hand" icon while in real time pan mode.

Transparent Panning.

A pan can be performed transparently while another command is in progress. To do this, enter **'Pan** at any non-text prompt. The same restrictions for transparent pans apply as for transparent zooms.

SCROLL BARS

Notice that AutoCAD has a pair of scroll bars, like other Windows software. The scroll bars allow you to pan the drawing horizontally and vertically (see Figure 7.59) but not diagonally.

To pan with a scroll bar, position the cursor over a scroll bar and click. AutoCAD transparently pans the drawing. There are three ways to use the scroll bars:

Click the arrow at either end of the scroll bar. This pans the drawing in one-tenth increments of the viewport size.

Click and drag on the scroll bar button. This pans the drawing interactively as you move the button. This is also the way to pan by a very small amount.

Click anywhere on the scroll bar, except on the arrows and button. AutoCAD pans the drawing by 80 percent of the distance.

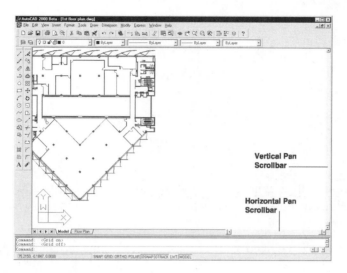

Figure 7.59 *Panning Scroll Bars*

AERIAL VIEW

AutoCAD has an alternative to the **Zoom** and **Pan** commands and scroll bars for changing the view. Called Aerial View, this window lets you see the entire drawing at all times in an independent window. This is sometimes called the bird's-eye view. To display the Aerial View, type **DsViewer** at the 'Command:' prompt.

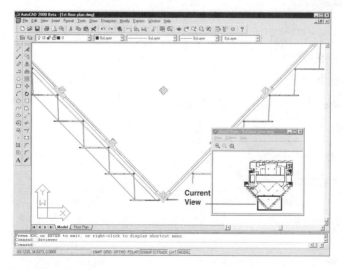

Figure 7.60 *Aerial View Window*

After the drawing appears in the Aerial View window, zooming in is as simple as with the Zoom Window command: pick two points. To instantly zoom to another area, pick another two points.

To pan, first choose the hand icon. Then, simply move the rectangle to the new location.

The Aerial View can also work in reverse. In **Global** mode, it shows a magnified view. This is sometimes known as the spyglass view. To use Global view, first set the level of magnification. Choose **Options** from the **Aerial View** menu bar, then **Locator Magnification**. When the **Magnification** dialog box appears, change the value from the default of 1 to 2 or 8 or 16. Choose the **OK** button.

DRAWING AIDS AND MODES

AutoCAD contains many commands that make drawing more accurate, efficient, and easy. Earlier in this chapter, you learned about object snap and tracking. In this section, you learn to use the following commands and modes:

- **Grid** command

- **Snap** command

- **Ortho** command

- Direct Distance Entry and Tracking

- **XLine** command

- **Ray** command

PLACING A GRID

The **Grid** command is used to display a grid of dots with a specified spacing. You can determine the spacing of the dots. You can specify the X and Y spacing separately.

The grid dot is for reference purposes and is not part of the drawing. You cannot erase the grid markers; the grid does not print or plot. To display a grid, choose the **GRID** button on the status bar, press F7, or press CTRL+G.

You can specify options for the grid through a dialog box. To access the dialog box, right-click the **GRID** button on the status line, and choose Settings from the cursor menu. Autodesk displays the **Snap and Grid** tab of the **Drawing Settings** dialog box.

Figure 7.61 *The Snap and Grid Tab*

The options are as follows:

Grid On:

The Grid On check box turns on the grid. The grid spacing and style is used. When the grid is on, you see the word GRID on the status line at the bottom of the screen. You can press F7 or CTRL+G at any time to turn the grid off and on.

Grid *X* and *Y* Spacing

The grid spacing of 0.5 is the default. When you enter a numerical value in the Grid Spacing text entry boxes, that value becomes the grid spacing.

You can enter a different value for the *X* spacing and *Y* spacing. This is called an *aspect* grid. The effect is shown by Figure 7.62.

Grid Snap:

Sets the grid spacing equal to the current snap setting. When the snap resolution changes, the grid value automatically changes to match.

Rectangular or Isometric Snap:

The **Rectangular Snap** displays the standard grid. When **Isometric Snap** is selected, the grid is suitable for use with isometric drawing. Isometric drafting is discussed in Chapter 21 "Constructing Isometric Drawings."

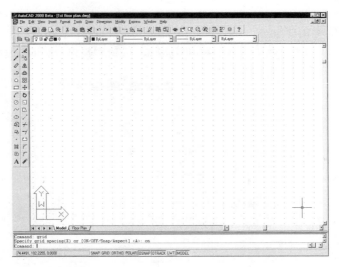

Figure 7.62 *Grid with Aspect*

> **Polar Snap:**
>
> The **Polar Snap** option turns on polar tracking mode, as discussed earlier in this chapter. The cursor snaps along polar alignment angles. You can also choose the **POLAR** button on the status bar.

If the grid spacing is too close to display properly, this message appears:

> Grid too dense to display

This can also happen when you zoom out the drawing.

 Note: Many designers use a grid when laying out or "sketching" design ideas. Setting the grid spacing to a convenient interval allows better visualization of scale.

SNAP MODE

Snap mode is useful for setting a drawing resolution. For example, if you set the snap spacing to 1", you are forced to draw to the nearest inch. When a point is not exactly aligned with the snap spacing, the point is forced to the nearest snap point, as shown by Figure 7.63.

You can turn snap mode on and off by several methods. Choose the word **SNAP** on the status bar. (Press F9 or CTRL+B to turn snap on and off at any time.)

You can specify different *X* and *Y* spacing for the snap spacing. The snap spacing can be set to coincide with the grid spacing. Unless the snap spacing is set with the grid spacing and the grid is currently displayed, the snap settings are invisible.

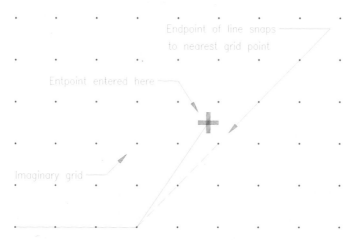

Figure 7.63 *Applying Snap Mode*

You set the snap options through the dialog box shown by Figure 7.61.

Snap On:

> Causes the Snap mode to turn on. The default snap value is determined by the prototype drawing. When snap is on, you can see the word **SNAP** depressed the status line at the bottom of the screen. You can press F9 or CTRL+B at any time to turn snap on and off.

Snap X & Y Spacing:

> The snap spacing of 0.5 is the default. When you enter a numerical value, that value becomes the snap spacing. It is not necessary to first enter a letter to designate the snap spacing option. The value must be non-zero and positive; you can enter a different value for **X Spacing** and **Y Spacing**.

Angle:

> The **Angle** option allows you to rotate set a rotation angle for the snap. This can be useful when you work on a rotated portion of a drawing.

X & Y Base:

> By default, the snap spacing starts at the origin 0,0. You can change the base point from which the snap is measured. This also affects crosshairs, rotating them.

> **Note:** The **Angle** and **Base** options affect the rotation of

> • Snap spacing

> • Hatch patterns

- Grid markings

- Ortho angle

DRAWING WITH ORTHOGONAL MODE

When you use a mouse or digitizer to draw in AutoCAD, placing lines at true horizontal and vertical orientation requires extra effort. AutoCAD's ortho mode (short for orthogonal) assures that all cursor movement is orthogonal— either horizontal or vertical. Many drawings consist of lines at 0 and 90 degrees; having ortho mode turned on greatly speeds up your drafting.

You can turn on the ortho mode either by pressing F8, by selecting the word **ORTHO** on the status line, by pressing CTRL+L, or by entering the **Ortho** command:

Command: **Ortho**

ON/OFF: **ON**

When ortho is on, the status line at the bottom of the screen displays the word **ORTHO** depressed. Turn off Ortho mode by reentering the command and responding to the prompt with off, or by pressing 8 again.

When ortho mode is on, all the lines are forced either horizontal or vertical. If a point is entered that is not true to either, the point is forced to the nearest vertical or horizontal point. That is, if the point is more nearly vertical, the line is forced vertical.

Ortho mode does not need to constrain cursor movement to the horizontal (0 and 180 degrees) and vertical (90 and 270 degrees). The **Snap** command's **Angle** option controls the rotation of the cursor. For example, to draw in ortho mode at a 45-degree angle, set the snap rotation to 45.

Now your drawing is constrained to the 45, 135, 225, and 315-degree angles, as shown in Figure 7.64. To return to "normal" ortho mode, change the snap angle back to 0.

 NOTE: Note that you turn on either ortho or polar mode; AutoCAD does not allow you to have both modes turned on at the same time.

Use ortho mode when you need the cursor constrained to just 0 and 90 degrees; use polar mode when you need more constraints, such as every 15 or 45 degrees.

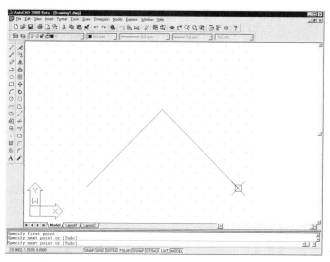

Figure 7.64 *Rotating Ortho, Grid, Snap, and Cursor by 45 Degrees*

SHORTCUT KEYS

AutoCAD provides three ways to "toggle" modes at any time. Toggle means to turn on and off, like a light switch. You can press a function key, or hold down [Ctrl] and press another key, or select the word on AutoCAD's status line:

Mode	Function Key	Ctrl Key	Status Line
Object Snap	F3	CTRL+F	OSNAP
Grid	F7	CTRL+G	GRID
Ortho	F8	CTRL+L	ORTHO
Snap	F9	CTRL+B	SNAP
Osnap Tracking	F10	CTRL+W	OTRACK
Polar Tracking	F11	CTRL+U	POLAR

DIRECT DISTANCE ENTRY & TRACKING

In addition allowing you to type coordinates or pick a point on the screen, AutoCAD has two more tools for specifying coordinates: direct distance entry and tracking. As we'll explain later, they are somewhat trickier to learn to use than regular coordinates, although you may already be familiar with the two if you use Autodesk's Generic CADD or AutoCAD LT for Windows.

DIRECT DISTANCE ENTRY

Direct distance entry is an alternative to entering polar or relative coordinates. To show direction, you move the mouse; then you type the distance. When ortho or polar modes are turned on, direct distance entry is an efficient way to draw lines.

Direct distance entry can be used any time a command prompts you to specify a point, such as the "Specify first point:" and "Specify next point:" prompts. Here is how to draw a line 10 units long using direct distance entry:

> Command: **line**
>
> Specify first point: *(Pick a point)*
>
> Specify next point: *(Move the mouse and you see a rubber band line. At the keyboard, type 10 and press* ENTER.)
>
> Specify next point: *(press* ENTER)

AutoCAD draws a line segment ten units long in the direction you moved the mouse. You can use direct distance entry for drawing polylines, arcs, multilines, and most other objects. Direct distance entry does not make much sense with some commands like **Circle** and **Donut**.

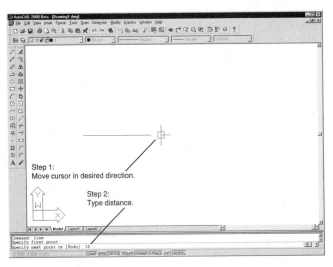

Figure 7.65 *Direct Distance Entry*

TRACKING

Whereas direct distance entry lets you *draw* relative distances, tracking lets you *move* relative distances within a command.

There are four ways to enter tracking mode. You can type **Tracking**, **Track**, or **Tk** at the "Specify first point:" and "Specify next point:" prompts. As an alternative, hold down the CTRL key, press the right mouse button, and choose **Temporary Track Point** from the cursor menu. Strictly speaking, **Tracking** is not a command but a command option; you can only use tracking within another command.

Tracking automatically switches on ortho mode and prompts for the "First tracking point:". (If polar mode is on, AutoCAD switches it off and turns on ortho mode.) Pick a point or type a coordinate pair. Tracking then prompts you, "Next point (Press ENTER to end tracking):" You continue showing points or press ENTER to return to the command currently in effect. Once you exit tracking mode, AutoCAD changes ortho and polar their states before you began tracking.

Some commands keep prompting you for additional points, such as the **Line** and **Pline** commands. With these, you can go in and out of tracking mode as often as you like. Here is an example with the **Line** command:

> Command: **Line**
>
> Specify first point: **tk**
>
> First tracking point: *(Move the mouse and click.)*
>
> Next point (Press Enter to end tracking): *(Move the mouse in another direction and click.)*
>
> Next point (Press Enter to end tracking): *(Press* ENTER.*)*
>
> To point: **2,3**
>
> To point: **10,5**
>
> To point: **tk**
>
> First tracking point: **5,10**
>
> Next point (Press Enter to end tracking): *(Press* ENTER.*)*
>
> To point: **3,2**
>
> To point: *(Press* ENTER.*)*

When you entered tracking mode, did you notice that ortho mode was turned on? The word **ORTHO** is depressed black on the status line, and cursor movement is constrained to 90-degree increments.

If you want tracking to move in angles other than 90 degrees, use the **Rotate** option of the **Snap** command to change the angle. Since the **Snap** command is transparent, you can change the tracking angle in the middle of tracking, as follows:

> Command: **Line**
>
> Specify first point: *(Pick.)*
>
> Specify next point: **tk**
>
> First tracking point: *(Pick.)*
>
> Next point (Press Enter to end tracking): **'snap**
>
> >>Specify snap spacing or [ON/OFF/Aspect/Rotate/Style/Type] <0.5000>: **R**
>
> >>Specify base point <0.0000,0.0000>: *(Press ENTER.)*
>
> >>Specify rotation angle <0>: **45**
>
> Resuming LINE command.
>
> Next point (Press Enter to end tracking): *(Pick.)*
>
> Next point (Press Enter to end tracking): *(Press ENTER.)*
>
> Specify next point: *(Pick.)*
>
> Specify next point: *(Press ENTER.)*

 Note: Tracking normally assumes you want to switch direction each time you use it. For example, if you first move north (or south), AutoCAD assumes you next want to move east (or west).

However, it is not easy to back up or track forward in the same direction. For example, you track north and then want to track north some more. Instead, you find the cursor wanting to move east or west, but not north or south.

To make the tracking cursor continue in the same direction, you need to move it back to its most recent starting point. Then you can move in the direction you want.

CONSTRUCTION LINES

AutoCAD has the ability to create *construction lines*. These lines are infinitely long; you snap to construction lines to help you create a drawing. Creating a grid of construction lines is better than using the grid dots, since you can use object snap modes on the construction lines. AutoCAD has two kinds of construction lines, which are created with the **Ray** and **Xline** commands.

THE RAY COMMAND

The ray is a semi-infinite construction line. The ray starts at a point you specify; the other end of the ray is in "infinity": no matter how far you zoom out, the ray always appears (whereas other objects in the drawing grow smaller and eventually disappear). Rays are created with the **Ray** command. From the menu bar, choose **Draw | Ray**.

Command: **Ray**

Specify start point: *(Pick point.)*

Specify through point: *(Pick point.)*

Specify through point: *(Press* ESC.*)*

The **Ray** command prompts you for the starting point of the ray, and then for a point that the ray passes through. The **Ray** command automatically repeats itself, using the "Specify through point:" prompt for new rays until you press ESC.

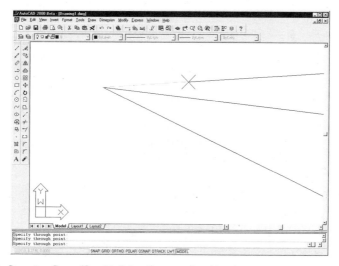

Figure 7.66 *Drawing Rays (Semi-infinite Construction Lines)*

THE XLINE COMMAND

The "xline" is an infinite construction line. The xline starts in one part of infinity, passes through the drawing area, and continues on into infinity. Like the ray, the xline always appears in the drawing, no matter how far out you zoom or whether you pan to AutoCAD's limit.

Xlines are created with the **Xline** command, which has six options. Like the **Ray** command, the **Xline** command keeps repeating its prompts until you press ESC.

Default Option.

The default option prompts you to pick a pair of through points (an Xline can't actually have a "from point" since it doesn't start anywhere), as follows:

> Command: **Xline**
>
> Specify a point or [Hor/Ver/Ang/Bisect/Offset]: *(Pick point.)*
>
> Specify through point: *(Pick point.)*
>
> Specify through point: *(Press ESC.)*

Pick the "from" and "through" points, then press ESC to end the **Xline** command.

Horizontal Option.

The **Hor** option draws a horizontal xline through a single point you pick, as follows:

> Command: **Xline**
>
> Specify a point or [Hor/Ver/Ang/Bisect/Offset]: **h**
>
> Specify through point: *(Pick point.)*
>
> Specify through point: *(Press ESC.)*

Vertical Option.

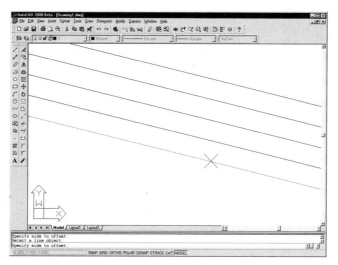

Figure 7.67 *Drawing Xlines (Infinite Construction Lines)*

The **Ver** option draws a vertical xline through a single point, as follows:

> Command: **Xline**
>
> Specify a point or [Hor/Ver/Ang/Bisect/Offset]: **v**
>
> Specify through point: *(Pick point.)*
>
> Specify through point: *(Press* ESC*)*

Angled Option.

The **Ang** option draws an xline at a specified angle through a picked point, or by a reference, as follows:

> Command: **Xline**
>
> Specify a point or [Hor/Ver/Ang/Bisect/Offset]: **a**
>
> Enter angle of xline (0) or [Reference]: *(Pick point.)*
>
> Specify second point: *(Pick point.)*
>
> Specify through point: *(Pick point.)*
>
> Specify through point: *(Press* ESC.*)*

Bisector Option:

The **Bisect** option draws an xline that bisects (half-way between) a pair of intersecting lines, as follows:

> Command: **Xline**

Specify a point or [Hor/Ver/Ang/Bisect/Offset]: **b**

Specify angle vertex point: *(Pick point.)*

Specify angle start point: *(Pick point.)*

Specify angle end point: *(Pick point.)*

Specify angle end point: *(Press* ESC.*)*

Offset Option.

The **Offset** option draws an xline parallel to an existing line. You specify either the off-set distance or pick a point the xline should go through, as follows:

Command: **Xline**

Specify a point or [Hor/Ver/Ang/Bisect/Offset]: **o**

Specify offset distance or [Through] <1.0000>: *(Enter distance.)*

Select a line object: *(Pick line.)*

Specify side to offset: *(Pick point.)*

Select a line object: *(Press* ESC.*)*

UNDOING & REDOING COMMANDS

AutoCAD allows you to undo your work, then to redo it! This is useful if you have just performed an operation that you wish to reverse. After you undo an operation, you can use the **Redo** command to reverse the undo. Let's take a look at how to do these interesting operations.

UNDOING OPERATIONS

The **U** command reverses the effect of the most recent command. You can execute a series of "undos" to backup through a string of changes. The **U** command should not be confused with the more complex **Undo** command, although it functions the same as **Undo 1**.

Undoing a command restores the drawing to the state it was in before the command was executed. For example, if you erase an object and then execute the **U** command, the object is restored. If you scale an object and then undo it, the object is rescaled to its original size. The **U** command lists the command that is undone to alert you to the type of command that was affected. For example, executing the **U** command after the **Scale** command results in the following:

Command: **U**

SCALE

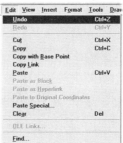

Undo   Redo

As an alternative to typing **U**, you can also press CTRL+Z or choose **Edit | Undo** from the menu bar.

There are several commands cannot be undone. **Save**, **Plot**, and **Wblock**, for example, are not affected because AutoCAD cannot "unsave" or "unplot". If you attempt to use the **U** command after these commands, the name of the command is displayed, as above, but the command's action is not undone.

Undoing a just-completed **Block** command restores the block, and deletes the block definition that was created, leaving the drawing exactly as it was before the block was inserted.

REDOING A DRAWING OPERATION

The **Redo** command is the antidote for the **U** command.

Redo "undoes" the undo. The **Redo** command must be used immediately after undo commands. However, **Redo** is limited to restoring a single undo, so use the **U** command with restraint!

To use **Redo**, choose **Edit | Redo** from the menu bar, press CTRL+Y, or type

> Command: **Redo**

EXERCISES

1. Let's draw a box using the Line command.

> Command: **Line**
>
> Specify first point: *(Enter point 1.)*
>
> Specify next point: *(Enter point 2.)*
>
> Specify next point: *(Enter point 3.)*
>
> Specify next point: *(Enter point 4.)*
>
> Specify next point: *(Enter point 1.)*

Figure 7.68 *Using the Line Command*

You may have noticed that the line "stretched" behind the crosshairs. This is called rubber banding.

2. Draw lines on grid paper, connecting the points designated by the following absolute coordinates.

   Point 1: **1,1**

   Point 2: **5,1**

   Point 3: **5,5**

   Point 4: **1,5**

   Point 5: **1,1**

3. Use Figure 7.69 to fill in the missing absolute coordinates. Each side of the shape is dimensioned. Place the answers in the boxes provided.

4. List the length of each side of the object in Figure 7.70. (Calculate from the absolute coordinates given.)

5. Use the following coordinates to draw an object.

   Point 1: **0,0**

   Point 2: **@3,0**

   Point 3: **@0,1**

   Point 4: **@–2,0**

   Point 5: **@0,2**

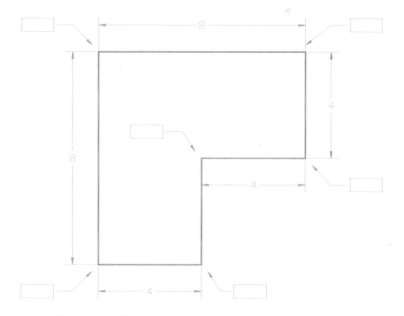

Figure 7.69 *Fill in the Coordinates*

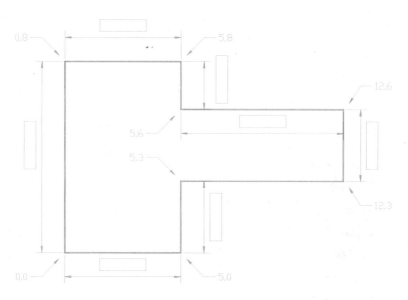

Figure 7.70 *List the Length of Each Side*

Point 6: **@–1,0**

Point 7: **0,0**

6. List the coordinates used to draw the following object. Use relative coordinates.

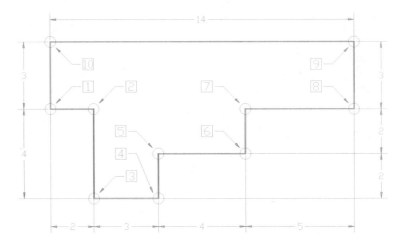

Figure 7.71 *List the Relative Coordinates*

7. Use the following coordinates to draw an object.

Point 1: **0,0**

Point 2: **@4<0**

Point 3: **@4<120**

Point 4: **@4<240**

8. List the coordinates used to construct the following object. Use polar coordinates whenever possible.

Point 1: **0,0**

To point 2: _____

To point 3: **@3<45**

To point 4: _____

To point 5: _____

To point 6: _____

To point 7: _____

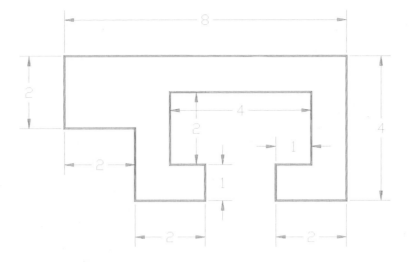

Figure 7.72 *List the Polar Coordinates*

To point 8: _____

9. Write a list of the coordinates used to construct the following objects. You can use any combination of absolute, relative, or polar coordinates.

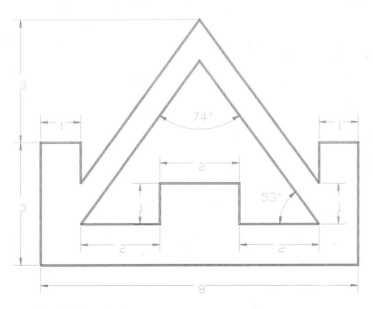

Figure 7.73 *List the Coordinates*

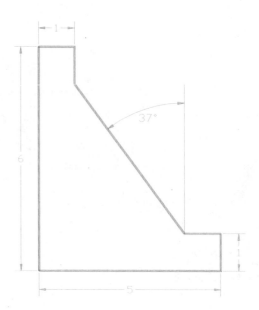

Figure 7.74 *List the Coordinates*

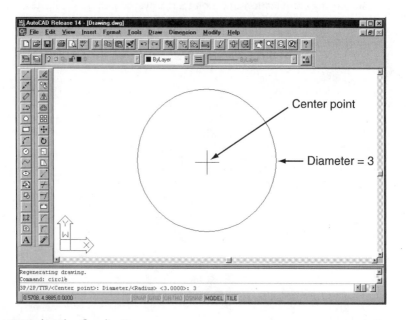

Figure 7.75 *List the Coordinates*

10. Let's set up a point type of 66 and place it in our drawing:

Command: **Pdmode**

New value for PDMODE <default>: **66**

Each point entered after a new **Pdmode** is set will be the type specified. Point objects entered before the change to the new setting are updated at the next command that causes a regeneration. To place the point we defined in the drawing, enter

Command: **Point**

Current point modes: **PDMODE=66 PDSIZE=0.0000**

Specify a point: *(Pick a point.)*

11. Let's construct a circle with a radius of 5. From the menu bar, choose **Draw**. From the **Draw** menu, choose **Circle**. Choose **Center, Radius**.

Command: _circle Specify center point for circle or [3P/2P/Ttr (tan tan radius)]: *(Pick a point.)*

Specify radius of circle or [Diameter]: **5**

12. Let's suppose you want to construct a circle using a center point and a diameter of 3.

Command: **Circle**

Specify center point for circle or [3P/2P/Ttr (tan tan radius)]: *(Enter point 1.)*

Specify radius of circle or [Diameter]: **D**

Specify diameter of circle <5.0000>: **3**

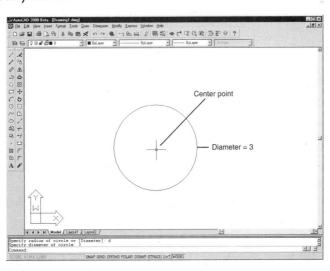

Figure 7.76 *Circle Drawn by Diameter*

13. Start a new drawing. Draw a line, circle, and arc of any size on the screen.

Choose the **Line** command from the **Draw** menu.

Hold down the SHIFT key and press the right mouse button to access the Object Snap menu.

Choose **Endpoint** from the menu. Notice the aperture box at the intersection of the crosshairs.

Place the aperture over one end of the line and click. The new line should "snap" precisely to the endpoint of the existing line.

> Before placing the endpoint of the line, choose **CENter** from the object snap menu (remember to display the object snap menu by pressing and right mouse button).
>
> Place the aperture over any part of the circle circumference and click. Notice how the line snapped to the center of the circle.
>
> Repeat the same procedure, except select a part of the arc. The line will snap to the center point of the arc.

14. Draw several more lines, circles, and arcs. Use each of the object snap modes to capture the parts of the entities. Be sure to use **TANgent** object to construct tangent lines with circles.

15. Right-click **GRID** on the status line and choose **Settings** from the cursor menu. When the **Snap and Grid** dialog box appears, choose **Grid On** to turn on grid mode. Set the **Grid X** and **Y Spacing** to **1.0**. Choose **OK**. The grid should be visible on the screen.

16. Press F7. This key toggles the grid on and off.

17. Return to the **Snap and Grid** dialog box. Set the X grid spacing to a value of 1, and the Y grid spacing to a value of 0.5. Choose **OK**. Notice the grid has changed.

18. Let's set up a snap spacing, with a snap resolution of 0.25. Right-click the word SNAP on the status line and choose **Setting** in the **Snap and Grid** tab and choose **Snap On** to turn on snap mode.

Change the **Snap X** and **Y Spacing** to **0.25**. Choose **OK**.

Next, choose the **Line** command and enter the line endpoints on the screen. Did you notice the crosshairs "snapping" to a point?

19. Right-click **GRID** on the status line and choose **Settings**. Set the grid spacing to 0.25 spacing. Use the **Line** command to set some endpoints. Notice how the crosshairs line up with the grid points.

20. Use F9 to turn the snap mode on and off. Draw two boxes; one with snap mode on, and one with snap mode off. Notice how the points are easier to line up with snap mode on.

21. Use the **Zoom Window** command to zoom in on one of the boxes. Move the crosshairs around the area of the box with the snap mode on (use F9 to turn snap mode on and off). Notice how the movement of the crosshairs is more exaggerated with the closer zoom. This is because the "screen distance" between the snap points is larger when zoomed in. The actual snap distance (resolution), however, remains unchanged.

22. Display the **Snap and Grid** dialog box again, and enter values as shown:

    Snap *X* spacing: 0.25

    Snap *Y* spacing: 0.50

    Grid *X* spacing: 0.25

    Grid *Y* spacing: 0.50

    Notice how the spacing for the vertical and horizontal snap resolution is different. It is not necessary for the vertical and horizontal values to be the same. Try drawing another box with the snap on.

23. Let's try rotating the snap points and the crosshairs. Choose the **Snap and Grid** dialog box again.

    Angle: **45**

    Your display should look like the one in Figure 7.77. Notice how the snap resolution and aspect are maintained. The entire snap grid and crosshairs are rotated. Draw a box with the rotated snap grid and crosshairs. This is an excellent method of drawing objects that have many angular lines.

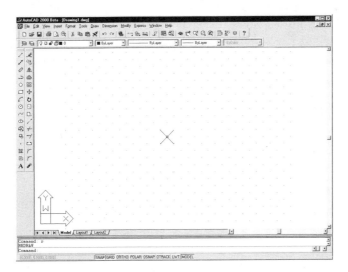

Figure 7.77 *Snap with Aspect and Rotate*

EXERCISE

24. Start a new drawing. Choose the **Line** command and draw a four-sided box with horizontal and vertical lines. Do not use the ortho mode. Now press F8 to turn on ortho mode and draw another box. Notice how much easier it is to create straight lines with the ortho mode turned on.

25. Choose the **Line** command again and enter the first point. Now move the crosshairs around the first point and notice how AutoCAD forces the line to be horizontal or vertical, depending on the location relative to the first end-point.

26. Use the **Snap and Grid** dialog box command to set a snap increment of 0.25. Be sure ortho is on and draw another box. Notice how the combination of snap and ortho makes drawing the box very easy.

27. Start a new drawing. Choose the Line command and draw a line segment. Use ESC to clear the command line. Enter U from the keyboard and press ENTER. Did the line segment disappear?

28. Use the **Line** command to draw several line segments. Choose the **Line** command and draw more line segments. Use the **U** command to undo the lines drawn with the last line command. Which segments were undone? Press [U] to repeat the **U** command. What happened?

29. Use the **Line** command to draw two line segments. Before entering the last point, use 6 to turn on ortho mode. Notice the **ORTHO** button on the status line. Now use the **U** command to undo the sequence. Is the ortho mode on now?

30. Use F8 to turn on ortho. Notice the listing on the status line. Now use the **Line** command to draw a line segment. Next, use the **U** command to undo the sequence. Is ortho mode still turned on? What is the difference between this and the last sequence you performed?

31. Use the **Lin**e command to draw several line segments. Use the **U** command to undo the lines. Now use the **Redo** command. Did the lines reappear?

32. Enter the **Redo** command again. What does the prompt line say? Why?

33. Use the commands you have learned to draw the following objects.

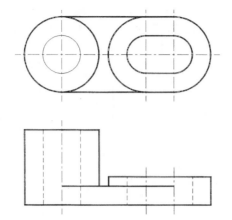

Figure 7.78

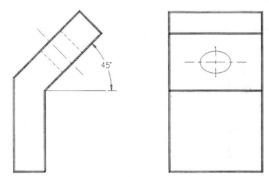

Figure 7.79

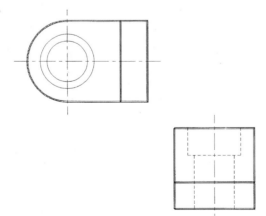

Figure 7.80

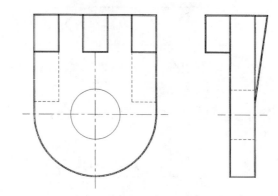

Figure 7.81

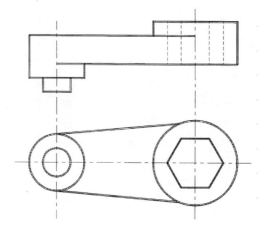

Figure 7.82

CHAPTER REVIEW

1. What command allows you to enlarge and reduce the display size of your drawing?

2. When you draw a circle and you are prompted for a diameter or radius, a numerical value can be given, but can the distance be shown by screen pointing as well?

3. Explain the function of the **Close** option under the **Line** command.

4. Why does AutoCAD offer so many methods of arc construction?

5. What is the purpose of the **Pan** command?

6. What is the default method of arc construction?

7. What command can be used to change the design and size of points?

8. Why is the **Close** option of the **Line** command a more accurate method of closing lines than attempting to line up the endpoints manually?

9. Can the **U** command be used to undo a sequence of commands?

10. **Pdmode** and **Pdsize** are *retroactive*. What does this mean?

11. What is the **Realtime** option under the **Zoom** command used for?

12. When a real-time zoom is in progress, can you override it? How?

13. How might you make a backup file of the drawing in progress without ending the drawing?

14. How is the **Line** command terminated?

15. There are six methods of constructing circles. What four circle properties are used in various combinations to comprise these methods?

16. What is the *chord* of an arc?

17. Can the **U** command be entered from the menu bar as well as from the keyboard?

18. How do **Pan** and **Zoom** commands differ?

19. When a dynamic **Zoom** is invoked, what information appears on the screen in reference to your drawing?

20. What is a *transparent* command? How is this option invoked?

21. If the **Undo** command is used while you are in a **Line** command sequence, does the entire sequence disappear or is each endpoint stepped through backwards?

22. How do **Redraw** and **Regen** differ?

23. What option allows you to enlarge or reduce the original drawing size by a numerical factor for viewing?

24. Relative and polar coordinates use the last point entered as a reference point. What reference point are absolute coordinate entries relative to?

25. In what type of coordinate entry is the @ symbol required?

26. Using Figure 7.83, list the coordinate entry necessary to enter the second point, first using absolute coordinates, then relative coordinates, finally using direct distance entry.

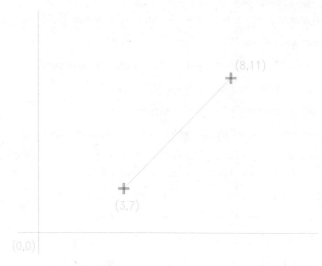

Figure 7.83 *List the Coordinates*

27. AutoCAD uses a default angle setting that can be altered if the user desires. Where would these setting changes be made?

28. When using polar coordinates, from what point is the distance and angle measured?

29. What method of point entry is entered in the *X,Y* format?

30. What is *tracking* used for?

31. What direction is negative in AutoCAD's default angle specification?

32. List the differences in the five types of coordinate entry, using illustrative examples of each.
    1.

    2.

    3.

    4.

    5.

33. Name two ways to create construction lines:
    1.

    2.

34. At what point on the coordinate axis is the lower left corner of the AutoCAD drawing screen normally located?

35. Where is the zero angle located in AutoCAD's default angle specification?

36. Given the first point of a line at 12,37, what coordinate entry would you use to locate an endpoint 15 units and 79 degrees from the first point?

37. Using AutoCAD's default angle settings, in which direction does the angle measurement increase?

38. How is the format different for *absolute* coordinate entry and *relative* coordinate entry?

39. What is the *aperture*?

40. What object snap mode would you choose to snap on to a point entity?

41. Can you snap on the midpoint of an arc?

42. When does AutoSnap come into effect?

43. Can different X and Y spacing be given to the snap grid? Can it coincide with the grid spacing?

44. What advantage do **Ray** and **Xline** have over grid?

45. What mode can provide great accuracy in creating true horizontal and vertical lines?

46. When can **APPint** mode *not* be used? What purpose does it serve?

47. Can you override the running object snap mode?

48. When does the **Redo** command have to be used?

49. How can the results of the last command executed be removed?

50. Although the grid is not a part of the drawing, can it be plotted?

51. Will the **U** command undo any command just executed?

52. What is the primary difference between tracking and direct distance entry?

53. Where does a ray start? End?

54. Name the three functions of the Aerial View:
    1.

    2.

    3.

CHAPTER 8

Editing the Drawing

Editing drawings is one of the strengths of CAD. AutoCAD 2000 provides an exceptional toolbox of editing functions. After completing this chapter, you will be able to

- Demonstrate a variety of methods of selecting the parts of the drawing to be edited
- Modify the way AutoCAD selects objects
- Understand some of AutoCAD's basic editing commands
- Use AutoCAD's edit commands with work problems

INTRODUCTION

So far you have learned how to use AutoCAD to create a simple drawing with lines, circles, arcs, and points.

From time to time you will have reason to change parts of your work. AutoCAD supplies edit commands to achieve this. In this chapter you will learn about

Object Selection Options

Provides a variety of methods of selecting objects to be edited.

SELECT

Preselects objects to be edited.

QSELECT

Selects objects by type.

GROUP

Creates a named group of objects.

ERASE

Deletes one or more objects from the drawing.

OOPS

Restores to the drawing erased objects.

BREAK

Removes a portion of an object.

MOVE

Moves objects in the drawing.

COPY

Copies objects, one or more times.

FILLET

Makes smooth and perfect corners with lines, arcs, and polylines.

Most of these commands are found in the **Modify** drop-down menu.

The ability to edit your work electronically is a very powerful feature of CAD drafting. Last-minute changes, correction of mistakes or change for any other reason can be accomplished quickly and accurately.

In order to edit a drawing, you must first determine the following:

Which object (or objects) would you like to edit?

How would you like to edit them?

After determining the objects you would like to edit, you use the object selection process to isolate them.

USING OBJECT SELECTION

To identify which objects are to be edited, you use the method of identification called *object selection*. Object selection is the standard method of object identification in AutoCAD, and it is used with most edit commands.

To efficiently edit objects, it is essential that you understand the options provided by object selection. As you will see, the edit commands are very easy to use; efficiency results from innovative use of the methods to select objects. This is especially true of intricate drawings. You can use any combination of selection methods, or a single method several times, to build the selected set of objects.

METHODS OF SELECTING OBJECTS

You can select objects for editing by a variety of selection methods. Let's look at each method, as described by the following list (uppercase letters indicate the option's abbreviation):

Standard selection modes

> **[Pick]** selects a single object.
>
> **ALL** selects all objects on thawed layers.
>
> **Window** selects all objects completely inside a rectangle defined by two points.
>
> **Crossing** selects objects within and crossing an area defined by two points.
>
> **BOX** selects all objects inside, or crossing a rectangle specified by two points.
>
> **AUto** selects a single object by pointing, or by the Box method.
>
> **Fence** selects all objects crossing a selection fence.
>
> **WPolygon** selects objects completely inside a polygon defined by points.
>
> **CPolygon** selects objects within and crossing a polygon defined by specifying points.

Special selection modes

> **SIngle** selects the first object(s) and refrains from prompting for further selections.
>
> **Multiple** specifies multiple points without highlighting the objects.
>
> **Previous** selects the most recent selection set.
>
> **Last** selects the most recently created object visible.
>
> **Group** selects all objects within a specified group.

Selection modifiers

> **Remove** removes objects from the current selection set.
>
> **Add** adds selected objects can be added to the selection set.
>
> **Undo** reverses the selection of the most recently added object.

STANDARD SELECTION MODES

To use a selection mode, type its abbreviation at the "Select objects:" prompt. For example, here is the **Copy** command with the [pick] and Window selection modes.

> Command: **copy**
>
> Select objects: *(Pick an object.)*
>
> I found Select objects: **w**
>
> Specify first corner: *(Pick a point.)*
>
> Specify opposite corner: *(Pick another point.)*
>
> 3 found (I duplicate), 3 total
>
> Select objects: *(Press* ENTER.*)*
>
> Specify base point or displacement, or [Multiple]: *(Etc.)*

Notice that the "Select objects:" prompt repeats until you press ENTER. This means you can keep selecting (and deselecting) objects until you are satisfied with the selection set.

Notice, too, that AutoCAD reports the number of objects it finds (as in "3 found"), as well as letting you know if you pick the same object more than once (as in "1 duplicate"). As AutoCAD finds each object you selected, it highlights them with a broken line pattern.

[Pick] (Select by Pointing)

When you first select an edit command, the cursor is replaced by a small box that we refer to as the *pickbox*. If you place the pickbox over the object and click (press the left mouse button), the drawing is scanned, and the object covered by the pickbox is selected.

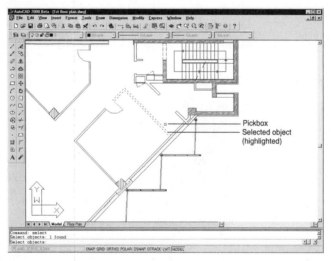

Figure 8.1 *Pickbox*

The pickbox can be changed in size to assist in working in drawings of differing complexities. The **Selection** tab of the **Options** dialog box allows you to change the pickbox size (covered later in this chapter).

 Note: Don't place the pickbox at an intersection of two objects. This will give unpredictable results, since AutoCAD won't know which object you want to choose.

After you select the item(s) to be edited, you will notice that AutoCAD highlights the items that are selected. Each selected item becomes dotted instead of solid. This helps you to see which items are selected. (The method of highlighting may be different on some display systems.)

ALL (Select All Objects)

The **All** option selects all the objects in the drawing— except those on frozen and locked layers. Some commands, such as **CopyClip**, select only all objects *visible* on the screen.

 Note: If most of the objects in a drawing are to be selected for edit, first designate all the drawing objects with the **All** option, then use the **Remove** (explanation follows) option to "deselect" the few exception items.

W (Select with a Window)

One of the most popular forms of object selection, the **W** option (short for **Window**) places a rectangle around the objects to be selected. You define the rectangle by picking its opposite corners, as follows:

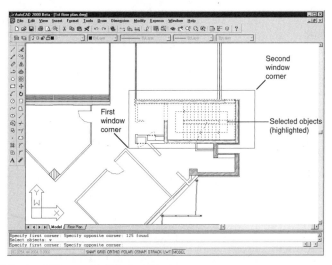

Figure 8.2 *Selecting Objects with Window Mode*

> Select objects: **w**
>
> Specify first corner: *(Pick a point.)*
>
> Specify opposite corner: *(Pick another point.)*

Any object *entirely* in the window rectangle is selected. If an object crosses the rectangle, it is not selected (see Crossing mode). Only objects that are currently visible on the screen are selected.

Note: You can select certain objects by placing the box so that all parts of those you want to choose are contained in the box and those which you don't want chosen are not entirely contained. With this method, you can make certain selections in areas of your drawing where objects overlap.

C (Select with a Crossing Window)

The **C** option (short for **Crossing**) is similar to the **Window** option, with one difference: objects crossing the rectangle are included in the selection set—as well as those objects entirely within the rectangle. You define the rectangle by picking its opposite corners.

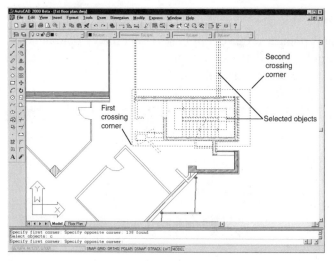

Figure 8.3 *Selecting Objects with Crossing Mode*

Note that the Crossing window is dashed (or highlighted in some other fashion). This distinguishes it from the Window rectangle, which is made of a solid line.

Box (Select with a Box)

The **B** option (short for **Box**) allows you to use either the Crossing or Window rectangle to select objects. You define the rectangle by picking its opposite corners.

After you pick the first corner of the box, you can move the cursor either to the right or to the left. If you move to the right, the result is the Window selection; AutoCAD selects objects completely within the selection rectangle.

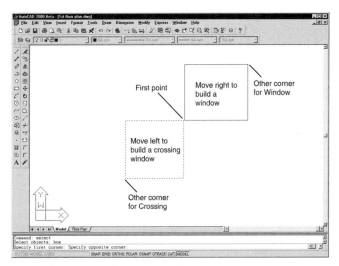

Figure 8.4 *Selecting Objects with Box or Auto Mode*

If you move to the left, the result is the Crossing selection; AutoCAD selects objects within the selection rectangle, as well as those crossing it. In addition, the rectangle is drawn of a dashed line.

AU (Select with the Automatic Option)

The **AU** option (short for **AUtomatic**) is a combination of the **[Pick]** and **Box** options. After you enter **AU** in response to the "Select objects" prompt, a pickbox is initially used for selection. If you select a point with the pickbox and an object is found, the selection is made.

If an object is not found, the selection point becomes the first corner of the Box option. Move the box to the right for Window, or the left for Crossing. The **AU** option is excellent for advanced users who wish to reduce the number of modifier selections that they input.

F (Select with a Fence)

The **F** option (short for **Fence**) uses a *polyline* (a series of connected line segments) to select objects. The fence line is displayed as a dashed line; all objects that cross the fence lines are selected. You can construct as many fence segments as you wish and can use the **Undo** option to undo a fence line segment.

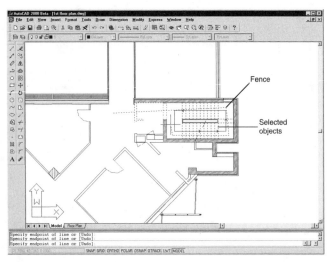

Figure 8.5 *Selecting Objects with Fence Mode*

> Select objects: **f**
>
> First fence point: *(Pick a point.)*
>
> Specify endpoint of line or [Undo]: *(Pick a point.)*
>
> Specify endpoint of line or [Undo]: *(Pick another point.)*

WP (Select with a Windowed Polygon)

The **WP** option (short for **Windowed Polygon)** selects objects by placing a polygon window around them. The polygon window selects in the same manner as the **Window** option: all objects completely within the polygon are selected. Objects crossing, or outside the polygon, are not selected. The difference is that the **W** option creates a selection rectangle, while the **WP** option creates a multisided window. Let's look at a sample command sequence:

> Select objects: **wp**
>
> First polygon point: *(Pick a point.)*
>
> Specify endpoint of line or [Undo]: *(Pick a point.)*
>
> Specify endpoint of line or [Undo]: *(Pick another point.)*

After you enter the first polygon point, you proceed to build the window by placing one or more endpoints. The polygon rubber bands to the cursor intersection, always creating a closed polygon window.

You undo the last point entered by entering a **U** in response to the prompt. Pressing ENTER closes the polygon window and completes the process.

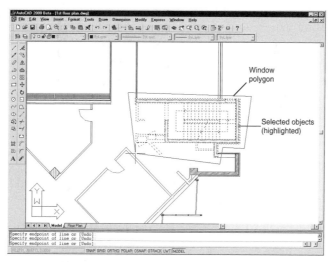

Figure 8.6 *Selecting Objects with Window Polygon Mode*

Note that the polygon window must not cross itself or be placed directly on a polygon object. If it does, AutoCAD warns "Invalid point, polygon segments cannot intersect" and refuses to place the vertex.

CP (Select with a Crossing Polygon)

The **CP** option (short for **Crossing Polygon**) works in the same manner as the **WP** option, except that the polygon functions in the same manner as a crossing window. All objects within or crossing the polygon are selected.

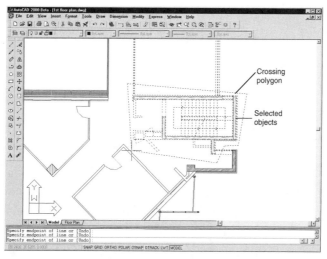

Figure 8.7 *Selecting Objects with Crossing Polygon Mode*

The crossing polygon is displayed as a dashed line similar to a crossing window.

SI (Select a Single Object)

The **SI** option (short for **SIngle**) forces AutoCAD to issue a single "Select objects:" prompt. (All other selection modes repeat the "Select objects:" prompt until you press Enter.) You can use another selection option in conjunction with **SI** mode, such as the Crossing selection shown in this example:

> Select objects: **si**
>
> Select objects: **c**
>
> Specify first corner: *(Pick a point.)*
>
> Specify opposite corner: *(Pick a point.)*
>
> 6 found

The single option can be used in menu strings for an efficient single object selection operation, since it deletes the requirement of a *null* response (ENTER) to end the object selection process.

M (Select Through the Multiple Option)

Each time you select an object, AutoCAD scans the entire drawing to find the object. If you are selecting an object in a drawing that contains a large number of objects, there can be a noticeable delay.

The **M** option (short for multiple) forces AutoCAD to scan the drawing just once. This results in shorter selection time. Press ENTER to finish the object selection and begin the scan. Note that AutoCAD does not highlight the objects you select until you press ENTER.

> Select objects: **m**
>
> Select objects: *(Pick three times.)*
>
> 3 selected, 3 found

SPECIAL SELECTION MODES

P (Select the Previous Selection Set)

The **P** option (short for **Previous**) uses the previous selection set. This very useful option is allows you to perform several editing commands on the same set, without needing to re-select them.

L (Select the Last Object)

The **L** option (short for **Last**) selects the last object drawn *still visible on the screen*. If the command is repeated and **Last** is used a second time, the selection chooses the same last object and reports "1 found (1 duplicate), 1 total." This option is useful for drawing an object, then immediately editing it.

G (Select the Group)

The **G** option (short for **Group**) adds the members of a named group to the selection set. (Groups are covered later in this chapter.) The **Group** option prompts:

> Enter group name: *(Enter a group name.)*

CHANGING THE ITEMS SELECTED

You can add or remove objects from the group of selected objects by using modifier commands. Modifier commands are entered after you have selected at least one object with the object selection process, but before you press ENTER to accept the designated objects.

U (Undo the Selected Option)

The **U** option (short for **Undo**) removes the most recent addition to the set of selections. If the undo is repeated, you will step back through the selection set. This shortcut replaces the two-step process of using the **Remove** option, then remembering the objects to pick.

R (Remove Objects from the Selection Set)

The **R** option (short for **Remove**) allows you to *remove* objects from the selection set. You can remove objects from the selection set by any object selection method.

> Select objects: **r**
>
> Remove objects: *(Pick an object.)*
>
> I found, I removed, 2 total

Note: Many new CAD operators think of object selection only in terms of adding objects to a selection set. It is often the case that you wish to choose a large number of objects, with the exception of one or two objects that are located within the area of the other objects.

A (Add Objects to the Selection Set)

The **A** option (short for **Add**) add objects to the set. The **Add** option is usually used after the **Remove** option. Add changes the prompt back to "Select objects:" so you may add objects to the selection set.

Canceling the Selection Process.

Pressing ESC at any time cancels the selection process and removes the selected objects from the selection set. The prompt line returns to "Command:".

SELECTING OBJECTS BY TYPE (QSELECT)

In the previous section, you learned how to select objects by picking them on the screen, either by actual picks or by a variety of windowing options. AutoCAD 2000 provides a second way to select objects through the **QSelect** command. The **Quick Select** dialog box lets you select objects based on their properties.

You would use this dialog box to select, for example, all circles with a radius larger than one inch. In Figure 8.8, the **Quick Select** dialog box is being used to select all blocks named FNCOMP (desktop computers symbols). In the screen image, you can see the 27 selected desktop computers (shown by dashed lines).

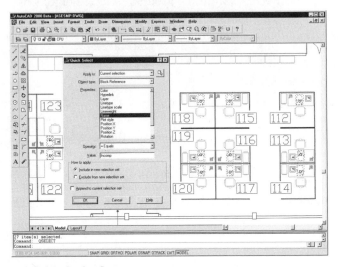

Figure 8.8 *Quick Select Dialog Box*

To display the dialog box, type the **QSelect** command. Alternatively, right-click and select **Quick Select** from the cursor menu.

The **Apply To** list box allows you to apply the selection to the entire drawing or a currently selected set of objects. The **Select** Objects button allows you to select objects, if you forgot to do it before starting the **QSelect** command.

The **Object Type** list box allows you to select a specific type of objects (such as a line, block, or circle). This list box lists only the objects found in the current drawing. As an alternative, you can pick the **Multiple** option, which applies to more than one type of objects.

The object type you select affects the list shown in the **Properties** list. This list contains all properties that AutoCAD can search for, such as color, layer, and linetype. You can select only one property from this list.

The **Operator** list lets you narrow the range AutoCAD will search for. Depending on the property you select, the choice of operator may include = Equals, <> Not Equal

To, > Greater Than, < Less Than, and *Wildcard Match. The Wildcard Match operator is used with text fields that can be edited, such as the names of blocks. The **Value** field goes together with the Operator list: if AutoCAD can determine values, it lists them here; if not, you type the value.

You can have AutoCAD include or exclude these objets from an existing selection set. Choose **OK** to exit the dialog box. AutoCAD highlights the objects specified.

SELECTION OPTIONS

AutoCAD allows you to set different modes of object selection. The previous sections covered the basic object selection methods. As you gain more proficiency, you will want to use other modes of selection.

SETTING SELECTION OPTIONS (DDSELECT)

Use the **DdSelect** command to set selection modes (from the menu, select **Tools | Options | Selection**). The dialog box contains several selection options. Turn on each option by selecting the check box next to it. Let's look at how each option functions.

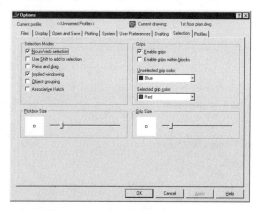

Figure 8.9 *Selection Tab of the Options Dialog Box*

Noun/Verb Selection

So far, you have learned to edit objects by: (1) first selecting the edit command and (2) then selecting the objects to modify with that command. An analogy is to choose the *verb* (the action represented by the edit command), then the *noun* (the object of the action represented by the selection set).

As an alternative, AutoCAD allows you to reverse this procedure by: (1) first picking the objects to edit and (2) then selecting the edit command to use. This is called *noun/verb selection*.

After you turn on the **Noun/Verb Selection** option in the **Selection** dialog box, AutoCAD places a pickbox at the intersection of the crosshairs. (This option is turned on by default.) This pickbox is used to select objects.

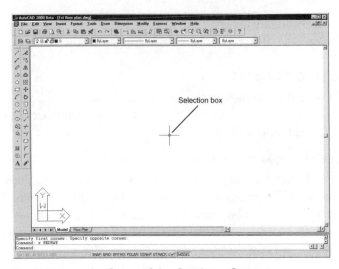

Figure 8.10 *The Pickbox at the Center of the Crosshairs Cursor*

To edit objects, first select the object(s) to be edited and then choose the edit command. Let's look at how the command sequence would look if we used this method to erase a single object.

Command: *(Select the Object.)*

Command: **Erase**

I found

The following editing commands can be used with the noun/verb selection method.

Align	List
Array	Mirror
Block	Move
Change	Properties
Chprop	Rotate
Copy	Scale
Dview	Stretch
Erase	Wblock
Explode	

Use Shift to Add

When using the selection process, you normally select the object(s), then press Enter when you are finished. As you choose each object, it is automatically added to the selection set.

As an alternative, you can turn on the **Use Shift to Add** option. When this option is on, you must hold down the SHIFT key to add objects to the selection set; this is similar to the method used by many Windows programs. (If you make more than one selection without holding the SHIFT key, the previous selections are removed from the selection set.)

Press and Drag

The traditional AutoCAD method of using a window to select objects is to click on one corner, then move the cursor to the other corner, and click again.

As an alternative, you can turn on the **Press and Drag** option. You build a window by clicking one corner, then *dragging* (holding down the mouse button) the cursor to the other corner. Like the **Shift to Add** option, this is also similar to the windowing method used by many Windows programs.

Implied Windowing

You previously learned how to use a selection window or a crossing window to select objects. All you have to do is enter either a **W** or a **C** to invoke the window mode.

The **Implied Windowing** option allows you do this automatically (this option is turned on by default). You "imply" a window by clicking on an empty area of the drawing. When AutoCAD does not find an object within the area of the pickbox, then it assumes that you want to use a window. If you move the cursor to the right, Window mode is entered. If you move to the left, Crossing mode is entered.

Since the first point entered describes the first corner of the window, be sure to select a desirable position.

Object Grouping

When **Object Grouping** is turned on, AutoCAD selects the group when just one object of that group is picked. When off, just the picked object is selected. You can toggle object grouping mode by pressing CTRL+A.

Associative Hatch

When **Associative Hatch** is turned on, AutoCAD also selects boundary objects when you pick an associative hatch pattern. When off, just the hatch is selected.

CHANGING THE PICKBOX SIZE (PICKBOX)

The pickbox size is set from the same **Selection** dialog box (select **Tools | Options | Selection** from the menu bar). The pickbox can be as small as zero pixels and as large as 50 pixels across; the default size is 3 pixels wide.

To change the size, drag the slider bar. Move it to left to decrease the size; move it to the right to increase the size. The window to the right of the slider bar shows the actual size of the pickbox as you change it.

If you prefer a more precise approach, use the **Pickbox** system variable, as follows:

Command: **pickbox**

Enter new value for PICKBOX <3>: **5**

Note: The pickbox size can aid in building the selection set. A larger pickbox makes selection less tedious, while a smaller pickbox allows you to select objects in a crowded area without the necessity of enlarging the drawing area with the **Zoom** command. Change the pickbox size as the conditions warrant.

SELECTING CLOSELY SPACED OBJECTS

A common problem in AutoCAD is in trying to select one object from several that are spaced closely together. One solution is to use transparent **'Zoom Window** command to get a closer look.

A second alternative is to select hold down CTRL and click repeatedly on the closely spaced objects. As you do this, AutoCAD highlights the objects, one by one. Cycle through the objects until AutoCAD highlights the objects you want. Press ENTER to end object selection.

PRESELECTING OBJECTS (SELECT)

Use the **Select** command to *preselect* objects for editing. To execute the **Select** command, enter

Command: **Select**

Select objects: *(Select objects to be edited.)*

Use one of the selection options, as described at the beginning of this chapter.

The next time you use an edit command, you simply enter the **P** option (short for **Previous**) to choose the preselected objects.

Note: The **Select** command is especially helpful when edits of several types must be performed on the same group of objects. You can select the target group of objects by executing the **Previous** option when using each successive edit command.

CREATING GROUPS OF OBJECTS (GROUP)

A selection set lasts only until a new selection set is created. When you select a group of circles, then later select a group of lines, AutoCAD forgets about the group of circles you selected first.

To overcome this limitation, AutoCAD lets you create any number of "groups." Each group has a name and consists of any selection set of objects. The **Group** command displays the **Object Grouping** dialog box (there is no menu selection).

Figure 8.11 *The Object Grouping Dialog Box*

Group Identification

The first step in creating a group is to give it a name. In the **Object Grouping** dialog box, you must type in the **Group Name** field a descriptive name, such as "Linkage." The name can consist of up to 255 letters, numbers, and blank spaces. You can also give the group a description up to 448 characters long, such as "The left end of the linkage."

The **Unnamed** check box determines whether the group is named. When selected (turned on), AutoCAD automatically assigns a name to the group. The first assigned name is *A0; the next unnamed group is assigned the name *A1; and so on. When **Unnamed** is not selected (turned off), you must give the group its name.

Create Group

The second step to creating a group is to select the objects that will be part of the group. Choose the **New** button, then use any object selection modes to select the objects.

After the group is created, you can remind yourself of the objects that belong to a particular group by: (1) selecting a group name and (2) choosing the **Highlight** button. The dialog box disappears and AutoCAD highlights the objects in the group.

The **Selectable** check box is very important. When selected (turned on), selecting one object in the group selects the entire group. When it is not selected (turned off), selecting an object in the group selects only the object, not the group. (You can toggle selectable mode on and off, while drawing, by pressing CTRL+A.)

Change Group

After the groups are created, you can change the description and the objects in the group, as follows:

> The **Remove** button lets you remove objects from the group.
>
> The **Add** button lets you add objects to the group.
>
> The **Rename** button lets you change the group's name.
>
> The **Re-order** button lets you change the "order" of the objects in the group. AutoCAD numbers the objects in the group as you add them (the first object in the group has number 0). Changing the order of objects in the group can be important in numerical control and analysis operations.
>
> The **Description** button lets you change the group's description.
>
> The **Explode** button gets rid of the group and name.
>
> The **Selectable** button changes the selectable setting.

Figure 8.12 *The Order Group Dialog Box*

REMOVING OBJECTS (ERASE)

Use the **Erase** command to remove objects from the drawing. From the menu bar, select **Modify | Erase**. The command sequence is as follows:

> Command: **Erase**

> Select objects:

You will notice that the crosshairs cursor is now a pickbox. The pickbox can be used to select an object (point method). If you wish to use one of the other object selection options (such as **Window**), enter the letter for the option at the "Select objects:" prompt.

RESTORING ERASED OBJECTS (OOPS)

The **Oops** command restores the objects that were last erased from the drawing. (**Oops** is not available from the menu bar.)

> Command: **Oops**

You cannot always oops backward through a drawing to return objects erased several commands back. Most of the time you will be required to oops back objects immediately after you erased them. Don't press your luck by executing any other command between the two!

If you find yourself unable to oops back erased objects, remember that you can undo the erase with the **U** command.

PARTIAL ERASING (BREAK)

The **Break** command is used to erase a part of a line, circle, arc, trace, ellipse, donut, spline, or 2D polyline. Unlike some other editing commands, **Break** operates on only one object at a time. From the menu bar, select **Modify | Break**.

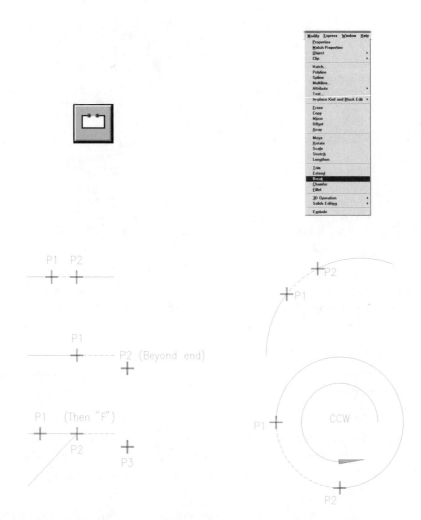

Figure 8.13 *Applying Break to Objects*

Command: **Break**

Select object: *(Pick object.)*

Specify second break point or [First point]:   *(Pick point.)*

To "break" an object, you select the two points on the object. The portion of the object between the two pick points is erased. By default, the point at which you select the object becomes the first break point.

If you wish to redefine the first break point, enter **F** in response to the prompt. You can then select another point to be the first break point, as follows:

Command: **Break**

Select object: *(Pick object.)*

Specify second break point or [First point]: **F**

Specify first break point: *(Pick point.)*

Specify second break point: *(Pick point.)*

 Note: The ability to redefine the first point is useful if the drawing is crowded or the break occurs at an intersection where pointing to the object at the first break point might result in the wrong object being selected.

A common problem is trying to break a line or arc completely—perhaps without knowing that you have indicated the complete object. **Break** must "leave something behind." If you think **Break** is not working, check to see if you have a (short) object. The **Break** command affects objects in different ways:

Line

The **Break** command removes the portion of line between the pick points. If one point is on the line and the other point is off the end of the line, the line is "trimmed back" to the first break point.

Trace

Traces are broken in the same manner as lines. The new endpoints of the trace are trimmed square.

Arc

An arc is broken in the same manner as a line.

Circle

A circle is broken into an arc. The unwanted piece is removed by going counter-clockwise from the first point to the second point.

Ellipse

An ellipse is broken in the same manner as a circle. The result is an elliptical arc.

Polyline

Polylines of non-zero width are cut square (similar to traces). Breaking a closed polyline creates an open polyline.

Viewport

Viewport borders cannot be broken.

MOVING OBJECTS (MOVE)

The **Move** command allows you to move one or more objects to another location. Move objects by showing a point to start from and a point to move to. The selected object(s) then move relative to the specified displacement. From the menu bar, select **Modify | Move**.

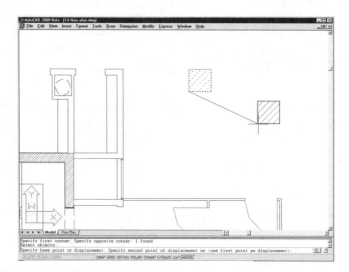

Figure 8.14 *Moving an Object*

Command: **Move**

Select objects: *(Pick object(s).)*

Select objects: *(Press Enter.)*

Specify base point or displacement: *(Pick point.)*

Specify second point of displacement or <use first point as displacement>: *(Pick point.)*

Note: Your first point need not be on the object to be moved; however, using a corner point or other convenient point of reference on the object makes the displacement easier to visualize.

MAKING COPIES OF OBJECTS (COPY)

The **Copy** command makes copies of existing objects in the drawing. Use the object selection process to choose the object to be copied. The prompt line then asks for the displacement from the original object to the location of the new object. From the menu bar, select **Modify | Copy**.

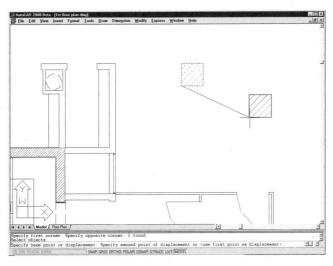

Figure 8.15 *Copying an Object*

Command: **Copy**

Select objects: *(Pick object(s).)*

Select objects: *(Press* ENTER.*)*

Specify base point or displacement, or [Multiple]: *(Pick point.)*

Specify second point of displacement or <use first point as displace-ment>: *(Pick point.)*

 Note: Use the **Copy** command freely to repeat items "on the go" as you draw.

Making Multiple Copies

The **Multiple** option of the Copy command allows you to place multiple copies. (See Figure 8.15).

Command: **Copy**

Select objects: *(Pick object(s).)*

Select objects: *(Press* ENTER.*)*

Specify base point or displacement, or [Multiple]: **M**

Specify base point: *(Pick point.)*

Specify second point of displacement or <use first point as displace-ment>: *(Pick point.)*

Specify second point of displacement or <use first point as displace-ment>: *(Pick point.)*

Specify second point of displacement or <use first point as displace-ment>: *(Press* ENTER.*)*

The "Second point of displacement" prompt repeats until you cancel the command. The copy originates from the originally selected object, using the base point you first selected.

ROUNDING CORNERS (FILLET)

The **Fillet** command is used to connect two lines or polylines with a perfect intersection or with an arc of a specified radius. Fillet can also be used to connect two circles, two arcs, a line and a circle, a line and an arc, or a circle and an arc.

The two objects do not have to touch in order to perform a fillet; this includes parallel lines. This property allows you to join two non-touching lines. In the case of parallel lines, a 180-degree arc is drawn between their ends.

From the menu bar, select **Modify | Fillet**.

Command: **fillet**

Current settings: Mode = TRIM, Radius = 0.5000

Select first object or [Polyline/Radius/Trim]: *(Pick object.)*

Select second object: *(Pick second object.)*

The command has three options:

Polyline

The **Polyline** option fillets the vertices of a polyline (instead of constructing a fillet between two objects). After specifying at the "Select first object" prompt, you select a single polyline.

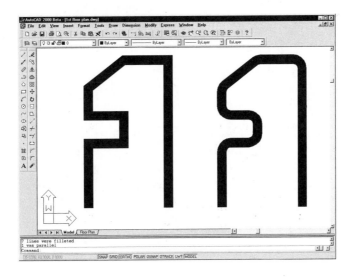

Figure 8.16 *Apply Fillet to a Polyline (at right)*

Radius

The **Radius** option determines the radius of the fillet arc (the default value is 0.5). When set to 0, the **Fillet** command is useful for ensuring two lines match precisely.

Trim

The **Trim** option determines what happens to the trimmed bits. When on (the default), the **Fillet** command trims away the selected edges, up to the fillet arc's endpoint.

FILLETING TWO LINES

The two lines are trimmed back (or extended, as necessary) so that an arc can be fit between them. A zero radius fillet connects two lines with a perfect intersection. Two parallel lines fillet with a radius equal to their offset distance.

FILLETING POLYLINES

You can fillet an entire polyline in one operation when you select the **P** option (short for Polyline). The fillet radius is constructed at all vertices of the polyline. If arcs exist at any intersections, they are changed to the new fillet radius. Note that the fillet is applied to one continuous polyline.

Figure 8.16 illustrates a fillet on a polyline.

FILLETING ARCS, CIRCLES, AND LINES TOGETHER

Lines, arcs, and circles can be filleted together. When you fillet such objects, there are often several possible fillet combinations. You can specify the type of fillet you want by the placement of points when you select the objects. AutoCAD attempts to fillet the endpoint that the selection point is closest to.

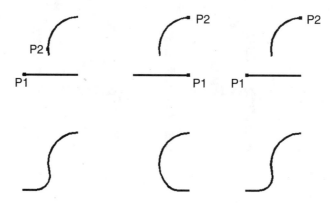

Figure 8.17 *Apply Fillet to Lines and Arcs (at right)*

Figure 8.17 shows several combinations between a line and arc. Observe the placement of the points used to pick the objects in the middle row, and the resulting fillet shown in the bottom row.

If you select two objects for filleting and get undesirable results, use the **U** command to undo the fillet and try respecifying the points closer to the endpoints you want to fillet.

The former prohibition against filleting a polyline with a line has been lifted with AutoCAD 2000. When the line and a polyline are filleted, all three objects (the line, the fillet arc, and the polyline) are converted to a single polyline.

FILLETING CIRCLES

As with lines and arcs, the result of filleting two circles depends on the location of the two points you use to select the circles. Figure 8.18 shows three possible combinations, each using different selection points.

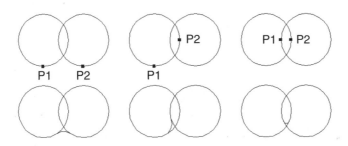

Figure 8.18 *Apply Fillet to Circles*

 Note: If you have several filleted corners to draw, construct your intersections at right angles, then fillet each one later. This allows you to continue your Line command, without interruptions, and results in fewer commands to be executed.

You can "clean up" line intersections by setting your fillet radius to zero and filleting the intersections.

Changing an arc radius by fillet is cleaner and easier than erasing the old arc and cutting in a new one. Let AutoCAD do the work for you!

EXERCISES

1. Open the CD-ROM drawing named "EDIT1." Use the **Erase** command to delete some of the objects. Next, issue the **Oops** command. Did the objects return?

2. Open the drawing named "EDIT2" from the CD-ROM. This exercise is a jigsaw puzzle. When you start the drawing, it appears as in Figure 8.19. Use the **Move** command to move the pieces into position, leaving a small space between them. The most effective method is to move the pieces roughly into position, then zoom in and fine position the pieces using object snap.

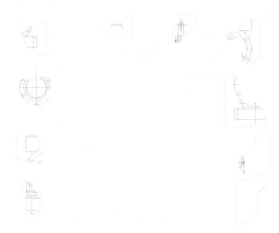

Figure 8.19 *EDIT2.DWG*

3. From the CD-ROM, open the drawing named "EDIT3." Figure 8.20 shows the drawing. Use the **Copy** command to copy the windows from the left side to the right side. Then copy all the windows on the lower level (including those you just copied) to the upper level. When you copy, use the object selection options you think will work best.

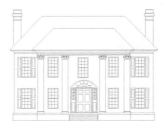

Figure 8.20 *EDIT3.DWG*

4. From the CD-ROM, open the drawing named "EDIT4." The drawing is a site plan, as shown in Figure 8.21. Use the **Copy** command, with the **Multiple** option, to copy the landscaping items (trees, shrubbery) and create your own landscape scheme.

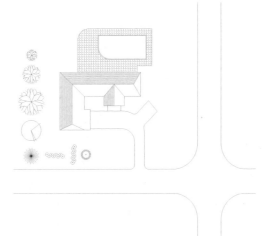

Figure 8.21 *EDIT4.DWG*

5. From the CD-ROM, open the drawing named "EDIT5." Figure 8.22 shows the drawing. Use the **Break** command, as shown completed in Figure 8.23, to break each of the objects in the drawing, achieving the results shown.

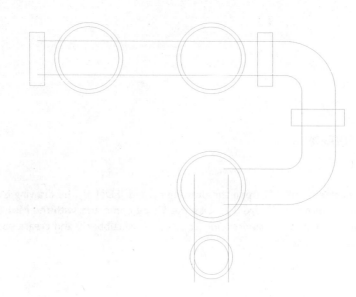

Figure 8.22 *EDIT5.DWG*

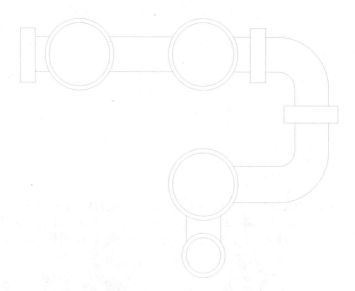

Figure 8.23 *Completed EDIT5 Exercise*

6. Let's look at an example of connecting two lines with a "zero" radius intersection. Refer to the following command sequences and Figure 8.24. First, draw lines similar to those in the illustration. Now let's set the fillet radius to zero.

Command: **fillet**

Current settings: Mode = TRIM, Radius = 0.5000

Select first object or [Polyline/Radius/Trim]: **r**

Specify fillet radius <0.5000>: **0**

Now, issue the **Fillet** command again and select the two lines.

Command: **fillet**

Current settings: Mode = TRIM, Radius = 0.0000

Select first object or [Polyline/Radius/Trim]: *(Pick one line.)*

Select second object: *(Pick other line.)*

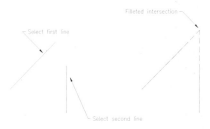

Figure 8.24 *Filleted Intersection*

Notice how the two lines are now connected in a perfect intersection. Let's continue and connect the same two lines with a radiused fillet. We will connect the lines with an arc with a radius of 0.15. Let's first set the fillet radius to 0.15.

Command: **fillet**

Current settings: Mode = TRIM, Radius = 0.0000

Select first object or [Polyline/Radius/Trim]: **r**

Specify fillet radius <0.0000>: **0.15**

The default radius is now set to 0.15 and will remain until it is changed to another value.

Now, select the **Fillet** command again. Refer to the following command sequence and Figure 8.25.

Command: **fillet**

Current settings: Mode = TRIM, Radius = 0.1500

Select first object or [Polyline/Radius/Trim]: *(Pick one line.)*

Select second object: *(Pick other line.)*

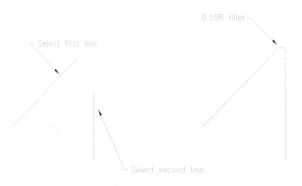

Figure 8.25 *Fillet with Radius 0.15*

7. Open the drawing named "EDIT6" from the CD-ROM. Perform fillets on the objects to achieve the results shown in Figure 8.26.

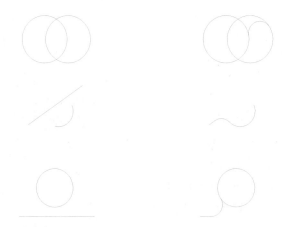

Figure 8.26 *EDIT6.DWG*

Additional drawing exercises:

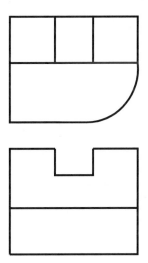

Figure 8.27

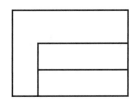

Figure 8.28

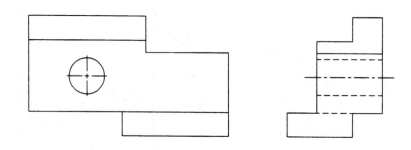

Figure 8.29

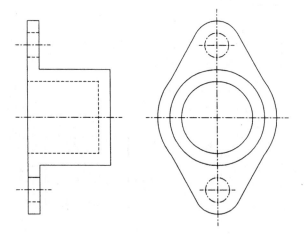

Figure 8.30

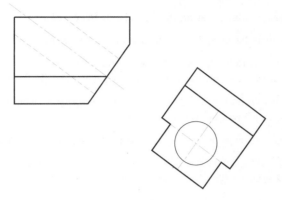

Figure 8.31

CHAPTER REVIEW

1. When an edit command is invoked, must a selection option, such as Single or Window, be entered before an object is selected?

2. What two changes can a fillet make to an intersection?

3. When choosing a set of objects to edit, other than using the U command, how can you alter an incorrect selection without starting over?

4. If you wish to edit an item that was not drawn last, but was the last item selected, would the "Last" selection option allow you to select the item?

5. How do you increase and decrease the size of the pickbox?

6. When moving an object, must the base point of displacement be on the selected object?

7. What objects are affected by the **Break** command?

8. When using the **Multiple** option of the **Copy** command, is each copy relative to the point of the last copy made, or the first base point entered?

9. In breaking an object, what happens if you do not enter an **F** for selection of the first point?

10. Can elements of different types be filleted (such as a line and an arc), or must they be alike?

11. When entering **BOX** in response to the "Select object:" prompt, how are you then allowed to choose objects?

12. What command restores objects just erased?

13. What makes it evident that an item has been selected?

14. How can objects be completely removed from the drawing?

15. When a group of items are selected with the **Window** option, does an item become a part of the selection set as long as it is partially inside the window?

16. What happens when you choose the **U** option during a selection process? Can you use this option more than once in a row?

17. In the object selection process, how is the box option different from the automatic option?

18. How does a group differ from a selection set?

TUTORIAL: SELECTING OBJECTS

Let's use the **Erase** command to delete some objects from a drawing. We will use a drawing named "EDIT1" from the CD ROM. Open the drawing. You should see the following on your screen.

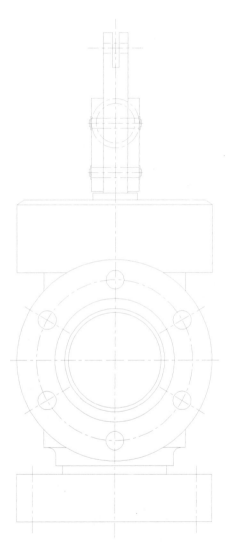

Figure 8.32 *EDIT1.DWG Exercise*

Using a Pickbox

Let's use the **Erase** command to delete some of the objects.

From the **Modify** menu, select **Erase.** You should now see a pickbox on the screen.

Place the pickbox over the bottom line of the part as shown in Figure 8-34 and click. The line should now be highlighted.

Now select the remainder of the lines as shown in the figure in the same manner.

Finally, press ENTER.

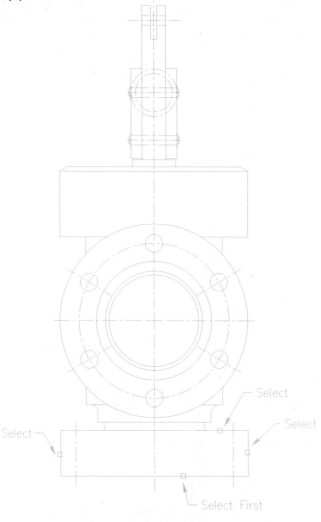

Figure 8.33 *Selecting Entities to Erase Using a Window*

Let's continue.

First, use the **U** command to undo the erase. Let's now use some of the object selection options.

Select **Erase** again and enter a **W** in response to the prompt. Refer to Figure 8.34 for the points referenced in the following command sequence.

Command: **Erase**

Select objects: **W**

Specify first corner: *(Select point 1.)*

Specify other corner: *(Select point 2.)*

58 found

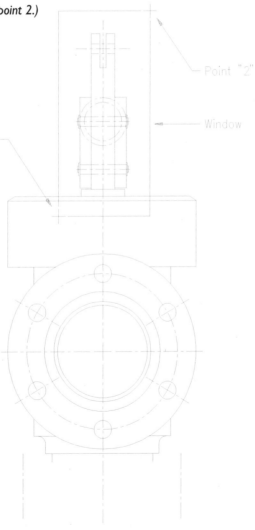

Figure 8.34 *Erasing with a Window*

All the items contained in the window were selected. The selected items are high-lighted. Notice that objects that extended into the window area, but were not wholly contained within the window, were not selected.

Press ENTER and the selected objects are deleted.

Using a Crossing Window

Use the **U** command to undo the erase.

Select **Erase** again, entering **C** (for Crossing) as the option. Refer to the following command sequence and Figure 8.35.

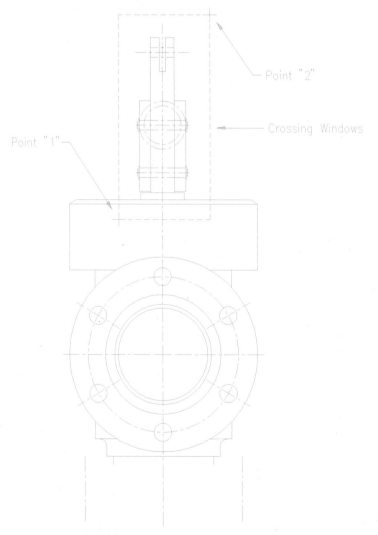

Figure 8.35 *Using a Crossing Window*

Command: **Erase**

Select objects: **C**

Select first corner: *(Select point 1.)*

Select other corner: *(Select point 2.)*

61 found

Select objects:

Notice that all the objects that were touched by the window were selected (as noted by the highlighting).

Removing Objects from the Selection Set

Let's remove some of the objects. After you placed the crossing window, AutoCAD again asked you to "Select objects:" (see the previous command sequence).

Enter **R** for Remove. The command sequence will continue:

Remove objects: *(Select one of the horizontal lines.)*

1 found, 1 removed

Remove objects: *(Select the other horizontal line.)*

1 found, 1 removed

Remove objects: *(Press ENTER.)*

The objects you removed from the object selection set were not erased.

If you exit the drawing, discard the changes so any edits are not recorded. If you want to practice with some of the edit commands, you can restore the drawing by using the **U** command to undo the edits.

TUTORIAL: USING EDITING COMMANDS

Let's try some edit commands with a drawing on the CD-ROM project named "EDIT7". Figure 8.36 shows the drawing.

Figure 8.36 *EDIT7.DWG*

Removing the Lower Circles

Suppose that a design change has been initiated and you have been instructed to remove the lower circle from the drawing.

> In the **Modify** menu, select **Erase**. The prompt line should now say "Select objects:".
>
> Move the pickbox until it intersects on the lower circle.

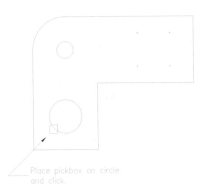

Figure 8.37 *Erasing a Circle*

> Click on the circle. Notice that the circle is now highlighted. The prompt shows that AutoCAD is ready for you to select more objects, but let's stop with this one now.
>
> Press ENTER to tell AutoCAD to execute the **Erase** command on the highlighted object. Watch the object disappear!

Removing the Points

> Now, remove the four points on the object by using a window selection:
>
> Command: *(Press spacebar to repeat the Erase command.)*
>
> Erase Select objects: **W**
>
> Specify first corner: *(Select point "1".)*
>
> Specify other corner: *(Select point "2".)*
>
> 4 found
>
> Select objects: *(Press ENTER.)*

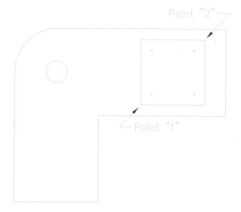

Figure 8.38 *Erasing the Points*

Notice how the window stretched out and followed the crosshairs! You can "stretch" the window to any size you wish. It is only "set" when you enter the second corner location.

Restoring the Points

Let's put back the four points you just erased. You could redraw them. Or you could say "Oops! I made a mistake!"

Command: **Oops**

Execute it and watch the points return.

Moving a Circle

Now that you have erased the lower circle on your drawing, move the remaining circle to the middle of the object. Refer to Figure 8.39

Command: **Move**

Select objects: *(Select the circle.)*

Select objects: *(Press* ENTER*.)*

Specify base point or displacement: *(Select point 1.)*

Specify second point of displacement or <use first point as displacement>: *(Select point 2.)*

Copying a Circle

Let's suppose that you have now been directed to add another circle to your drawing, which is identical to the remaining circle. (Remember erasing the larger circle earlier?)

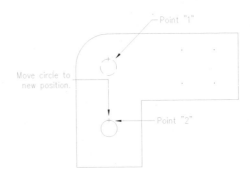

Figure 8.39 *Moving a Circle*

Command: **Copy**

Select objects: **P**

Select objects: *(Press* ENTER.*)*

Specify base point or displacement, or [Multiple]: *(Select point 1.)*

Specify second point of displacement or <use first point as displacement>: *(Select point 2.)*

You now have an exact copy of the first circle.

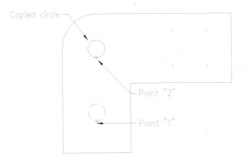

Figure 8.40 *Copying a Circle*

Adding a Notch

Let's add a notch to the object by breaking a line to receive the notched-in area. Refer to Figure 8.41.

Command: **Break**

Select objects: *(Select the bottom line.)*

Enter second point (or F for first point): **F**

First point: *(Select point 1.)*

Second point: *(Select point 2.)*

Now, using the **Line** command, draw in the notch, as shown in Figure 8.42.

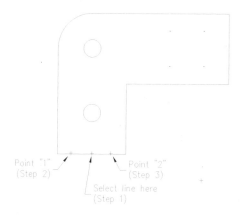

Figure 8.41 *Breaking a Line*

Radiusing the Corners

Let's add a radius to each of the two right corners of the object. We will set the radius to 0.5.

Command: **Fillet**

Current settings: Mode = TRIM, Radius = 0.0000

Select first object or [Polyline/Radius/Trim]: **r**

Specify fillet radius <0.0000>: **.5**

Now repeat the **Fillet** command again.

Command: **Fillet** *(or press Enter to repeat the command.)*

Current settings: Mode = TRIM, Radius = 0.0000

Select first object or [Polyline/Radius/Trim]: *(Select point 1.)*

Select second object: *(Select point 2.)*

The corner now has an arc with a radius of 0.5.

Now repeat the fillet on the lower corner.

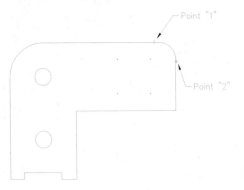

Figure 8.42 *Radiusing a Corner*

Use the **Quit** command to leave the drawing unchanged.

CHAPTER 9

Plotting Your Work

The intended end product of most CAD drawing products is a hard copy of the work. In this chapter, you learn how to plot your drawings. After completing this chapter, you be able to

- Set up AutoCAD to work with a plotter.
- Use AutoCAD's **Plot** command
- Understand the relationship of plot scales and limits
- Arrange a drawing for a plot of a specified scale
- Work with AutoCAD's batch plotting utility

INTRODUCTION

You have learned so far how to create and edit drawings using AutoCAD commands, but wouldn't you really like to see them on paper? After all, the end product is usually a finished drawing on paper. In this chapter, you learn to make a hard copy of the AutoCAD drawing by plotting. You learn these commands:

PlotterManager

Sets up AutoCAD for a new plotter.

Plot

Displays the Plot Control dialog box.

Preview

Displays a preview of the drawing before plotting.

Batchplt

Allows you to set up more than one drawing for plotting.

SETTING UP THE PLOTTER

Before you can plot the drawing, you may need to set up AutoCAD for the plotter attached to your computer. You do not need to go though the setup process if you plan to plot your drawings on standard office printers, such as laser printers or inkjet printer.

LOCAL AND NETWORK PRINTERS

In many cases, you can use AutoCAD with any printer already attached to your computer—either directly or through a network. When the printer is connected directly to your computer through a parallel port, a serial port, or a USB port, it is known as a *local printer*.

A computer can have up to four parallel ports (although one is common), up to four serial ports (although two are common), and up to 128 USB ports, although one or four is common. Parallel ports are most commonly used for printers, although, as this book is being written, USB ports are just starting to become accepted since they are faster and configure automatically. (USB is short for "universal serial bus.") Serial ports are rarely used for printers anymore, since they are slower and difficult to configure.

When the printer is connected to your office network, then it is known as a network printer. A *network printer* is connected either: (1) directly to the network, in which case it has a network card installed; or (2) indirectly to the network. In the latter case, the printer is connected to a computer (through parallel, serial, or USB port), and the computer is connected to the network.

In both cases, if your computer is connected to the network, you can print to any network printer—provided your computer has been given permission to access the network printer. This works in reverse, too: you can give other networked computer permission to use your computer's printer.

The primary difference between a local and a network printer is for the need of Windows. During printer setup, you tell Windows whether the printer is a Local or Network printer, so that Windows knows to search your computer's local ports or hunt along the network.

The other difference between a local and a network printer is that your local printer is typically more available. A network printer might be inaccessible for several reasons, such as the network being down or too many other computers are sending files to the printer. On the other hand, the advantage to networking printers is that your office can share printers.

SYSTEM PRINTERS

AutoCAD 2000 works "out of the box" with any printer connected to your computer and your network. That's because AutoCAD simply checks for all *system printers* registered with Windows 95/98/NT. (The term "system printer" refers to all local and network printers recognized by your computer.) To see the list of system printers, choose the **Start** button on the task bar. From the **Start** menu, select **Settings | Printers**. Windows opens the **Printers** window, which lists your computer's system printers.

Figure 9.1 *The Printers Window*

Figure 9.1 shows the **Printers** window for the author's computer. From left to right, the icons indicate:

Add A Printer: Double-click this icon to add a new printer to your computer. You only need to do this when Windows does not automatically recognize a new printer or plotter added to your computer (this automatic recognition is called "Plug and Play").

Canon: Notice the "wires" attached to the bottom of the icon. These indicate a **network printer**. This printer is attached to another computer on the author's computer network. Windows, unfortunately, does not have a special icon for printers connected directly to the network.

HP LaserJet: This icon represents a **local printer**. To allow others to access your computer's printer(s) over the network, right-click one of the local printer icons and select **Sharing**. In the **Properties | Sharing** dialog box, select the **Shared As** radio button and then choose **OK**.

Lexmark Optra: Notice the check mark on this icon. This indicates the **default printer**. AutoCAD and other Windows programs will use this printer, unless you select another printer. To select a different default printer, right-click one of the other printer icons and select **Default Printer** from the cursor menu.

Microsoft Fax: This icon represents a "non-traditional printer," as described in the next section.

Phantom AutoCAD: These two icons are used by AutoCAD to interface with the Windows printing system. The **ADI** icon is for AutoCAD Release 13 and 14; the **HDI** icon is for AutoCAD 2000.

You change the properties of your system printers by right-clicking their icon and selecting **Properties**. The content of the **Properties** dialog box varies, depending on the printer's capabilities. Commonly, however, you can select the default resolution, paper size and source, color management, and other settings.

NON-TRADITIONAL PRINTERS

Windows also works with non-traditional printers. The most common example is the fax (I am assuming your computer has fax capability, which is typically included with today's modems). Windows allows you to fax from any software program, including AutoCAD. For you, the process is as simple as selecting the fax as the printer.

The drawback to faxing from AutoCAD is that you are typically limited to squeezing a large drawing onto an A- or A4-size of paper (roughly 8"x11" in size).

AUTOCAD-SPECIFIC PRINTERS

As mentioned earlier, AutoCAD can plot to any printer connected to your computer. This includes *large-format* plotters, which are often used by engineering and architectural offices for creating D- and E-size plots (roughly three or four feet across).

Windows 95/98/NT works with these plotters as system printers—just like any other printer. The drawback is that the device drivers provided by Microsoft are not accurate and flexible enough for CAD output. For this reason, AutoCAD includes its own set of improved drivers, known as HDI (short for "HEIDI Device Interface;" Heidi is an acronym meaning "HOOPS Extended Immediate Mode Drawing Interface;" and HOOPS is yet another acronym meaning "Hierarchical Object-Oriented Picture System" – whew!)

Before you can plot a drawing with the HDI drivers, you need to run the **Plotter Manager**.

1. From AutoCAD's menu bar, select **File | Plotter Manager**. Notice that AutoCAD displays the **Plotters** window, which displays several icons, including **Add-A-Plotter Wizard**.

Figure 9.2 *The Plotters Window*

2. Double-click the **Add-A-Plotter Wizard** icon. Notice that AutoCAD displays the **Introduction Page** of the "wizard."

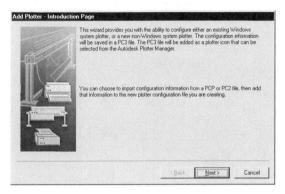

Figure 9.3 *The Introduction Page*

3. Choose **Next**. Notice the **Begin** page. You can set up AutoCAD with a local printer (called "My Computer" by the wizard), a Network Plotter Server, or a System Printer (an existing printer).

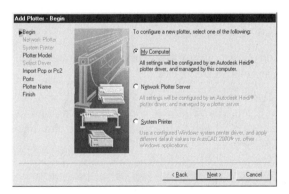

Figure 9.4 *The Begin Page*

4. Choose **My Computer** and **Next**. AutoCAD displays a list of printer and plotter drivers provided by AutoCAD 2000.

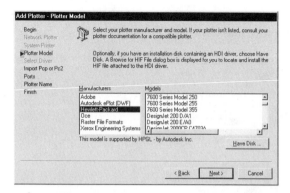

Figure 9.5 *The Plotter Model Page*

5. In the **Manufacturer** list, select the brand name of the plotter. As an alternative, you can plot the drawing to a file in a specific raster **format**: select **Raster File Formats**.

 Note: If you do not see your plotter manufacturer's name, check the plotter documentation. Often, it lists the brand names and model numbers of compatible plotters. If you cannot find this information, try selecting **Adobe** for PostScript printers, or **Hewlett-Packard** for large-format inkjet plotters.

If your new plotter comes with AutoCAD-specific drivers on a diskette or CD-ROM, choose the **Have Disk** button.

6. In the **Models** list, select the specific model number. If you selected Raster File Format, select a file format.

7. Choose **Next** to see the **Import PCP or PC2** page. This page is only important if you created plotter configuration files with AutoCAD Release 13 (PCP) and 14 (PC2). Here, you can import those files for use with AutoCAD 2000 (as PC3 files).

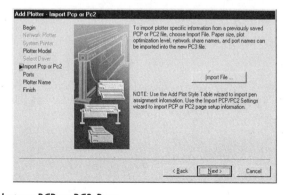

Figure 9.6 *The Import PCP or PC2 Page*

8. Choose **Next**. The **Ports** page may be the most difficult step for some users. Here you select the port that the plotter is connected to—or no port at all. Your choices are:

Plot to a Port: AutoCAD sends the drawing to the plotter through a local or network port, such as a parallel port (designated as LPT), a serial port (designated as COM). Select this option when you want the drawing plotted by the printer or plotter.

Plot to a File: AutoCAD sends the drawing to a file of disk. Select this option when you want the drawing saved to disk in another file format, such as a raster format.

AutoSpool: AutoCAD sends the drawing to a file in a specific folder (defined by AutoCAD's **Options** dialog box), where another program processes the file. Select this option only if you know what you are doing.

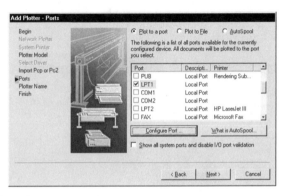

Figure 9.7 *The Ports Page*

9. Some ports have further configuration options. Choose the **Configure Port** button, if necessary. Parallel ports allow you to specify the transmission retry time, serial ports allow you to change their communications settings, and network ports have nothing to configure.

10. Choose **Next** to give the plotter configuration a name with the **Plotter Name** page.

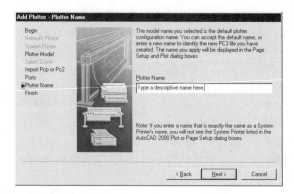

Figure 9.8 *The Plotter Name Page*

11. Type a descriptive name, then choose **Next**.

12. The **Finish** page has three buttons:

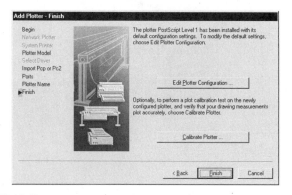

Figure 9.9 *The Finish Page*

Edit Plotter Configuration: Choose this button to display AutoCAD's Plotter Configuration Editor dialog box. This dialog box allows you to specify options, such as media source, type of paper, type of graphics, and initialization strings.

Calibrate Plotter: Choose this button to enter another "wizard," which allows you to calibrate your plotter. This allows you to confirm that a, say, ten-inch line is indeed plotted 10.0000000 inches long.

Finish: Choose this button to complete the plotter configuration.

13. Choose **Finish**.

This completes the process of setting up a new plotter configuration for AutoCAD 2000. When you next use the **Plot** command, you will see this configuration in the list of available plotters.

PLOTTING THE DRAWING

To plot a drawing, enter the **Plot** command at the command line, or select **File | Plot**, or press CTRL+P, or choose the printer icon. The following dialog box is displayed:

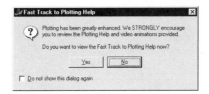

Figure 9.10 *The Fast Track Dialog Box*

AutoCAD 2000 completely changes the **Plot** dialog box. For this reason, an initial dialog box asks if you would first like to see a tutorial on how to use the new plotting features. Choose **Yes** to view the tutorial; choose **No** to continue with the **Plot** command.

 Note: AutoCAD does not plot frozen layers and the Defpoints layer. If part of your drawing does not plot, it could be you drew on the Defpoints layer.

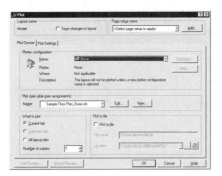

Figure 9.11 *The Plot Dialog Box's Plot Device Tab*

The **Plot** dialog box consists of two tabs: **Plot Device** and **Plot Settings**. The **Plot Device** tab is entirely new to AutoCAD 2000. Here you select the plotter and the *plot style table* (more on plot styles later in this chapter). Think of this tab as "where to plot."

The **Plot Settings** tab may be familiar if you have used other versions of AutoCAD. Here you select the plot scale, media size, orientation, and other parameters. Think of this tab as "what to plot."

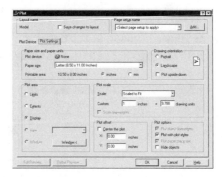

Figure 9.12 *The Plot Dialog Box's Plot Settings Tab*

Before plotting a drawing, you must first determine several aspects of the plot. Let's take these in recommended order.

Step 1: Select the Plotting Equipment

AutoCAD stores settings for many plotters and printers. For example, your computer may have two different plotters available. The **Plotter Configuration** list box of the **Plot Device** tab contains the available plotters. If the plotter you want to use is not listed here, read the "Setting Up the Plotter" section earlier in this chapter.

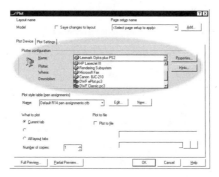

Figure 9.13 *Selecting the Plotter*

To change properties specific to the printer, such as its resolution or color management, choose the **Properties** button.

Step 2: Specify the Plot Style

Plotters are capable of plotting with several "pens." The term refers to an earlier age when plotters used actual pens to plot with. Today, the term carries over to refer to varying widths, colors, and shades of gray that imitate pens. Previous releases of AutoCAD used object colors to assign pens. For example, you would specify that all objects in the drawing that are red should be plotted with a certain pen, while those that are blue are plotted with a different pen. The pens need not be red or blue; they could be 0.1" or 0.05" wide.

With AutoCAD 2000, pens are assigned through *plot styles*. If you have drawings set up to plot by matching object colors with pens, select **Default R14 Pen Assignments** from the **Name** list. Otherwise, select one of the pre-assigned plot styles from the list.

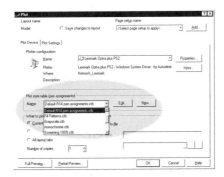

Figure 9.14 *Selecting the Plot Style*

Step 3: Select the Tab to Plot

You can specify the layout of the drawing that you wish to plot. The choices are contained in the **What to Plot** area of the dialog box, together with the number of copies of make of the plot.

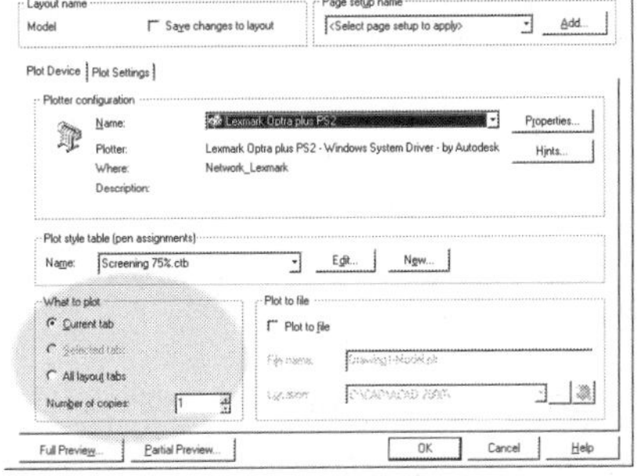

Figure 9.15 *Selecting What to Plot*

Step 4: Select the Media Size

Select the **Plot Settings** tab. The **Paper Size and Paper Units** area lets you select the size of media. AutoCAD knows the media sizes that the selected printer can work with. Make your selection from the **Paper Size** list box.

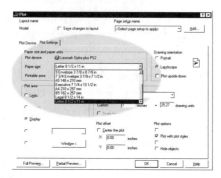

Figure 9.16 *Selecting the Media Size*

Step 5: Select the Orientation

CAD drawings are often plotted in "landscape" mode, with the long edge of the page laid horizontal. To do this, choose the **Landscape** button in the **Drawing orientation** area.

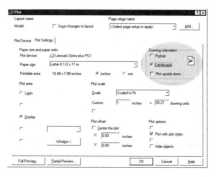

Figure 9.17 *Selecting the Media Orientation*

Step 6: Select the Plot Area

In most cases, you want to plot the entire drawing, also called the "extents" of the drawing. In some cases, however, you may want to plot a specific area of the drawing. In those cases, AutoCAD offers the following options:

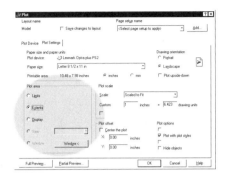

Figure 9.18 *Selecting the Plot Area*

Limits. Plotting the *limits* of a drawing uses the drawing's limits as the border definition of the plot. The limits are set by the **Limits** command. This option plots all areas of the drawing bounded by the limits, including "blank" areas.

Display. Plotting the *display* area of the drawing results in a plot of the current screen view. This is a "what you see is what you get" plot.

Extents. Plotting the *extents* plots every part of the drawing that contains objects, eliminating "blank" areas. Everything is plotted, even if outside the current viewport, except objects on frozen layers. The plot is the same as performing the **Zoom Extents** command and then plotting with the **Display** option.

View. Plotting the *view* plots a named view. For this option to be available, the drawing must contain at least one named view created with the **View** command. If you wish to plot a view, choose the **View** button. AutoCAD prompts you for the name of the view through a dialog box, which displays all the named views that exist in the drawing. Select a view name and choose **OK**.

Window. Plotting the *window* means that AutoCAD plots a rectangular area identified by a window. When you choose the **Window** button, AutoCAD prompts you to pick the two corners of a rectangle:

Specify first corner: *(Pick a point or type X,Y coordinates.)*

Specify opposite corner: *(Pick another point or type X,Y coordinates.)*

Step 7: Set the Plot Scale

For a fast plot, such as for draft purposes, select **Scaled to Fit** as the **Scale**. Together with a plot area of **Extents**, this ensures your entire drawing fits whatever size of media you selected.

If, instead, you want the drawing plotted to a specific scale, select one of the predefined scale factors provided in the **Scale** list box. These range from 1:1 to 1:100 and 1/128"=1'.

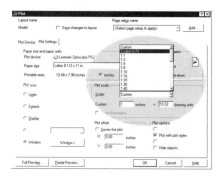

Figure 9.19 *Selecting the Plot Scale*

The third choice is **Custom**, which is a scale factor that you work out yourself. You determine the plot scale by setting a specified number of inches or millimeters on the paper the number of drawing units to be plotted on. For example, a scale of 1" = 8'-0" means that 1" of paper contains 8'-0" (96") of drawing. This is the same, of course, as the scale 1/8" = 1'-0". (These examples assume drawing units are set to architectural units.)

To set the custom scale, enter 1" next to **Custom** and 9' 6" next to **Drawing Units**.

If you are working in metric, choose the **mm** radio button in the **Paper** size and paper units section, just above. AutoCAD can provide metric conversion for you: type the number of inches in **Drawing** units and then choose the **mm** button.

Many CAD operators initially calculate the proper plot scale. As unusual as it might seem, proper scaling begins at the time you first prepare the drawing. The limits, paper size, and the scale at which you intend to plot must all match for a properly scaled plot.

If you begin a drawing using traditional drafting techniques, you must know the size of paper on which you are drawing (of course), and the scale at which you are drawing. The same is true for CAD drafting. But how do all these work together?

Let's first look at how the plotter translates the scale factor you give it. When you give the plot ratio, you are stipulating how many units will be contained in one inch on the paper that is being plotted on (or millimeters, if you choose that option). For example, if you intend to plot the drawing at 1/4" = 1'-0", and you have stipulated that one unit equals one foot, you can also say that 1/4" = 1 unit. (One unit = 1 foot, remember?) We might take this further and say that 1" = 4 units, or there are four units in one inch. Let's review this process:

If 1/4" = 1'-0" & one unit = 1'

then 1/4" = 1 unit

then 1" = 4 units

Let's take this one step further. If 1" = 4 units, then each inch on the paper will contain 4 units. If you intend to plot your drawing on paper that is 36" wide by 24" high, you can multiply the number of inches in each direction of the paper by the number of units in one inch.

36" x 4 units/inch = 144 (X limit)

24" x 4 units/inch = 96 (Y limit)

Therefore, assuming the following parameters:

Intended scale: 1/4" = 1'-0"

1 unit = 1 foot

Paper size = 36 x 24

the limits would be 0,0 and 144,96.

From this, we can derive a formula that determines the proper limits when we have the other necessary information.

(No. of units/inch) x (paper width) = X limit

(No. of units/inch) x (paper height) = Y limit

When you are using architectural units, one unit = one inch.

If 1/4" = 1'-0" & 1 unit = 1"

then 1/4" = 12 units

then 1" = 48 units

therefore,

36" x 48 units/inch = 1728 (X limits)

24" x 48 units/inch = 1152 (Y limits)

You could also say that 1728 inches is equal to 144 feet and 1152 inches is equal to 96 feet. Therefore, you could enter either 1728,1152 or 144',96' as the upper right limits.

As you can see, calculating limits requires some basic mathematics skills. In fact, any type of drafting requires these skills. With practice, the process becomes easier.

The appendix has a chart that makes things quite a bit easier. This chart lists plot scales, paper sizes, and limits. You can use the chart to set up a drawing for a specific plot scale and paper size.

Step 9: Preview the Plot

You should preview the plot to see the area of the paper on which the plot appears. AutoCAD's plot preview allows two types of preview: partial and full. Let's look at each.

 Note: Use plot previews to discover plotting results before committing time, paper, and pens to plotting a drawing that might have been set up incorrectly for the plot.

The **Partial Preview** shows the effective plotted area on the page size you selected. On color displays, the plotted area is shown as a blue hatched area, and the paper area as a blue dashed rectangle. A partial preview does not show the drawing, but rather the position of the plot on the paper.

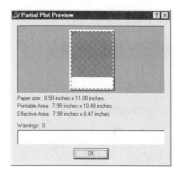

Figure 9.20 *Partial Plot Preview*

The plotted area also shows a red rotation icon. This triangular icon represents the lower left corner of the drawing as it is positioned on the drawing screen. When the drawing is rotated to angle 0 on the paper, the icon appears at the lower left. At 90 degrees rotation, it is at the upper left.

If the plotted area is entirely within the paper size, the plot area is slightly offset from the lower left corner so the plot area lines and the paper lines do not lie on top of each other. If the plotted area and the paper size are the same, the lines are shown dashed, with the dashes alternating red and blue. If the plot origin is offset so that the plotted area extends outside the paper area, AutoCAD displays a green line at the clipped side and displays the following message:

Effective area clipped to display image

AutoCAD also uses the partial preview box to display warning messages such as "Plotting area exceeds maximum."

The **Full Preview** shows the drawing as it appears as the final plot on paper. When you select full preview, the drawing is regenerated to display the plot. Since the regeneration takes time, AutoCAD displays a meter showing the percent of regeneration completed in the lower right corner of the Plot dialog box.

When the full preview is displayed, you can right-click to display a pop-up menu.

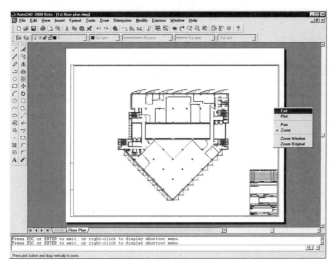

Figure 9.21 *Full Plot Preview*

The menu has these choices: **Exit, Plot, Pan, Zoom, Zoom Window,** and **Zoom Original.** The pan and zoom are used to examine parts of the drawing to be plotted, and they work in the same way as the **Zoom** and **Pan** commands. When you select **Pan** or **Zoom,** the cursor changes to indicate zoom (magnifying glass) or pan (hand). Hold down the left mouse button and as you move the mouse around, the view moves (pan) or changes size (zoom).

To end the preview, select the **Exit** button. You are returned to the **Plot** dialog box.

Note: You can get to Print Preview directly with the **Preview** command or by selecting **File / Plot Preview** from the menu or by choosing the Preview icon.

Step 10: Save the Settings

You have spent some time saving setting up the plot parameters for this drawing. If you plan to use the same parameters again, it makes sense to save them for future use. You do this by choosing the **Add** button in the **Page** setup name section (at the top of the **Plot** dialog box.) AutoCAD displays the **User Defined Page Setups** dialog box.

Figure 9.22 *The User Defined Page Setups Dialog Box*

Type a descriptive name, and choose **OK**. The only other purpose for this dialog box is to rename and delete named plot setups.

From the **Page setup name** list box, you select a saved plot setup.

Figure 9.23 *The Page Setup Name List Box*

Step 11: Plot the Drawing

When you have completed all the steps outlined, make sure the plotter is ready and choose the **OK** button at the bottom of the **Plot** dialog box.

SELECTING OTHER PLOT PARAMETERS

The steps listed above describe the most important parts of setting up an AutoCAD plot. After selecting the portion of the drawing to plot, you can choose additional parameters that affect the plot. Let's look at the choices. These are found in the **Plot Offset** and **Plot Options** areas.

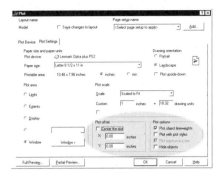

Figure 9.24 *Additional Parameters*

Plot Offset

For most plotters, the plot origin is at the lower left corner of the media just as it is for the AutoCAD drawing. Some plotters have their origin at the center of the media; AutoCAD normally adjusts for that.

If, for any reason, you need to shift the plot on the media, you can use the **Plot Offset** section of the **Plot** dialog box; you can enter negative values to shift the plot to the left and down. As an alternative, you can have AutoCAD center the plot on the media by choosing **Center the Plot**.

 NOTE: Caution: changing the plot's offset may result in a clipped plot (where part of the drawing is not plotted because it is positioned off the edge of the media).

Plot Object Lineweights / Plot with Plot Styles

The **Plot with Lineweights** and **Plot with Plot Styles** are toggles: you have one or the other turned on. The first option plots lineweights (if any) as defined by the drawing. The second option applies plot styles as defined in the plot style table.

Plot Paperspace Last

AutoCAD normally plots paper space objects before plotting model space objects. This option reverses the order.

Hide Objects

This option removes hidden lines in paper space (hidden-line removal of model space controlled by the **Viewports Hide** option found in the **Object Property Manager**). You can see the effect of hidden-line removal in the plot preview. Note that removing hidden lines from complex drawings can take a period of time and slows down the plotting process.

Note: Using the **Hide** command to remove hidden lines on the drawing screen does not carry over to the **Plot** command. The hidden lines must be removed at the time of the plot, even if the display shows the hidden lines removed.

Plot to File

AutoCAD can output the plot of your drawing to a file. The plot file can then be used to import the file into desktop publishing software or other software. To create a plot file:

In the **Plot Device** tab, **Plot to File** section, select the **Plot To File** check box. You can type a file name, or choose the ... (browse) button and select the file name from the file dialog box. (Choosing the **Browse the Web** button lets you save the file to a location on the Internet.)

The plot file name should be entered *without* a file extension. AutoCAD appends a file extension of .plt for plot files.

A plotter file can be used with a plot file utility program supplied by a third party to plot the file, or you can import the plot file into another program. For example, HPGL is a useful format for desktop publishing, while TIFF format is good for raster editing programs.

AutoSpool

If you set up the plotter with the AutoSpool option, this lets AutoCAD spool the plot. *Spooling* is an old term, better known today as buffering or by the Windows term, Print Manager. In fact, since you are using AutoCAD 2000 with Windows 95/98/NT, you are probably (unknowingly) using spooling anytime you plot an AutoCAD drawing to the Windows System Printer. The AutoSpool feature is primarily designed for plotters.

AutoCAD allows you, however, to use independent software for plot spooling. Before you can use the plot spool software, you must set up AutoCAD as follows:

Step 1

AutoCAD needs a subdirectory into which to put the spooled files. The default is either: (1) the subdirectory specified by the Temporary Drawing File Location preference; or (2) Windows system temporary directory. If you want to specify a different subdirectory, select **Tools | Options | Files | Print File, Spooler, and Prolog Section Names**. Specify the plot spool subdirectory and the spooler program name. Choose **OK**.

Step 2

Select the plotter set up for spooling by selecting **File | Plotter Manager**, and working through the wizard.

Step 3

Select the spooler **Plotter Configuration** from the **Plot** dialog box.

BATCH PLOTTING

Batch plotting is another way to plot AutoCAD drawings. *Batch* means to line up more than one drawing for plotting. Today, most plotters are capable of producing more than one plot unattended, because they have continuous rolls of paper or sheet feeders. Thus, if you have many drawings to plot, it can be more efficient to let Batchplt, AutoCAD, and your plotter do the work overnight, rather than tie up the plotter during the day.

For this reason, AutoCAD includes the Batchplt utility (short for batch plot), which runs outside AutoCAD. Batchplt puts together a list of drawings and then uses AutoCAD to go through the list and plot each drawing. Batchplt uses the **Plot** command's **Display** option, so be sure to save each drawing with the view you want plotted.

If you want to plot multiple views from a single drawing, you must first set up one surrogate drawing for each view. The surrogate drawings use AutoCAD's Xref command to externally reference views in the base drawing.

Batchplt lets you specify a different plotter and plotter setup with each drawing. Specify a PC3 (plotter control) file for the drawing.

To create a list of drawings for plotting, follow these steps:

Step 1

Load each drawing into AutoCAD and ensure you see the view you want plotted. Save each drawing.

Step 2 (optional)

If you plan to use more than one plotter configuration, create PC3 files.

Step 3

Double-click the Batchplt icon to start the program. If you cannot find the Batchplt.Exe program, it is located in subdirectory \Acad2000\Support\Batchplt. As an alternative, choose the **Start** button, then select **Programs | AutoCAD 2000 | Batch Plot Utility**.

Step 4

When the **AutoCAD Batch Plot Utility** dialog box appears, select **Add Drawing**. Select the drawings you want to plot. You can select more than one drawing in a folder by holding down the Shift key while you select file names.

Step 5

To change the plotter configuration, right-click a drawing name. The cursor menu displays five options:

> **Layouts:** Select a pre-defined layout from the list presented by the **Layouts** dialog box.

> **Page Setups:** Select a pre-defined page setup from the list presented by the **Page Setups** dialog box.

> **Plot Devices:** Select a plotter device from the list presented by the **Plot Devices** dialog box. These are the same plotters as listed by the **Plot** dialog box inside AutoCAD.

> **Plot Settings:** Select plot settings from the dialog box, which is an abbreviated version of the **Plot** dialog box.

> **Layers:** Toggle the display of layers.

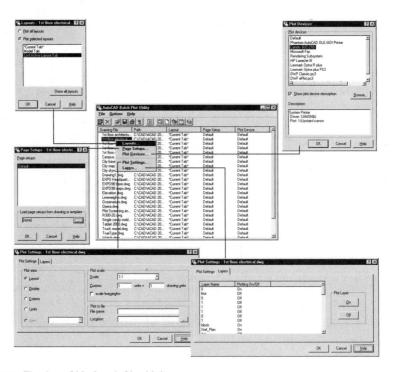

Figure 9.25 *The AutoCAD Batch Plot Utility*

Step 6

Check that the plotter is turned on, is set online, and is filled with paper. If you wish, select **File | Plot Test**, which performs a dry run on all of the drawings. This loads each drawing into AutoCAD but does not plot it.

Finally, select **File | Plot.** As each drawing is processed by AutoCAD and sent to the plotter, the Batch Plot Progress dialog box reports the progress of the batch plotting.

EXERCISES

1. If you have a printer or plotter hooked up to your computer or network, use the **Plot** dialog box to verify the plotter is configured for use.

   Use the chart in the appendix to set the limits for an A-size drawing. Draw a simple drawing and save your work. Plot the drawing, using "Scaled to Fit" to scale the drawing.

2. Plot the drawing again, using the scale you configured from the chart in the appendix. After the drawing is plotted, use a draftsman's scale to check the scale of the drawing at a known dimension length. Did your drawing plot to scale?

3. Plot the drawing again, rotating the plot through the landscape or portrait option.

4. Enter the drawing you constructed. Select the **Plot** command. Plot the drawing, using a **Window**. Place a window around any part of the drawing. Set the plot units to **Scaled to Fit**. Plot the drawing.

5. Select the **Plot** command again. Set the plot for **Limits**. Select **Scaled to Fit**. Set the plot file name to "MYPLOT." Plot to file.

CHAPTER REVIEW

1. What is the purpose of a plot?

2. What part of the drawing would be plotted if you plotted the extents of the drawing?

3. How would you plot only a portion of a drawing?

4. How would you rotate the plot on the paper?

5. Why would you write a plot file to a disk?

6. What are two ways to execute a plot in AutoCAD?

7. What is the plot origin?

8. How would you offset the plot origin?

9. What units does the plot origin use?

10. When does AutoCAD need to know the plotter pen width? Why?

11. What drawing limits would you set if you set up a drawing to be plotted in architectural units, with a scale of 1/4 in. = 1 ft.-0 in., and on architectural C-sized paper?

12. What is batch plotting best for?

CHAPTER 10

Applying Text, Fonts, and Styles

Drawings are a means of communication that uses both graphics and text. The proper use and placement of text in a drawing is an important aspect of constructing effective drawings. After completing this chapter, you will be able to

- Manipulate AutoCAD's text capabilities
- Create text styles that contain fonts and make modifications to those fonts
- Utilize the methods of placing text in a drawing

INTRODUCTION

In this chapter, you learn to place text and modify the look of the text using these commands:

Text

Interactively places one or more lines of text.

MText

Places paragraph text within a boundary.

Properties

Changes the properties of text objects.

DdEdit

Edits text.

Style

Changes the look of fonts.

QText

Redraws text as a rectangle for faster display speed.

USING TEXT IN DRAWINGS

Graphics are used in a drawing to convey information. The total description of a project, however, requires the written word, or text. Traditionally, placing text in a paper drawing by hand is a laborious manner. In contrast, text is easily placed in your drawing by AutoCAD.

TEXT STANDARDS

The use of text in drafting is usually governed by standards. Many North American industries use the American National Standards Institute (ANSI) style of letters and numbers. European companies use ISO (International Organization for Standardization) lettering. Other companies and countries set standards of their own.

The most important aspect of lettering is that it is *clear and concise*. The standards must be consistent. In general:

> Headings should be 3/16-inch high.
>
> Note text should be 1/8-inch high.
>
> Text is typically left-justified. This means that each line of text is aligned at its left edge.
>
> The clear font typically used is called "Simplex" by AutoCAD.

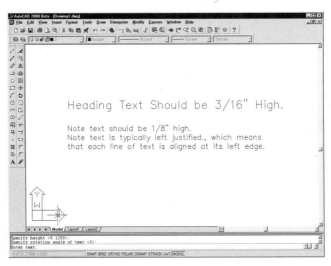

Figure 10.1 *Text Size Standards*

AUTOCAD TEXT COMPONENTS

AutoCAD provides several tools for text. You can apply many different styles; create text in any size; place the text slanted, compressed, and rotated; and even tell AutoCAD how long any text line length should be.

Text can be a variety of different designs; these designs are called *fonts*. A font is the "design" of the text letter. Gothic, Swiss, Script, and ISO are examples of fonts. See Figure 10.2. AutoCAD 2000 uses any TrueType font found on your computer, as well as its own font format called SHX. Many (but not all) fonts can be stretched, compressed, obliqued, mirrored, or drawn in a vertical stack.

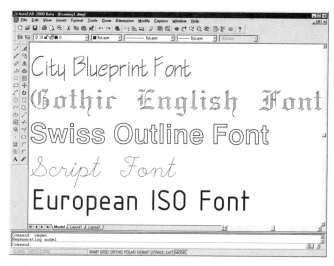

Figure 10.2 *Examples of Text Fonts*

The text font, the text height, and its modifications are saved as a *style*. The style includes the font and font modifications. The style is created with the **Style** command *prior* to text placement. Once created, many styles can be used many times in the same drawing. Figure 10.3 shows how a single font can be represented by many styles.

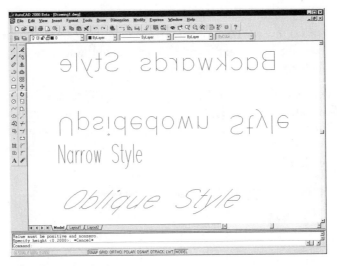

Figure 10.3 *Examples of Text Styles*

PLACING TEXT (TEXT)

The **Text** command is used for placing lines of text in the drawing. The **MText** command is used to place paragraphs of text. (The **DText** command, found in AutoCAD Release 14 and earlier, has been combined with the **Text** command in AutoCAD 2000; when you type the **Text** command, it now acts like the **DText** command.)

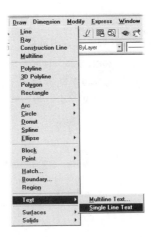

To start the **Text** command, select **Draw | Text | Single-line Text** from the menu bar. Or, type **Text** at the command prompt. You see the following prompt:

> Command: **text**
>
> Current text style: "Standard" Text height: 0.2000
>
> Specify start point of text or [Justify/Style]: *(Pick a point.)*
>
> Specify height <0.2000>: *(Press* ENTER*)*
>
> Specify rotation angle of text <0>: *(Press* ENTER.*)*
>
> Enter text: *(Type a line of text and press* ENTER.*)*
>
> Enter text: *(Type another line of text.)*
>
> Enter text: *(Press* ENTER.*)*

As you type the first *string* of text, notice that the text is displayed in the drawing as you type it. ("String" is another way of saying a line of text.) After you have placed one text string, you can move the cursor to a new location, click (press the left mouse button), and start a new text string. You can place several strings of text at different positions in the drawing without exiting and reentering the **Text** command repeatedly. To exit the **Text** command, press ENTER twice.

As you place text in the drawing, notice this feedback from AutoCAD:

- As you place text, an *I-beam* cursor is displayed at the text location. The height of the I-beam represents the height of a character in the current text style.

- The second time you use the Text command, the last text string entered is highlighted. When you press ENTER at the "Specify start point of text:" prompt (instead of picking the text position), the text box is placed on the next line below the last text, as though you had not previously exited **Text**. This is useful when you want to go back and continue placing multiple text lines.

- You can press BACKSPACE to erase—one character at a time—the text placed in that operation (even on previous lines).

- If you use a text alignment code other than Starting Point, the text does not appear to be properly aligned until you finish the command by pressing ENTER twice.

- If you cancel the **Text** command at any time during the operation, all text placed during that operation is erased, not just the current line.

- All menu selections, tablet picks, and command options are "locked out" during the **Text** command's "Enter text" prompt. Only keyboard entry is permitted while you place text.

TEXT CONTROL CODES

AutoCAD allows for adding some extra characters to your text. You use *control codes* to create text that is underlined, that includes the degree and plus/minus symbols, and other text notations. For the **Text** command, this is accomplished with a code consisting of a pair of percent characters (%%).

The following table shows the control codes and their function:

Control Code	Meaning
%%o	Toggle overscore mode on and off
%%u	Toggle underscore mode on and off
%%d	Degrees symbol: 27°
%%p	Plus/minus tolerance symbol: ±65
%%c	Circle diameter dimensioning symbol: 22Ø
%%%	Percent sign (think about it): %
%%nnn	Draw a special character designated by ASCII code nnn.

Table 10.1

Let's look at an example. The following text string:

> If the piece is fired at 400%%dF for %%utwenty%%u hours, it will achieve %%p95%%% strength.

would be drawn as:

> If the piece is fired at 400°F for twenty hours, it will achieve ±95% strength.

Notice that you toggle the underscore on and off by typing **%%u** to start the underscore, and again to end the underscoring. If you don't turn off underscoring, AutoCAD automatically turns it off at the end of the line.

You can also "overlap" the symbols by using, for example, the degrees symbol between the underscore symbols.

When you use text control codes, the codes are initially displayed instead of the effect. For example, when you enter:

> %%UTHIS IS UNDERLINED%%U

the text codes %%u are shown in the drawing as you are entering the text. When the Text command is completed, the text is redrawn, showing the effect.

<u>THIS TEXT IS UNDERLINED</u>

Special Characters

The special characters refer to the ASCII (American Standard Code for Information Interchange) character set. This character set uses a number code for different symbols. Table 10.2 shows the ASCII character set. To use one of the characters, enter two percent signs (%%) followed by the ASCII character code. For example, to place a tilde (~) in your text, enter %%126.

PLACING JUSTIFIED TEXT

The **Text** command's **Justify** option allows you to specify how the text is aligned. You can respond to the prompt "Specify start point of text or [Justify/Style]:" by picking a point, or typing *X,Y* coordinates. This becomes the starting point for the line of text, which is placed to the right. This is known as "left-justified" text, since the lines of text start at their left edge.

To see all of AutoCAD's text justification options, respond with a **J** (short for Justify) at the "Specify start point of text or [Justify/Style]:" prompt. AutoCAD prompts for the alignment options.

> Command: **text**
>
> Current text style: "Standard" Text height: 0.2000
>
> Specify start point of text or [Justify/Style]: **j**
>
> Enter an option
>   [Align/Fit/Center/Middle/Right/TL/TC/TR/ML/MC/MR/BL/BC/BR]:

Let's stop and look at how the text alignment options work and see that the abbreviations are easy to remember.

Figure 10.4 shows the alignment modes for text. Notice that you can align text by both the vertical and horizontal alignments.

Figure 10.4 *Text Alignment Options*

32		space	64	@		96	'	left apost.
33	!		65	A		97	a	
34	"	doublequote	66	B		98	b	
35	#		67	C		99	c	
36	$		68	D		100	d	
37	%		69	E		101	e	
38	&		70	F		102	f	
39	'	apostrophe	71	G		103	g	
40	(		72	H		104	h	
41	)		73	I		105	i	
42	*		74	J		106	j	
43	+		75	K		107	k	
44	,	comma	76	L		108	l	
45	-	hyphen	77	M		109	m	
46	.	period	78	N		110	n	
47	/		79	O		111	o	
48	0		80	P		112	p	
49	1		81	Q		113	q	
50	2		82	R		114	r	
51	3		83	S		115	s	
52	4		84	T		116	t	
53	5		85	U		117	u	
54	6		86	V		118	v	
55	7		87	W		119	w	
56	8		88	X		120	x	
57	9		89	Y		121	y	
58	:	colon	90	Z		122	z	
59	;	semicolon	91	[		123	{	
60	<		92	\	backslash	124	|	vert. bar
61	=		93	]		125	}	
62	>		94	^	caret	126	~	tilde
63	?		95	_	underscore			

Table 10.2 *ASCII Character Set*

AutoCAD uses a single-letter abbreviation to designate each alignment. Refer to Table 10.3.

Alignment	Abbreviation
Top	T
Middle	M
Bottom	B
Left	L
Center	C
Right	R

Table 10.3 *Alignment Abbreviations*

Notice the series of two-letter options. To designate an alignment, use two letters: one to describe the vertical alignment, and one to designate the horizontal alignment.

If you know the type of alignment you want, it is not necessary to type the **J** option; just enter the two letters that describe the alignment you desire: Vertical alignment followed by horizontal alignment, such as TL for Top Left.

Standard Alignment Options

Several alignment modes listed under the **Justify** option are shown as full words. These are the modes you are most likely to use. They are Align, Fit, Center, Middle, and Right. Let's take a closer look at each type of alignment.

Placing Aligned Text

Align mode requires you to select two points that the text will fit between. AutoCAD adjusts the text height so that the baseline of the text fits perfectly between the two points. Note that the two points can be placed at any angle in relation to each other (shown in Figure 10.5).

Command: **TEXT**

Current text style: "Standard" Text height: 0.2000

Specify start point of text or [Justify/Style]: **j**

Enter an option
[Align/Fit/Center/Middle/Right/TL/TC/TR/ML/MC/MR/BL/BC/BR]:
a

Specify first endpoint of text baseline: *(Pick a point.)*

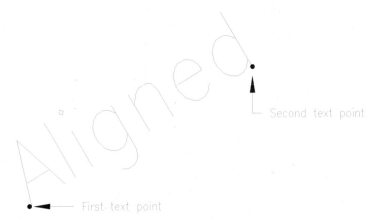

Figure 10.5 *Aligned Text*

> Specify second endpoint of text baseline: *(Pick another point.)*
>
> Enter text: *(Type text.)*
>
> Enter text: *(Press* ENTER.*)*

If you were to pick the two points in reverse order, AutoCAD would draw the text upside down.

Placing Centered Text

The **Center** option is used to center the baseline of the text on the specified point.

Figure 10.6 *Centered Text*

> Command: **TEXT**
>
> Current text style: "Standard" Text height: 0.2000
>
> Specify start point of text or [Justify/Style]: **j**

Enter an option
    [Align/Fit/Center/Middle/Right/TL/TC/TR/ML/MC/MR/BL/BC/BR]:
    c

Specify center point of text: *(Pick a point.)*

Specify height <0.2000>: *(Press* ENTER.*)*

Specify rotation angle of text <0>: *(Press* ENTER.*)*

Enter text: *(Type text.)*

Enter text: *(Press* ENTER.*)*

Fitting Text to a Specified Distance

The **Fit** option is similar to the **Aligned** option. You are prompted for two points to place the text between. With **Fit**, however, you are prompted for a text height. AutoCAD calculates only the text width and adjusts the width to fit perfectly between the two entered points.

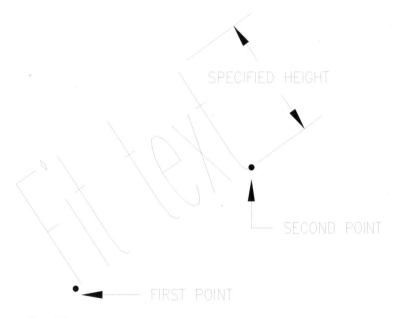

Figure 10.7 *Fitted Text*

Command: **TEXT**

Current text style: "Standard" Text height: 0.2000

Specify start point of text or [Justify/Style]: **j**

Enter an option
[Align/Fit/Center/Middle/Right/TL/TC/TR/ML/MC/MR/BL/BC/BR]:
f

Specify first endpoint of text baseline: *(Pick a point.)*

Specify second endpoint of text baseline: *(Pick a point.)*

Specify height <0.2000>: *(Press* ENTER.*)*

Enter text: *(Type text.)*

Enter text: *(Press* ENTER.*)*

When you pick the second point to the left of the first, AutoCAD draws the text upside down. Fitted text is often used inside constrained areas, such as for labeling a small closet.

Placing Middle Aligned Text

Middle is used as a shortcut when you center text both horizontally and vertically. It achieves the same result as the **MC** option.

Figure 10.8 *Middle Aligned Text*

Command: **TEXT**

Current text style: "Standard" Text height: 0.2000

Specify start point of text or [Justify/Style]: **j**

Enter an option
[Align/Fit/Center/Middle/Right/TL/TC/TR/ML/MC/MR/BL/BC/BR]:
m

Specify middle point of text: *(Pick a point.)*

Specify height <0.2000>: *(Press* ENTER.*)*

Specify rotation angle of text <0>: *(Press* ENTER.*)*

Enter text: *(Type text.)*

Enter text: *(Press* ENTER.*)*

Middle-aligned text is often used in title blocks.

Placing Right Aligned Text

Using right-justified text is similar to using Start point, except the text ends at the reference point instead of beginning at it.

Figure 10.9 *Right-Justified Text*

Command: **TEXT**

Current text style: "Standard" Text height: 0.2000

Specify start point of text or [Justify/Style]: **j**

Enter an option
[Align/Fit/Center/Middle/Right/TL/TC/TR/ML/MC/MR/BL/BC/BR]:
r

Specify right endpoint of text baseline: *(Pick a point.)*

Specify height <0.2000>: *(Press* ENTER.*)*

Specify rotation angle of text <0>:*(Press* ENTER.*)*

Enter text: *(Type text.)*

Enter text: *(Press* ENTER.*)*

Right-justified text is often used for text placed to the left of a leader line.

SETTING THE TEXT STYLE

The **Style** option of the **Text** command is used to change between defined text styles. *Note that you must first create a text style before you can use it.* (The only exception is the **Standard** style, which is predefined in every new drawing.) Assuming the new drawing contains a style called "Bigtxt" (styles are covered in detail later in this chapter), the following command sequence changes from the default style "Standard" to the user-designed style named "Bigtxt":

Command: **TEXT**

Current text style: "Standard" Text height: 0.2000

Specify start point of text or [Justify/Style]: **s**

Enter style name or [?] <Standard>: **bigtxt**

Current text style: "Bigtxt" Text height: 0.2000

Specify start point of text or [Justify/Style]:

 Note: The **Style** option of the **Text** command and the **Style** command are not the same. The Style option selects a style, while the Style command creates the style.

You can enter a ? at the "Enter style name" prompt to see a list of style names. The list looks like this:

Enter style name or [?] <Standard>: **?**

Enter text style(s) to list <*>: *(Press* ENTER *to see all styles.)*

Text styles:

Style name: "Standard"   Font files: txt

    Height: 0.0000 Width factor: 1.0000 Obliquing angle: 0

    Generation: Normal

Current text style: "Standard"

And the **Text** command resumes with its "Specify start point of text" prompt.

ROTATING THE TEXT

Text can be placed at any angle in your drawing. As you may have noticed, the previous command sequences included the prompt line:

Specify rotation angle of text <0>:

You can specify the angle at which the text is drawn by designating the angle. Figure 10.10 shows text placed at several angles.

Press ENTER to accept the default angle, 0 degrees in this case. Note that the zero-angle direction set with the **Units** command affects which direction is zero. Unless changed, the default zero-angle in AutoCAD is to the right (East). Thus, a text angle of zero would place text that runs from the left to right.

Once the text angle is set, the angle remains the default angle until it is changed again.

PLACING PARAGRAPH TEXT (MTEXT)

The **Text** command places one or more lines of text. The **MText** command (short for multiline tex.), in contrast, creates paragraphs of text. It fits the text into an invisible boundary defined by you (the boundary is not printed or plotted). The **MText** com-

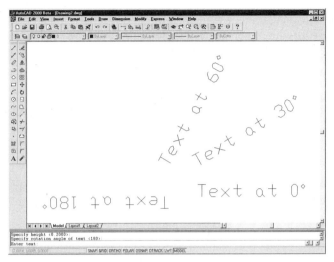

Figure 10.10 *Text Placed at Several Rotation Angles*

mand permits more text enhancements than the **Text** command. For example, text placed by the **MText** command can have varying heights and colors.

To write the text, you use the text editor built into AutoCAD. (You can specify another text editor, such as **NotePad**, with the system variable **MtextEd**.)

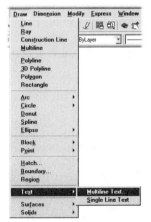

To place a paragraph of text, select **Draw | Text | Multiline Text** from the menu bar or type the **MText** command. The **MText** command has the following options:

> Command: **mtext**
>
> Current text style: "Standard" Text height: 0.2000

Specify first corner: *(Pick a point.)*

Specify opposite corner or [Height/Justify/Line
spacing/Rotation/Style/Width]: *(Pick another point.)*

(Text editor appears.)

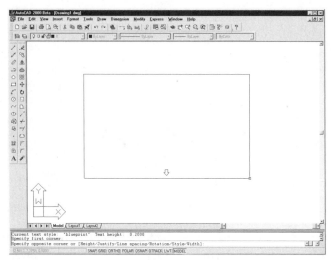

Figure 10.11 *Rectangle Contains Paragraph Text*

The default options prompt you to pick two points on the screen. The two points define the rectangle that contains the paragraph text. Then AutoCAD loads the text editor. Enter the text and then choose **OK**. AutoCAD places the text in the rectangle.

The other options at the command line allow you to control the text, as follows:

Height Option

The **Height** option allows you to change the height of the text used by the **MText** command. (This option is also available in the **Character** tab of the **Multiline Text Editor** dialog box.) The option prompts you:

Specify height <0.2000>: *(Type a number or indicate a height.)*

Each time you respond to an option (except the **Width** option), AutoCAD redisplays the **MText** prompt until you pick the opposite corner:

Specify opposite corner or [Height/Justify/Line
spacing/Rotation/Style/Width]: *(Pick.)*

Justify Option

The **Justify** option lets you specify the justification and positioning of text within the boundary rectangle. (This option is also available in the **Properties** tab of the **Multiline Text Editor** dialog box.)

Enter justification [TL / TC / TR / ML / MC / MR / BL / BC / BR]
<TL>: *(Type a two-letter option or press* ENTER.*)*

Unlike the **Text** command, the **MText** command's **Justify** option determines two kinds of justification: (1) text justification and (2) flow justification. *Text justification* is left, center, or right-justified relative to the left and right boundaries of the rectangle. *Text flow* positions the block of text top, middle, and bottom relative to the top and bottom boundaries of the rectangle. The abbreviations have the following meaning:

Justification	Meaning
TL	Top Left
TC	Top Center
TR	Top Right
ML	Middle Left
MC	Middle Center
MR	Middle Right
BL	Bottom Left
BC	Bottom Center
BR	Bottom Right

Linespacing Option

The **Linespacing** option allows you to change the spacing between lines of text. (This option is also available in the **Linespacing** tab of the **Multiline Text Editor** dialog box.) The option has two ways of specifying the line spacing: At least and Exactly. The **At least** option specifies the minimum distance between lines, while the **Exactly** option specifies the precise distance between lines of text.

Enter line spacing type [At least/Exactly] <At least>: **a**

Enter line spacing factor or distance <1x>: *(Enter a value, including the x suffix.)*

The value you enter is a multiplier of the standard line spacing distance, which is defined in the font.

Rotation Option

The **Rotation** option lets you specify the rotation angle of the text rectangle. (This option is also available in the **Properties** tab of the **Multiline Text Editor** dialog box.)

Specify rotation angle <default>: *(Type a number or indicate the angle.)*

This option rotates the entire block of text. For example, you might specify 90 degrees to have text placed sideways at the edge of a drawing.

 Note: When you use the mouse to show the rotation angle, AutoCAD calculates the angle as follows:

- The start of the angle is the X-axis.

The end of the angle is the line anchored by the "Specify first corner:" pick; the other corner is now in line with the indicated angle.

Style Option

The **Style** option lets you select a text style to use for the multiline text. (This option is also available in the **Properties** tab of the **Multiline Text Editor** dialog box.)

Enter style name (or '?') <Standard>: *(Type a name, or press ? to get a list of styles, or press* ENTER.)

When you type ? in response to this prompt, AutoCAD displays the names of text styles defined in the current drawing. You create a new text style with the Style command (as described later in this chapter) or by borrowing a style from another drawing using the **Xbind** command (covered in detail later in this book).

Width Option

The **Width** option lets you specify the width of the multiline text rectangle, instead of picking the second point. (This option is also available in the **Properties** tab of the **Multiline Text Editor** dialog box.)

Specify width: *(Type a value or pick a point to show the width.)*

A width of 0 (zero) has special meaning. AutoCAD draws the multiline text as one long line (no word wrap), as if there were no bounding rectangle at all.

USING THE MTEXT DIALOG BOX

The **Multiline Text Editor** dialog box lets you import text, set the style of the text insert symbols, and stack fractions. The dialog box has four tabs:

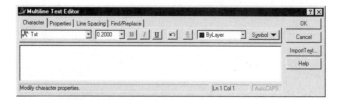

Figure 10.12 *The Multiline Text Editor Dialog Box*

- **Character** for controlling the look of the text

- **Properties** for mimicking the command-line prompts of style, justification, width, and rotation

- **Line Spacing** for controlling the distance between lines of text

- **Find/Replace** for changing text within the block

The Character Tab

Select the **Character** tab to display the character controls. (Some of the controls are unlabeled in the dialog box; pause your cursor over an item to display a tooltip describing the feature.) From left to right, the controls are

Font

Select a font from the drop list. The fonts listed are those installed on your computer. The list will probably include TrueType fonts included with Windows and fonts installed by AutoCAD 2000. TrueType fonts have a small TT logo in front of them; AutoCAD fonts are prefixed by the Autodesk logo. (Printer fonts, PostScript fonts, and other fonts are not listed since AutoCAD does not support them.)

Since you can use several different fonts within the MText block, you must highlight the text before selecting a font. (To highlight text, click and drag.)

To select all text, right-click the text and choose **Select All** from the pop-up menu. You can then change the case to all UPPERCASE or all lowercase.

Font Height

If you want some (or all) text at a different height, highlight the text, then type a text height in the list box.

B (Bold)

Highlight text and choose the **B** button to make the text boldface. Choose the button a second time to "unboldface" the text, making it regular text. When the **B** on the button is gray, text in the current font cannot be boldfaced.

I (Italic)

Highlight text and choose the **I** button to make the text italic.

U (Underline)

Highlight text and choose the **U** button to make the text underlined.

Undo

Choose the **Undo** button to reverse the last operation. This dialog box is limited to a single undo. Choose the button a second time to "undo" the undo (a redo command in disguise).

a/b (Stacked/Unstack)

The **a/b** button allows you to stack fractions. Instead of using the side-by-side format, like 11/32, this button places the 11 over the 32. You use it like this:

> Step 1: Type the fraction, such as 11/32.
>
> Step 2: Highlight the entire fraction.
>
> Step 3: Choose the **a/b** button.

AutoCAD replaces the numbers with a stacked fraction. Repeat steps 2 and 3 to "unstack" stacked fractions.

This function is not limited to numbers. AutoCAD will stack any combination of text, numbers, and symbols. Whatever is in front of the slash goes on top; whatever is behind the slash goes under.

Text Color

Highlight a portion of the text and select a color from the **Text Color** list box. You can choose from Bylayer, Byblock, the seven basic AutoCAD colors, or AutoCAD's other 248 colors.

Symbol

Choose the **Symbol** button to insert a symbol. You can choose from three basic symbols or a symbol from any Windows TrueType font (every Windows system includes the WingDings font, which contains 255 symbols). The three basic symbols are the most commonly used in drafting: degree, plus–minus, and diameter.

A fourth symbol is invisible and is called the **Non-breaking space**. When you place it between two words, it prevents AutoCAD from using that space for word wrapping. For example, if your text includes the phrase "six inches," you might want a non-breaking space between the "six" and "inches" to keep the two words together.

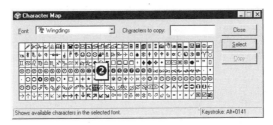

Figure 10.13 *Character Map Dialog Box*

Import Text

The **Import Text** button lets you bring text into the drawing from a file stored on disk. The file must be in plain ASCII format (also known as DOS text or Text document); you cannot import text saved in a word processing format, such as Write WRI, Word DOC, or WordPerfect files. Import is limited to files 16KB in size.

To import a text file, choose the **Import Text** button. AutoCAD displays the **Open** dialog box. Select the name of the file to import, then choose the **Open** button. AutoCAD places the text in the MText boundary.

Properties Tab

The **Properties** tab lets you specify some of the properties of the text and the bounding box.

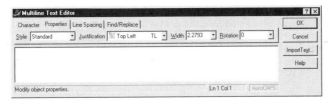

Figure 10.14 *The Properties Tab*

For the text itself, you can select the text style. The **Style** list box lists styles only defined in the current drawing. The **Justification**, **Width**, and **Rotation** options are the same as the command-line options, discussed earlier. For example, if you set the rotation angle to 17 degrees, the entire block of text is rotated by that amount. Changing the width causes AutoCAD to reflow the text.

The Line Spacing Tab

The **Linespacing** options are the same as the command-line options, discussed earlier.

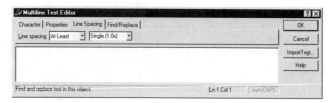

Figure 10.15 *The Line Spacing Tab*

The Find/Replace Tab

The **Find/Replace** tab lets you find text in the current text block (not in the entire drawing). To find text, click in the text box, type the text, and choose the binocular button.

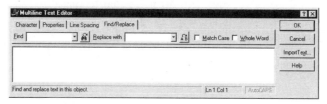

Figure 10.16 *The Find/Replace Tab*

To replace text, fill out the **Find** and **Replace** text boxes, then choose the **AB** button.

MTEXT CHARACTER CODES

In addition to the character codes discussed earlier for the **Text** command, you can use a set of codes in paragraph text created with the **MText** command. Using these codes allows you to format text with a text editor or word processor external to AutoCAD. (You may even be able to write a set of macros that convert your word processor's codes to those recognized by AutoCAD.)

MText codes consist of a backslash symbol, followed by a character. (Codes are *case-sensitive*, which means you must use an uppercase P and not a lowercase p in \P.) Table 10.4 shows the codes and their meanings.

Code	Meaning
\P	New line
\S	Stacked fractions
^	Removes the previous character
\L	Start underline
\l	Stop underline
\O	Start overscore
\o	Stop overscore
\W	Width of the font
\Q	Obliquing
\F	Change font
\H	Height of font
\An	Alignment value
	\A0; = bottom alignment
	\A1; = center alignment
	\A2; = top alignment
\Tn	Tracking (spacing between characters)
	\T3; = results in wide spacing
\~	Prevent breaks (forces words to stay together)
\Cn	Color change
	\C1; red
	\C2; yellow
	\C3; green
	\C4; cyan
	\C5; blue
	\C6; magenta
	\C7; white
{	Opening brace to separate a function so that the entire text isn't affected by a change.
\}	Closing brace, used with \{
\\	Literal backslash (used when you need to include a backslash in the text)

Table 10.4 *MText Codes*

Let's look at some examples. Normally, MText automatically wraps text to fit a rectangle. If you want to force a line break in the text, you use the \P code (short for paragraph). For example, AutoCAD reads the following line:

> This is the first line of text\P and the second line.

and places it in the drawing as:

> This is the first line of text
>
> and the second line.

The opposite is to force text to remain together, which you do as follows:

> This\~is\~text\~that\~is\~forced\~to\~remain\~together.

To underline text, use the \L and \l codes, as follows:

> This is \Lunderlined\l text.

Use the curly brackets, { and }, to specify a change in text for a specific part. For example, to write the following warning in white (color #7) and red (color #1) text:

> {\C7;Do not} {\C1;change} {\C7;this drawing.}

Notice how the semicolon is required when the MText code consists of more than a single character:

- \P   does not require the semicolon

- \C7; requires the semicolon

 Note: Use the **Leader** command to place a leader and annotation in the drawing. Use the \P code to create line breaks in the text. When done, the text is an MText object.

CHANGING TEXT (DDEDIT, PROPERTIES)

AutoCAD provides several ways to edit text and change its properties.

EDITING TEXT

The **DdEdit** command is meant for editing the wording of the text. From the menu bar, select **Modify | Text** and select a line of text, as follows:

> Command: **ddedit**
>
> Select an annotation object or [Undo]: *(Select.)*

If the text is regular text, **DdEdit** displays the **Edit Text** dialog box. Make your changes, and choose the **OK** button. AutoCAD returns to the **DdEdit** prompt:

Figure 10.17 *The Edit Text Dialog Box*

> Select an annotation object or [Undo]: *(Press* ESC.*)*

Press ESC to exit the DdEdit command.

If the text was created by the **MText** command, AutoCAD instead brings up the **Multiline Text Editor** dialog box, which lets you edit text and change the properties. The command displays the familiar **Multiline Text Editor** dialog box with the **Properties** tab showing.

CHANGING TEXT PROPERTIES

The **Properties** command is a powerful way to change the properties of text. Select the text, then type **Properties** or select **Modify | Properties** from the menu bar. AutoCAD displays the **Properties** window (see Figure 10.18).

Figure 10.18 *The Properties Window*

The **Properties** window lets you change every aspect of the text, including the color, layer, insertion point of the MText, and linetype. You change the elevation of the text by changing the Z value of the insertion point (called **Geometry**).

The **Contents** box displays the text of the text object. (Note that the \P character indicates line breaks for MText objects.) You can change the text in the **Contents** box. If the text is paragraph text, choose the **[...]** button, which causes AutoCAD to load the familiar **Multiline Text Editor** dialog box.

CREATING TEXT STYLES (STYLE)

A *style* is a collection of modifiers applied to the font of your choice. For example, you may want your text to be slanted, or perhaps condensed, so that it takes up less space than normal. The style is stored by a name of your choice.

A text style is composed of the following information:

- A style name, which can be up to 255 characters in length.

- A font file that is associated with the style. The font can be either TTF (TrueType) or SHX (AutoCAD) format.

- A height, which is the height of the text. If you want to specify the height each time you place text, enter a 0 for this value.

- A width factor, which is a multiplier that makes the text wider or narrower. A width of 1 is the standard width factor; a width factor of .5 produces text at one half the width.

- An obliquing angle that determines the slant. A positive angle is a forward slant; a negative angle is a backward slant.

- A backward text indicator.

- An upside down text indicator.

- An orientation indicator to determine whether the text will be vertical (stacked), or horizontal (not available for TrueType fonts and some SHX fonts).

After you create a style, you refer to it whenever you want that particular set of parameters. You do not have to go through a series of questions each time you want to define the "look" of your text!

CREATING A TEXT STYLE

The **Style** command creates a named text style, which allows multiple text "looks" on the drawing at a time. Note that the **Style** command is not the same as the **Style** option under the **Text** command. The **Style** *command* creates a text style; the **Style** *option* under the **Text** command selects a previously created text style for use.

To create a text style, type the **Style** command or select **Format | Text Style** from the menu bar. The **Text Style** dialog box is displayed. Let's look at each of the settings.

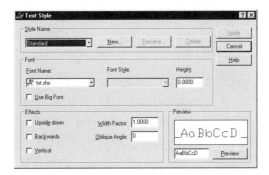

Figure 10.19 *The Text Style Dialog Box*

Style Name

This **Style Name** list box tells you the names of styles already in the drawing. You use this list to change the characteristics of an existing style.

New

Choose the **New** button to create a new style. AutoCAD displays the **New Text Style** dialog box. Type a name of up to 255 characters in length and choose **OK**. AutoCAD adds the name to the **Style Name** list.

Figure 10.20 *The New Text Style Dialog Box*

Rename

Choose the **Rename** button to rename a text style. AutoCAD displays the **Rename Text Style** dialog box. Type a different name of up to 255 characters in length and choose **OK**. AutoCAD changes the name in the **Style Name** list. You cannot rename the STANDARD text style.

Figure 10.21 *The Rename Text Style Dialog Box*

Delete

Choose the **Delete** button to erase a text style from the drawing—but only if it is unused. If the style is being used by text placed in the drawing, AutoCAD pops up this warning: "Style is in use, can't be deleted."

Font Name

There are many font files available for AutoCAD. The program comes with an excellent selection. Specialty fonts for map making (draws symbols instead of letters), cursive writing, and other applications are included with AutoCAD. Fonts with an Autodesk logo are AutoCAD's own vector format, known as SHX. Fonts with the TT logo are TrueType fonts.

The font file should be appropriate for the application. Most mechanical applications use the ANSI type lettering. The "Simplex" font (also known as "RomanS" — roman, single stroke) closely approximates this style. Architectural drawings typically use "hand-lettered" fonts such as City Blueprint. Engineering applications often use the Simplex font since it closely resembles a traditional font constructed with a lettering template.

Font Style

Some fonts let you choose between Regular, Boldface, Italic, or Bold Italic. Not all font styles are available for all fonts.

Height

The text height represents the height, in scale units, that the text will be drawn. For example, if the drawing is plotted at 1/4"=1'0" and the text height is 6, the plotted text will be 1/8" high on the paper. Because of this, you must first determine the scale of the drawing (see Chapter 6 "Setting Up a New Drawing") before you can set the final plotted text height.

You may find it preferable to leave the height at the default of 0.000. This gives you the flexibility to change the height in the drawing after the text has been placed. This applies particularly if you are not sure of the scale until the end of the drawing process.

 Note: You can use a standard architect's or engineer's drawing board scale to "see" the size of the plotted text. Place the scale on a piece of paper, using the numerical scale that represents the final plotted scale and mark the height of the text. See Figure 10.22.

Create a text style to use with dimensions. Make sure the text height is set to 0. This allows the dimension style to adjust the text height.

Figure 10.22 *Measuring Scaled Text*

Use Big Font

The "big font" class of fonts is AutoCAD's way of being able to handle more than 256 characters per file. A big font is primarily used for Chinese and other languages with thousands of characters.

Width Factor

The width factor determines how wide the text will be drawn. A width factor of 1.0 can be thought of as a "standard" width. A decimal value, such as 0.85, draws text that is narrower, resulting in condensed text. Values greater than 1 create expanded text. Figure 10.23 illustrates the same text font with different scale factors applied.

Figure 10.23 *Text Width Factors*

Oblique Angle

A slant can be applied to a font with the oblique angle setting. A zero obliquing angle draws text that is "straight up." Positive angles, such as 15 degrees, draw text that slants forward, while negative angles draw backward slants.

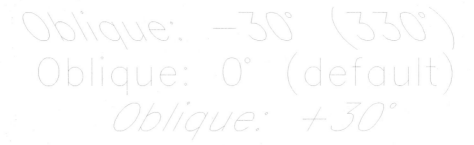

Figure 10.24 *Text Oblique Angles*

Backwards

Text can also be drawn backwards. Backwards text is useful if you want to plot your drawing on the back side of film media. Check the **Backwards** box.

Figure 10.25 *Backwards Text*

Upside Down

If you want the text drawn upside down, check the **Upside down** box.

Figure 10.26 *Upside-down Text*

Vertical

Text can be oriented so it draws vertically. This is *not* the same as text that is rotated 90 degrees. Vertically oriented text is drawn so that each letter within the text string is vertical and the text string itself is vertical. TrueType and some AutoCAD fonts cannot be drawn vertically.

Figure 10.27 *Vertical Text*

Note: Obliquing, underscoring, or overscoring should not be used with vertical text, since the result will look incorrect.

Preview

After applying effects, such as a width factor, choose the **Preview** button to see the result.

USING TEXT STYLES

The concept of fonts and styles is sometimes confusing to learn in the beginning. Remember that a style is a modification that is applied to a particular font (such as height, width factor, and slant factor). You must first create a style in order to use a font. The first step is to choose the particular font that you would like to use in your drawing and then determine the modifications you would like to apply to it by using the **Style** command parameters.

You can have different styles applied to the same font. Just store them with a different style name of your choice.

Note: "Fancy" styles (those created with multistroke fonts) are slow to regenerate and redraw. To speed your drawing operations, you can avoid such fonts, place the text in your drawing last, or use the QText command (covered in the next section) to make redraws and regenerations faster.

Multistroke fonts are identified as duplex, complex, AutoCAD names fonts, such as "RomanC," with the "D" denoting duplex (double stroke), "C" complex, and "T" triple (triple stroke).

REGENERATING TEXT FASTER (QTEXT)

Text strings redraw, regenerate, and plot relatively slowly. As your drawings become more complex, the time required to handle text can slow down the drawing process. AutoCAD provides the **QText** (short for quick text) command to speed handling of text.

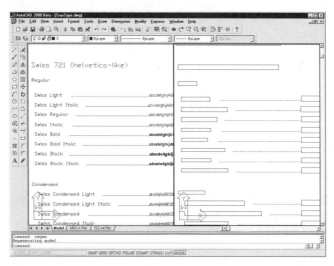

Figure 10.28 *Normal Text (at left) and Quick Text (at right).*

Quick text replaces text with a rectangle of the same size. It is turned on and off by the **QText** command, followed by the **Regen** command.

> Command: **Qtext**
>
> ON/OFF <OFF>: **ON**
>
> Command: **Regen**

QText uses rectangular boxes to represent the height and length of the text. The text boxes handle more quickly when AutoCAD performs redraws, regens, and plots. The change to text boxes does not take place until the next regeneration.

The drawing plots either text or quick text boxes. The display setting at the time of the plot determines the type of plot. If you wish to plot the actual text, turn off the quick text, then regenerate the drawing with the **Regen** command before plotting.

Note: When a drawing contains a large amount of text, several techniques can be used to minimize the time required to display and plot the text.

- Place the text in the drawing last. This eliminates having to regenerate text while you are constructing the drawing elements.

- Create a special layer for text. Freeze the layer when you are not using the text (see Chapter 14).

- Use QText if you want the text placements shown but still need fast regenerations.

- Plot check . drawings with the **QText** option on. The text plots as rectangles, greatly reducing plot time.

- Whenever possible, avoid multistroke "fancy" fonts or True Type fonts. These fonts regenerate and plot more slowly than single stroke fonts. If you wish to use fancy fonts, the quick text option reduces the regeneration time.

EXERCISES

1. We want to place the text "PART A" so that the text will be centered both vertically and horizontally on a selected point. We might choose this method of placement if we wanted to place text in the center of a space. Let's use the text command to place the text. Refer to the following command sequence and Figure 10.29.

Figure 10.29 *Middle-Center Alignment*

Command: **text**

Current text style: "Standard" Text height: **0.2000**

Specify start point of text or [Justify/Style]: **mc**

Specify middle point of text: *(Select point A.)*

Specify height <0.2000>: *(Move cursor up to show height.)*

Specify rotation angle of text <0>: **0**

Enter text: **Part A**

Enter text: *(Press Enter.)*

Notice how you entered simply **MC** instead of selecting the **Justify** option. It is not necessary to specify **Justify** unless you want to see the options. We entered **MC** because we wanted to place the text vertically in the Middle and horizontally in the Center (see Figure 10.4 and Table 10.3).

For the height, we moved up the cursor and picked another point to "show" AutoCAD the height. We could have entered a value. The height of text is in scale units, not actual size.

2. Figure 10.30 shows several text strings. The placement point is marked with a solid dot. Use the **Text** command to place the text as shown. If you need, refer back to Figure 10.4 and Table 10.3 to identify the alignment mode abbreviations.

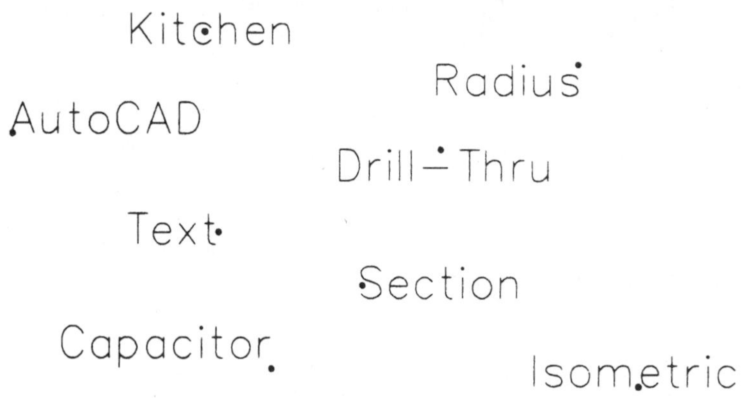

Kitchen

Radius

AutoCAD

Drill−Thru

Text

Section

Capacitor

Isometric

Figure 10.30 *Text Alignment Exercise*

3. Using the text angle prompt, place the following line of text at angles of 0, 45, 90, 135, and 180 degrees.

THIS IS ANGLED TEXT

Use the same text starting point for each line.

4. Let's place some multiple lines of text. We will use the following lines of text:

THIS IS THE FIRST LINE OF TEXT

THIS IS THE SECOND & CENTERED LINE OF TEXT

Refer to Figure 10.31. Let's look at the command sequence used to do this.

EXERCISE

THIS IS THE FIRST LINE OF TEXT
THIS IS THE SECOND & CENTERED LINE OF TEXT

Figure 10.31 *Multiple Text Lines*

Command: **text**

Current text style: "Standard" Text height: 0.2000

Specify start point of text or [Justify/Style]: **c**

Specify center point of text: *(Select point 1.)*

Specify height <0.2000>: *(Move cursor up to show height.)*

Specify rotation angle of text <0>: *(Press* ENTER.*)*

Enter text: **THIS IS THE FIRST LINE OF TEXT**

Enter text: **THIS IS THE SECOND & CENTERED LINE
OF TEXT**

Enter text: *(Press* ENTER.*)*

Repeat this procedure for as many lines of text as you want.

5. Use the **Text** command to place a line of text. Press ENTER and place a second line of text.

   Press BACKSPACE to "erase" the text on the second line, then continue to use BACKSPACE to erase text on the first line.

   Place two more lines of text and then press ENTER twice to end the **Text** command.

   Press ENTER to repeat the **Text** command.

   Press ENTER again and notice how the text box is placed on the next line below the last text string. Enter another line of text.

   When you reach the end of the line, move the crosshairs and enter a new text location. Type a new line of text at this location.

6. Use text codes to construct the following text string in your drawing.

   The story entitled MY LIFE is ±50% true.

7. Start a new drawing and use the **MText** command to place several lines of text on the screen. Now turn on the **QText** command.

   Command: **qtext**

   ON/OFF <OFF>: **ON**

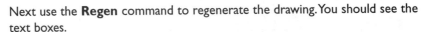

Next use the **Regen** command to regenerate the drawing. You should see the text boxes.

8. If you have a word processor capable of creating an ASCII file (or know how to use Notepad) create a text file that contains your name, address, city, and state (or province). Save the file in TXT format. Import the text file into your drawing with the **MText** command.

9. Use AutoCAD's text fonts to create styles you like and design your own business card.

10. Using the same font, create five different styles that represent very different text appearances, such as wide text, slanted, etc.

11. Use a combination of text and drawing commands to design a logo for a graphic design office.

CHAPTER REVIEW

1. How can you see a listing of text styles stored in the drawing?

2. What is a *font*? How does a *style* differ from a font?

3. What text characteristics are altered by means of the **Style** command?

4. There are several methods of placing text. List them with a brief explanation of each.

5. When an underscore or overscore is added to text, is an entire string altered, or can segments be treated separately? Explain.

6. When a text height is asked for, is a numerical entry required? Explain.

7. List the alignment modes under the **Text** command.

8. Can text be rotated at any angle? How does oblique angle differ from text angle?

9. What is *justified* text?

10. How can the text height be altered each time you enter text without redefining the style?

11. How can you compress or expand text?

12. What format must text be before it is imported into AutoCAD?

TUTORIAL

Let's place some simple text in a drawing.

1. Start AutoCAD and open a new drawing.

2. Next, type the **Text** command at the command prompt.

   When you enter the **Text** command, you will see the command prompt:

   Command: **text**

   Current text style: "Standard" Text height: 0.2000

   Specify start point of text or [Justify/Style]: *(Pick.)*

   We will use the default style that is already contained in the drawing.

3. At this point, move your cursor into the drawing area and place a point on the screen.

4. The command prompt continues and AutoCAD will ask you for the text height:

   Specify height <0.2000>: *(Pick two points.)*

   Let's just "show" AutoCAD how high the text should be. Move the crosshairs up from point you entered. The point will rubber band to the crosshairs.

5. Enter a point approximately 1/4 inch (actual distance) above the first point you entered.

6. AutoCAD continues and prompts you:

   Specify rotation angle of text <0>: **0**

   Enter zero (0) in response to the prompt.

7. Now it's time to type the text. Type your name. If you make a mistake, use BACKSPACE to erase back and start again. Notice how the text shows up on the command line, but not on the screen. When you are finished typing your name, press ENTER. You should now see your name on the screen!

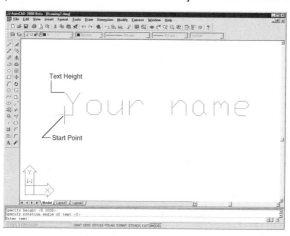

CHAPTER 11

Working with Layers

Professional CAD drawings are constructed on different layers that can be turned on and off, or changed. This chapter covers the methodology of using layers in your drawings. After completing this chapter, you will be able to

- Use the concept of CAD layering

- Apply layers to your drawings

- Perform layering with proficiency

INTRODUCTION

In this chapter, you learn to create and control layers. That includes giving layers their names, setting their color and linetype, sorting layers, and toggling the visibility of layers. All this is performed with a single command, with the assistance of two other commands:

Layer

Displays the Layers Properties Manager dialog box.

AI_MOLC

Allows you to select an object to make its layer current.

CLAYER

Allows you to type the name of the layer to make current.

LAYERS

Traditional drafting techniques often include a method of drawing called *overlay drafting*. This consists of sheets of transparent drafting media that are overlaid so the drawing below shows through the top sheet. Items placed on the top sheet line up with the drawing below. Both sheets are blueprinted together, resulting in a print that shows the work on both sheets.

The bottom sheet is typically referred to as the *base drawing*. Each additional sheet is used to place different items.

As an example, if you draw a set of floor plans, you must prepare a separate drawing for the to-be-removed plan, the plumbing plan, the electrical plan, and so forth. Since most of your drawing time is spent redrawing the floor plan, you spend a great deal of time doing repetitive tasks. The floor plan can be thought of as the base drawing, with each discipline, such as electrical and plumbing, placed on overlay sheets.

AutoCAD provides capabilities that eliminate the repetition in redrawing the base drawing. You can use drawing layers to place different parts of your drawings.

You can think of layers as transparent sheets of glass that are stacked on top of each other. You can draw on each layer and see through all the layers so that all the work appears as though it were on one drawing.

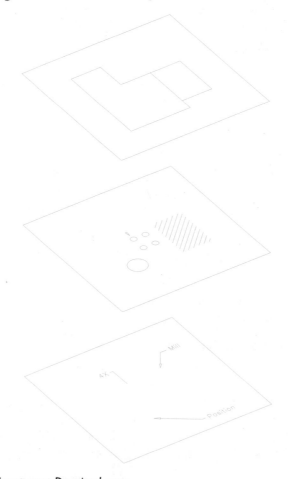

Figure 11.1 *Transparent Drawing Layers*

AutoCAD goes one step further. You can turn on or off each layer so that it is either visible or invisible!

USING THE LAYER COMMAND

You use the **Layer** command to manipulate the layers in your drawing. Let's look at the **Layer Properties Manager** dialog box. There are four ways to display the dialog box.

- Type the **Layer** command at the command line.

- Use the transparent command **'Layer** during a current command.

- Select **Layer** from the **Format** drop-down menu.

- Select **Layers** button from the **Object Properties** toolbar.

Figure 11.2 shows the Layer Properties Manager dialog box.

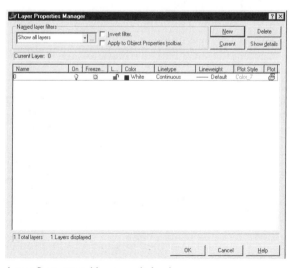

Figure 11.2 *The Layer Properties Manager dialog box*

Let's look at the methods used to manipulate layers with the dialog box.

Creating a New Layer

To create a new layer, choose the **New** button. AutoCAD creates a new layer with the default name of "Layer1." The layer is turned on, is colored white, and has the continuous linetype. To rename the layer to something more meaningful, click Layer1 twice and then edit the name.

Deleting a Layer

AutoCAD lets you delete only empty layers. Layer 0, DefPoints, and externally referenced layers cannot be deleted. Select one or more layers and choose **Delete**. If the layer cannot be erased, AutoCAD displays the following dialog box:

Figure 11.3 *The AutoCAD Message Box*

Turning Layers On or Off

To turn one or more layers on or off, first highlight the target layer(s) by clicking on the layer name and then choose the lightbulb icon. The "on" column will immediately reflect the change. If a layer is on, the lightbulb is yellow. If off, the lightbulb is blue-gray.

Figure 11.4 *Turning a Layer Off and On*

Freezing and Thawing Layers

To freeze or thaw one or more layers, highlight the target layer(s), then choose the sun or snowflake icon. A frozen layer is indicated by a snowflake. A non-frozen or thawed layer is indicated by a sun in the **Freeze in All Viewpoints** column.

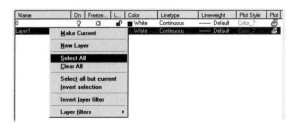

Figure 11.5 *Freezing and Thawing a Layer*

 Note: To select all layers in the layer dialog box, right-click on a layer name. AutoCAD displays a small pop-up menu. Choose **Select All** to highlight all layer names. Choose **Clear All** to deselect all layers.

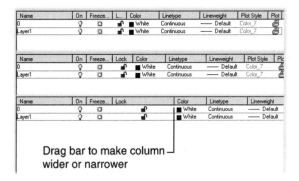

 Note: To change the width of the columns in the layer dialog box, simply grab the black bar separating column tiles and drag left or right. In Figure 11.6, dragging the bar to the right makes the column **Lock wider**.

Drag bar to make column wider or narrower

Figure 11.6 *Dragging a Column Wider*

Locking and Unlocking Layers

To lock or unlock layers, select the target layer(s) and then choose the padlock icons in the **Lock** column. A layer that is not locked shows an open padlock.

Figure 11.7 *Locking and Unlocking Layers*

Setting the Layer Color

To set the layer color, first select the layer(s) to be modified, then choose the color name, such as White. The **Select Color** dialog box is displayed as shown in Figure 11.8.

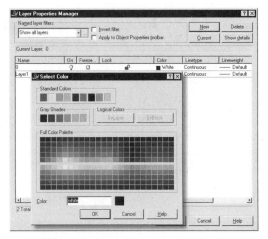

Figure 11.8 *The Select Color Dialog Box*

To set the color, select a color from the color display chart. Note that this chart can display 255 colors.

The **Color** edit field displays either the name or number of the selected color. The first seven colors are listed by name, with the remaining colors listed by number. You can also enter the color name or number manually from the keyboard in the **Color** edit field. After you have selected the color, choose the **OK** button to return to the layer dialog box.

Setting the Layer Linetype

You set the linetype for all the objects residing on a layer. First select the layer(s) for which you wish to set a linetype, then choose the linetype, such as Continuous. The **Select Linetype** dialog box is displayed, as shown in Figure 11.9.

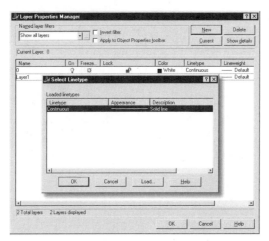

Figure 11.9 *The Select Linetype Dialog Box*

The dialog box displays the linetypes that have been loaded. Select a linetype and choose **OK**. (To load more linetypes, choose the **Load** button.)

Setting the Layer Lineweight

You set the lineweight for all the objects residing on a layer. First, select the layer(s) for which you wish to set a lineweight, then choose the lineweight, such as Default. The **Lineweight** dialog box is displayed, as shown in Figure 11.10.

The dialog box displays the lineweights that are available. Select a lineweight and choose **OK**.

Setting the Layer Plot Style

You set the plot style for all the objects residing on a layer. First, select the layer(s) for which you wish to set a plot style, then choose the plot style, such as Color_7. Select a plot style from those available and choose **OK.**

Setting the Layer Print Toggle

You can specify that some layers do not print. Choose the printer icon for the layer(s) you don't want to print or plot. To allow the layer to print, simply choose the icon a second time.

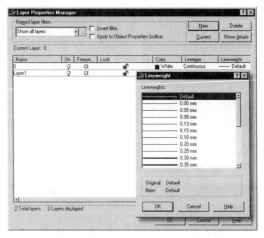

Figure 11.10 *The Lineweight Dialog Box*

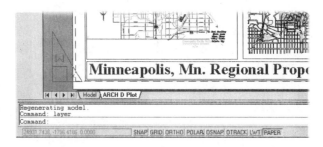

Figure 11.11 *Toggling the Print Status of a Layer*

Layout Modes

In layout mode, two more columns appear in the layer dialog box: **Active VP Freeze** and **New VP Freeze**. To see these columns, switch to a layout mode by selecting any tab except **Model.** Open the **Layer Properties Manager** dialog box and scroll the layer listing all the way to the left.

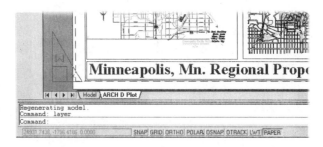

Figure 11.12 *Switching to Layout Mode*

As you learn in Chapter 23 "Working with Viewports and Layouts," AutoCAD can work with tiled or floating viewports. You can freeze layers in floating viewports—something you cannot do in model view—to display different sets of layers in different viewports. AutoCAD has two controls over layer visibility in floating viewports: (1) freeze active viewport and (2) freeze new viewports.

The first option automatically freezes the specified layers in the current viewport. Select the layer name(s), then choose the icon in the **Active VP Freeze** column. When the icon is a shining sun, the layer is thawed; when the icon is a snowflake, the layer is frozen. When the icon is gray, AutoCAD cannot display floating viewports because the drawing is not in layout mode.

The second option freezes specified layers when a new viewport is created. Select the layer name(s), then choose the icon in the **New VP Freeze** column. As before, when the icon is a shining sun, the layer is thawed; when the icon is a snowflake, the layer is frozen.

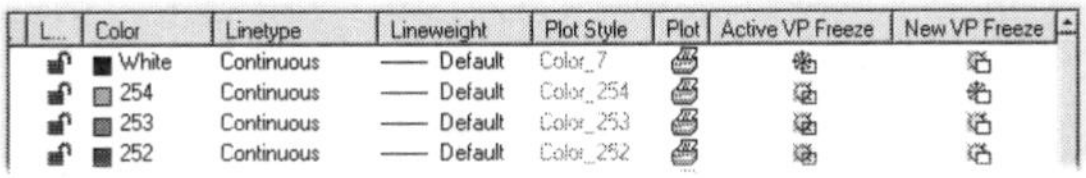

Figure 11.13 *Toggling the Paper Space Status of a Layer*

Selecting the Layer Listings

Some types of drawings may have many, many named layers. In fact, an AutoCAD drawing is capable of having thousands and thousands of layers. Manipulating through the layer listings can become tedious. AutoCAD allows you to select the listings so that only certain types of listings are displayed.

There are these ways to sort layers:

- Sort by columns
- The Show list
- By filters

Sort by Columns

The names of the columns—such as Name, On, Freeze, Lock, etc.—are actually buttons. Click once on the **Name** column button and the column sorts in alphabetical (A - Z) order; click a second time and the column sorts in reverse order (Z - A).

Choosing the **Color** column button sorts the colors by color number (Red is Color #1). Choosing a second time reverses the color numbers, starting with the highest color number used by drawing.

Click to sort alphabetically

Choosing the **On, Freeze,** and other toggle column titles sorts the layers by icons. Click once, and the column sorts by light bulb, sun, or open padlock. Click a second time and the column reverses the sort by dim bulb, snowflake, or closed padlock.

Show List

The list box under **Named Layer Filters** lets you see logical groups of layers. Your choices include:

Show all layers

Displays all layer names in the drawing. This is the default.

Show all used Layers

Displays all layers being used, that is, all layers with at least one object on the layer.

Show all Xref dependent layers

Displays all layers found in externally referenced drawings.

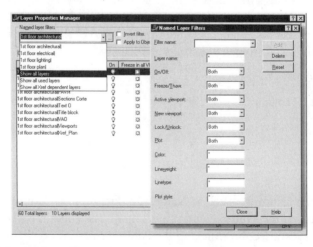

Figure 11.14 *Layer Name Filters*

Invert filter

Displays all layers not displayed by the named filters, as listed above. For example, when **Invert Filter** is on and you select **All in Use**, the layer dialog box displays all layers *not* in use.

(...) Named Layer Filter

AutoCAD's filter option lets you decrease the number of layers displayed by filtering—after all, a drawing can contain more than 32,000 layers! Choose the ... button next to the drop-down list and AutoCAD displays the **Named Layer Filter** dialog box.

For example, you can filter the listing so that only thawed layers are displayed. The filter capabilities can be used to display or suppress layers that are on or off, frozen or thawed, locked or unlocked, and whether the layer is frozen in the current viewport.

You can also filter by name, color, and linetype. With these filters, you can use the wild-card characters of question mark (?) and asterisk (*) to designate multiple layers.

Some of the filters shown (see Figure 11.14) use pop-up boxes to specify either property listed, or both properties to be listed. For example, when you select the **Freeze/Thaw** pop-up box, the selections are **Both**, **Frozen**, and **Thawed**.

The other selections require you to enter the filter type in the text box. The default listing is an asterisk (*), which means that all layer names will be listed.

Let's assume that you wanted to list all layers that started with the letters LEVEL and ended with any two characters. The filter would read:

LEVEL??

This would list layers that are named LEVEL21, LEVEL_A, and LEVEL2C. It would not list ILEVEL, LEV21, or 3RD_LEVEL. If you do not remember how wildcard characters work, you might want to review Chapter 3 "The Windows Operating System" earlier in this book.

You can reset the values so that all the layer names are listed when you choose the "Reset" button.

CONTROLLING LAYERS WITH THE TOOLBAR

When the **Object Properties** toolbar is available (as it usually is), you select the current layer without using the **Layer** command. When you click on the down arrow at the right end of the layer name on the toolbar, AutoCAD displays a drop-down box, as shown in Figure 11.15.

Figure 11.15 *Layer Control on the Toolbar*

The drop-down box lists the names of all layers in the drawing, as well as a quintet of icons. Each icon has two states:

- Lightbulb on or off: Layer is on (default) or off.

- Sun or snowflake: layer is thawed (default) or frozen. Frozen layers show their names colored gray.

- Padlock open/closed: layer is unlocked (default) or locked.

- Printer: layer is printable or not.

- Square: color assigned to the layer.

 Note: AutoCAD uses some layer names for its own, special purposes. Two layers you must be acquainted with: Layer 0 and Layer Defpoints.

Layer 0: Every new drawing contains one layer, called 0 (zero). This layer cannot be removed. Layer 0 has a special property for creating blocks, which you learn about later in this book.

Layer Defpoints: The first time you draw a dimension in a drawing, AutoCAD creates layer Defpoints. This layer contains data that AutoCAD needs to keep its dimensioning associative. This layer cannot be removed.

Anything you draw on this layer, accidentally or otherwise, AutoCAD will not plot. For this reason, some students are frustrated to find part of their drawing will not plot, because they accidentally drew on layer Defpoints. (You can use the No Print toggle to make any layer non-printing.)

MAKE LAYER CURRENT

Before you can work on a layer, you need to switch to that layer. The working layer is called the current layer. AutoCAD can have only one current layer at a time. Current layers cannot be frozen.

One easy way to switch to another layer is with the **Make Object's Layer Current** button. Suppose you are interested in switching to the layer holding a green dotted line whose name you're not sure about. The **MOLC** command switches you to the layer when you select the green dotted line. You use the feature as follows:

1. Choose the **Make Object's Layer Current** button. AutoCAD prompts you, "Select object whose layer will become current:"

2. Select the green dotted line. AutoCAD reports, "*layername* is now the current layer."

Check the toolbar and you will see that layer "*layername*" is now current.

Note: If you prefer the keyboard to dialog boxes, then **CLayer** is the fastest way to switch layers. **Clayer** is a system variable that holds the name of the current layer. You use **Clayer** as follows:

1. When you type **Clayer** at the command prompt, AutoCAD prompts "New value for CLAYER <default>:"

Command: **Clayer**

New value for CLAYER <default>:

2. Type the name of the layer and press ENTER. AutoCAD immediately makes that layer current.

EXERCISES

1. Create a layer named "MYLAYER." Set "MYLAYER" as the current layer. Do you see the layer name in the toolbar at the top of the screen?

2. Set the layer color to green. Draw some objects on the layer.

3. Set the current layer to "0." Freeze the layer named "MYLAYER." Did the objects you drew disappear? Select the **Layer** dialog box from the drop-down menu and Thaw "MYLAYER." Did the objects reappear?

4. Create seven layers at the same time with the **Layer** command, using these names:

ToBeRemoved

Landscape

Roadways

Hydro

StormSewer

CableTV

Building

5. Change the color of layer "HYDRO" from white to blue using the **Layer** command.

EXERCISE

6. Change the linetype of layer "HYDRO" from Continuous to Gas_line using the **Layer** command.

7. Using the Layer dialog box, change the "TOBEREMOVED" layer from white to red.

8. Change the linetype of the "TOBEREMOVED" layer from Continuous to Dashed using the Layer dialog box.

CHAPTER REVIEW

1. Why would a CAD drafter use layers?

2. What layer option would you use if you wanted to create a new layer?

3. How do you turn on a frozen layer?

4. What is the difference between turning *off* a layer and *freezing* a layer?

5. How would you obtain a listing of all the layers in your drawing?

6. Can you have objects of more than one color on the same layer? Explain.

7. What do the following layer symbols mean?

Snowflake:

Open lock:

Lightbulb glowing:

Printer:

Colored square:

CHAPTER 12

Introduction to Dimensioning

The ability to place dimensions in a drawing is one of the most essential elements of CAD drawing. In this chapter, you will learn the basics of AutoCAD dimensioning. After completing this chapter, you will be able to

- Understand the components of a dimension
- Apply the basic elements of dimensioning
- Manipulate some aspects of dimensions

DIMENSIONING IN AUTOCAD

AutoCAD constructs dimensioning *semi-automatically*. That is, it constructs the dimensioning lines and measures the distances for you. All it needs from you is some basic information, such as where to start and end the dimension.

AutoCAD's dimensions are *associative*. That means they automatically update themselves when stretched. This is a very powerful feature. If mastered, it can be a great time-saver and provide professional results.

DIMENSIONING COMPONENTS

A dimension is made up of several parts. Before we try some dimensioning, you should know these parts. Let's look at each.

DIMENSION LINE

The *dimension line* is the line with the arrows or "ticks" at each end. You place the dimension text in this line, thus dividing it into two parts, or over the line.

Figure 12.1 *Dimension Line with Arrows*

When you use angular dimensioning, the dimension line is an arc instead of a straight line.

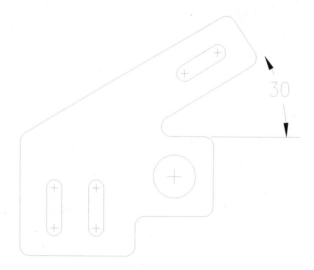

Figure 12.2 *Angular Dimension*

ARROWHEADS

If you wish, the arrows at the end of the dimension line can be replaced with tick marks or other symbols. Figure 12.3 shows dimension line with tick marks. AutoCAD includes many different arrowheads, and you can define your own.

Figure 12.3 *Dimension Line with Ticks*

EXTENSION LINES

The *extension lines* (sometimes called "witness lines") are the lines constructed perpendicular to the dimension line and extending to the points that are being dimensioned to.

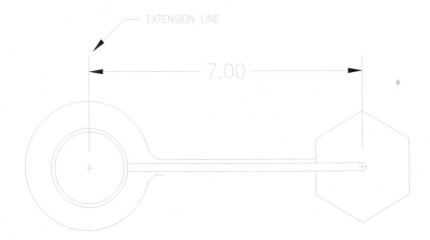

Figure 12.4 *Extension Lines*

DIMENSION TEXT

The dimension text is the text that appears at the dimension line. You can allow AutoCAD to measure the distance and enter the text, you can specify different text, or you can suppress the text entirely by entering a space in place of the text.

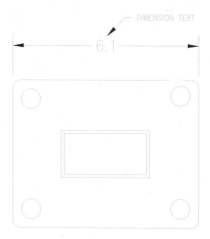

Figure 12.5 *Dimension Text*

The dimension text is drawn with the text style and font specified by the dimension style. You cannot use the current text style to specify the look of the dimension text; you must specify a text style name through the **DimText** system variable, or in the **Dimension Style Manager** dialog box.

 Note: You must use a horizontal text font for dimension text. You should not use a vertical text style for placing dimensions.

The format of the dimension units is determined by the current dimension style. Dimension text does *not* follow the format set by the **Units** command.

DIMENSION TOLERANCES

The dimension tolerances are the plus and minus amounts that are appended to the dimension text. These tolerances are added to the text that AutoCAD generates automatically. You specify the plus and minus amounts. They can be equal or unequal. If they are equal, they are drawn with a plus/minus symbol. If they are unequal, the tolerances are drawn one above the other. The following figure shows examples of equal and unequal tolerances.

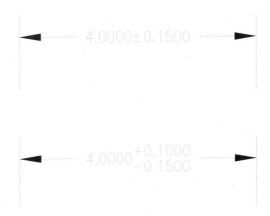

Figure 12.6 *Dimension Tolerances*

DIMENSION LIMITS

Instead of showing the tolerances, you can have them applied to the dimension. The example shown in Figure 12.7 is a measurement of 4.0000 units with a tolerance of ±0.15.

Figure 12.7 *Dimension Limits*

ALTERNATE DIMENSION UNITS

Alternate units are used to show two systems of measurement simultaneously. For example, you can show English and metric units on the same dimension line.

Figure 12.8 *Alternate Dimension Values*

LEADER LINES

Leader lines are "arrowed lines" with text at the end. Leaders are often used to point out a specific part of a drawing to be noted. The leader line can be straight, angled, or curved.

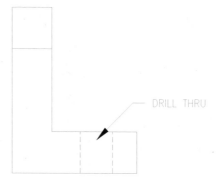

Figure 12.9 *Dimension Leader*

CENTER MARKS AND CENTERLINES

A center mark is a cross designating the center of a circle or arc. Centerlines are lines that cross at the center of the circle or arc and may intersect the circumference. Examples of each are shown in Figure 12.10.

Figure 12.10 *Center Marks and Centerlines*

DIMENSION VARIABLES

Dimension variables determine how the dimensions are drawn. Some variables are values and some are simply turned on and off. You can change the variables to change the dimension "look" and function. Dimension variables are covered in Chapter 13 "Setting Dimension Styles and Variables."

DIMENSION MENU

You begin dimensioning by selecting **Dimension** from the menu bar. When you dimension, you first select the type of dimensioning you wish to perform, and then you use the appropriate commands to draw the dimensions.

AutoCAD dimensioning commands can be grouped into four basic types of commands. These are

Dimension *drawing* commands

Dimension *editing* commands

Dimension *utility* commands

Dimension *style* commands (covered in Chapter 13)

Let's now look at each of the types of dimensioning commands.

DIMENSION DRAWING COMMANDS

Dimension drawing commands are used to draw the different types of dimension lines. The following is a listing of dimension drawing commands.

Linear

Draws a dimension with a horizontal, vertical, or rotated dimension line.

Aligned

Draws a dimension line parallel to an object or to the dimension extension line origins you specify, often slanted.

Ordinate

Dimensions the X or Y coordinate, referenced to a specified point, of an object.

Radius

Constructs a dimension line that shows the radius of an arc or circle.

Diameter

Dimensions the diameter of a circle or arc.

Angular

Dimensions the angle between two nonparallel lines. The dimension line is constructed as an arc, with the dimension value shown as the angle between the lines.

Baseline

Continues a dimension line in reference to the first extension line origin of the first dimension line in the string. The dimension line is offset to avoid conflict with the previous dimension line.

Continue

Continues a dimension line after you have placed the first dimension from the second extension line of the previous dimension. The continuing string uses the first dimension to determine the correct positioning of the next dimension.

Leader

Places text attached to a line and arrowhead.

Tolerance

Displays a dialog box that lets you select a tolerancing symbol.

Center Mark

Places a center mark, or alternately, centerlines, at the center point of a circle or arc.

QDim

Quickly dimensions in continuous mode.

Let's look at each command and learn how to use each to place a dimension.

PLACING A LINEAR DIMENSION

The **DimLinear** dimension command creates the most common types of dimension lines. Select **Dimension | Linear** from the menu bar. The dimension line is drawn horizontally (as shown in Figure 12.11), or vertically, depending on how you move the cursor.

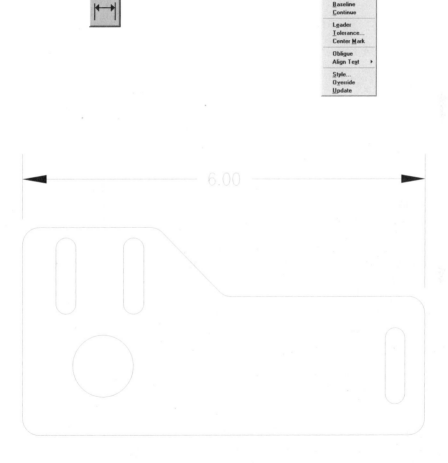

Figure 12.11 *Horizontal Dimension Line*

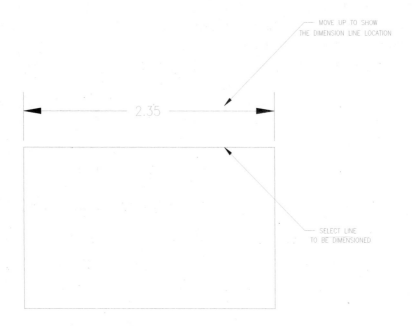

Figure 12.12 *Placing a Horizontal Dimension Line*

When placing a linear dimension, you can either specify an object to dimension, or select the starting and ending points for the extension lines.

Let's place a horizontal dimension line. Draw a box as shown in Figure 12.12. Use the following command sequence and follow the instructions at the prompts.

> Command: **dimlinear**
>
> Specify first extension line origin or <select object>: *(Press* ENTER.*)*
>
> Select object to dimension: *(Select the top line of the box.)*
>
> Specify dimension line location or
>
> [Mtext/Text/Angle/Horizontal/Vertical/Rotated]: (Move the cursor up to locate the dimension line.)
>
> Dimension text = 2.35

Note that you can also enter the extension line origin points. When AutoCAD prompts you for the first extension line origin, enter the actual point instead of pressing Enter. We use this method in a later exercise.

 Note: Use object snap to set exact points for accurate dimensioning.

If you choose to select the object to be dimensioned by pressing ENTER instead of specifying the extension line points, you can save the steps necessary for entering each beginning point.

This method works on lines, arcs, or circles and can be used with the **DimLinear**, **DimAligned**, and **DimRotated** dimension commands. **DimLinear** dimensions individual line and curve segments of polylines, etc. Figure 12.13 shows typical pick points for dimensioning three objects using this method.

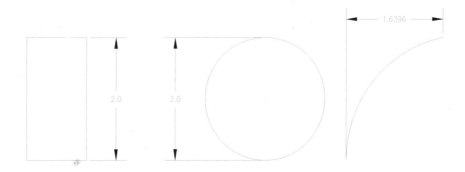

Figure 12.13 *Selecting an Object to Dimension*

DIMENSIONING SYMBOLS

AutoCAD uses two symbols for special purposes in dimensioning text.

< *and* > *Symbols*

The < > (closed angle brackets, with no space between) are a shorthand notation for the default dimension text. For example, if you place a dimension with a distance of 2.54, AutoCAD displays < > in the dialog box (or at the command prompt, depending on the dimensioning command). The < > acts as a placeholder.

You can add text in front and behind the < > symbol. For example, you could add the words "Distance is" in front and "miles (approx.)" behind, as follows:

> Distance is < > miles (approx.)

AutoCAD would then display the dimension text as "Distance is 2.54 miles (approx.)".

Another possibility is to replace the < > symbol entirely. In that case, AutoCAD displays your text, not the 2.54 dimension text. For example, if you erase the < > and replace the place holder with:

> Two miles

AutoCAD displays the dimension text as "Two miles".

[and] Symbols

AutoCAD displays alternate units, such as for metric [imperial], in square brackets. Using the 2.54 from above,

> < >[1.00]

AutoCAD displays the dimension text as "2.54[1.00]".

DIMENSION OPTIONS

Many of the dimensioning commands include a number of options for placing the text and the dimension itself. The typical prompt reads:

> Specify dimension line location or
> [Mtext/Text/Angle/Horizontal/Vertical/Rotated]:

The options have the following meaning:

Dimension line location

Pick a spot on the drawing where you want the dimension line located. AutoCAD automatically stretches or shrinks the extension lines to compensate.

Mtext

Type **M** to display the **Multiline Text Editor** dialog box. You can edit the text by adding a prefix and suffix to the < > symbols, or change it entirely by erasing the < > symbols. The setting of units, tolerance, and current text style determine the display of text.

Text

Type **T** and AutoCAD displays the prompt, "Dimension text <default>:". You can change the text or press ENTER to accept the default.

Angle

Type **A** and AutoCAD prompts you, "Enter text angle:". When you provide a number, such as 45, AutoCAD rotates the dimension text (but not any other part of the dimension) by 45 degrees.

Horizontal

To force a linear dimension to be horizontal, type **H**. AutoCAD prompts you, "Dimension line location (Mtext/Text/Angle):", which are the same options as above.

Vertical

To force a linear dimension to be vertical, type **V**. AutoCAD prompts you, "Dimension line location (Mtext/Text/Angle):".

Rotated

To force a linear dimension to be rotated, type **R**. AutoCAD prompts you, "Dimension line angle <0>:", which creates the rotated dimension by rotating the dimension line.

CONTINUING THE DIMENSION

The **DimContinue** dimension command is used to place a dimension segment that follows the first dimension you placed. The continued dimension is constructed with the last extension line of the previous dimension being used as the first extension line point of the new dimension.

Continued dimensions work with previously placed linear, angular, and ordinate dimensions. If you haven't placed a dimension immediately before using the **DimContinue** command, AutoCAD prompts you to select a dimension it can work with.

Construct the simple object shown in Figure 12.14. Place a horizontal dimension as shown.

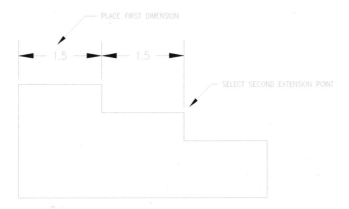

Figure 12.14 *Continuing a Dimension String*

Select **Dimension** from the menu bar, and then select **Continue**. AutoCAD prompts:

> Command: **Dimcontinue**
>
> Specify a second extension line origin or [Undo/Select] <Select>:
> **Int**

Let's use object snap to capture the upper right intersection. Type **INT** from the keyboard and press ENTER. AutoCAD prompts you for the intersection point.

> of *(Select the intersection.)*
>
> *(Press* ESC *to end the command.)*

 Note: There is no requirement for the dimension text to be a numerical value. You can type in any characters from the keyboard. For example, you may want to place a horizontal dimension and use the dimension text:

FIVE EQUALLY SPACED NOTCHES

PLACING A VERTICAL DIMENSION

The **DimLinear** dimension command works equally as well for vertical dimensions as for horizontal dimensions, except that the dimension line is drawn vertically. Again, select **Dimension | Linear** from the menu bar.

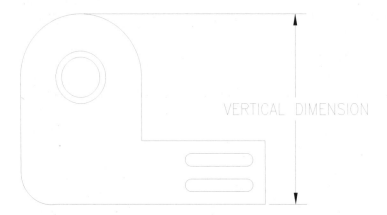

Figure 12.15 *Vertical Dimension Line*

To place vertical dimensions, either select an object to dimension or select the extension line origin points using object snap modes, such as ENDpoint or INTersection. Use Exercise 1, at the end of this chapter, to place vertical dimension lines.

DIMENSIONING ANGLED SURFACES

Many times you are required to place dimensions or objects that are constructed at an angle. AutoCAD provides two commands for placing angled dimensions. Let's see how each one works.

Drawing Aligned Dimensions

The **DimAligned** dimension command draws the dimension line parallel to the extension line origin points. Select **Dimension | Aligned** from the menu bar. See Figure 12.16.

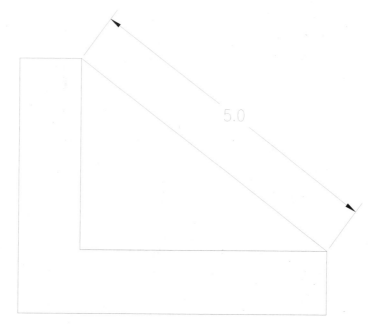

Figure 12.16 *Aligned Dimension*

To draw an aligned dimension, either select the object that you want the dimension line to be parallel to or manually select the extension line origins. The exercise at the end of the chapter guides you through the methodology of constructing an aligned dimension.

Constructing Rotated Dimensions

The **DimLinear** command functions in the same way as the aligned command, except that you must first specify the rotated option. This is especially useful in situations where the extension line origins do not accurately describe the desired dimension line angle. The illustration in Figure 12.17 shows such a situation.

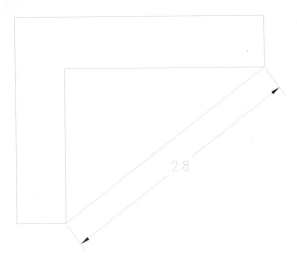

Figure 12.17 *Rotated Dimension*

When you select the **Dimension | Linear** from the menu bar, AutoCAD prompts:

Command: **dimlinear**

Specify first extension line origin or <select object>: *(Pick point.)*

Specify second extension line origin: *(Pick point.)*

Specify dimension line location or
    [Mtext/Text/Angle/Horizontal/Vertical/Rotated]: **R**

Specify angle of dimension line <0>: 45

Specify dimension line location or
    [Mtext/Text/Angle/Horizontal/Vertical/Rotated]: *(Pick point.)*

Dimension text = 2.8

You tell AutoCAD the angle you want to use in two ways. The first is to enter the angle from the keyboard. The angle is specified in AutoCAD's standard angle notation, with the zero angle to the right (east) direction.

The second way is to use two points to "show" AutoCAD the angle. To do this, respond with two points when prompted for the dimension line angle. AutoCAD measures the angle between the two points and uses that angle to construct the rotated dimension.

CREATING BASELINE DIMENSIONS

The **DimBaseline** dimension command creates continuous dimensions from the first extension line. The first extension line acts as the baseline from which the dimensions originate. AutoCAD offsets each new dimension line to avoid drawing on top of the preceding dimension line.

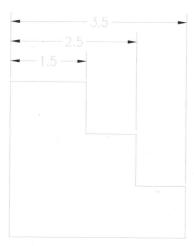

Figure 12.18 *Baseline Dimension*

To construct a baseline dimension (see Figure 12.19), first construct an initial dimension line, then select **Baseline** from the **Dimension** menu. When AutoCAD prompts for the second extension line origin, select the next point. AutoCAD constructs the second dimension line, using the extension line origin of the first dimension line and the point you entered as the second extension line origin to describe the second dimenson line. AutoCAD also offsets the new dimension line so that it is not constructed on top of the first.

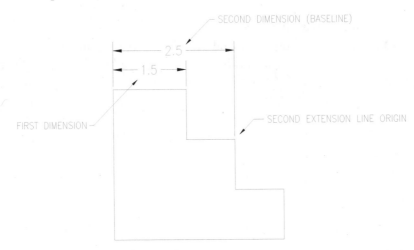

Figure 12.19 *Constructing a Baseline Dimension*

Press ENTER to end the command.

DIMENSIONING ANGLES

The DimAngular command is used to dimension angles. There are four ways to specify an angular dimension:

- Two non-parallel lines: the intersection is the vertex; the two lines define the extension lines.

- An arc: the arc's center is the vertex; the endpoints define the extension lines.

- A circle: the circle's center is the vertex.

- Three points: define the (1) angle vertex, (2) first extension line, and (3) second extension line.

Figure 12.20 *Angular Dimensions*

When you select **Angular** from the **Dimensioning** menu, AutoCAD prompts:

> Command: **dimangular**
>
> Select arc, circle, line, or <specify vertex>:

DIMENSIONING CIRCLES AND ARCS

Diameter and radius dimensioning of circles and arcs is easy with AutoCAD. The **DimDiameter** and **DimRadius** commands are used to place these types of dimensions. Let's see how each works.

Dimensioning the Diameter

The **DimDiameter** dimension command draws dimensions of the diameter of either a circle or an arc. Select **Dimension | Diameter**. An example of each is shown in Figure 12.21.

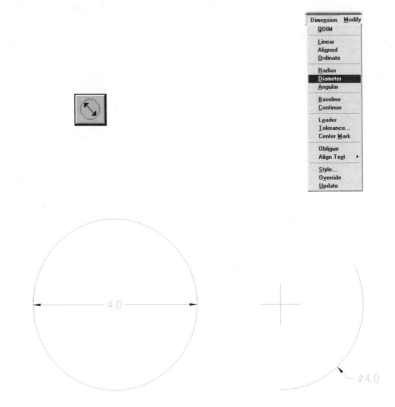

Figure 12.21 *Diameter Dimension*

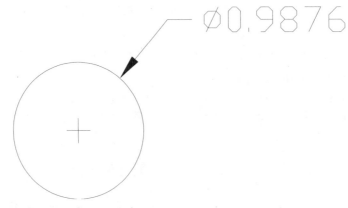

Figure 12.22 *Diameter Dimension Line*

The dimension is placed initially according to the point you pick when selecting the arc or circle; you can adjust the dimension location after picking the object. The dimension line intersects that point, passing through the center point of the arc or circle.

If the dimension line and text are too large to fit within the arc or circle, AutoCAD places a leader line outside the circle or arc. Figure 12.22 shows an example of such a leader line.

You indicate the length of the leader line.

> Select arc or circle: *(Pick object.)*
>
> Dimension text = 12.0000
>
> Specify dimension line location or [Mtext/Text/Angle]: *(Pick point.)*

Dimensioning the Radius

The **DimRadius** command draws the dimension of the radius from the center of a circle or arc. Select **Dimension | Radius** from the menu.

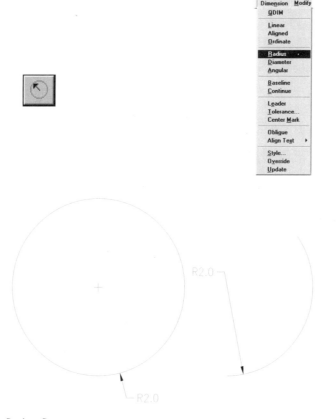

Figure 12.23 *Radius Dimension*

The dimension is placed initially according to the point you pick when selecting the arc or circle; you can adjust the dimension location after picking the object.

Placing Center Marks

You can place a center mark at the center point of an arc or circle with the **DimCenter** command. Center marks are drawn as an intersection cross as shown Figure 12.24. The center mark size is controlled by the **DimCen** system variable:

- DimCen=0: No center marks or lines are drawn.

- DimCen=0.09: Center marks are drawn 0.9 units long.

- DimCen=-0.09: Centerlines are drawn.

The absolute value specifies the size of the center mark.

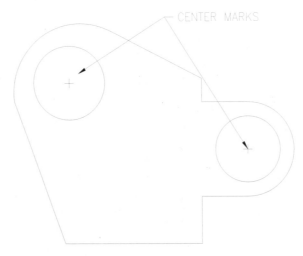

Figure 12.24 *Center Marks*

To place a center mark, select **Center Mark** from the **Dimension** menu.

> Command: **Dimcenter**

> Select arc or circle: *(Pick object.)*

ORDINATE DIMENSIONING

The **DimOrdinate** command places dimensions that are relative to a *reference point*. The reference point used is the current User Coordinate System (UCS) origin (Chapter 32 describes UCS in detail). Values for ordinate dimensions are specified as either X or Y coordinates relative to the origin.

The UCS origin is located, by default, in the lower left corner of the screen (the 0,0 coordinate point). In most cases, you want to relocate the UCS origin point so the ordinate dimensions will be referenced to a specific point. You can move the UCS origin by using the UCS command with the Origin option.

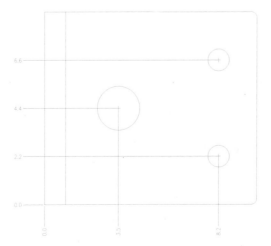

Figure 12.25 *Ordinate Dimensioning*

Command: **UCS**

Enter an option
    [New/Move/orthoGraphic/Prev/Restore/Save/Del/Apply/?/
    World] <World>: **o**

Specify new origin point <0,0,0>: *(Select the new origin point.)*

Ordinate dimension lines are shown as simple leader lines. AutoCAD asks for a feature location. This is the location to which the dimension is measured. You can either stipulate whether the point is to be measured along the X coordinate or the Y coordinate. When you select **Ordinate** from the **Dimension** menu, AutoCAD prompts:

Command: **Dimordinate**

Specify feature location: *(Pick the dimension start.)*

Specify leader endpoint or [Xdatum/Ydatum/Mtext/Text/Angle]: *(Pick
    the leader endpoint.)*

Dimension text = 0.5000

You can "shortcut" the process by entering a point on the drawing instead of specifying either an X or Y point. If you do this, AutoCAD determines whether the measurement is along the X or Y direction by measuring between the two points.

QUICK DIMENSIONING

The **QDim** command quickly creates a series of baseline or continued dimensions. It can also quickly dimension a series of circles and arcs. From the menu bar, select **Dimension | QDIM**.

Command: **qdim**

Select geometry to dimension: *(Select the objects to dimension.)*

Select geometry to dimension: *(Press* ENTER.*)*

Specify dimension line position, or
  [Continuous/Staggered/Baseline/Ordinate/Radius/Diameter/
  datumPoint/Edit] <Continuous>: *(Select an option.)*

As you move the cursor, notice how AutoCAD ghosts the dimensions it plans to place. You can press the keys representing options, such as C for Continuous and S for Staggered. As you press the keys, AutoCAD displays the effect of the option.

Specify dimension line position, or

[Continuous/Staggered/Baseline/Ordinate/Radius/Diameter/datum
  Point/Edit] <Continuous>: *(Pick a point.)*

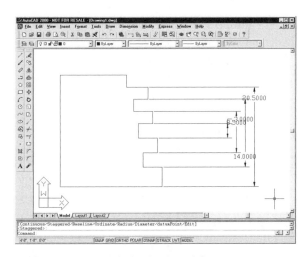

Figure 12.26 *Quick Dimensioning with the Staggered Option*

The options create different kinds of continuous dimension. As you can see, this one command replicates the features of all other dimension drawing and editing commands.

Baseline

Creates a series of baseline dimensions. The dimensions are horizontal or vertical.

Continuous

Creates a series of continued dimensions.

Staggered

Creates a series of staggered dimensions.

Diameter

Creates a series of diameter dimensions.

Radius

Creates a series of radius dimensions.

Ordinate

Creates a series of ordinate dimensions.

datumPoint

Sets a new datum point for baseline and ordinate dimensions.

Edit

Adds and removes dimensioning points. It displays the following prompt:

Indicate dimension point to remove, or [Add/eXit] <eXit>:

AutoCAD displays a small x at every location where it plans to place an extension line. You can add or remove the x markers. In **Remove mode,** choose an x to remove it; similarly, in **Add mode**, choose a point to add an x marker.

DIMENSION EDITING COMMANDS

Dimension editing commands are used to alter a dimension after it has been placed in your drawing. AutoCAD's dimension editing commands are:

DimEdit

Restores dimension text to its original position, changes existing dimension text, and forces extension lines to a specified angle.

DimTEdit

Changes placement and orientation of dimension text, and rotates dimension text.

Let's explore each function.

RESTORING DIMENSION TEXT POSITION

If the dimension has been moved to a new location, it can be restored to its default position with the **DimEdit** command's **Home** option. Select **Dimension | Align Text | Home** from the menu bar.

Command: **dimedit**

Enter type of dimension editing [Home/New/Rotate/Oblique]
    <Home>: **H**

Select objects: *(Select one or more dimensions.)*

I found Select objects: *(Press* ENTER.*)*

CHANGING DIMENSION TEXT

In the normal course of dimensioning, some dimension values or dimension text are changed. Instead of erasing the dimension and replacing it with a new dimension containing the desired value or text, you can use the **DimEdit** command's **New** option

Command: **dimedit**

Enter type of dimension editing [Home/New/Rotate/Oblique]
    <Home>: **n**

AutoCAD displays the **Multiline Text Editor** dialog box. Make changes and choose **OK**.

Select objects: *(Select one or more dimensions.)*

I found Select objects: *(Press* ENTER.*)*

If you press ENTER when prompted for the new dimension text, type <> in the mtext dialog box.

As an alternative, you can select **Modify | Text** from the menu bar, then select the dimension text. AutoCAD displays the **Multiline Text Editor** dialog box.

OBLIQUING EXTENSION LINES

In some situations, the standard orientation of dimension extension lines can conflict with other dimensions. You can remedy the conflict by obliquing the extension lines. Figure 12.27 shows an object with oblique extension lines.

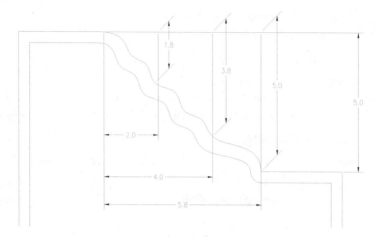

Figure 12.27 *Oblique Dimensioning*

Extension lines can be obliqued through the DimEdit command's **Oblique** option. Select **Dimension | Oblique** from the menu bar.

> Command: **dimedit**
>
> Enter type of dimension editing [Home/New/Rotate/Oblique]
> <Home>: **o**
>
> Select objects: *(Select one or more dimensions.)*
>
> 1 found Select objects: *(Press* ENTER.*)*
>
> Enter obliquing angle *(press* ENTER *for none)*: **45**

Note that the dimension must first exist before the **Oblique** command can be used.

RELOCATING DIMENSION TEXT

Some situations require that you move the dimension text. For example, dimensions that cross sometimes have overlapping dimension text. You can move the dimension text by using the **DimTEdit** (short for dimension text edit) command. When you select **Dimension | Align Text** from the menu bar, AutoCAD displays a submenu with five choices: **Left**, **Center**, **Right**, **Home**, and **Angled**.

As an alternative, type the **DimTEdit** command, as follows:

> Command: **Dimtedit**
>
> Select dimension: *(Select a single dimension.)*
>
> Specify new location for dimension text or
> [Left/Right/Center/Home/Angle]:

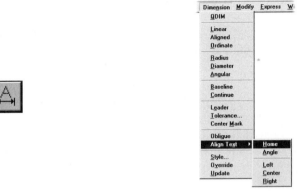

After selecting the dimension, moving the cursor dynamically moves the dimension text. Move the text to the desired location, click, and the text is repositioned. Let's look at how each of the options work.

Figure 12.28 *Move Dimension Text Left*

Left

Positions the dimension text to the left side of the dimension line.

Right

The reverse of **Left**, positions the dimension text to the right side of the dimension line.

Figure 12.29 *Move Dimension Text Right*

Center

Moves the text to the center of the dimension line. In many cases, this restores the text to its default position.

Angle

Allows you to set the angle for the dimension text. You can set the angle in two ways. Entering an actual angle from the keyboard will cause the text to rotate to that angle. The default angle, with text reading from left to right, is 0. You can also enter two points to show AutoCAD the angle to which you wish to rotate the text.

Figure 12.30 *Changing Dimension Text Angle to 45 Degrees*

ROTATING DIMENSION TEXT

The **DimEdit** dimension command has an option that allows you to rotate dimension text to a specified angle. The **Rotate** option functions similarly to the **Angle** option of the **DimTEdit** command. The difference is that {**mEdit** is used to rotate the dimension text of several dimensions. (The command is not available from the menu bar.)

> Command: **dimedit**
>
> Enter type of dimension editing [Home/New/Rotate/Oblique] <Home>: **r**
>
> Specify angle for dimension text: **45**
>
> Select objects: *(Select one or more dimensions.)*
>
> I found Select objects: *(Press* ENTER.*)*

You can either enter the actual angle or show two points to specify the angle. The "Select objects:" prompt allows you to select several dimensions in one object selection process.

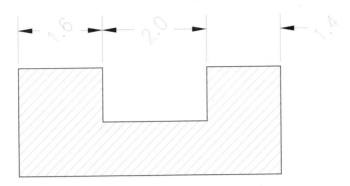

Figure 12.31 *Dimensions with Text Rotated by 45 Degrees*

Note: Use **Dimedit** to change all the dimensions in your drawing by selecting the entire drawing with a window.

DIMENSION UTILITY COMMANDS

When you type **Dim** at the command prompt, the "Command:" prompt changes to "Dim:". This is called the dimension mode. Type **Exit** to get out of the dimension mode and return to the "Command:" prompt:

> Command: **dim**
>
> Dim: **exit**
>
> Command:

The dimension utility commands are used within the dimension mode for special purposes. Let's look at each of the utility commands.

DISPLAYING THE DIMENSION STATUS

When you enter **Status** while in the dimension mode, AutoCAD displays the current dimension variable settings in the Text window. Press 0 to return to the drawing screen.

> Dim: **status**

```
AutoCAD Text Window - Drawing1.dwg
Edit

Command: dim

Dim: status

DIMASO      Off             Create dimension objects
DIMSTYLE    Standard        Current dimension style (read-only)

DIMADEC     0               Angular decimal places
DIMALT      Off             Alternate units selected
DIMALTD     2               Alternate unit decimal places
DIMALTF     25.4000         Alternate unit scale factor
DIMALTRND   0.0000          Alternate units rounding value
DIMALTTD    2               Alternate tolerance decimal places
DIMALTTZ    0               Alternate tolerance zero suppression
DIMALTU     2               Alternate units
DIMALTZ     0               Alternate unit zero suppression
DIMAPOST                    Prefix and suffix for alternate text
DIMASZ      0.1800          Arrow size
DIMATFIT    3               Arrow and text fit
DIMAUNIT    0               Angular unit format
DIMAZIN     0               Angular zero supression
DIMBLK      ClosedFilled    Arrow block name
DIMBLK1     ClosedFilled    First arrow block name

Dim:
```

BACKGROUND TO AUTOCAD DIMENSIONS

AutoCAD uses some special objects to help it draw dimensions. These include text styles, defpoints, and blocks.

DIMENSION TEXT STYLE

By default, the dimension text is drawn using the current text style. That's because the **DimText** system variable is set to the default text style, called "Standard." You cannot change the dimension text style using the **Style** command, as you do for regular text. Instead, you must change the value of **DimText**.

DEFINITION POINTS

When you create dimensions, AutoCAD places *definition points* (or defpoints, for short) in the drawing. These points are placed at the object end of extension lines. If you look closely at a dimensioned drawing, you can see a tiny dot. The points are placed on a layer named DEFPOINTS. This layer never plots, so never draw on this layer!

AutoCAD uses defpoints as a reference for some commands, such as the **Stretch** command. When stretching an dimensioned object, be sure to include the dimensions in the selection set.

ARROW BLOCKS

Many disciplines require specific symbols at the end of the dimension line. AutoCAD provides arrows, tick marks, solid dots, and other shapes for your use. Your discipline, however, may specify another type of symbols.

You can specify *blocks* to be used in place of the dimensioning arrows or ticks. The block reference must already exist in the drawing. (Blocks are discussed in detail in Chapter 18 "Learning Intermediate Draw Commands.")

The following rules should be used when you prepare a block for use as an arrowhead:

- The block should be prepared as the *right* arrow of a horizontal dimension line.

- The *insertion point* (base point) should be placed at the point that would normally be the *tip* of the arrow.

- The dimension line stops a distance from the tip of the arrow. Draw a short tail line to the left so it will connect to the dimension line.

- For the arrow block to scale properly, draw the block exactly *one drawing unit* from the tip to the end of the tail line.

To select the block for use in dimensioning, use the **Ddim** command's **Modify** button to select the **Lines and Arrows** tab, then select the **Arrowheads 1st** list box, and select **User Arrow**. Enter the name of the block. When you place a dimension, your arrowhead block will be used in place of arrows or ticks.

Figure 12.32 *Dialog Box for Setting Arrowhead Properties*

To create a separate arrow block at each end of a dimension line, select **User Arrow** for the **1st** and **2nd User Arrow**. Next, enter the names of the first and second arrow blocks in the edit box.

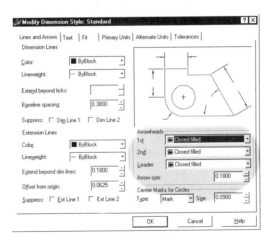

Figure 12.33 *Creating Separate Arrowheads*

EXERCISES

1. Draw the following objects and place the dimensions at the locations shown.

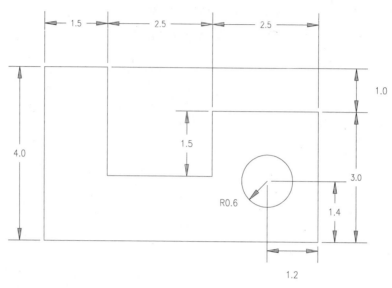

Figure 12.34

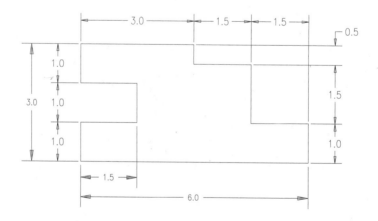

Figure 12.35

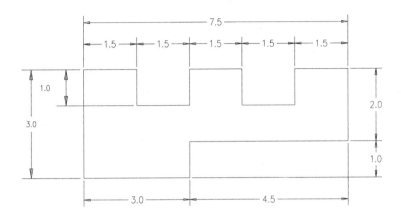

Figure 12.36

2. Draw the following objects and use the dimension commands to place the vertical dimensions at the locations shown.

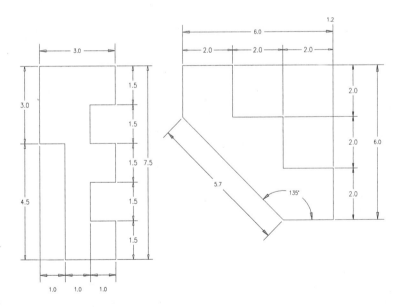

Figure 12.37

3. Construct the following drawing and place all the dimensions as shown.

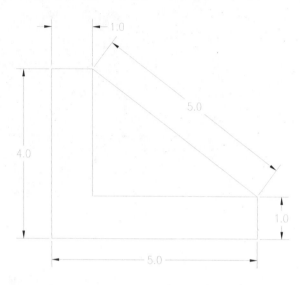

Figure 12.38

4. Construct the following drawing and place all the dimensions as shown.

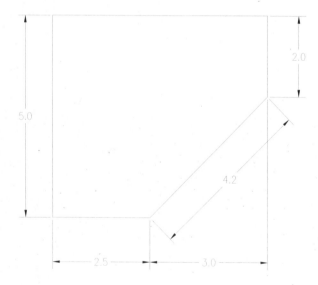

Figure 12.39

5. Construct the following object and place the baseline dimensions.

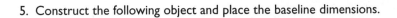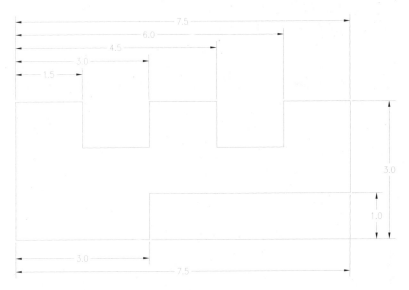

Figure 12.40

6. Let's place an angular dimension. Draw two lines as shown in Figure 12.41. Let's select **Dimension | Angular** from the menu bar. Use the following command sequence and Figure 12.41.

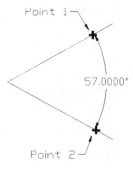

Figure 12.41 *Placing an Angular Dimension*

Command: **dimangular**

Select arc, circle, line, or press ENTER: *(Select "POINT 1".)*

Second line: *(Select "POINT 2".)*

Dimension arc line location (Mtext/Text/Angle): *(Move the dimension arc into place and click.)*

Dimension text = 57 *(Press* ENTER*)*

7. Draw four circles on the screen. From the menu bar, select **Dimension | Diameter**. When AutoCAD prompts for the arc or circle, select any part on the first circle. Accept the default dimension.

8. Draw the following object and use radius dimensioning to place the dimensions shown.

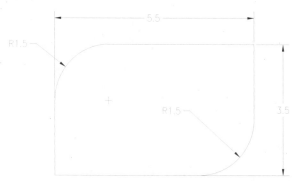

Figure 12.42

9. Draw a square approximately one-third the size of the screen. From the **Dimension** menu, select **Linear**.

Command: **dimlinear**

First extension origin or press ENTER to select:

Use object snap INTersection to capture the lower left corner of the box you drew. The command sequence will continue.

Second extension line origin:

Use object snap intersection again to capture the lower right corner. The command sequence will respond:

Dimension line location
(Mtext/Text/Angle/Horizontal/Vertical/Rotated):

Enter a point on the screen approximately 1/4-inch actual distance from the bottom line of the square. You have just placed a dimension.

10. Repeat the same procedure for the upper line of the square. Instead of specifying the corners, however, just press ENTER when prompted and select the upper line to dimension. The command sequence will appear as follows:

Command: **dimlinear**

First extension line origin or press ENTER to select: *(Press* ENTER.*)*

Select object to dimension: *(Select the line.)*

Dimension line location
(Mtext/Text/Angle/Horizontal/Vertical/Rotated): **M**

(The **Multiline Text Editor** dialog box appears. Type in your name. The "< >" characters represent the default dimension measurement. Delete them if you do not want them to show. Choose the **OK** button when finished.)

Dimension line location
(Mtext/Text/Angle/Horizontal/Vertical/Rotated): *(Enter a point where you want the dimension line.)*

Notice how AutoCAD measured the selected line length, placing extension lines in the proper positions. For the dimension text, it entered your name. Notice that you can accept the default measurement, enter your own numerical value, or enter a text string (such as your name).

11. Select **Linear** from the **Dimension** menu and dimension one of the vertical sides of the box you drew. The vertical dimension works the same as the horizontal dimension.

12. Draw a triangle. Let's use the **DimAligned** command to dimension one of the angled sides. Refer to the following command sequence.

Command: **dimaligned**

First extension line origin or press ENTER to select: *(Press* ENTER.*)*

Select object to dimension: *(Select an angled side.)*

Dimension line location (Mtext/Text/Angle): *(Select a point.)*

AutoCAD draws a dimension line parallel to the selected side.

13. Clear the screen and draw one horizontal dimension line. Accept the default dimension measurement. Type **DimEdit** and select the **New** option. Delete the "< >" characters in the editor and enter 250. Choose **OK**. AutoCAD will ask you to "Select objects:". Select the dimension text of the dimension line you drew. The text will update to "250."

14. Select **Dimension | Align Text**. Then select **Angle**. Select the dimension line. When prompted for the angle, enter 45 for 45 degrees. Did the text turn to a 45-degree angle?

15. Repeat but select **Left** from the **Dimension | Align Text** menu.

16. Now select **Home** from the **Dimension | Align Text** menu. The dimension text should return to the horizontal position.

EXERCISE

17. Use the dimension commands to dimension the following objects, which are on the CD-ROM.

Figure 12.43 *File 15-42.Dwg*

Figure 12.44 *File 15-43.Dwg*

Figure 12.45 *File 15-44.Dwg*

Figure 12.46 *File 15-45.Dwg*

Figure 12.47 *File 15-46.Dwg*

CHAPTER REVIEW

1. What is the dimension line?

2. What is the difference between the extension line and the witness line? Draw an example of a baseline dimension.

3. Name the types of linear dimensions that the **DimLinear** command draws.

4. What does the **DimDiameter** dimension command do?

5. What command would you use to tilt the dimension text?

6. How would you apply new changes to an existing dimension?

7. What layer are the dimension definition points placed on?

8. What are "arrow blocks"?

9. What is the purpose of the **QDim** command?

TUTORIAL

1. Open the Dimen.Dwg drawing from the CD-ROM. You will dimension this drawing using several of the dimensioning commands you have learned. The finished drawing will look like Figure 12.48.

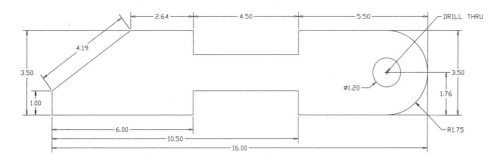

Figure 12.48 *Dimen.Dwg Drawing*

Let's start with linear dimensioning.

Linear Dimensioning

2. From the menu bar, choose **Dimension**.

3. From the **Dimension** menu, choose **Linear**. Respond to the "First extension line origin:" prompt by picking a point using INT object snap at the corner shown as point 1 in Figure 12.49.

4. Respond to the "Second extension line origin:" prompt by picking a point using INT object snap at the corner shown as point 2 in Figure 12.49.

5. You should now be prompted with "Dimension line location:". This is the distance above the object that the dimension line will be placed. Enter a point at the location of point 3.

    AutoCAD now has measured the distance and calculated it for you. It is displayed as "Dimension text value." Isn't that easy? From now on, we will simply refer to the points labeled in the illustrations for each entry.

6. Let's continue this dimension line. From the **Dimension** menu, select **Continue.**

    Respond to "Second extension line origin:" with INT object snap at point 4.

7. Repeat this procedure by selecting point 5 as the extension line origin with QUA object snap. (Don't forget to press ENTER twice.)

352

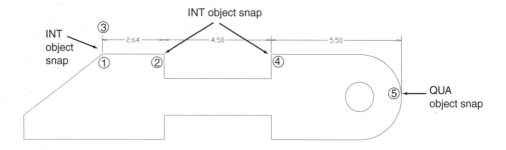

Figure 12.49 *Linear Dimensioning (Continue)*

You have now completed a horizontal and a continuing dimension line.

Baseline Dimensioning

Let's now do some baseline dimensioning.

8. This dimensioning method starts out the same way that continuing dimensioning does. Use INT object snaps. From the **Dimension** menu, choose **Linear**, and then press ENTER so you can select an object to dimension. When prompted to select an object, select the line defined by endpoints 6 and 7 in Figure 12.50. Choose point 8 for the dimension line location.

9. Now, this is where we create the baseline dimensioning style. Choose **Baseline** from the **Dimension** menu and enter point 9 as the second extension line origin. Press ENTER twice.

10. Repeat the procedure by choosing **Baseline** again, entering point 10 as the second extension line origin with QUA object snap.

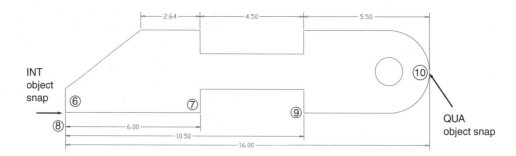

Figure 12.50 *Linear Dimensioning (Baseline)*

Notice that AutoCAD offsets each dimension line the proper distance.

Vertical Dimensioning

Now let's draw the vertical dimensions.

11. Choose **Linear** from the **Dimension** menu. As shown by Figure 12.51, choose point 11 as the first extension line origin, point 12 as the second extension origin and point 13 as the dimension line origin.

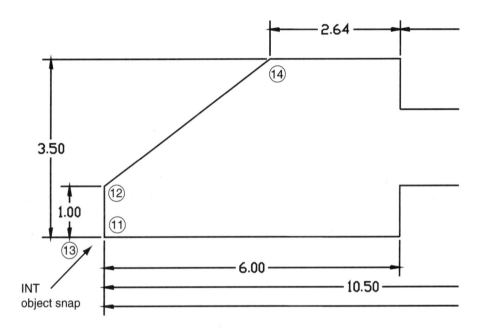

Figure 12.51 *Linear Dimensioning (Vertical)*

12. From the **Dimension** menu, choose **Baseline**, and enter point 14 as the second extension line origin.

13. Repeat the procedure for the opposite end, using points 15 through 18 (Figure 12.52) as the entered points. You can do this one on your own.

    If you mess up, use the **Undo** command to eliminate the last dimension, or cancel if the dimension line is not yet drawn.

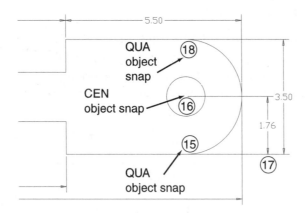

Figure 12.52 *Completing the Vertical Dimension*

Aligned Dimensioning

Let's construct the dimension line that describes the angle at the left side of the part.

14. Choose **Aligned** from the **Dimension** menu, then press ENTER. Select the angled line on the object. When prompted for the dimension line location, select point 19 of Figure 12.53.

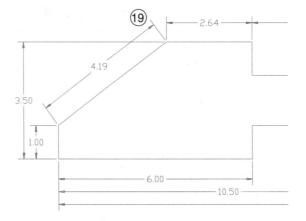

Figure 12.53 *Linear Dimensioning (Aligned)*

AutoCAD constructed the dimension line in alignment with the two extension line origin points.

Diameter Dimensioning

15. Select **Diameter** from the **Dimension** menu. You are prompted "Select arc or circle:".

16. Enter a point on the circle (point 20 of Figure 12.54) to identify the object you want dimensioned.

17. Press ENTER to accept AutoCAD's measurement. Show point 21 as the distance for the leader line.

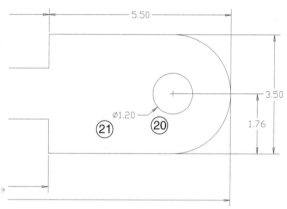

Figure 12.54 *Diameter Dimensioning*

Notice that AutoCAD used the point you entered on the circle as one of the endpoints of the dimension line. The other endpoint is positioned by the dimension line drawn from this point through the center of the circle to the opposite circumference. The extended text position is referenced beside the first point entered.

Radial Dimensioning

Now, let's dimension the arc that closes the right end of the object.

18. Select **Radius** from the **Dimension** menu.

19. When prompted, choose the point on the arc that you want the dimension line to intersect (point 22 of Figure 12.55). Move the crosshairs to place the dimension leader.

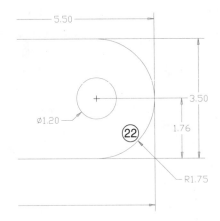

Figure 12.55 *Radius Dimensioning*

Leader Construction

20. Let's construct a leader line and some text. Select **Leader** from the **Dimension** menu.

21. Enter point 23 (see Figure 12.56) as the leader start and point 24 as the "to" point.

22. For the dimension text, let's put in some words (remember, we can use words instead of numerals for any dimension text). Press ENTER to stop drawing the leader.

23. Type in "DRILL THRU" and press ENTER twice.

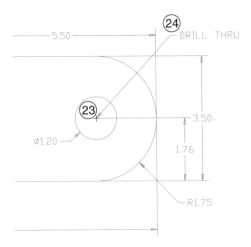

Figure 12.56 *Constructing a Leader Line with Text*

Your dimensioned drawing should look like the one in Figure 12.57.

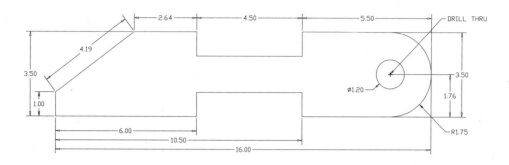

Figure 12.57 *Completed Dimensioned Drawing*

CHAPTER 13

Setting Dimension Styles and Variables

AutoCAD is rich in dimensioning capabilities. To obtain professional results, you must understand how to set the many dimension variables that control the look of the dimension. After completing this chapter, you will be able to

- Manipulate different types of variables used to control the look and function of AutoCAD's dimensions

- Control the dialog boxes to set the many variables

INTRODUCTION

How dimensions look and act depends on the settings of *dimension variables* (called "dimvars," for short). The variables control characteristics such as whether the dimension text is placed within or above the dimension line, whether arrows or ticks are used, site of the dimension and many other options.

After you set the variables for a dimension, you store the settings in a *style*. Creating dimension styles (or "dimstyles," for short) is similar to creating text styles. In both cases, you can specify a few settings, then save the settings with a name. Just as every AutoCAD drawing has a text style named STANDARD, every drawing also has a dimension style called STANDARD.

To effectively use AutoCAD's dimensioning capabilities, you must become familiar with the dimension variables. The following sections show you what options are available and how to set and store them.

THE DIMENSION STYLE MANAGER

AutoCAD's dimension variables are set through a dialog box. To see the **Dimension Style Manager** dialog box, type **DDim** at the command line, or select **Style** from the **Dimension** menu.

The **Dimension Style Manager** consists of an initial dialog box, which leads to several other dialog boxes. These dialog boxes control almost all dimension variables, which in turn affect the look of dimensions in your drawings. After you customize the look of the dimensions, you save that look by giving it a name, called a dimension style. You can use many dimstyles, each with its own name, in a drawing. This means you can create many customized sets of dimensions—one for each client and project, if need be.

There are two dimvars not set by **DDim**. These are **DimSho** (toggles whether dimensions are updated while dragged) and **DimAso** (determines whether dimensions are drawn as associative or non-associative). By default, both have a value of 1 (are turned on). You can change the value of these two dimvars at the "Command:" prompt, but there isn't any good reason to do so, since both dimvars exist only for compatibility with old versions of AutoCAD.

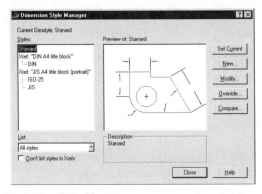

Figure 13.1 *The Dimension Style Manager*

The dialog box is called a "manager" because it tells you about the dimension styles available in the drawing. The preview window provides a graphical overview of the current dimstyle. The buttons lead to other dialog boxes, as described in the following sections. (The Release 14 concept of a dimstyle *family* has been removed from AutoCAD 2000. It has been replaced by the **Use for** drop-down list, which is found in the **Create New Dimension Style** dialog box.)

SET THE CURRENT DIMSTYLE

The **Set Current** button sets a dimension style as "active." Select a dimstyle name from the list under **Styles** and then choose the **Set Current** button. The current dimension style name is stored in dimvar **DimStyle**.

Every new drawing contains at least one dimension style named "STANDARD." For this reason, you might see only a single dimstyle name the first time you use the **DDim** command.

There are several ways to obtain additional dimension styles. You can get dimstyles by creating your own, opening a template drawing, using the **XBind** command on an externally-referenced drawing, copying dimstyles through the AutoCAD DesignCenter, or importing a dimstyle with **DimIm** command (found in the **Express | Dimension | Dimstyle Import** menu item).

Be careful, though, since it is possible to overwrite existing dimstyle values with different values. You cannot set as current an xrefed dimstyle.

Create Your Own Dimstyle

To create your own dimension style, choose the **New** button, as described below.

Dimstyles in Template Drawings

When you create a new drawing based on a template drawing, then it contains a specific dimstyle. Template drawings are found in the **\Acad 2000\Template** folder and have the DWT file extension. For example, the **ISO AO title block.dwt** template file contains a dimstyle named "ISO" that conforms to the dimensioning standards of the International Organization for Standardization.

Other template drawings contain additional dimension styles, as shown in the following table:

DWT Prefix	Dimension Style	Comments
Acad.dwt	Standard	Default template drawing.
ANSI*.dwt	Standard	American National Standards Institute
DIN*.dwt	DIN	German Industrial Standards
ISO*.dwt	ISO-25	International Organization for Standardization
JIS*.dwt	JIS	Japanese Industrial Standards

Table 13.1

Binding Xrefed Dimstyles

As you will learn in Chapter 25 "Attaching External Drawings," you can view (but not edit) other drawings at the same time as the current drawing. This is done with the **XRef** command. Once a drawing has been *attached* with the **XRef** command, you can access its dimension styles.

Notice the **Don't list styles in Xrefs** check box below the **List** drop-down list. When this check box is off, you can see the dimstyle names of xrefed drawings.

You add the xrefed drawing's dimstyles to the current drawing with the **XBind** command. This command, in effect, copies the dimstyles from the xrefed drawing into the current drawing.

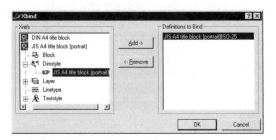

Figure 13.2 *The Xbind Dialog Box Copies Dimstyles*

Adding Dimstyles with AutoCAD DesignCenter

You can use the AutoCAD DesignCenter to copy dimension styles from one drawing to another. (You learn more about DesignCenter in Chapter 26 "AutoCAD DesignCenter.") Here are the steps you take:

> Open both drawings. Make current the drawing you wish to copy the dimstyles *to*.
>
> Select **Tools | AutoCAD DesignCenter** from the menu bar.
>
> In the tree view, open the Dimstyles of the drawing you wish to copy the dimstyles *from*.
>
> Right-click the dimstyle name and select **Add Dimstyle(s)** from the cursor menu.

Importing Dimstyles

If the menu bar of your AutoCAD 2000 contains the Express item, then you can import and export dimension styles. This is useful if you do not have easy access to the drawing containing the dimension style you require.

Before you can import a dimstyle, it must first have been exported from another drawing with the **DimEx** command (found in the **Express | Dimension | Dimstyle Export** menu item). This command exports the dimension styles to a DIM file. You can receive this file by email or diskette – it is very small, typically 2 KB.

Figure 13.3 *The Dimension Style Export Dialog Box*

Once the dimstyle has been exported, you can import it with the **DimIm** command (found in the **Express | Dimension | Dimstyle Import** menu item).

Figure 13.4 *The Dimension Style Import Dialog Box*

CREATE A NEW DIMSTYLE

The **New** button leads to the **Create New Dimension Style** dialog box. Creating a new style is as easy as typing a name and choosing **OK**. There are, however, some nuances you should be aware of.

Figure 13.5 *The Create New Dimension Style Dialog Box*

The standard way to create a new dimension style is to make a copy of an existing style, make changes to some dimension variables, and then save the result. For this reason, AutoCAD displays "Copy of Standard" in the **New Style Name** text field.

The **Start With** text field lets you select which dimstyle to make a copy of— provided the drawing contains two or more dimstyles. This allows you to start with an existing dimstyle (such as the DIM, JIS, and ISO styles stored in the template drawings) and modify it to your needs.

The **Use for** drop-down list lets you apply changes to a limited group of dimensions:

- All dimensions
- Linear dimensions
- Angular dimensions
- Radius dimensions
- Diameter dimensions
- Ordinate dimensions
- Leaders and tolerance

(The **Use for** list replaces the concept of *family* of dimensions in Release 14.)

To create a new dimension style:

1. Select an existing dimension style from the **Start With** drop-down list. Most likely, you have no choice: most drawings contain a single dimension style.

2. Decide if the changes apply to all dimensions or just a subgroup. Select the group from the **Use for** drop-down list.

3. Type a name for the style in **New Style Name** field. The name can be any descriptive name up to 255 characters long.

4. Choose **Continue**. Notice that the new style name appears in the **Styles** list.

5. Now choose the **Modify** button to make the changes to the dimension style.

RENAME AND DELETE DIMSTYLES

It is not immediately apparent how to rename or delete a dimension style in the **Dimension Style Manager**. Here's how:

1. Select a dimension style name in the **Styles** list.

2. Right-click to display the cursor menu.

3. Select **Rename** to rename the dimstyle; select **Delete** to erase the dimstyle from the drawing.

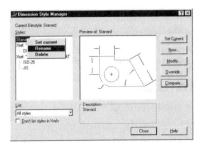

Figure 13.6 *Renaming and Deleting Dimension Styles*

The "Standard" dimstyle can be renamed but it cannot be deleted. Xrefed dimstyles cannot be renamed or deleted.

MODIFYING A DIMSTYLE

The **Modify** button leads to the **Modify Dimension Style** dialog box, which contains several tab views listing most dimvars and their options. Here you modify the values of dimvars to create new dimstyles. These options are discussed in detail later in this chapter.

Xrefed dimstyles cannot be modified or overridden.

OVERRIDING A DIMSTYLE

The **Override** button leads to the **Override New Dimension Style** dialog box, which looks identical to the **Modify** dialog box (mentioned above) but is meant to allow you to override the settings of a current dimstyle. These options are discussed in detail later in this chapter.

COMPARE TWO DIMSTYLES

The **Compare** button leads to the **Compare New Dimension Style** dialog box, which lets you see the differences in dimvar values between two dimstyles.

Figure 13.7 *The Create New Dimension Style Dialog Box*

The table shows the differences in dimension variables between any two styles. Choose the button at the right to copy the list to the Windows Clipboard. (The data is in tab-delimited ASCII format.)

DIMENSION VARIABLES

The **Modify** and **Override** buttons lead you to a group of tabbed dialog boxes that allow you to change the settings of dimension variables. Changing a dimvar changes the look of the dimension. Be forewarned: dimvars are AutoCAD's most confusing aspect, since there are so many dimvars.

Lines and Arrows

The **Lines and Arrows** tab allows you to select the properties for dimension lines, extension lines, arrowheads, and center marks.

Text

The **Text** tab allows you to specify the format, placement, and alignment of dimension text.

Fit

The **Fit** tab allows you to determine the placement of dimension text, arrowheads, leader lines, and dimension line.

Primary Units

The **Primary Units** tab allows you to set the format of primary dimension units, as well as set the prefix and suffix of dimension text.

Alternate Units

The **Alternate Units** tab allows you to choose the format of units, angles, dimension, and scale of alternate measurement units.

Tolerances

The **Tolerances** tab allows you to control the format of dimension text tolerances. This tab is not for formatting Tolerance dimensions.

Each dialog box contains a preview window that gives you a rough idea of the effect of your changes on the look of the dimensions. Let's now go on a tour of each tab.

LINES AND ARROWS

The **Lines and Arrows** tab allows you to select the properties for dimension lines, extension lines, arrowheads, and center marks. (Note for Release 14 users: the overall scale factor has been moved to the **Fit** tab.)

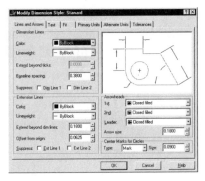

Figure 13.8 *The Lines and Arrows Tab*

Dimension Lines

The **Dimension Lines** area of the dialog box lets you change the color of the dimension line from its default color of white (or black, when the drawing background is white). Choose the drop-down list and select a standard color. Choose **Other** to display the standard Select Color dialog box. The color is stored in the dimension variable named **DimClrd** (short for "DIMension CoLoR Dimension").

To see the effect of changing a variable in the Preview box, select the color blue. Notice how the dimension line's color changes from black to blue.

To change the lineweight of the dimension line, choose the drop-down list next to **Lineweight.** Select a predefined width. The dimension line lineweight is stored in dimension variable **DimLwd.**

The **Extend beyond ticks** option only applies when tick marks are used in place of arrowheads; this value is the distance that the extension line extends beyond the dimension line (stored in **DimDle**), as shown by Figure 13.9.

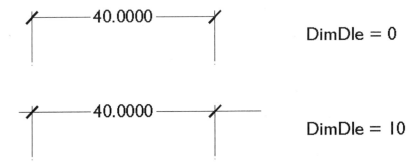

Figure 13.9 *Dimension Line Extended Beyond Extension Lines*

The **Baseline spacing** option determines the distance between dimension lines when they are automatically stacked by the **DimBaseline** and **QDim** commands (see Figure 13.10). The distance is stored in system variable **DimDli**.

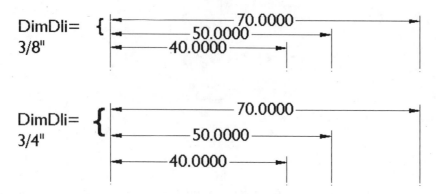

Figure 13.10 *The Distance Between Stacked Dimensions*

AutoCAD can suppress either or both dimension lines. To do so, select the check boxes to suppress the first (the value is stored in system variable **DimSd1**) or second dimension line (stored in system variable **DimSd2**). The effect is shown in Figure 13.11.

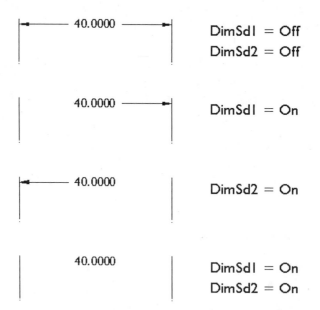

Figure 13.11 *Supressing Dimension Lines*

Extension Lines

As with dimension lines, you can specify the color (**DimClre**) and lineweight (**DimLwe**) of the extension lines.

The **Extend beyond dim lines** value is the distance the extension line protrudes beyond the dimension line (stored in **DimExe**). See Figure 13.12.

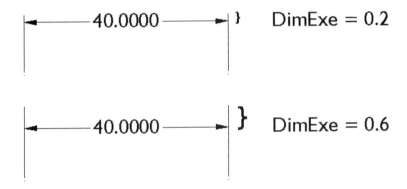

Figure 13.12 *Extending the Extension Line Beyond the Dimension Line*

The **Offset from origin** value is the distance the extension line begins away from the pick point or the object being dimensioned (**DimExo**). See Figure 13.13.

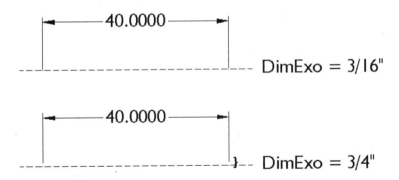

Figure 13.13 *Distance Between Pick Point and Extension Line*

You can opt to suppress the first (**DimSe1**) or second (**DimSe2**) or both extension lines (see Figure 13.14.)

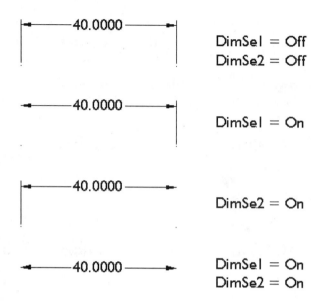

Figure 13.14 *Extension Line Suppression*

Arrowheads

You can select from 20 predefined arrowhead styles (stored in **DimBlk**), or select your own arrowhead that you have previously customized. You have the option of specifying a different arrowhead at each end of the dimension line (**DimBlk1** and **DimBlk2**) and for the end of a leader dimension (**DimLdrBlk**). When you use a different arrowhead at each end, **DimSah** is set to 1.

In addition to the standard arrowhead, AutoCAD includes arrowheads (some are shown in Figure 13.15) such as the tick, the dot, open arrowhead, open dot, right-angle arrowhead, or no head at all.

None	Architectural Tick	Dot Small
Closed	Open	Boxed Filled
Dot	Origin Indication	Box
Closed Filled	Right-Angle	Datum Triangle Filled
Closed Blank	Open 30	Datum Triangle
Oblique	Dot Blanked	Integral

Table 13.2

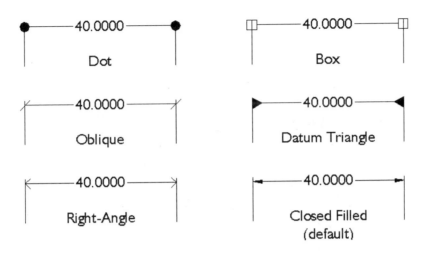

Figure 13.15 *Some of the Arrowheads Provided with AutoCAD*

The **User Arrow** (customized arrowhead) is defined by the creation of an object scaled to unit size and then saved as a named block. The name of the arrowhead block is stored in system variable **DimBlk**. See Figure 13.16 for an example of customized arrowheads.

Figure 13.16 *User-Defined Arrowheads*

The **Arrow** size is the distance from left end to right end; for custom arrowheads, AutoCAD scales the unit block to size (0.18 units, by default). See Figure 13.17. The value is stored in **DimAsz**.

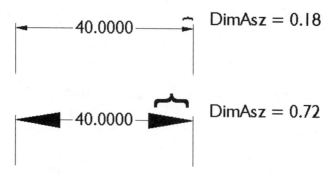

Figure 13.17 *Length of the Arrowhead*

Center Marks for Circles

AutoCAD lets you mark the centers of circles and arcs with the **DimCenter** command, which places a center mark (see Figure 13.18), or a center mark with extending lines, or no mark at all. The value is stored in **DimCen**, as the table below shows:

DimCen Value	Meaning
positive (such as 0.09)	Center mark
negitive (such as −2.5)	Center mark with lines
zero	No mark or lines

Table 13.3

Select the **Type** (None, Mark, or Line) from the drop-down list, then enter the value next to **Size.**

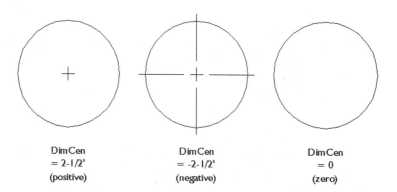

Figure 13.18 *Center Marks and Lines*

Choose the **OK** or **Cancel** button to return to the manager portion of the dialog box.

TEXT

The **Text** tab allows you to specify the format, placement, and alignment of dimension text.

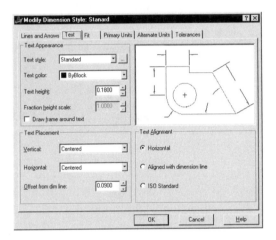

Figure 13.19 *The Text Tab*

Text Appearance

The **Text Appearance** area lets you make the dimension text look different from other text used in the drawing. By default, AutoCAD uses the Standard text style for dimension text. Choose the **[...]** button to display the Text Style dialog box, which lets you create or modify a text style.

You can select a text style name (**DimTxsty**), the color of the text (**DimClrt**), a fixed height (**DimTxt**), and the size of fractional dimension text (**DimTfac**).

To draw a box around the dimension text, select the **Draw** frame around text check box (**DimGap**). The distance is controlled lower in the dialog box, with the **Offset from dim line** option.

Text Placement

The **Vertical drop-down** list box lets you select the vertical placement of dimension text (**DimTad**). See Figure 13.20.

- **Centered:** The dimension text is centered between the extension lines and on the dimension line (the default).

- **Above:** Places text above the dimension line.

- **Outside:** Places text "below" the dimension line (on the side furthest away from the dimension pick points).

- **JIS:** Places text as per the Japanese Industrial Standards for dimensions.

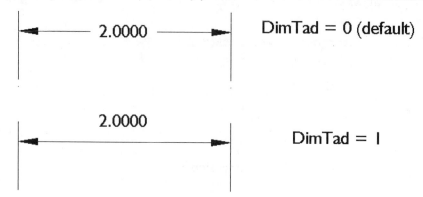

Figure 13.20 *Vertical Justification*

The **Horizontal** drop-down list box lets you select the horizontal placement of dimension text (**DimJust**). See Figure 13.21.

- **Centered:** Dimension text is centered between the extension lines and on the dimension line (the default).

- **At Ext Line 1:** Text is left-justified against the first extension line (usually the left-hand extension line).

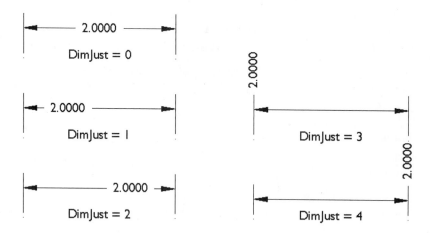

Figure 13.21 *Horizontal Justification*

- **At Ext Line 2:** Text is right-justified against the second extension line.

- **Over Ext Line 1:** Text is positioned vertically over the first extension line.

- **Over Ext Line 2:** Text is positioned vertically over the second extension line.

The **Offset from dim line** specifies the gap between dimension line and text (**DimGap**).

Text Alignment

The **Text Alignment** area lets you force text outside the extension lines.

The **Horizontal** radio button (dimvar **DimTih** and **DimToh**) controls whether text inside or outside extensions is horizontal. The **Aligned with dimension line** button forces text to be aligned with the dimension line.

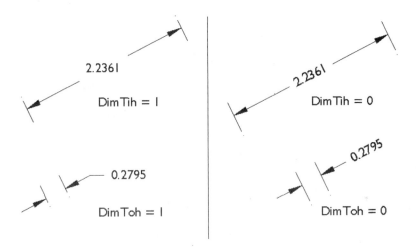

Figure 13.22 *Text Locations*

The third alternative matches the **ISO Standard** for text alignment, which aligns text with the dimension line when text is inside the extension lines, and aligns it horizontally when text is outside the extension lines.

Choose the **OK** or **Cancel** button to return to the dimension style manager.

FIT

The **Fit** tab allows you to determine the placement of dimension text, arrowheads, leader lines, and dimension line.

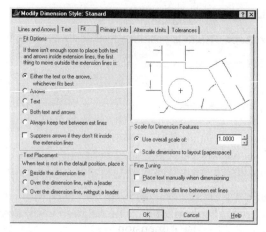

Figure 13.23 *The Fit Tab*

Fit Options

The radio buttons in the **Fit Options** section let you control where elements are placed when the distance between extension lines is too narrow (**DimAtFit**). See Figure 13.24.

- **Either the text or the arrows, whichever fits best:** Dimension elements are placed where there is room to fit.

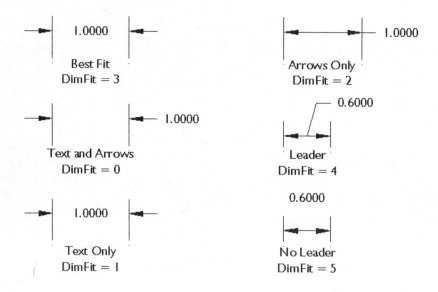

Figure 13.24 *Fitting Text and Arrows*

- **Arrows:** When space is available for arrowheads only, places them between extension lines.

- **Text:** Dimension text is placed between extension lines, while arrowheads are placed outside—when space is lacking.

- **Both text and arrows:** Both text and arrows are forced between the extension lines.

- **Always keep text between ext lines**: Self-explanatory (**DimTix**).

- **Suppress arrows if they don't fit inside the extension lines:** no arrowheads are drawn (**DimSoxd**)

Text Placement

When AutoCAD cannot place the dimension text in its normal position, you have these options for alternate locations (**DimTmove**):

- **Beside the dimension line:** Text is placed beside the dimension line.

- **Over the dimension line, with a leader:** When there isn't enough room for the text, AutoCAD draws a leader line between the dimension line and the text.

- **Over the dimension line, without a leader:** No leader is drawn when text does not fit.

Scale for Dimension Features

This important part of the dialog box controls the overall scale of dimensions when they are placed in the drawing (**DimScale**). It does not change the size of dimensions already in the drawing. Entering a value of **2**, for example, doubles the arrowheads and text.

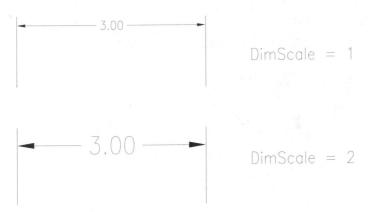

Figure 13.25 *Changing the Scale of Dimension Elements*

Selecting the check box next to **Scale dimensions to layout (paperspace)** scales dimensions to a factor based on the scale between model and paper space.

Fine Tuning

The **Place text manually when dimensioning** check box (formerly **User Defined** in Release 14) is one of the most important. When selected (turned on), all dimension commands prompt you for the position of the dimension text, allowing you to change the text position, if necessary (**DimUpt**).

Select the check box next to **Always draw dim line between ext lines** (formerly **Force Line Inside** in Release 14) to force AutoCAD to always draw the dimension lines inside the extension lines (**DimTofl**). The arrowheads and text are placed outside, if there isn't enough room (see Figure 13.26).

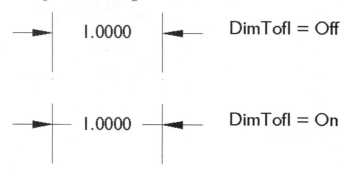

Figure 13.26 *Determining the Location of the Dimension Lines*

Choose the **OK** or **Cancel** button to return to the manager portion of the dialog box.

PRIMARY UNITS

The **Primary Units** tab allows you to set the format of primary dimension units, as well as set the prefix and suffix of dimension text.

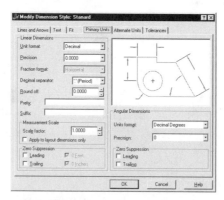

Figure 13.27 *The Primary Units Tab*

Linear Dimensions

Here you can specify the dimensioning units, and whether the dimension text will have a prefix or suffix. This section has features similar to the dialog box displayed by the **Units** command. Whereas the **Units** command controls the units used by AutoCAD for everything in the drawing, the **Primary Units** tab overrides the global units. The values you select here apply only to the dimension text.

Since AutoCAD does not work with units (other than architectural), you may need to specify the units used by dimensions, such as mm. Choose the **Units format** button to select a unit format. You can select from decimal (the default), architectural, engineering, fractional, and scientific style of units (**DimUnit**).

The **Fraction format** option (**DimFrac**) is only available when you select a unit with fractions, such as Architectural. Fractions are stacked horizontally, diagonally, or not stacked.

Decimal separator (**DimDsep**) allows you to select a decimal separator other than period (.) in Decimal units. This setting is typically used in countries outside North America.

Round off lets you specify (**DimRnd**) how decimals are rounded, such as to the nearest 0.5; note that this setting is different from Precision, which truncates decimal places.

Enter any alphanumeric value in the **Prefix** or **Suffix** text boxes. For example, to prefix every dimension with the word Verify, enter that in the **Prefix** box (see Figure 13.28).

Figure 13.28 *Dimension Text with "VERIFY" Prefix*

To suffix every dimension with "(TYPICAL)," enter that in the **Suffix** box (see Figure 13.29). Prefix and suffix text are stored in dimvar **DimPost**.

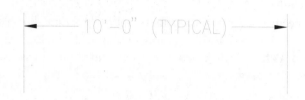

Figure 13.29 *Dimension Text with "(TYPICAL)" Suffix*

Measurement Scale

The **Measurement Scale** area lets you set a scale factor for the value of all linear dimensions. (**DimLfac**). This lets you convert, for example, an imperial drawing to metric units.

The **Apply to layout dimensions only** check box applies the linear scale factor to dimensions created in layouts (paper space). When it is turned on, the scale factor is stored as a negative value in **DimLfac**.

Zero Suppression

You can specify whether leading and trailing zeros and the zero feet or inches are suppressed (**DimZin**).

Angular Dimensions

This section lets you specify the format of angular dimensions, including the **Units** format (**DimAunit**), the **Precision** (**DimAdec**), and **Zero Suppression** (**DimAzin**).

The **Precision** drop-down list specifies the number of decimal places or fraction precision. Select from decimal degrees (DDD.dddd, the default), Degrees/Minutes/Seconds (DD.MMSSdd), grads (DDg), radians (DDr), or surveyors units (NDDE). The value is stored in **DimAUnit**.

Choose the **OK** or **Cancel** button to return to the manager portion of the dialog box.

ALTERNATE UNITS

AutoCAD allows you to place dimensions with double units: primary units, plus a second or alternate units. The alternate units are within square brackets. This is particularly useful for drawings that must show imperial and metric units.

The **Alternate Units** tab allows you to choose the format of units, angles, dimension, and scale of alternate measurement units. Select the **Display alternate units** check box (**DimAlt**) to turn on alternate units. The options are the same as for the **Primary Units** tab.

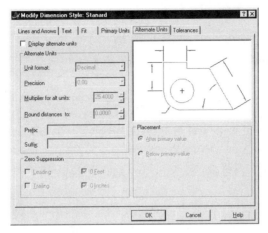

Figure 13.30 *The Alternate Units Tab*

Alternate units are a second set of units display in [square brackets] behind the primary units (a.k.a. dimension text). See Figure 13.31.

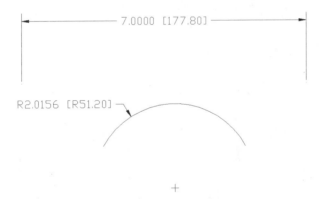

Figure 13.31 *Alternate Units*

Placement

The **Placement** options determine where the alternate units are placed: either after or below the primary value (**DimAPost**).

DimAltd

Alternate unit decimal places.

DimAlttd:

Alternate tolerance unit decimal places.

DimAltRnd

Rounding of alternate units.

DimAlttz

Alternate tolerance units zero suppression.

DimAltu

Alternate units format, except for angular dimensions.

DimAltz

Alternate unit zero suppression

Choose the **OK** or **Cancel** button to return to the manager portion of the dialog box.

TOLERANCES

The **Tolerances** tab allows you to control the format of dimension text tolerances. This tab is not for formatting Tolerance dimensions.

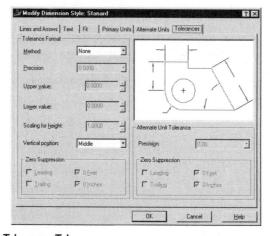

Figure 13.32 *The Tolerances Tab*

Tolerance Format

The **Method** drop-down list allows you to select from five different styles of tolerance text (controlled by **DimTol** and **DimLim**). **DimTol** appends the tolerances to the dimension text. **DimLim** displays dimension text as limits. Turning on **DimLim** turns off **DimTol**.

None

The default is none. No tolerance text is placed.

Symmetrical

The **Symmetrical** option displays a single plus/minus notation (see Figure 13.33).

Figure 13.33 *Dimension Text with Tolerance*

Deviation

The **Deviation** option displays a pair of plus/minus notations (see Figure 13.34)

Figure 13.34 *Dimension Text with Uneven Tolerance*

Limits

The **Limits** option draws a pair of dimension texts (see Figure 13.35).

Figure 13.35 *Dimension Text with Limits Applied*

Basic

The **Basic** option draws a box around the dimension text (see Figure 13.36). The distance between the text and the box is stored in **DimGap**.

Figure 13.36 *Boxed Dimension Text*

In the remainder of the **Tolerance Format** area, you specify the value of the upper (**DimTp**) and lower tolerance values (**DimTm**).

The **Scaling for height** value determines the size of the tolerance text relative to the main dimension text (**DimTfac**). A value of 1 indicated each character of tolerance (or limits) text is the same size as the primary dimension text. A scale of 0.5 reduces the text by half.

The **Vertical position** option determines the relative placement of tolerance text (**DimTolj**): aligned to the top, middle, or bottom of the main dimension text.

Figure 13.37 *The Gap Between Dimension Line and Text*

The **Zero Supression** areas are identical to that found in the **Primary Units** tab.

The **Alternate Unit Tolerance** item (**DimAltTd**) specifies the number of decimal places for the tolerance values in the alternate units of a dimension.

Choose the **OK** or **Cancel** button to return to the manager portion of the dialog box.

CHAPTER 14

Applying Dimensioning Practices

Establishing dimensions practices ensures clear and concise graphic instruction.

After completing this chapter, you will be able to

- Perform standard dimensioning practices
- Use dimensioning practices for 3D mechanical components

DIMENSIONING PRACTICES

The construction of an object can be shown pictorially by drawing methods. In order to construct an object, you must also describe it in terms of its dimensions. The combination of pictorial representation and dimensional information provides complete detail that can be used for construction of the object.

Proper dimensioning provides the necessary distances and notes to completely describe the object. The distances needed to draw the object are not necessarily those required for construction. Because of this, you must carefully select the dimensions you provide in your drawings. To properly dimension an object of any discipline, you should be familiar with the construction techniques used to build the object. It is helpful to mentally construct the object using the process that is common for building that object. Provide the dimensions that you would use in building.

PLACING DIMENSION INFORMATION IN THE DRAWING

There are two methods of placing dimensional information in a drawing:

- A dimension
- A note

Dimensions stipulate distances between two points. The extension lines of the dimension designate the points to which the dimension is applied. The dimension line shows the direction of the dimension, the arrows or ticks the extent of the dimension, and the dimension text conveys the actual distance in numerical terms.

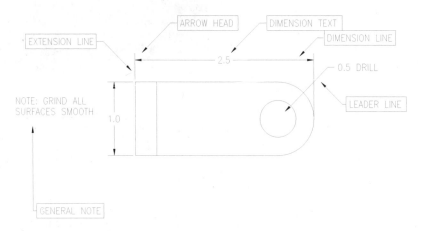

Figure 14.1 *The Components of a Dimension*

Notes indicate explanatory information that cannot be properly conveyed with a dimension. Notes that are specific to a part of the object are indicated with a dimension leader. The arrow of the leader points to the part of the object to which the note applies. General notes drawn without a leader are used to give information that is applied to the drawing as a whole.

CONSTRUCTING DIMENSION COMPONENTS

Many disciplines use drawing as a means of conveying information. The dimensioning practice of each discipline varies. The objective, however, is always the same: provide clear, concise information that can be used to construct the object. Because the information must be conveyed in such a concise manner, the methodology of dimensioning must be carefully considered. Let's look at some of the principles of dimensioning.

Using Dimension Arrows

Arrowheads are used to designate the extent of the dimension line. The length of the arrowhead will vary, depending on the scale of the drawing. Generally, arrowheads are 1/8 inch in length when used in small drawings, and 3/16 inch in larger drawings. Arrowheads that are too large or small are either distracting or difficult to read.

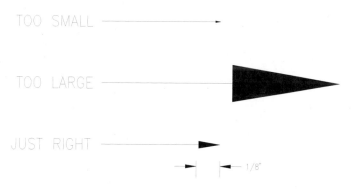

Figure 14.2 *Dimension Arrow-size*

Constructing Extension Lines

Extension lines designate the points to which the dimension refers. The extension lines should not touch the points they reference. The normal offset from the reference points is 1/16 inch. Extension lines should extend approximately 1/8 inch beyond the dimension line.

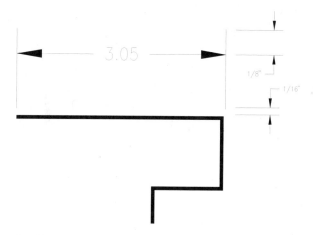

Figure 14.3 *Extension Lines*

Placing Dimension Lines

Extension lines and their dimension lines should be drawn outside the object whenever possible. Dimensions should only be drawn within the object when no other option is available.

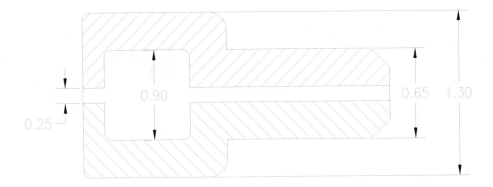

Figure 14.4 *Placing Dimensions*

Constructing Dimension Text

Dimension text is used to provide the actual distance described by the dimension components. Text used in dimensions is generally 1/8 inch for small drawings and 5/32 inch for larger drawings. The location of dimension text is dependent on the discipline. In architectural and structural drawings, the dimension is placed on top of the dimension line. In mechanical drawings, the dimension is usually placed within the dimension line.

When distances are stipulated, the text designating feet and inches is separated by a dash, such as 5'-4". If there are no inches, a zero is used: 6'-0". If dimensions are stipulated strictly in inches, such as 72", the inch mark should be used to avoid confusion.

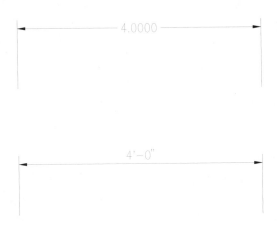

Figure 14.5 *Dimension Text Practices*

Using Fractions

Fractions are given either as common fractions, such as 1/2, 3/4, etc., or as decimal fractions such as 0.50 and 0.75. Normally, inches and common fractions are stipulated without a dash between them. In CAD, however, many text fonts do not have "stacked" fractions. When fractions must be constructed from standard numerical text, you should use a dash to avoid confusion, such as 3-1/2".

DIMENSIONING 3D OBJECTS

Three-dimensional objects are generally dimensioned by stipulated rules. Let's look at some 3D objects and the accepted methodology of dimensioning each.

Dimensioning Wedges

Wedges are dimensioned in two views. The three distances that describe the length, width, and height are dimensioned in the two views.

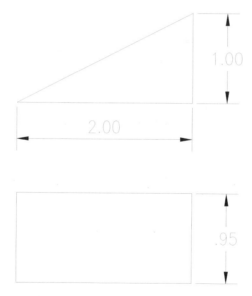

Figure 14.6 *Dimensioning a Wedge*

Dimensioning Cylinders

Cylinders are dimensioned for diameter and height. The diameter is typically dimensioned in the non-circular view. If a drill-through is dimensioned, it is described by a diameter leader.

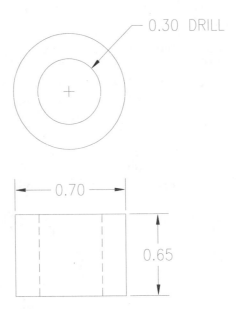

Figure 14.7 *Dimensioning a Cylinder*

Dimensioning Cones

Cones are dimensioned at the diameter and the height. Some conical shapes, such as the truncated cone shown below, require two diameter dimensions.

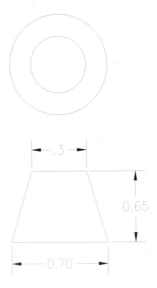

Figure 14.8 *Dimensioning a Cone*

Dimensioning Pyramids

Pyramids are dimensioned in a manner similar to cones.

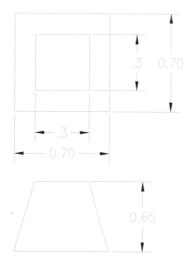

Figure 14.9 *Dimensioning a Pyramid*

Dimensioning Arcs and Curves

An arc is dimensioned as a radius. The dimension line should be placed at an angle, avoiding placement as either horizontal or vertical. The letter R, designating that it is a radius dimension, should prefix the dimension text designating the radius value.

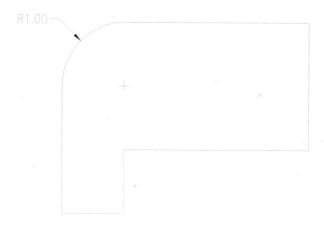

Figure 14.10 *Dimensioning an Arc*

If a circle is dimensioned by its diameter, the dimension text is prefixed by the diameter symbol: Ø4.25.

An object constructed of several arcs is dimensioned in two stages: (1) locate the center of the arcs with horizontal or vertical dimensions and (2) showing their radii with radius dimensioning.

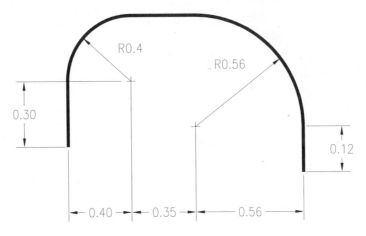

Figure 14.11 *Dimensioning Multiple Arcs*

The dimensions of irregular curves are offset, as shown in the figure.

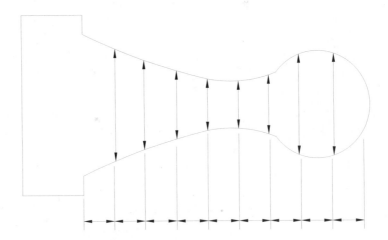

Figure 14.12 *Dimensioning Irregular Curves*

DIMENSIONING MECHANICAL COMPONENTS

The following sections illustrate how to dimension mechanical components. Note that many techniques vary due to individual interpretation.

Dimensioning Chamfers

A chamfer is an angled surface applied to an edge. A leader dimensions chamfers of 45 degrees, with the leader text designating the angle and one (or two) linear distances.

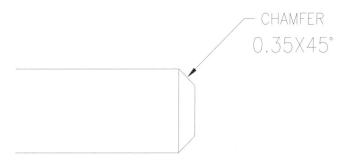

Figure 14.13 *Dimensioning a Chamfer*

If the chamfer is not 45 degrees, dimensions showing the angle and the linear distances describe the part.

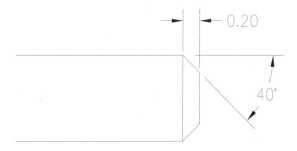

Figure 14.14 *Dimensioning a Chamfer by Angle*

Enlarging Parts for Dimensioning

A portion of a part may be too small to properly dimension. To properly show the dimensions, a segment can be enlarged.

Figure 14.15 *Enlarging a Detail for Dimensioning*

Dimensioning Holes

Holes can be drilled, reamed, bored, punched, or cored. It is preferable to dimension the hole by note, giving the diameter, operation, and (if there is more than one hole) the number. The operation is used to describe such techniques as counter-bored, reamed, and countersunk.

Standards dictate that drill sizes be designated as decimal fractions.

Whenever possible, point the dimension leader to the hole in the circular view.

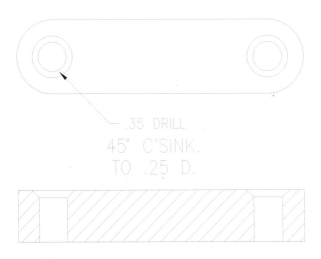

Figure 14.16 *Dimensioning a Hole*

Holes that are made up of several diameters can be dimensioned in their section.

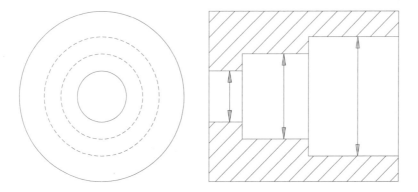

Figure 14.17 *Dimensioning Holes of Several Diameters*

Dimensioning Tapers

A taper can be described as the surface of a cone frustum. Dimension tapers by giving the diameters of both ends and the rate of taper, given as the distance of taper per foot.

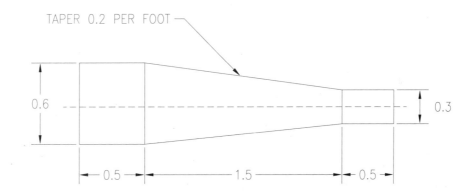

Figure 14.18 *Dimensioning a Taper*

EXERCISES

Dimension the following drawings.

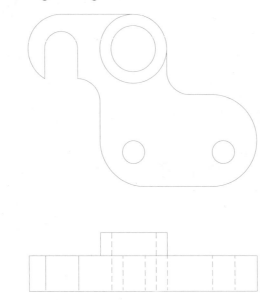

Figure 14.19

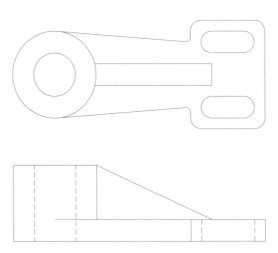

Figure 14.20

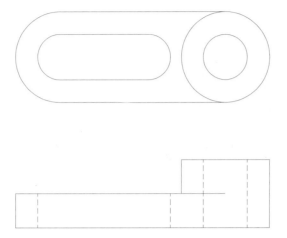

Figure 14.21

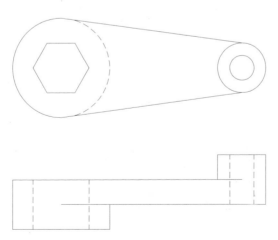

Figure 14.22

Figure 14.23

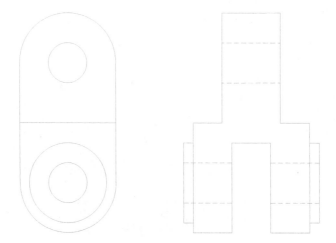

Figure 14.24

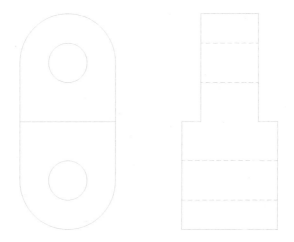

Figure 14.25

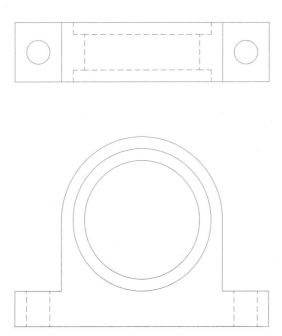

Figure 14.26

Figure 14.27

Figure 14.28

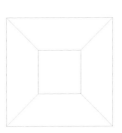

Figure 14.29

Figure 14.30

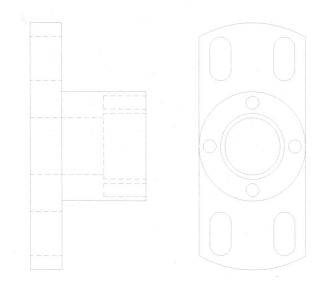

Figure 14.31

Figure 14.32

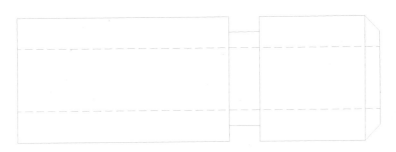

Figure 14.33

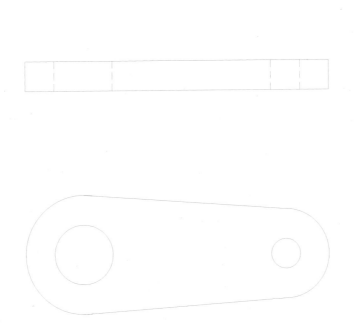

Figure 14.34

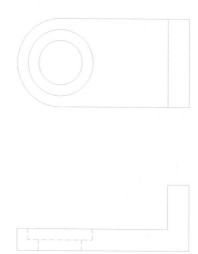

Figure 14.35

CHAPTER 15

Constructing Multiview Drawings

Three-dimensional objects are commonly described by multiview drawings. A multiview drawing shows more than one view of an object, typically three-dimensional. In this chapter, you learn about using AutoCAD for constructing multiview drawings. After completing this chapter, you will be able to

- Perform the fundamentals of multiview drawings
- Manipulate the commands that draw multiview drawings
- Use AutoCAD methodologies to construct multiview drawings

INTRODUCTION

So far, you have learned how to use AutoCAD's most basic drawing, editing, and dimensioning commands. In this—and the next—chapter, you start applying those commands to multiview and sectional drawings. In this chapter, you learn these new commands:

Linetype

Causes objects take on a dotted or dashed look (linetypes).

LtScale

Changes the scale of the linetype.

MULTIVIEW DRAWINGS

The description of three-dimensional objects by the use of flat, two-dimensional (2D) drawings is a common drafting practice. You accomplish an accurate 2D description by drawing the object from several directions—thus, multiple views.

ORTHOGRAPHIC PROJECTION

An *orthographic projection* is a view of an object that is created when a single view is projected onto an imaginary projection plane. Let's look at an example. The face of an object in Figure 15.1 is projected onto the viewing plane.

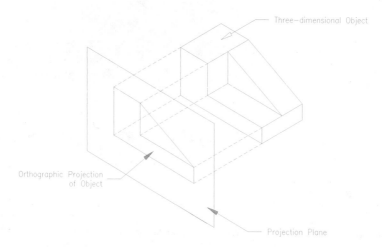

Figure 15.1 *Orthographic Projection*

The image that is projected onto the projection plane represents the true lengths of the edges on the object. The projection plane in this example is parallel to a viewing face on the object. This is referred to as a normal view of the object. Normal views are a more accurate method of viewing an object in orthographic projection.

One-View Orthographic Projections

Many thin, simple objects can be described by a single orthographic projection, such as the one in Figure 15.2.

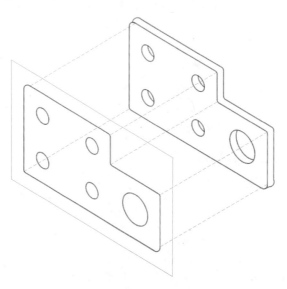

Figure 15.2 *One-View Orthographic Projection*

Two-View Orthographic Projections

More often, a single view cannot adequately describe a three-dimensional object. Faces that do not lie in the same plane can be projected onto the projection plane. The viewer may see the edge lines but cannot determine the location of the different planes.

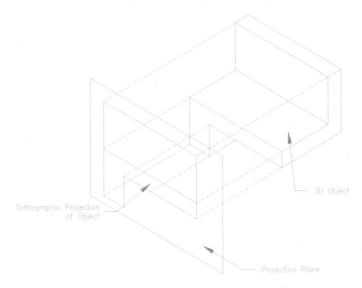

Figure 15.3 *Orthographic Projection of a Complex Object*

Two orthographic views can be used to accurately describe such an object. To do this, two projection planes must be used. The views should show the length, height, and width of the object.

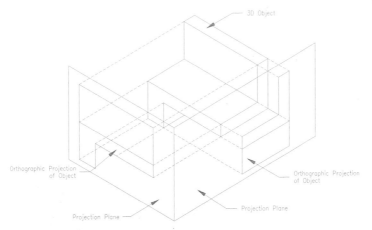

Figure 15.4 *Two-View Orthographic Projection*

Multiple-View Orthographic Projections

More complex objects may require multiple views to adequately describe them. These views can be projected from several sides of the object. The term *multiview* describes several views of an object.

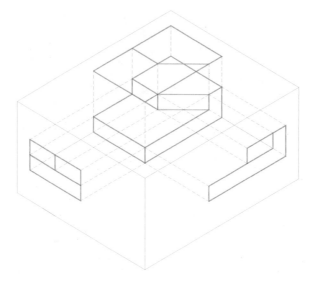

Figure 15.5 *Multiple-View Orthographic Projection*

Positioning Views

In most situations, three views adequately describe an object. The views are usually labeled the front, top, and side views. The *front view* is considered the primary view, with the *top view* positioned above it, and the *side view* to the right side of the front view (Figure 15.6).

With this arrangement, the dimensions for the side view are transferred from the top and front views. Figure 15.7 shows the previous example with imaginary transfer lines shown. Note how the transfer lines from the top view are reflected off a 45-degree miter line. This method is used as an alternative to constructing each line length.

AutoCAD is an excellent tool for constructing multiview drawings. The method of construction closely follows that used on the drawing board.

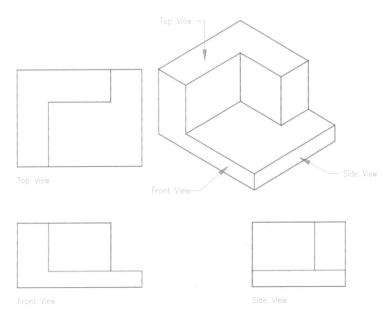

Figure 15.6 *Three Views of an Object*

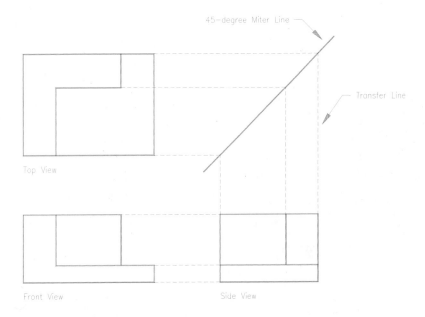

Figure 15.7 *Transferring Line Lengths*

AUXILIARY VIEWS

Objects sometimes have angular faces that are not parallel to the projection plane. The object shown in Figure 15.8 contains an angular face that is not truly represented in the top or side view.

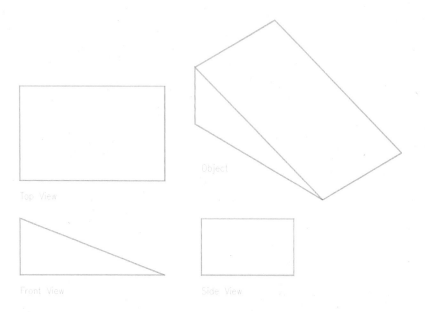

Figure 15.8 *Object with Angular Surface*

We show the *true size* of the angular face by adding a projection from the object called an *auxiliary view*. An auxiliary view is one that is projected onto a projection plane parallel to the angular surface. You can also think of this as the view you would see if you looked at the object from a point perpendicular to the angular face. Figure 15.9 shows the previous object with an auxiliary view added.

Auxiliary views serve three purposes that cannot be achieved by the normal three views:

- To show the true size of the angular surface

- To illustrate the true shape of the surface

- To aid in the projection of other views

Auxiliary views can be projected from any other view. The name of the view is determined from the view from which it is projected. For example, if you project the auxiliary view from the front view, it is named the *front view auxiliary*.

Figure 15.9 *Auxiliary View*

CONSTRUCTING AUXILIARY VIEWS WITH AUTOCAD

Constructing an auxiliary view with AutoCAD involves drawing the view at an angle parallel to the angular face. There are some tricks that make this process easier. Let's look at how we draw an auxiliary view.

SHOWING HIDDEN LINES IN MULTIVIEW DRAWINGS

Auxiliary views are line drawings. Solid lines are used to represent the edges of the object. In AutoCAD, solidly drawn lines are referred to as *continuous lines*.

Edges that are hidden from view are shown in a linetype referred to as *hidden*. This is a line that is constructed from a series of short line segments. Let's look at an example. Figure 15.10 shows an object containing edges that are hidden in some views. These edges are defined with the hidden linetype.

AutoCAD provides several linetypes for your use. Let's look at how to draw different types of lines.

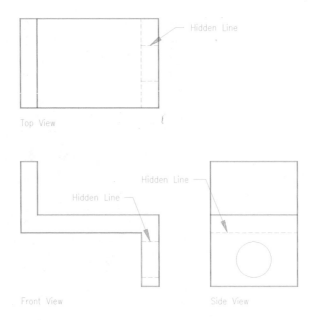

Figure 15.10 *Object with Hidden Lines Shown*

Using the Linetype Command

Different linetypes are used frequently in drawings. Figure 15.11 shows the linetypes provided in AutoCAD.

These linetypes can be added to your drawing through the **Linetyp**e command. Before you can use a linetype, you must first load that linetype. Let's look at how to use the **Linetype** command to load a linetype.

LOADING LINETYPES

You must load a linetype before using it. When you use the **Linetype** command's **Load** option, the linetype is loaded from Acad.lin when it is needed.

Figure 15.11 *AutoCAD Linetypes*

If you explicitly want to load a linetype into your drawing, you can also choose the linetype icon or select **Format | Linetype** from the menu bar:.

Command: **Linetype**

AutoCAD displays the **Linetype Manager** dialog box. It displays the linetypes currently loaded into the current drawing.

To load a linetype, choose the **Load** button. AutoCAD displays the **Load or Reload Linetypes** dialog box. The **File** field displays the name of the file in which the linetype is located. In most cases, you use the Acad.lin file. It is not necessary to enter the LIN file extension. Select the linetype you want to load, and then choose the **OK** button.

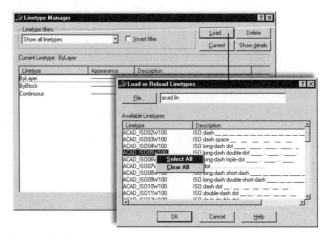

Figure 15.12 *Linetype Dialog Box*

You can use this process to reload current linetypes. If a regeneration is performed, the effect of the new linetype is immediately seen on the screen.

(To load *all* linetypes at once, right-click on any linetype. AutoCAD displays a small pop-up menu with two selections. Choose **Select All** and AutoCAD highlights all linetypes. Choose **OK** and AutoCAD loads all linetypes.)

Specifying Linetypes

Linetypes can be designated in many ways:

- The **Linetype** command is used to designate the current linetype, overriding the layer setting for linetype.

- The **Layer** command specifies a default linetype for a layer. This linetype is known as the BYLAYER linetype.

- The linetype can be assigned during the plotting setup to a specified object color.

Direct Linetype Editing

To change the linetype of an object, AutoCAD has a direct method. (For this method to work, ensure all linetypes are loaded into the drawing, as noted earlier.) The direct method is very fast because it takes just three screen selections:

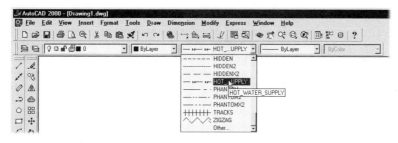

Figure 15.13 *Selecting Linetype From Toolbar*

Selection #1

Select the object (or objects) by clicking on it. AutoCAD highlights the object you select.

Selection #2

Choose the linetype list box. The list box drops down.

Selection #3

Select the new linetype, such as Hot Water Supply in Figure 15.13. AutoCAD changes the object's linetype.

The same process can be used to change the color and layer of an object.

Linetype Scales

Linetypes are constructed of line segments and dots. The line segments are of a specified length of 1. The scale of these segments can be adjusted by the use of the **LtScale** (short for "linetype scale") command:

> Command: **Ltscale**
>
> Enter new linetype scale factor <1.0000>: *(Type a value.)*
>
> Regenerating model.

Entering larger numbers results in longer line segments, while smaller numbers create shorter line segments.

So that the new linetype scale is displayed correctly, AutoCAD automatically performs a regeneration. If a regeneration does not occur after you change the linetype scale, you can force a regeneration with the **Regen** command.

If you change linetypes and the line still appears as a continuous line, change the **Ltscale** to a different setting. It is possible for the linetype scale to be too large or small to display. If lines take a longer time to draw, it is likely that the scale is too small.

Global and Object Linetype Scale

AutoCAD lets you specify the linetype scale in two ways: (1) as a *global* linetype scale for all objects in the drawing and (2) as an *object* linetype scale for individual objects. While it is poor drafting practice to use multiple linetype scales in a drawing, the object scaling is particularly useful for complex linetypes, such as Batting, Hot Water Supply, and ZigZag.

To set the global linetype scale, use the **Ltscale** command.

Linetype Scale by Dialog Box

Instead of the **Ltscale** command, you can change the linetype scale using the dialog box.

Select **Format | Linetype**. When the **Linetype Manager** dialog box appears, choose the **Details** button (see Figure 15.14). AutoCAD expands the dialog box downward to show more linetype details.

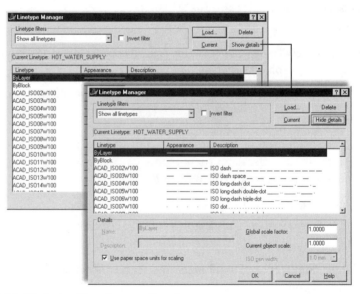

Figure 15.14 *Displaying More Linetype Details*

The **Details** section of the dialog box lets you set four different kinds of linetype scaling:

- **Global scale factor:** Sets the linetype scale for all objects in the drawing. This scale factor takes effect with the next drawing regeneration. (The value is stored in system variable **Ltscale**.)

- **Current object scale:** Sets the linetype scale for newly created objects. (The factor is stored in system variable **Celtscale**.)

- **ISO pen width:** Select from one of the ISO standard values, such as 1 mm. (ISO stands for International Organization for Standardization.)

- **Use paper space units for scaling:** Turn on this option when you want linetypes to scale the same in paper space and model space. (The toggle is stored in system variable **Psltscale**.)

EXERCISES

1. Use AutoCAD to draw three views of the following 3D objects.

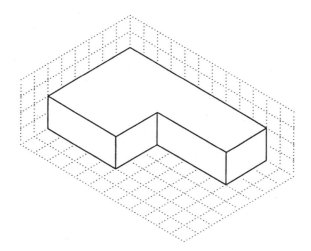

Figure 15.15

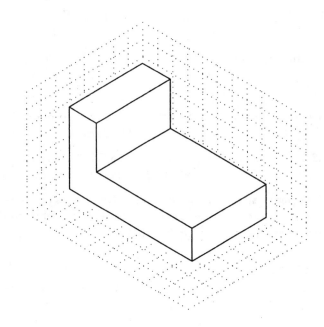

Figure 15.16

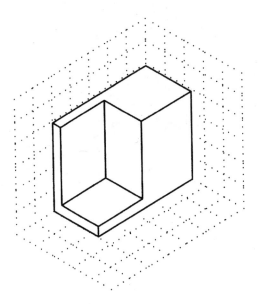

Figure 15.17

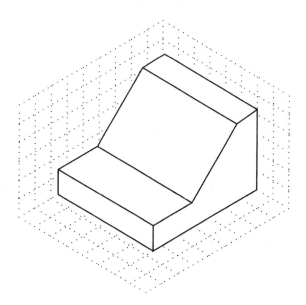

Figure 15.18

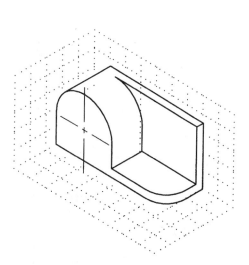

Figure 15.19

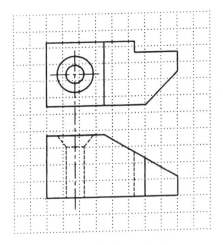

Figure 15.20

Figure 15.21

CHAPTER REVIEW

1. What is an orthographic projection?
2. What type of object can be described by only one projection?
3. What term is used to describe a drawing that contains several views of an object?
4. What are the three typical views used in orthographic projection?
5. What is an auxiliary view?
6. Why would you use an auxiliary view?
7. What do hidden lines show in a drawing?
8. Name two ways a linetype can be designated in AutoCAD.
9. How would you control the length of individual segments in a dashed line?
10. How do you load all linetypes into a drawing?
11. What is the difference between global and object linetype scaling?

TUTORIAL

Let's construct a simple three-view drawing with AutoCAD. Figure 15.22 shows the object you will use.

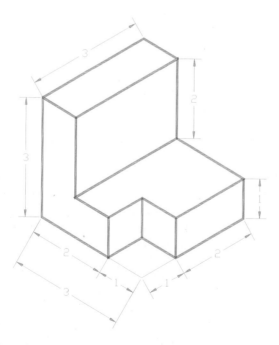

Figure 15.22 *3D Object (with Dimensions Shown)*

Let's begin a new drawing.

1. Select **File | New** from the menu bar. When the **Startup** dialog box appears, select **Start from Scratch**, then select **English**, and choose **OK**.

2. Set a snap increment so your crosshairs move one unit at a time.

   Command: **Snap**

   Specify snap spacing or [ON/OFF/Aspect/Rotate/Style/Type]
       <0.5000>: 1

3. Now use the **Line** command to draw the front view of the object. Figure 15.23 (left) shows the dimensions of the front view. Figure 15.23 (right) shows how your drawing should look after you have drawn the front view.

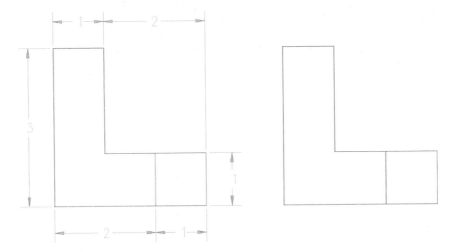

Figure 15.23 *Dimensioned Front View (left); Completed Front View (right)*

4. Now draw the top view. Use the **Line** command to draw transfer lines to transfer the widths of the object, then draw a line that serves as the upper edge of the top view as shown in the following illustration.

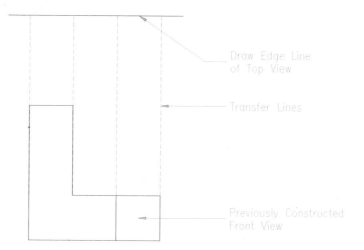

Figure 15.24 *Transferring Object Widths*

5. Next, you use a command called **Offset** to offset the 3-inch width of the object.

6. Following this, again use the **Trim** command to create the "notch." Use the following illustrated sequence.

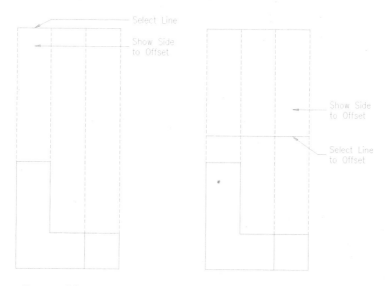

Figure 15.25 *Illustrated Sequence*

7. Now draw the side view. Start by placing a miter line so you can transfer the top view dimensions to the side view. Use the **Line** command for this.

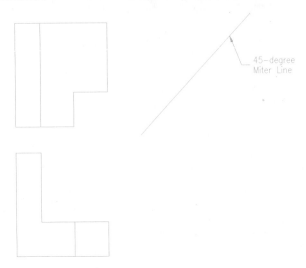

Figure 15.26 *Placement of Miter Line*

8. Now use the **Line** command to transfer the object's edges from the top and front views as shown in Figure 15.27. Using **Ortho** mode and object snap will aid you in constructing accurate transfer lines.

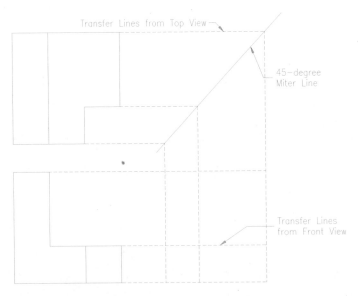

Figure 15.27 Constructing Transfer Lines

9. Now use the **Trim** command to trim away lines so your drawing looks like the one in Figure 15.28. Be sure to erase any remaining transfer lines.

Figure 15.28 *Completed Drawing*

Save your work with name "3FACE."

TUTORIAL

You will construct an auxiliary view for the following object. The CD-ROM contains a drawing named "AUX_VIEW" that can be used if you wish to follow along.

Figure 15.29 *3D Object with Angular Surface*

1. A simple trick to get started is to copy the angle line from a projected view into the position where the auxiliary view will be placed. Figure 15.30 illustrates this procedure.

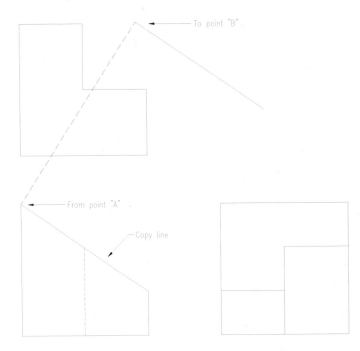

Figure 15.30 *Copying the Angle Line*

As an alternative, you can use the **Offset** command with the **Through** option. This ensures it is parallel and projected perpendicular.

2. Next, use the **Snap** command to rotate the snap grid. This not only rotates the snap grid, but also rotates the crosshairs so we can draw at the proper angle. The following command sequence and Figure 15.31 show how to do this.

Command: **snap**

Specify snap **spacing** or [ON/OFF/Aspect/Rotate/Style/Type] <1.0000>: **R**

Specify base point <0.0000,0.0000>: **end**

of *(Select point 1.)*

Specify rotation angle <0>: **end**

of (Select point 2.)

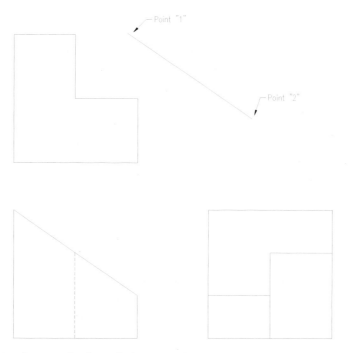

Figure 15.31 *Rotating the Snap Grid*

Notice how using object snap assists in obtaining greater accuracy. If you move the crosshairs around the screen, you will notice that it is rotated to the same angle as the line you copied.

3. Next, use the **Line** command to complete the drawing. Be sure to use the **Ortho** mode and the snap increment setting to assist in drawing the object. You may also want to construct some temporary transfer lines to assist in determining the intersections. Figure 15.32 shows the completed view with transfer lines shown dashed.

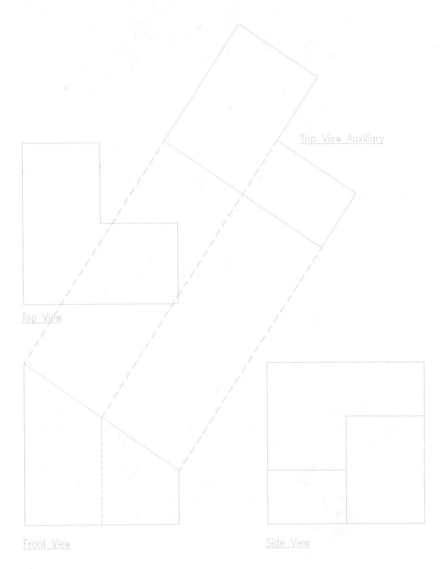

Top View Auxiliary

Top View

Front View

Side View

Figure 15.32 *Completed View with Transfer Lines Indicated*

CHAPTER 16

Constructing Sectional and Patterned Drawings

Sectional views are used to describe objects when other views are not sufficient.

Computer-aided design is an excellent tool for constructing sectional views. After completing this chapter, you will

- Understand the types of sectional views and their application
- Apply hatch patterns to sectional views
- Create boundaries
- Draw solid quadralaterals that can be used as part of sectional drawings
- Change the width of objects

INTRODUCTION

In the last chapter, you learned to apply drawing and editing commands to multiview drawings. In this chapter, you learn to use these new commands for creating sectional drawings:

BHatch

Applies crosshatch patterns to closed objects.

Boundary

Creates a boundary from several objects.

Solid

Draws solid-filled triangles and quadrilaterals.

Lineweight

Applies a width to objects.

SECTIONAL VIEWS

Many parts and assemblies cannot be fully described by the use of orthographic projection. It is often helpful to view the object as if it were cut apart. A view of an object or assembly that has been cut apart is called a *section*.

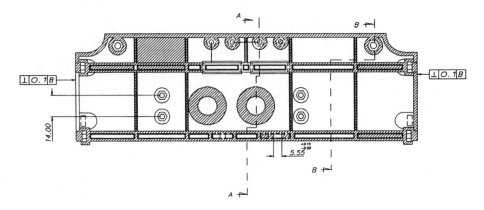

Figure 16.1 *Section*

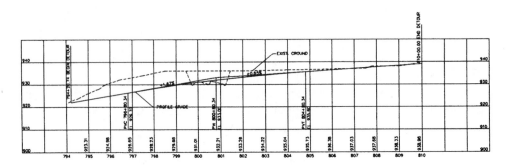

Figure 16.2 *Roadway Profile*

A section is used in many disciplines. The mechanical designer uses sections to show details that cannot be described in other ways.

Civil engineers detail roadway profiles by showing sections along the road.

Architects use sections frequently. Sections through entire structures show how a building structure is designed. Individual wall sections detail vertical measurements and delineate materials.

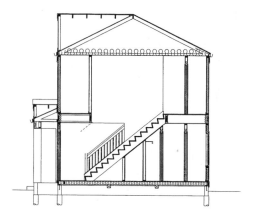

Figure 16.3 *Architectural Section*

TYPES OF SECTIONS

Depending on the desired information to convey, the designer uses different types of sections. Let's study each type of section and look at an example of each.

Full Sections

A *full section* is a section that is cut across the entire object. Full sections are usually cut through the larger axis of the object.

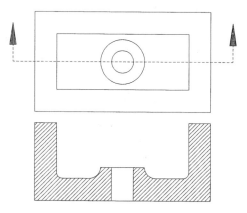

Figure 16.4 *Full Section*

Such a section cut along the longer axis is referred to as the longitudinal axis. If the section is cut along the minor axis, it is referred to as the latitudinal axis.

Parts of the object that are "cut" are shown with crosshatching.

Revolved Sections

It is often helpful to view a section of a part of an object that is transposed on top of the point where the section was cut. Such a section is referred to as a revolved section. Figure 16.5 shows a revolved section.

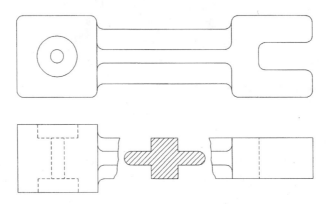

Figure 16.5 *Revolved Section*

Removed Sections

A removed section is similar to a revolved section, except the section is not placed at the point where the section was cut.

Figure 16.6 *Removed Section*

Offset Sections

Offset sections are sections that are cut along an uneven line. Offset sections should be used carefully; change the cutting plane only to show essential elements.

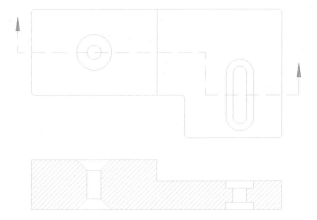

Figure 16.7 *Offset Section*

CROSSHATCHING SECTIONAL VIEWS

Crosshatching refers to placing a pattern within a boundary area. When drawing sections, it is customary to place crosshatching on the faces through the section cuts.

The spacing of the crosshatching should be relative to the scale of the section. The angle of the crosshatching should be oriented 45 degrees from the main lines of the cut area whenever possible.

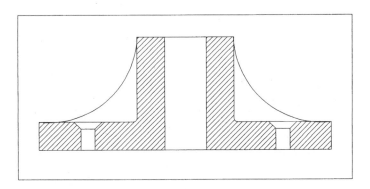

Figure 16.8 *Section with Crosshatching*

In most drafting applications, a simple crosshatch composed of parallel lines is used. In some applications, however, it is acceptable to show different materials with representative crosshatching. Figure 16.9 shows some standard hatch patterns for different materials.

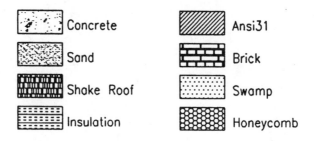

Figure 16.9 *Typical Material Hatches*

CREATING SECTIONAL VIEWS IN AUTOCAD

AutoCAD can be used to create sectional views effectively. You draw the section, then render the cut areas with hatch patterns that are provided with the AutoCAD program.

Hatches differ from other objects in that the entire hatch is treated as one object. If you identify one line of a hatch to be erased, the entire hatch will be erased. Likewise, an entire hatch can be erased by the **Erase Last** command.

Hatch patterns are placed with the **BHatch** command. Let's look at how the command is used.

USING THE BHATCH COMMAND

The **BHatch** (short for "boundary hatch") command creates a boundary around an area and then places an associative hatch pattern within that area. An *associative* hatch pattern is one that automatically updates when you change the hatch boundary. To use this command, select **Draw | Hatch** from the menu bar.

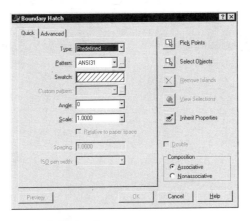

Figure 16.10 *Boundary Hatch Dialog Box*

The **BHatch** command displays the **Boundary Hatch** dialog box as shown in Figure 16.10. It has two tabs: **Quick** and **Advanced,** plus a common area. Let's look at how you use the dialog box to place a hatch.

Decide the Type of Hatch

The first step in placing a hatch pattern is to decide the type of pattern. AutoCAD supports three classes of patterns:

Predefined: Allows you to select a predefined hatch pattern stored in the Acad.pat and Acadiso.pat files. The ISO patterns allow you to control the ISO pen width as well.

User Defined: Allows you to specify a simple pattern of lines based on the angle and spacing. In addition, you can use linetypes to make unique patterns.

Custom: Allows you to select a pattern in any PAT file other than Acad.pat and Acadiso.pat files. Typically, third party or in-house developers provide these PAT files.

Select a Pattern Name or Swatch

The next step is to select the pattern you want to use from the **Pattern** drop-down list. The default is a hatch pattern made of diagonal lines called "ANSI31." To select a new pattern, click the down arrow and select the name of the available hatch patterns.

If you don't know the name, the alternative is to select the pattern next to **Swatch**. AutoCAD displays a tabbed dialog box of patterns grouped together, as shown in Figure 16.11. Select a pattern swatch and choose **OK**.

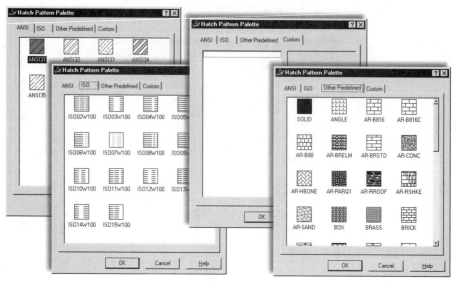

Figure 16.11 *Hatch Patterns*

 Note: The Solid pattern has some limitations not found in other patterns. The hatch boundary must not intersect itself. If the hatch area contains more than one loop, the loops must not intersect.

Specify the Angle and Scale

Hatch patterns can be drawn at any angle and scale. The angle of the hatch is specified in degrees and conforms to AutoCAD's angle specifications (see Chapter 6 "Setting Up a New Drawing"). The angle of the hatch pattern samples shown in the dialog box is 0 degrees. This is true even if the lines of the hatch are drawn at an angle. To change the angle, select the **Angle** box. Select one of the preset angles or use the keyboard to type the different angle.

The scale of the hatch pattern is multiplied by a numerical factor. The default scale is 1. Setting the factor to 2, for example, draws a hatch twice as large, while setting a factor of 0.5 creates a hatch half the size. To change the scale, select the **Scale** box. Select one of the preset scales, or use the keyboard to type the different scale factor.

Scaling a hatch pattern is the same as scaling text and linetypes. You must enter an inverse scale factor in the **Scale** box. For example, if the drawing is to be plotted at a scale of 1:100, then the hatch pattern must be applied at a scale of 100.

Scale: 0.5   Scale: 1.0   Scale: 2.0

Figure 16.12 *Hatch Scale Factors*

 Note: AutoCAD normally draws the hatch pattern based on 0,0 as the starting point, even if your hatch pattern does not physically pass through the origin (0,0). If it is critical that the hatch pattern start at a different location, such as a brick pattern on a wall, use the **Snap** command's **Rotate** option to change the base point.

User Defined Parameters

If you selected **User Defined** as the hatch type, then you have no choice of pattern. You work with a single pattern consisting of simple, parallel lines. You can specify three parameters: the angle, the spacing between lines, and double hatching. The **Angle** option is the same as above.

The **Spacing** option lets you specify how far apart the parallel lines are drawn. It is similar to the Scale option.

The **Double** check box causes AutoCAD to draw the pattern a second time, at 90 degrees to the first pattern, creating a crosshatch. Figure 16.13 shows examples of using linetypes together with user-defined hatch patterns. In two areas, the Batting hatch pattern was used, once with double hatching.

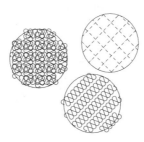

Figure 16.13 *User Defined Hatch Patterns*

Other Hatch Parameters

There are a several other options in the **Quick** tab of the **Boundary Hatch** dialog box that are used less often. The **Relative to paper space** option is only available when you start the **BHatch** command while in layout mode (a.k.a. paper space). It matches the hatch scale to the paper space scale.

The **ISO** pen width option is only available when you select an ISO hatch pattern. This option allows you to specify the width of the lines making up the ISO pattern.

The **Composition** area lets you decide whether you want associative or non-associative hatch patterns. In almost all cases, you would keep the default of **Associative**. Associative hatch patterns stretch when their boundary is stretched. In addition, associative hatch patterns store data about themselves. This provides two advantages: you can copy hatch parameters through the **Inherit Properties** button and you can change the parameters of an existing hatch.

Nonassociative hatch patterns do not update themselves, nor do they know their own parameters. The pattern is applied as a block. The advantage is that you can explode the block (reduce the pattern to its composite lines and points), then edit them. This may prove useful in some limited cases.

SELECTING THE HATCH BOUNDARY

A hatch must be contained within a *boundary*. The boundary is made up of one or more objects that form a closed polygon. Objects such as lines, polylines, circles, and arcs can be used to form the boundary.

Before you can place a hatch, you must identify the objects that form the boundary. This might be one of the trickiest aspects to CAD. If the boundary is not "closed," the hatch pattern will "leak" out.

There are several ways to identify the boundary:

> Select the object(s) that make up the boundary.
>
> Select a point within the hatch area and let AutoCAD find the boundary.
>
> Remove islands inside the boundary.

Let's look at each.

Defining the Boundary by Selecting Objects

You can specify the boundary for the hatch pattern by selecting the objects that surround the area to be hatched. Let's look at a simple example. Figure 16.14 shows a shape constructed with four lines.

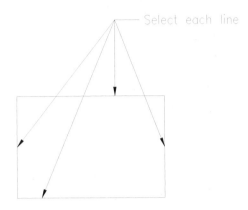

Figure 16.14 *Selecting the Hatch Boundary*

To place a hatch in the rectangle, start the **BHatch** command. In the **Boundary Hatch** dialog box, choose the **Select Objects** button. Don't be alarmed when the dialog box disappears from the screen. Look at the command prompt area:

Select objects:

AutoCAD is prompting you to select the objects that will form the boundary. Place the pickbox over each object and select.

 Note: Use the window or crossing option to select all the sides of the boundary in one operation.

When you are finished selecting objects, press ENTER. You have selected the objects that bound the area to be hatched.

Previewing the Hatch Pattern and Boundary Objects

Choose the **Preview** button to see what the hatch will look like. The dialog box disappears again and AutoCAD prompts you:

<Hit enter or right-click to return to the dialog>

Press ENTER or right-click to return to the **Boundary Hatch** dialog box. You can make changes to the hatch parameters, such as the scale factor.

If the boundary is inadequate to contain the hatch pattern, AutoCAD reports on the command line:

Unable to hatch the boundary.

Press ENTER or right-click to return to the **Boundary Hatch** dialog box. To review the objects selected for the boundary, choose **View Selections**. The dialog box again disappears and AutoCAD highlights the boundary objects. AutoCAD prompts you:

<Hit enter or right-click to return to the dialog>

Press ENTER or right-click to return to the **Boundary Hatch** dialog box. If you need to, you can make changes to the boundary objects. Choose the **Select Objects** button. When the dialog box disappears, you can select additional objects.

Choose **OK** to accept your settings and add the hatch to the drawing.

Defining the Boundary by Selecting an Area

A simpler way to define the hatch area is to show AutoCAD the area. You show AutoCAD by selecting a point within the area to be hatched.

Let's look at an example. Figure 16.15 shows an object with different areas that could be hatched. Start the **BHatch** command. In the dialog box, choose the **Pick Points** button. When the dialog box disappears from the screen, place the crosshairs in the area shown in Figure 16.15 (left).

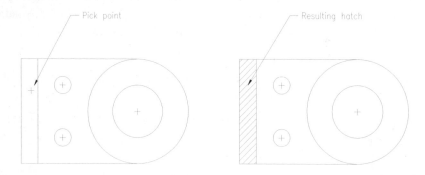

Figure 16.15 *Picking Inside the Boundary*

AutoCAD prompts you:

> Select internal point: *(Pick point.)*

Click anywhere inside the area. AutoCAD reports:

> Selecting everything...
>
> Selecting everything visible...
>
> Analyzing internal islands...

You can think of those statements as AutoCAD mumbling to itself as it searches for the boundary. AutoCAD analyzes the area, looking for leakage. Unknown to you, AutoCAD then places a polyline boundary around the inside of the area.

If you select a point in an area that is not contained, AutoCAD displays the warning: "Valid hatch boundary not found." You must reconsider the area you want hatched; perhaps you need to draw another line to close off the area to be hatched.

You can select more than one area. When finished picking areas, press ENTER. The dialog box is again displayed. Choose **Preview** to see how the hatch will turn out. When satisfied, choose **OK** and the hatch is completed.

Removing Islands

Sometimes the area to be hatched is not as straight forward as the rectangle hatched in Figure 16.15. It is common for the hatch area contains *islands*. An island is another closed object within the hatch boundary. In Figure 16.15, the two small circles are considered islands if the area around them were to be hatched.

AutoCAD assumes you do not want the islands hatched. If you want AutoCAD to draw the hatch pattern right through the islands, choose the **Remove Island** button. The dialog box disappears and AutoCAD prompts you to pick the islands to be removed:

> Select island to remove: *(Pick object.)*

After you select an object, AutoCAD prompts you to continue:

> <Select island to remove>/Undo:

Or, you can type **U** to deselect the object. When done removing island, press ENTER. Back in the dialog box, choose the **Preview** button to see the effect of having removed the islands.

ADVANCED HATCH OPTIONS

As you have read, the **Pick Points** button pretty much automates the process of finding the hatch area. In some cases, the hatch area might be complicated. The **Advanced** tab of the **Boundary Hatch** dialog box provides additional parameters for determining the hatch area.

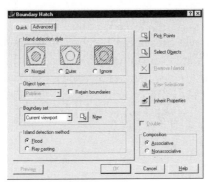

Figure 16.16 *Advanced Options Dialog Box*

Let's look at each part of the dialog box, starting with island detection.

Island Detection Style

When the area to be hatched contain other objects, you can choose the manner in which the hatch behaves. To define complex areas to be hatched, a *island detection* is used to control the hatch. Let's look at how boundary styles work. Figure 16.17 shows an area called "PART A."

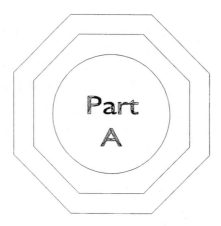

Part
A

Figure 16.17 *Part "A" Drawing*

Normal

The default style of hatching is called "Normal." The normal style hatches inward from the outermost boundary, skips the next boundary, and hatches the next. Notice how the text is hatched around in Figure 16.18. With Normal Style hatching, an invisible window protects text; this ensures that the text is not obscured by the hatch pattern.

Outer

The "Outer" hatching style causes the hatch pattern to hatch only the outermost enclosed boundary. The hatch continues until it reaches the first inner boundary and continues no further. Text is not hatched.

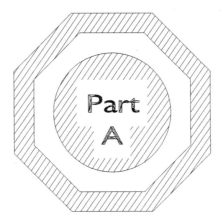

Figure 16.18 *Normal Style*

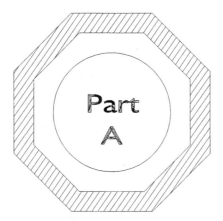

Figure 16.19 *Outermost Style*

Ignore

The "Ignore" style hatches all areas that are defined by the outer boundary—with no exceptions. This style also hatches through text.

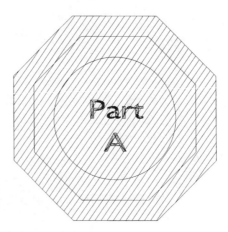

Figure 16.20 *Ignore Style*

Object Type

The boundary is drawn as a polyline or a region. When the **Retain Boundary** option is turned on, you can request that AutoCAD use a polyline for compatibility with older versions of AutoCAD; otherwise, use the region.

AutoCAD normally erases the boundary after the BHatch command is finished. If you wish to retain the boundary as a polyline in the drawing, select the **Retain Boundaries** box.

Boundary Set

When you use the **Pick Points** option to define a boundary, AutoCAD analyzes all objects visible in the current viewport. You can, however, change the set of objects AutoCAD examines. In large drawings, reducing the set lets AutoCAD operate faster.

The **Current Viewport** option is the default and examines all objects visible in the current viewport to create the boundary.

The **New** button prompts you to select objects from which you want to create the boundary set. (AutoCAD includes only objects that can be hatched.)

The **Existing Set** option creates the boundary from the objects selected with the **New** option. (You must use the **New** option before the **Existing Set** option.)

Island Detection Method

The **Flood** option includes islands as boundary objects.

The **Ray Casting** option runs an imaginary line from the point you pick to the nearest object. It then traces the boundary in the counterclockwise direction; it excludes islands as boundary objects.

FIXING LEAKY HATCHES

The objects that bound the area should be perfectly joined at their intersections. If, for example, a line extends beyond the intersection, the hatch may not operate smoothly. In a situation such as that shown in Figure 16.21, the line that runs past the intersection along the full length of the object can cause problems for the hatch.

To prevent incorrect hatching, the lines should be broken at the intersection. A good way to do this is to use the **Break** command with the **First** option to break the object at the intersection as shown in Figure 16.21.

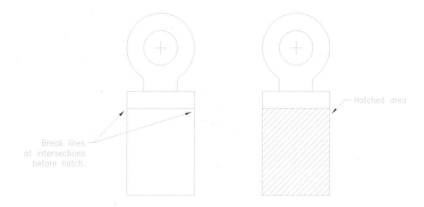

Figure 16.21 *Intersection Break for Hatch Boundary*

Now the objects that define the boundaries meet properly at their endpoints.

PATTERN ALIGNMENT

There may be times when you want adjacent hatches to line up. AutoCAD compensates for alignment problems by normally using the 0,0 point as the hatch origin for all hatches. This means that the hatches align properly. You can change the origin point by using the **SnapBase** system variable to change the base point.

 Note: Hatches can be handled more easily if they are put on their own layer. They can also be turned off and frozen to speed redraw time. Be sure that the layer linetype is continuous. Although the hatch pattern may contain dashed lines and dots, the linetype should be continuous to ensure a proper hatch.

CREATING A BOUNDARY

Earlier, you read how AutoCAD creates a temporary boundary out of a polyline or region to hold the hatching. To define an entire boundary as a single object, use the **Boundary** command. When you select **Draw | Boundary** from the menu bar, the **Boundary Creation** dialog box (Figure 16.22) is displayed.

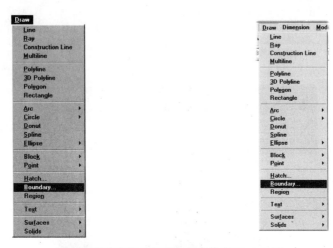

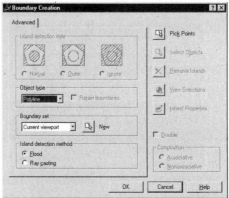

Figure 16.22 *Boundary Creation Dialog Box*

This dialog box is similar to the one displayed under **Advanced** tab of the **BHatch** command. The boundary is selected in the same manner. The difference is that no hatch is placed within the boundary; only the boundary is constructed as a polyline or region.

Note: The **Boundary** command has other options. Converting the outline of an object to a polyline allows you to select all parts of it as a single object when editing.

CREATING SOLID AREAS

The **BHatch** command has a solid hatch pattern that allows you to fill areas with a solid color. As an alternative, the **Solid** command allows you to fill triangular and quadrilateral areas with a solid color. Solids plot as solid ink areas.

 Note: The solid areas are displayed only if the **FillMode** system variable is on.

To use the **Solid** command, select **Draw | Surfaces | 2D Solid** from the menu bar.

Command: **Solid**

Specify first point: *(Pick point.)*

Specify second point: *(Pick point.)*

Specify third point: *(Pick point.)*

Specify fourth point or <exit>: *(Pick point.)*

Specify third point or <exit>: *(Pick point.)*

Specify fourth point or <exit>: *(Pick point.)*

Specify third point or <exit>: *(Pick point.)*

Specify fourth point or <exit>: *(Press* ENTER.)*

This sequence continues until you terminate it with ENTER. This allows you to create solid polygons with any number of sides.

In some cases, the **Solid** command won't work properly when the points are entered in the wrong sequence. For example, entering the points in the fashion shown in Figure 16.23 results in a "bow tie" rather than the solid you want.

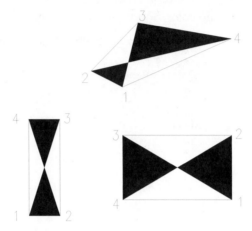

Figure 16.23 *Bow Tie Sequences for Solid Command*

The following examples show the sequence to enter points for a correct solid fill. Draw each shape and enter the points as shown.

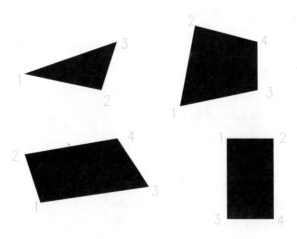

Figure 16.24 *Proper Sequences for Solid Command*

Turn off the fill (using system variable **FillMode**) after using the **Solid** command to speed up the regeneration process caused by zooms and pans, then turn the fill back on and **Regen** to redisplay filled solids.

 Note: To fill an arbitrary shape with a solid fill, use the **BHatch** command's solid hatch pattern.

APPLYING LINEWEIGHTS

In addition to filling areas with patterns and drawing solid-filled areas, you can give objects a width. Before AutoCAD 2000, only polylines and traces could have a width (see Chapter 22 "Understanding Advanced Operations"). All other objects are drawn one pixel wide, which means they are drawn as thin as the display or plotter allows. (Widths were assigned at plot time based on the object's *color*!)

As of AutoCAD 2000, almost any object can have any width. Or, as it is called, *weight*. Using a variety of weights (heavy and thin lines) helps make a drawing clearer. Weights can be assigned to layers or on an object-by-object basis.

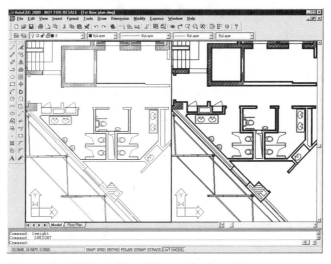

Figure 16.25 *Lineweights Off (at left) and Turned On (at right)*

Lineweights range from 0.05 mm (0.002") through to 2.11 mm (0.083"). AutoCAD displays the values in mm (millimeters). The lineweight of 0 is compatible with earlier versions of AutoCAD and displays as one pixel in model space. It is plotted at the thinnest width by the plotter.

When turned on, the weight appears when displayed and when plotted, with one exception: an object with lineweight 0.025 mm or less is displayed one pixel wide in model space. At plot time, objects with weight are plotted at the exact same widths. Objects copied to the Windows Clipboard (through Ctrl+C) retain their lineweight data.

 Note: Do not use lineweights to represent the actual width of objects. If a wall is 4 inches wide, use two lines (or a multiline) four inches apart. Do not draw a single line four inches wide.

Table 16-1 shows the default lineweight values used by AutoCAD and their equivalent values for industry standards. There are 25.4 mm per inch; 72.72 points per inch.

By default, lineweights are not displayed in the drawing. To turn on the display of lineweights, choose the **LWT** button on the status bar. Lineweights are on when the button looks depressed.

Figure 16.26 *The LWT Button on the Status Bar*

To apply a lineweight to an object, select the object, then select a lineweight from the lineweight list box on the **Object Properties** toolbar.

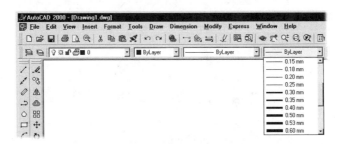

Figure 16.27 *Accessing Lineweights from the Object Properties Toolbar*

To display the **Lineweight Settings** dialog box, type the **LWeight** command, or right-click the **LWT** button on the status bar and select **Settings**, or select **Format | Lineweights** from the menu bar, or select **Lineweight Settings** in the **Display** tab of the **Options** dialog box.

Millimeters	Inches	Points	Pen size	ISO	DIN	JIS	ANSI
0.05	0.002						
0.09	0.003	1/4 pt.					
0.13	0.005				*		
0.15	0.006						
0.18	0.007	1/2 pt.	0000	*	*	*	
0.20	0.008						
0.25	0.010	3/4 pt.	000	*	*	*	
0.30	0.012		00				2H or H
0.35	0.014	1 pt.	0	*	*	*	
0.40	0.016						
0.50	0.020		1	*	*	*	
0.53	0.021	1-1/2 pt.					
0.60	0.024		2				H, F, or B
0.70	0.028	2-1/4 pt.	2-1/2	*	*	*	
0.80	0.031		3				
0.90	0.035						
1.00	0.039		3-1/2	*	*	*	
1.06	0.042	3 pt.					
1.20	0.047		4				
1.40	0.056			*	*	*	
1.58	0.062	4-1/4 pt.					
2.0	0.078			*	*		
2.11	0.083	6 pt.					

Table 16.1 *Lineweight Standards*

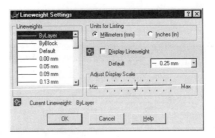

Figure 16.28 *The Lineweights Settings Dialog Box*

The **Lineweights** list is the same one as displayed by the lineweights list box on the **Object Properties** toolbar.

The **Units** for Listing area allows you to select between inches and millimeters (the default).

The **Display Lineweight** check box is identical in function to the LWT button on the status bar.

The **Default** list box lets you select the lineweight by which all objects are drawn, unless overridden by layer or by object override. AutoCAD includes three named lineweights:

- BYLAYER means that the object takes on the lineweight based on the value set for the layer.

- BYBLOCK means that the object takes on the lineweight of the block which it is a part of.

- DEFAULT is the value set here in the dialog box. The default is 0.25 mm.

The **Adjust Display Scale** slider controls the display scale of lineweights in model space. This slider is beneficial if your computer displays AutoCAD on a high-resolution monitor. Experiment with adjusting the lineweight scale to see if you get a better display of different lineweights. As you move the slider, notice that the widths of lines in the **Lineweights** list change.

Choose **OK** to dismiss the dialog box.

EXERCISES

1. Use the CD-ROM and open the drawing named "Solids.Dwg." The drawing shows several hot-air balloons. Use the BHatch command to place the solid hatch pattern areas in some of the balloon areas.

Figure 16.29 *Solids.Dwg Hatch Pattern Exercise*

2. Now let's try a variety of hatch patterns. Open the drawing on the student CD-ROM named "Solids.Dwg." This is the same drawing you used in the previous exercise. Use the **BHatch** command to place a variety of hatched areas in some of the balloons.

CHAPTER REVIEW

1. Explain what the BHatch command does.

2. If you did not want the solid filled areas in your drawing to plot, what could you do?

3. What is a hatch boundary?

4. When would you use the BHatch command instead of the Solid command?

5. How would you change the origin point of a hatch?

6. When does a hatch pattern obscure text?

7. Name one benefit to using lineweights.

454

TUTORIAL

Let's look at a complex hatching example. Figure 16.30 shows an object that contains three areas that could be hatched. Let's suppose that we want to hatch the square and the circle, but not the triangle. Create a quick drawing similar to the one in the illustration.

This is a good example of how boundary styles can be used to control how a hatch pattern behaves.

1. From the **Draw** menu, select **Hatch**.

2. In the **Pattern** list box, select **ANSI37**.

3. Now choose the **Advanced** button. Make sure that **Normal** appears in the **Style** list box, and the **Island Detection** method has **Flood** selected.

4. Select **Pick Points<**. Click inside the circle but outside the box. See Figure 16.30.

5. Click the right mouse button to return to the previous dialog box.

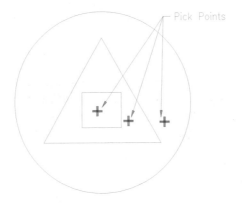

Figure 16.30 *Selecting Pick Points*

6. Choose the Preview button. Your hatch should look similar to the one in Figure 16-31. If not, make changes to the hatch options and/or your pick points.

7. When the preview hatch looks right, choose OK to apply the hatch pattern.

The **Island Detection** option eliminates the need to select multiple points when boundaries are nested inside boundaries. Try turning this option on and only selecting the point within the circle. The result should look like Figure 16.31.

Figure 16.31 *Hatch Drawing*

TUTORIAL

1. From the CD-ROM, open the **Hatch.Dwg** drawing.

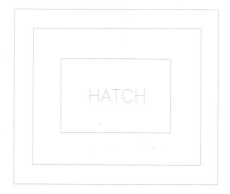

Figure 16.32 *HATCH1. DWG CD-ROM Drawing*

2. Start by using the **Normal** hatch style and the ANSI31 hatch pattern.

   Command: **bhatch**

   Pattern: **ANSI31**

   Scale: **1**

   Angle: **0**

   Select objects: **W**

3. Place the window around the object and press ENTER. Your drawing should look similar to the following illustration:

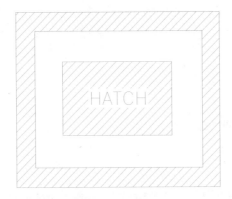

Figure 16.33 *Normal Island Detection*

4. Let's now hatch the same drawing using the **Outer** style.

   Command: **bhatch**

   Pattern: **ANSI31**

   Scale: **1**

   Angle: **0**

   Island Detection Style: **Outer** (Found in the **Advanced** tab.)

   Select objects: **W**

   Using the Outer style, your drawing should look similar to the following illustration:

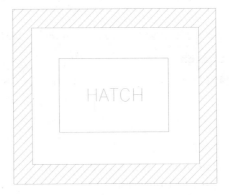

Figure 16.34 *Outer Island Detection*

5. Now use the **Ignore** style of island detection.

   Command: **bhatch**

   Pattern: **ANSI31**

   Scale for pattern: **1**

   Angle for pattern: **0**

   Island Detection Style: **Ignore**

   Select objects: **W**

   Notice how the hatch has ignored the boundaries and the text.

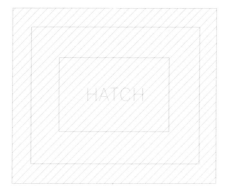

Figure 16.35 *Ignore Island Detection*

CHAPTER 17

Using Inquiry Commands

When drawing with CAD, you must be capable of obtaining information from the drawing. This chapter covers the inquiry and utility commands available to achieve this. After completing this chapter, you will be able to

- Determine coordinates, distances, and areas with the ID, Dist, and Area commands

- Obtain information about objects with the Property, List, and Dblist commands

- Use the time-keeping function with the Time command

INTRODUCTION

AutoCAD's inquiry commands allow you to obtain information and perform utility functions while working within your drawing. You learn about the following commands:

ID

Identifies the *X,Y,Z* coordinates of a point.

Dist

Computes the distance and angle between two points.

Area

Computes the area of an open or closed polygon.

Property

Displays the Property window with information about one or more objects.

List

Lists information on one or more objects in report format.

DbList

Lists information on all objects in the current drawing in report format.

Time

Displays elapsed time.

DRAWING COORDINATES (ID)

The **ID** command (short for "identification") allows you to identify the 3D coordinates of any point in the drawing. When you enter a location on the screen as the response to the prompt, the coordinates of that location are displayed on the command line.

The information displayed contains the X coordinate, the Y coordinate, and the current Z-value elevation.

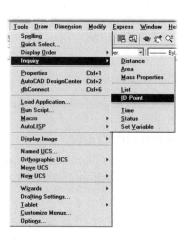

To execute the **ID** command, enter:

> Command: **ID**
>
> Specify point: *(Pick a point.)*
>
> X = 4.3361   Y = 1.7094   Z = 0.0000

You can also use the **ID** command to show a known point on the screen. If you enter a set of coordinates in response to the prompt, a marker blip is displayed at that point on the screen.

For example:

1. Type the **ID** command at the command prompt, or select **Tools | Inquiry | ID Point** from the menu bar.

2. Enter the coordinate value of 8,5:

   Command: **id**

   Specify point: **8,5**

3. Notice that a blip mark is displayed at the *X, Y*-coordinates of 8,5.

Note: Blipmode must be on before marker blips are displayed. To turn on the display of blip marks:

Command: **blipmode**

Enter mode [ON/OFF] <OFF>: **on**

COMPUTING DISTANCES

The **Dist** command (short for "distance") computes the distance between two points, the angle created by the relative position of the points, and the difference of the X and Y values of the points. Type the **Dist** command, or select **Tools | Inquiry | Distance** from the menu bar:

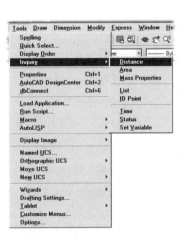

Command: **dist**

Specify first point: **(Pick first point.)**

Specify second point: **(Pick second point.)**

Distance = 7.6519, Angle in XY Plane = 335, Angle from XY Plane = 0

Delta X = 6.9378, Delta Y = -3.2279, Delta Z = 0.0000

Note: Use object snap modes to ensure an accurate distance.

Here is what all this data means:

Distance

The 3D distance between the two points in drawing units. This is the same as the shortest distance between two points.

Angle in XY Plane

The angle created by relative position of the points in the 2D X,Y plane.

Angle from XY Plane

The angle up from the page into the Z direction (used in 3D construction).

Delta X

The X distance between the two points.

Delta Y

The Y distance between the two points.

Delta Z

The Z distance between the two points.

You can use the **Dist** command to show a distance by specifying a relative coordinate at the "Second point:" prompt. For example:

1. Start the Dist command.

   Command: **dist**

2. Select a point in the drawing, using object snap modes, if required.

   Specify first point: *(Pick first point.)*

3. Respond with the relative coordinate of @10,0 for the second point.

   Specify second point: **@10,0**

Notice that AutoCAD displays a blip mark 10 drawing units to the right of the first point (in the positive X direction), as shown by Figure 17.1. **Blipmode** must be on before blip marks are displayed.

As an alternative, you can use polar coordinates with the **Dist** command. Respond to the "Specify second point:" prompt with @15<45, AutoCAD displays a marker blip at a distance of 15 drawing units and 45 degrees from the first point.

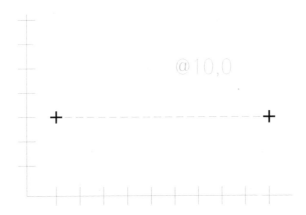

Figure 17.1 *Using Dist Command to Show a Relative Distance*

Figure 17.2 *Using Dist Command to Show a Polar Distance*

CALCULATING PERIMETERS & AREAS

The **Area** command is used to calculate the perimeter and area of a polygon—whether open or closed. The results are returned in current drawing units. This is useful for computing the land bounded by property lines or calculating the areas of floor plans.

The **Area** command permits different methods of calculating areas. The following explanations explain how to use the options.

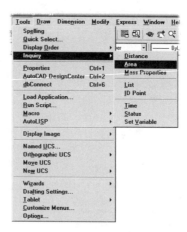

Pointing Method

One method of calculating the area of an object is to select points at the intersections of a polygon, preferably with the assistance of INTersection object snap. Figure 17.3 shows this method. First, use the **OSnap** command to turn on INT object snap. Then, type the **Area** command or select **Tools | Inquiry | Area** from the menu bar.

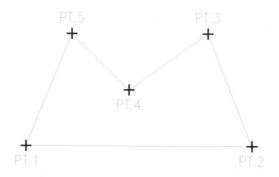

Figure 17.3 *Using the Area Command*

Command: **-osnap**

Enter list of object snap modes: **int**

Command: **area**

Specify first corner point or [Object/Add/Subtract]: *(Select point "1".)*

Specify next corner point or press ENTER for total: *(Select point "2".)*

Specify next corner point or press ENTER for total: *(Select point "3".)*

Specify next corner point or press ENTER for total: *(Select point "4".)*

Specify next corner point or press ENTER for total: *(Select point "5".)*

Notice that the "Specify next corner point:" prompt repeats until you press ENTER to tell AutoCAD that you have selected all points. You do not have to "close" the polygon by entering the last point on top of the first point entered. AutoCAD automatically closes from the last point to the first point entered.

After you press ENTER, you are shown the following information in the prompt area:

Area = 15.2178, Perimeter = 16.7360

Add and Subtract Areas

The **Add** option is used to calculate running totals of calculated areas. To use the **Add** option, select **Add** before calculating the areas. The prompt is reissued after each calculation and AutoCAD displays the running total.

The **Subtract** option is used to subtract an area from the running total. Select **Subtract** before selecting the area(s) to be subtracted from the total.

Object Method

When you select the **Object** option of the **Area** command, AutoCAD computes the area of closed objects, such as a circle, polyline, ellipse, polygon, spline, region, and 3D solids. It will also work with certain types of open objects, such as open polylines and arcs. The following prompt appears:

Command: **area**

Specify first corner point or [Object/Add/Subtract]: **o**

Select objects: *(Pick object.)*

Area = 15.2178, Perimeter = 16.7360

Although AutoCAD prompts you to "Select *objects*" (note the plural), you can select only one object. Attempting to use a selection mode such as Crossing or Window results in the error message, "*Invalid selection* Expects a point or Last."

AutoCAD calculates the area of closed objects in the following manner.

Circle, ellipse, 2D closed spine

Displays the area and the circumference.

Polyline and polygon

When the polyline is closed, the area and perimeter are calculated. A wide polyline is calculated from the centerline of the polyline.

When the polyline is open, the area and perimeter are calculated as though a line is drawn between the endpoints to close the polyline. The area for a polyline is valid only if the polyline can enclose an area. In other words, you cannot compute the area of a polyline with a single segment.

The perimeter of an open polyline is its length.

For example, draw an open polyline. Make a copy, and close the copied polyline. Both have the same area; the open polyline, however, has a shorter perimeter than the closed polyline.

Open object

AutoCAD "closes" the object with a straight line joining the start and endpoints. The area and perimeter are displayed.

Region

The area and perimeter of the region are displayed.

3D solid

The surface area and perimeter are displayed.

LISTING DRAWING INFORMATION

AutoCAD provides you with information about objects in the drawing in three ways:

1. Interactive reports through the **Object Properties** toolbar

2. Comprehensive reports through the **Properties** window

3. Database reports through the **List** and **DbList commands**

INTERACTIVE REPORTING

Any time you select an object, AutoCAD displays some of its properties in the **Object Properties** toolbar. Figure 17.4 shows the data available.

From left to right, the list boxes tell you

Layer status and name

Figure 17.4 *The Object Properties Toolbar*

Color

Linetype

Lineweight

Plot style

There are, however, many other pieces of information about objects that the **Object Properties** toolbar does not tell you. For this reason, we go on to the **Properties, List,** and **DbList** commands.

 Note: If the **Object Properties** toolbar is not visible, right-click any toolbar and select **Object Properties** from the cursor menu.

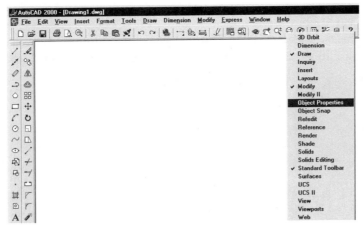

Figure 17.5 *Turning On the Object Properties Toolbar*

THE PROPERTIES WINDOW

 The **Properties** window displays all known information about an object that you select. In addition, it allows you to change nearly all of the object's properties—short of changing the object into something else.

To display the window, type the **Properties** command, or select **Tools | Properties** from the menu bar, or press CTRL+1. The data you see in the window varies depending on: (1) when no objects are selected, (2) when a single object is selected (and its object type), and (3) when more than one object is selected.

The **Properties** window is *non-modal.* This means it can remain open while you draw and edit. As you select objects, the window updates itself.

When values are shown in black, they can be modified. Choose the value box and then either select a value from the drop-down list or type in a new value. In some cases, an additional button appears, which displays a dialog box or allows you to select a point in the drawing. When values are shown in gray, they cannot be changed.

No Objects Selected

When no objects are selected, the **Properties** window displays a general set of properties common to all objects— sort of like the **Object Properties** toolbar.

Figure 17.6 *The Properties Window: No Objects Selected*

Color

Specifies the color of the object. Click to select another standard color, or choose **Other** to display the **Select Color** dialog box.

Layer

Specifies the layer of the object. Click the list to select another layer from among the layers created in current drawing.

Linetype

Specifies the linetype of the object. Click the list to select another linetype from among the linetypes loaded into the current drawing.

Linetype Scale

Specifies the linetype scale factor of the object. Click and type a new scale factor.

Lineweight

Specifies the lineweight of the object. Click the list to see all available lineweights in the current drawing.

Thickness

Sets the current extrusion distance in the Z direction.

Plot Style

Specifies the plot style of the object. Click the list to select NORMAL, BYLAYER, BYBLOCK, or another plot style.

Plot Style Table

Specifies the plot style table. Click the list to select another plot style.

Plot Table Attached To

Determines whether the current plot style table is attached to model space or layout (paper space).

Plot Table Type

Displays the type of plot style tables, if available.

View

Describes the view of the current viewport. The values are *read-only*, meaning that they cannot be changed; to change these values, use one of the view commands, such as **Zoom** and **Pan**.

The **Center X**, **Center Y**, and **Center Z** items describe the *X,Y,Z* coordinates of the center of the viewport. The **Height** and **Width** items describe the height and width of the current viewport. All are reported in the working units.

Misc

Describes data related to UCSs (user coordinate system). **USCicon On** can be turned on and off. **UCS icon at origin** can be turned **on** and off. **UCS per viewport** can be turned on and off. **UCS** name is the name of the named UCS.

Single Object Selected

When a single object is selected, the **Properties** window displays the most information. The information displayed, however, varies with each type of object. For example, select a line and the coordinates of the endpoints are shown. Select a circle, and its center point and radius are displayed instead.

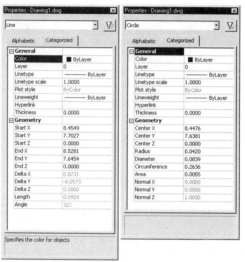

Figure 17.7 *The Properties Window: One Object Selected*

Two or More Objects Selected

When two or more objects are selected, the **Properties** window shows only the properties that the objects have in common.

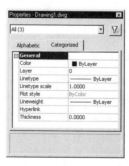

Figure 17.8 *The Properties Window: More Than One Object Selected*

You do, however, have the option to display the properties of each type of object. Notice the list box at the top of the **Properties** window. It probably reads "All." This indicates it is listing the properties in common with all selected objects.

Click the list box to reveal a list of object types. For example, if you select two lines and a circle, the drop-down list reads:

> All (3)
>
> Line (2)
>
> Circle (1)

This indicates that the selection consists of 3 objects: 2 lines and 1 circle. Select "Line (2)." The **Properties** window now lists the properties common to the two lines.

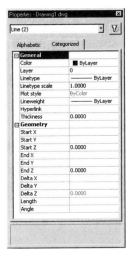

Figure 17.9 *The Properties Window: Two Objects Selected*

DATABASE REPORTS (LIST AND DBLIST)

The **List** command is used to display the data stored by AutoCAD on any object. The data is displayed in the Text window. Type the **List** command or select **Tools | Inquiry | List** from the menu bar.

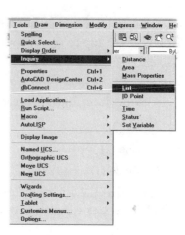

Command: **list**

Select objects: *(Select one or more objects.)*

Select objects: *(Press* ENTER.*)*

AutoCAD displays the Text window and list information about each object you selected. The information listed varies for each type of object. For example, if you select a circle, you obtain the layer the circle is drawn on, the radius, circumference, area, and center point of the circle. AutoCAD switches to the Text window to display the data.

If the object you select contains a lengthy amount of data, you terminate the listing by pressing ESC.

The following shows a sample listing of a circle:

```
CIRCLE        Layer: "0"
              Space: Model space
      Handle = 2F
      center point, X=  8.4476  Y=  7.6381  Z=  0.0000
      radius    0.0420
      circumference   0.2636
      area    0.0055
```

Press F2 to return to the AutoCAD graphics window.

Whereas the **List** command is used to obtain information on a selected set of objects in a drawing, the **DbList** command is used to display the data stored on *all* objects in the drawing. This listing can be very long! AutoCAD switches to the Text window and pauses the listing after every screen fill. See Figure 17.10. The listing begins scrolling again when you strike any key. You terminate the listing by pressing ESC.

Return to the graphics window by pressing F2.

```
AutoCAD Text Window - Drawing1.dwg                                    _ □ ×
Edit
Command: dblist
            REGION     Layer: "0"
                       Space: Model space
                Handle = 2C
                        Area: 15.2178
                   Perimeter: 16.7360
  Bounding Box: Lower Bound X = 3.5882   , Y = 2.7111   , Z = 0.0000
                Upper Bound X = 9.2848   , Y = 5.3825   , Z = 0.0000

            3DSOLID    Layer: "0"
                       Space: Model space
                Handle = 2D
  Bounding Box: Lower Bound X = 9.9792   , Y = 4.1350   , Z = -2.1539
                Upper Bound X = 14.2870  , Y = 8.4428   , Z = 2.1539

            LINE       Layer: "0"
                       Space: Model space
                Handle = 2E
       from point. X=   8.4549  Y=   7.7027  Z=   0.0000
       to point. X=    8.5281  Y=   7.6454  Z=   0.0000
   Length =   0.0929,  Angle in XY Plane =      322
           Delta X =   0.0731, Delta Y =   -0.0573, Delta Z =   0.0000

Press ENTER to continue:
            CIRCLE     Layer: "0"
                       Space: Model space
                Handle = 2F
       center point. X=   8.4476  Y=   7.6381  Z=   0.0000
       radius      0.0420
   circumference   0.2636
        area       0.0055

            LINE       Layer: "0"
                       Space: Model space
                Handle = 30
       from point. X=   8.4921  Y=   7.7207  Z=   0.0000
Command: |
```

Figure 17.10 *The DbList Command*

KEEPING TRACK OF TIME

The **Time** command uses the clock in your computer to keep track of time functions for each drawing. To obtain a listing of the time information for the current drawing, type the Time command or select **Tools | Inquiry | Time** from the menu bar.

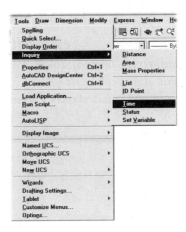

Command: **Time**

Figure 17.11 displays a text screen as shown. The explanation of each follows:

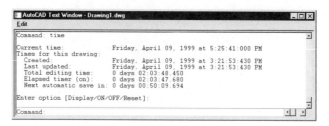

Figure 17.11 *The Time Command*

Date and time the current drawing was initially created. If the drawing was created with **WBlock**, the date of that execution is the creation time. If you edit a drawing created with a previous version of AutoCAD, the first edit time is used as the creation time.

Cumulative time spent on drawing. The **Save** command does not reset the time. If you exit the drawing by using **Quit**, the time does not count. (Plotter time is not added.)

Last time the drawing was updated and saved through the **Save** command.

The time when an automatic drawing save will occur.

"Stopwatch" timer. You can turn this timer on or off. It is independent of the other functions:

Display

Redisplays the time functions with updated times.

ON

Starts the stopwatch timer (the timer is initially on).

OFF

Stops the stopwatch timer, freezing the display at the accumulated time.

Reset

Resets the stopwatch timer to zero.

If you do not wish to execute any of the options, enter a null response (ENTER) or ESC (cancel). Press F2 to return to the AutoCAD drawing window.

EXERCISES

1. Start a new drawing. You use this drawing for all the exercises in this chapter. Draw a line, using the absolute endpoints of 2,2 and 6,6.

   Start the **ID** command, then specify object snap **ENDpoint.** Select the first endpoint of the line with the object snap aperture. The command line should show the ID point as 2,2. Repeat the procedure with the other endpoint location of 6,6.

2. Draw a line, circle, and arc on the screen. Use the **List** command to select the circle. You should obtain a listing of the circle similar to the one shown previously. Select each of the objects and notice the information displayed for each. The **List** command can be useful when you need to obtain information on a single object in your drawing.

3. Repeat with the **Properties** command. Select, in turn, the line, arc, and circle. Notice how the data displayed in the **Properties** window changes.

4. Select all three objects and notice how the **Properties** window displays a smaller amount of information.

5. Use the same drawing with the objects you used for the **List** exercise. Enter **DbList** at the command line. Press F2 and notice how the information scrolls on the Text window. You will need to use F1 to switch back to the graphics display.

6. Use the **Copy** command to copy all the objects in the drawing three times. Use the **DbList** command again and notice the length of the listing.

7. Select the **Dist** command. Using object snap ENDpoint, select two endpoints of one of the lines on the screen. Notice how the length of the line is displayed on the command line.

8. Draw two rectangles of different sizes with the **Rectang** command. Use the **Area** command to figure out the area of one of the rectangles. Be sure to use object snap INTersection to precisely capture the intersections of the rectangle. Repeat the same procedure on the second rectangle. Mentally add the areas together to obtain the total area.

9. Use the **Area** command again to obtain the area of the first rectangle. This time, use the **Add** option to add the area of the second rectangle. Did the two areas add up to the same area you computed before?

10. Open the drawing named Area.Dwg from the CD-ROM (see Figure 17.12). Compute the area of each object and write down the figures.

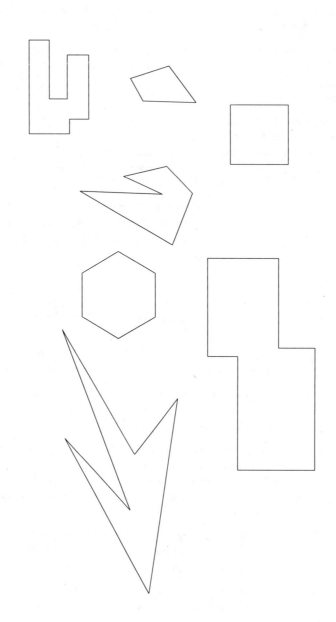

Figure 17.12 *Area.Dwg*

11. Use the **Time** command to check the current time in the drawing you are currently using.

12. Use the elapsed timer option in the **Time** command to time the period needed to draw a car.

CHAPTER REVIEW

1. If you want to identify the coordinates of a specific point in a drawing, what command would you use?

2. What is displayed when you use the **List** command?

3. What is the difference between the **Properties**, **List**, and the **DbList** commands?

4. What are the six distances returned by the **Dist** command?

5. What option under **Area** would you use if you wanted to calculate the area of a circle?

6. What values are displayed when a circle is selected with the **List** command?

7. After using the **Area** and **List** commands on the circle, which command do you find more useful?

8. Repeat the **Area** and **List** commands on a circle and a rectangle. Which command is more useful for finding the total area of both objects?

9. How would you determine the last time a drawing was updated?

CHAPTER 18

Learning Intermediate Draw Commands

After you have become proficient with the basic draw commands in AutoCAD (see Chapter 7), the next step is to master the intermediate commands. Learning the use of these commands allows you to construct more complex drawings in less time. After completing this chapter, you will be able to

- Create special objects, such as ellipses and polygons

- Construct new objects from existing ones

- Operate one of AutoCAD's most powerful features: blocks.

INTRODUCTION

In this chapter, you learn the following intermediate drawing and modification commands:

Ellipse

Draws ellipses and elliptical arcs, as well as isometric circles.

Donut

Draws thick and solid-filled circles.

Polygon

Draws regular polygons from 3 to 1,024 sides.

Offset

Offsets any object.

Chamfer

Cuts off the corner.

Block

Groups several objects together into a single, named object.

WBlock

Writes the block to disk as a DWG file.

Insert

Places a block or another drawing in the current drawing.

Base

Redefines the insertion point for a drawing.

MInsert

Inserts a block as a rectangular array.

DRAWING ELLIPSES

Ellipses look like squashed circles. They represent circles viewed at an angle. The **Ellipse** command is used to construct an ellipse. (You also use the **Ellipse** command to place isometric circles in any of the three planes of an isometric drawing; see Chapter 21 "Constructing Isometric Drawings.")

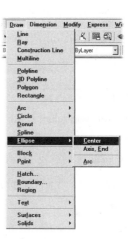

Before constructing an ellipse, let's look at the parts of an ellipse. A *major axis* and a *minor axis* define the ellipse. (Recall that a circle has a constant diameter.) The major axis measures the longest "diameter" of the ellipse; the minor axis measures the shortest "diameter" of the ellipse. The major and minor axes intersect at the center-point of the ellipse.

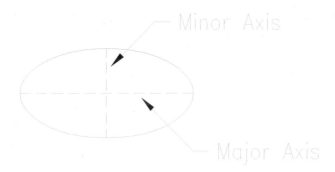

Figure 18.1 *Ellipse Axes*

AutoCAD provides you with several ways to construct an ellipse. Let's look at them now.

CONSTRUCTING AN ELLIPSE BY AXIS

In the first example, we construct an ellipse by showing AutoCAD: (1) the full length of one axis and (2) the half-length of the other axis. It doesn't matter if you first draw the major axis (the longer one) or the minor axis (the shorter one.).

Type the **Ellipse** command, or select **Draw | Ellipse | Axis, End** from the menu bar.

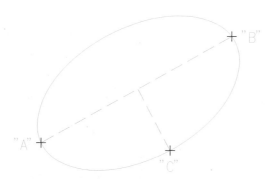

Figure 18.2 *Ellipse by Axis and Eccentricity*

Command: **Ellipse**

Specify axis endpoint of ellipse or [Arc/Center]: *(Pick point A.)*

Specify other endpoint of axis: *(Pick point B.)*

Specify distance to other axis or [Rotation]: *(Pick point C.)*

You can also specify point "C" by using a numerical distance.

The first two points entered can specify the major or minor axis. The distance specified for point "C" determines which axis becomes the major axis. (Remember, the major axis is the longer axis.)

CONSTRUCTING AN ELLIPSE BY AXIS AND ROTATION

When you respond to the "Specify distance to other axis or [Rotation]:" with **R**, you rotate the ellipse around the axis first specified. The ellipse is considered a circle, with the previously specified axis acting as the diameter line. The rotation takes place in the Z plane. That is, it will be as though you rotated the circle "into" the screen. The circle can be rotated into the Z plane at any angle, from 0 to 89.4 degrees. Specifying a zero angle results in a full circle being drawn.

You can also show AutoCAD the angle by drag specification. The angle is measured with the drag line from the midpoint of the ellipse:

> 0 degrees is parallel to the X axis.

> 90 degrees is parallel to the Y axis.

The following illustrations show ellipses of varying rotation angles. Points "A" and "B" designate the endpoints of the major axis.

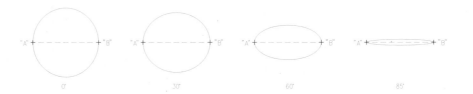

Figure 18.3 *Ellipse by Axis and Rotation*

CONSTRUCTING AN ELLIPSE BY CENTER AND TWO AXES

You can also define an ellipse by specifying its center point, the endpoint of one axis, and the half-length of the other axis.

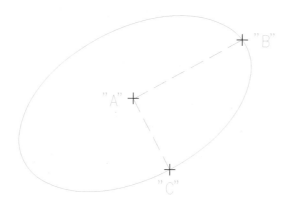

Figure 18.4 *Ellipse by Center and Two Axes*

Let's construct an ellipse by this method. Enter the Ellipse command:

> Command: **ellipse**
>
> Specify axis endpoint of ellipse or [Arc/Center]: **c**
>
> Specify center of ellipse: *(Pick point "A".)*
>
> Specify endpoint of axis:*(Pick point "B".)*
>
> Specify distance to other axis or [Rotation]: *(Pick point "C".)*

Notice that the location of point B determines the angle of the ellipse. The "Specify endpoint of axis:" prompt can be a numerical distance. This distance is the distance from the center point, and perpendicular to the first axis specified by points A and B. Note that a numerical distance represents one-half of the axis defined.

Figure 18.5 shows the results of ellipses constructed by the center and two axes method. Note that the angle and the major axis are determined by the placement and distance of the points.

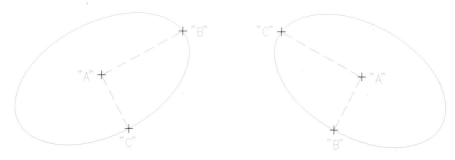

Figure 18.5 *Specify Ellipse Axis Angle*

You can also choose the **Rotation** option when constructing an ellipse using the center method. Figure 18.6 shows the effects of this type of construction.

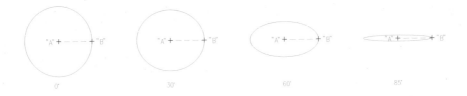

Figure 18.6 *Using Rotation Option with Ellipse*

CONSTRUCTING ISOMETRIC CIRCLES

The **Ellipse** command allows you to correctly place circles in isometric drawings. AutoCAD must be in *iso mode* (use the **Snap** command's **Style** option to set isometric mode). When you execute the **Ellipse** command, it displays an additional prompt: **Isocircle**.

Command: **ellipse**

Specify axis endpoint of ellipse or [Arc/Center/Isocircle]:

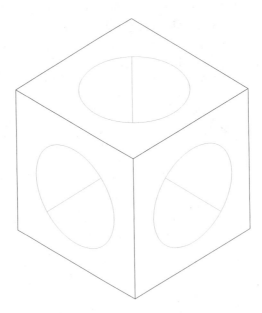

Figure 18.7 *Drawing Isometric Circles with the Ellipse Command in Iso Mode*

For more details on isocircles, see Chapter 21 "Constructing Isometric Drawings."

CONSTRUCTING AN ELLIPTICAL ARC

To construct an elliptical arc, use the **Arc** option of the **Ellipse** command, placing the axes as before. You also specify the starting and ending angles of the arc, as follows:

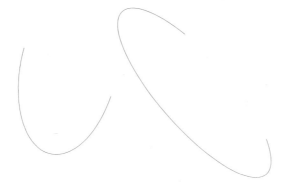

Figure 18.8 *Elliptical Arcs*

Command: **ellipse**

Specify axis endpoint of ellipse or [Arc/Center]: **a**

Specify axis endpoint of elliptical arc or [Center]: *(Pick point.)*

Specify other endpoint of axis: *(Pick point.)*

Specify distance to other axis or [Rotation]: *(Pick point.)*

Specify start angle or [Parameter]: *(Pick point.)*

Specify end angle or [Parameter/Included angle]: *(Pick point.)*

In addition to the method of selecting the start and end angle, there are two additional options for constructing an elliptical arc, **Parameter** and **Included**.

The **Parameter** option lets you specify the arc's angle by sweeping in real time, while the **Included** option lets you specify the included angle for the elliptical arc.

TWO TYPES OF ELLIPSES

AutoCAD can draw two kinds of ellipses: (1) the mathematically accurate ellipse made from the ellipse object introduced by AutoCAD Release 13 and (2) the less accurate ellipse made from a connected series of polyline arcs, which approximate an elliptical shape. The second type of ellipse is the kind drawn by AutoCAD Release 12 and earlier.

You cannot draw elliptical arcs with the polyline ellipse, although you can use the **Trim** command to "cut" the polyline ellipse into an arc. The only time you would use the second type of ellipse is if you need to exchange drawings with someone using an earlier version of AutoCAD (through the **SaveAs** command).

The system variable **PEllipse** determines which style of ellipse is drawn:

Pellipse	Meaning
0	True ellipse is drawn (default)
I	Ellipse is made up of polyline arcs

You change the value of **PEllipse** as follows:

Command: **pellipse**

New value for PELLIPSE <0>: I}

If you wish to edit the ellipse, use the **Trim, Extend, Offset,** and **Break** commands.

DRAWING SOLID-FILLED CIRCLES

The Donut command constructs solid-filled circles and donuts. A *donut* is a thick circle. Execute the **Donut** command by entering either **Donut** or **Doughnut**. The following explanation uses both forms of the command.

To construct a donut, you must tell AutoCAD both the inside and the outside diameters. If you wish to make a solid-filled circle, specify an inside diameter of zero.

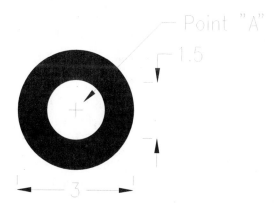

Figure 18.9 *Constructing a Donut*

Let's first make a regular doughnut. Enter the **Donut** command:

> Command: **donut**
>
> Specify inside diameter of donut <0.5000>: **1.5**
>
> Specify outside diameter of donut <1.0000>: **3**
>
> Specify center of donut or <exit>: *(Pick point "A".)*
>
> Specify center of donut or <exit>: *(Press* ENTER.*)*

Notice that the "Specify center of donut" prompt repeats to allow you to place several identical donuts. The default values are saved from the last values used.

Entering a value of zero (0) for the inside diameter results in a solid-filled circle. Let's construct a solid-filled circle using the **Donut** command.

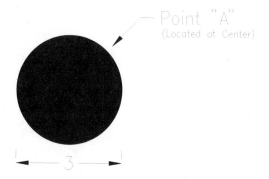

Figure 18.10 *Constructing a Solid-Filled Circle*

Command: **donut**

Specify inside diameter of donut <0.5000>: **0**

Specify outside diameter of donut <1.0000>: **3**

Specify center of donut or <exit>: *(Pick point "A".)*

Specify center of donut or <exit>: *(Press* ENTER.*)*

Donuts are constructed as a closed polyline with two wide arc segments. To edit a donut, use the **PEdit** (polyline editing) command or the edit functions.

The display of the solid fill in the donut is determined by the current setting of the system variable **Fill**. If **Fill** is on, the donut will be solid-filled. When **Fill** is off, only the outline of the donut will be displayed and plotted.

DRAWING POLYGONS

The Polygon command allows you to construct a regular polygon (a closed object with all edges of equal length) with a specified number of sides. You can choose any number of sides, from 3 to 1,024. The default is 4 sides, which constructs a square.

AutoCAD draws polygons from polylines; thus, you can give the lines of the polyline width with the **PEdit** command.

There are three methods of constructing polygons: inscribed, circumscribed, and by the specification of one edge of the polygon. Let's look at each method of constructing polygons.

INSCRIBED POLYGONS

Inscribed polygons are constructed inside a circle of a specified radius (the circle is not drawn). The center point of the circle is first specified, followed by any point on the circumference of the circle. A vertex point of the polygon starts on the point you selected on the circumference, establishing the angle of the polygon. Figure 18.11 shows a polygon constructed through the inscribed method.

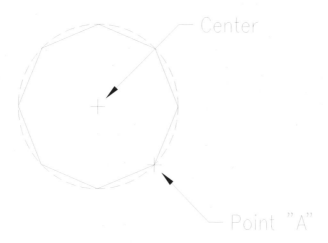

Figure 18.11 *Inscribed Polygon*

CIRCUMSCRIBED POLYGONS

Circumscribed polygons are constructed outside a circle of a specified radius. The center point of the circle is first specified, followed by any point on the circumference of the circle. The midpoint of one edge of the polygon will be placed on the point specified on the circumference, establishing the angle of the polygon. Figure 18.12 shows a polygon constructed using the circumscribed method.

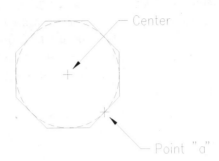

Figure 18.12 *Circumscribed Polygon*

EDGE METHOD OF CONSTRUCTING POLYGONS

You can also construct a polygon by specifying the length of one edge. Simply enter the endpoints of the edge. The angle of the two points specifies the angle of the polygon. Figure 18.13 shows construction of a polygon through the edge method.

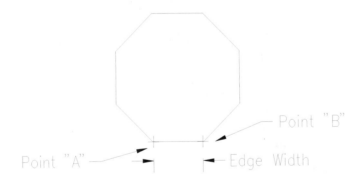

Figure 18.13 *Edge Method of Constructing a Polygon*

CONSTRUCTING POLYGONS

To construct a polygon using either the inscribed or circumscribed methods, enter the **Polygon** command:

> Command: **polygon**
>
> Enter number of sides <4>: *(Enter number of sides.)*
>
> Specify center of polygon or [Edge]: *(Pick center point.)*

> Enter an option [Inscribed in circle/Circumscribed about circle] <I>:
> *(Enter I or C.)*
>
> Specify radius of circle: *(Pick point.)*

To construct a polygon using the edge method, respond to the "Specify center of polygon or [Edge]:" prompt with **E**.

> Command: **polygon**
>
> Enter number of sides <4>: *(Enter number of sides.)*
>
> Specify center of polygon or [Edge]: **E**
>
> Specify first endpoint of edge: *(Pick first edge point.)*
>
> Specify second endpoint of edge: *(Pick second edge point.)*

The polygon is constructed of a closed polyline and you can edit it using **PEdit** or other edit commands.

Note: AutoCAD cannot snap to the center of a polygon. If you need this point, draw an inscribed circle and snap to the center of the circle.

OFFSETTING OBJECTS

The **Offset** command allows you to construct a copy parallel to an object or to construct a larger or smaller image of the object through a point.

CONSTRUCTING PARALLEL OFFSETS

Let's first look at how to construct a parallel object. First, construct a vertical line on the screen similar to the one shown in Figure 18.14. Now, enter the **Offset** command:

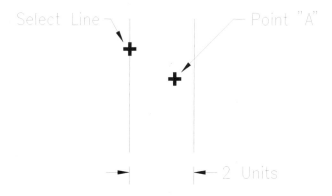

Figure 18.14 *Offsetting a Line*

Command: **offset**

Specify offset distance or [Through] <Through>: **2**

Select object to offset or <exit>: *(Select the line.)*

Specify point on side to offset: *(Pick point "A".)*

Select object to offset or <exit>: *(Press* ENTER.*)*

The default value shown in the brackets is either a numerical value or **Through**, whichever was last chosen.

You can choose to show AutoCAD two relative points to designate the offset distance, instead of entering a numerical value.

The "Select object to offset:" prompt repeats to allow you to offset as many copies as you wish. Enter a null response (press ENTER) to terminate the command.

The direction of the object's Z axis must be parallel to the current user coordinate system (UCSs are explained in Chapter 32). If it is not, AutoCAD displays the message:

That object is not parallel with the UCS.

The "Select object:" prompt is then repeated.

CONSTRUCTING "THROUGH" OFFSETS

The **Through** option allows construction of the offset through a point. Let's create an offset to a circle. Draw a circle as in the following illustration. Enter the **Offset** command.

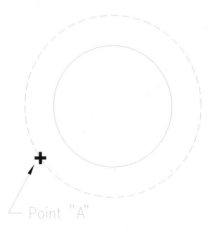

Figure 18.15 *Using a Through Point with the Offset Command*

Command: **offset**

Specify offset distance or [Through] <2.0000>: **t**

Select object to offset or <exit>: *(Pick the circle.)*

Specify through point: *(Pick point "A".)*

Select object to offset or <exit>: *(Press* ENTER.*)*

Note: You can choose only one object at a time to offset. If you choose to offset a complex object, it must be one object. Constructing objects with polylines allows you to do this.

CHAMFERING LINES AND POLYLINES

The **Chamfer** command is used to trim segments from the ends of two lines or polylines and draw a straight line or polyline segment between them.

You can specify the amount to be trimmed by entering either a numerical value or by showing AutoCAD the distance using two points on the screen.

The distance to be trimmed from each segment can be different or the same. The two objects do not have to intersect, but they should be capable of intersecting within the limits. If the limits are off, the segments should be capable of intersecting at some point.

The following example shows how the **Chamfer** command works. Figure 18.16 will be used:

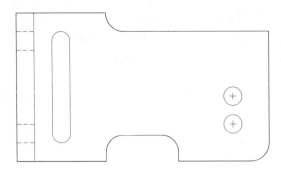

Figure 18.16 *Chamfer Example Drawing*

We will first set the **Chamfer** defaults:

> Command: **chamfer**
>
> (TRIM mode) Current chamfer Dist1 = 0.5000, Dist2 = 0.5000
>
> Select first line or [Polyline/Distance/Angle/Trim/Method]: **d**
>
> Specify first chamfer distance <0.5000>: **2**
>
> Specify second chamfer distance <2.0000>: **4**

You have now set the default chamfer distances. When you next enter the **Chamfer** command, the first line you choose will be trimmed by two units and the second by four units.

Reenter the command:

> Command: **Chamfer** *(Or press Enter.)*
>
> (TRIM mode) Current chamfer Dist1 = 2.0000, Dist2 = 4.0000
>
> Select first line or [Polyline/Distance/Angle/Trim/Method]:*(Select first line.)*
>
> Select second line: *(Select second line.)*

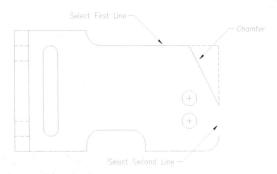

Figure 18.17 *Chamfering a Corner*

The default settings remain until you reset them or you enter a new drawing. The initial setting is determined by the prototype drawing.

> **Note: Chamfer** only works with line segments, such as line, 2D polylines, and traces. **Chamfer** does not work with arc segments, such as arcs, circles, and ellipses. In 3D drawings, you can use **Chamfer** on 3D solid models, such as boxes and other rectangular shapes. When the chamfer distances are both set to 0, AutoCAD performs the equivalent of a **Trim** or **Extend** command on the two line segments.

Polylines

Like filleting, chamfering a polyline is different from chamfering a pair of lines (fillets are discussed in Chapter 8). If you used the **Close** option to finish the polyline, then AutoCAD chamfers all corners of the polyline.

> Command: **chamfer**
>
> (TRIM mode) Current chamfer Dist1 = 2.0000, Dist2 = 4.0000
>
> Select first line or [Polyline/Distance/Angle/Trim/Method]: **P**
>
> Select 2D polyline: *(Pick.)*

If the polyline contains arcs, the arcs are converted to chamfers. If some parts of the polyline do not chamfer, it could be that the segment length is too short or two segments are parallel to each other. In that case, AutoCAD warns: "5 lines were chamfered. 1 was parallel. 2 were too short."

Angle

As an alternative to specifying a chamfer by two distances, AutoCAD lets you specify a distance and an angle (as shown in Figure 18.17).

> Command: **chamfer**
>
> (TRIM mode) Current chamfer Dist1 = 2.0000, Dist2 = 4.0000
>
> Select first line or [Polyline/Distance/Angle/Trim/Method]: **A**
>
> Specify chamfer length on the first line <1.0000>: *(Type the distance.)*
>
> Specify chamfer angle from the first line <0>: *(Type the angle.)*

Trim

Normally, **Chamfer** erases the line segments not needed by the chamfer process. However, the **Trim** option lets you select whether or not you want to keep the excess lines.

> Command: **chamfer**
>
> (TRIM mode) Current chamfer Dist1 = 2.0000, Dist2 = 4.0000
>
> Select first line or [Polyline/Distance/Angle/Trim/Method]: **T**

> Enter Trim mode option [Trim/No trim] <Trim>: *(Type T or N and press* ENTER.*)*

Method

The default for **Chamfer** is for you to specify two distances. If you prefer the distance-angle method, the **Method** option lets you change the default:

> Command: **chamfer**
>
> (NOTRIM mode) Current chamfer Dist1 = 2.0000, Dist2 = 4.0000
>
> Select first line or [Polyline/Distance/Angle/Trim/Method]: **M**
>
> Enter trim method [Distance/Angle] <Angle>: *(Type D or A and press* ENTER.*)*

BLOCKS AND INSERTS

One of the most valuable elements of CAD drafting is the ability to reuse parts of drawings. AutoCAD provides commands to store such symbol details and drawings. These commands are **Block**, **WBlock**, and **Insert**.

The **Block** and **WBlock** commands are used to create and store the symbols. The **Insert** command is used to place them in a drawing. Let's look at how to use the **Block** and **Insert** commands.

COMBINING OBJECTS IN A BLOCK

A block is a group of objects that have been identified by a name and combined into a single object. This grouping can be placed into a drawing. Blocks can be placed any place in the drawing, scaled, and rotated to your specifications.

A block is considered as a single object. Because of this, you can move or erase a block as though it were a single object.

Note: It is more efficient to use blocks than to copy the same objects over and over. The **Copy** command makes a complete copy each time you use it. With blocks, AutoCAD creates a single copy, then points to copies of the block you make with the Insert command. That helps reduce the drawing's file size and improves the display time.

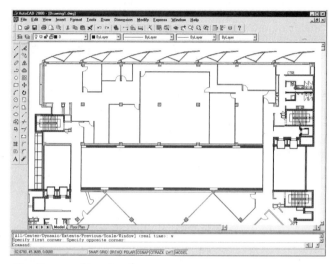

Figure 18.18 *This Drawing Uses Blocks*

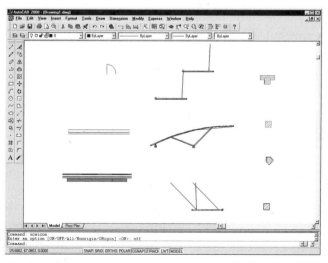

Figure 18.19 *Blocks Used in Drawing*

CREATING A BLOCK

To create a block, select **Draw | Block | Make** from the menu bar, or choose the **Make Block** icon on the **Draw** toolbar, or type **Block** at the command line. AutoCAD displays the **Block Definition** dialog box. There are three crucial pieces of information AutoCAD needs to know before it creates the block:

- A name for the block, so that you can identify it

- X,Y,Z coordinates for the insertion point, so that AutoCAD knows where to start placing the block

- The objects you want collected into a block

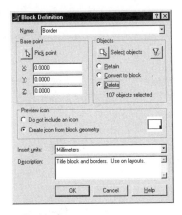

Figure 18.20 *Block Definition Dialog Box*

The **Block Definition** dialog box takes you through the stages in providing the information and gives you one option.

Name: You must give the block a name. The name can be anything from a single letter through to a word 255 characters long. AutoCAD stores blocks in the drawing in which they were made.

If you try to give the block a name that already exists in the current drawing, AutoCAD warns, "*Blockname* is already defined. Do you want to redefine it?" Most of the time you will choose **No** and change the name. Sometimes, however, you deliberately want to redefine an existing block. Perhaps you made an error in your first attempt at creating the block or you left out attribute definitions. In this case, choose **Yes**.

Figure 18.21 *Duplicate Block Name Warning*

List Block Names: If you need help seeing the names of blocks already defined in the drawing, click the down arrow in the **Name** field. AutoCAD displays the names of blocks in the current drawing. Block names that begin with * (asterisk), such as *X20, are blocks created by AutoCAD. These are called *anonymous* blocks.

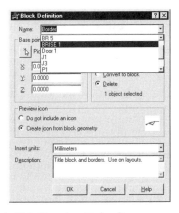

Figure 18.22 *Block Names in This Drawing Dialog Box*

Base Point: You can type the *X,Y,Z* coordinates or choose the **Pick Point** button to select the insertion point in the drawing. The dialog box disappears and AutoCAD prompts you, "Specify insertion base point:". I recommend you use an object snap mode to make the selection accurate, such as ENDpoint for the end of a line or CENter for the center of a circle or arc. After the selection, the dialog box reappears and AutoCAD fills in the values for the *X,Y,Z* values. Normally the *Z* value is 0, unless you have set the elevation or are selecting a point in a 3D object.

Objects: Choose the **Select Objects** button to select the objects that make up the block. The dialog box disappears and AutoCAD prompts you, "Select objects:". You can use any method of object selection, such as Window or Fence. After you are finished selecting objects, press ENTER and the dialog box reappears. AutoCAD reports the number of objects you selected.

Retain: AutoCAD normally erases the objects making up the block. The **Retain** option prevents this, like using the **Oops** command after the **Block** command (see below).

Two other options are **Convert to block** (which erases the objects and replaces them with the block) or **Delete** (which simply erases the objects).

QSelect: Displays the **Quick Select** dialog box, which allows you to create a selection set of objects to be converted to a block.

Preview Icon: It can be useful to include an icon with the block.

Insert Units: AutoCAD usually makes one drawing unit equal the current unit. With this list box, you can select another unit, ranging from light years to angstroms.

Description: You can include, if you are so inclined, a long description of the block.

OK: After you've given the new block its name, its insertion point, and constituent members, choose the **OK** button. AutoCAD creates the block, which you can now access with the **Insert** command.

Blocks and Layers

When inserted, a block retains its original layer definitions. That is, if an object was originally located on a layer named "PCBOARD," it will be on that named layer when inserted.

The exception is an object that was originally on layer 0. Objects created on layer 0 in the block are assigned to the layer on which the block is inserted.

Nested Blocks

Blocks can be *nested*. That is, you can place one block inside another. Consider the following example. Each object is a block and was inserted into another block. If you wish, you can "block" the entire part (made up of other blocks) and have a new block that contains the original and the nested blocks together.

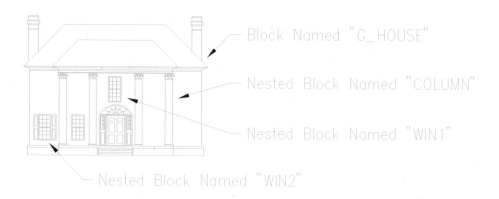

Figure 18.23 *Nested Blocks*

CREATING A DRAWING FILE FROM A BLOCK

The **WBlock** command can be used to capture a portion or all of a drawing and write it to the disk as a separate drawing.

After you execute the **WBlock** command, AutoCAD 2000 displays the **Write Block** dialog box, which is very similar to the **Block** command's dialog box. The primary difference is that you have a choice of *what* to write to disk in the **Source** area:

Block: Saves a block to a DWG file on disk. The block must already exist in the drawing.

Entire Drawing: Saves the entire drawing to disk with one exception: unused layers and blocks are not saved. This option is often used to quickly "clean" a drawing, which helps reduce its size.

Objects: Allows you to select the objects to save to a DWG file.

In the **Destination** area, you can specify the file name and location (drive and folder.)

The **WBlock** method of storage is one way to use the block in other drawings. A separate drawing can be placed in any drawing. The other method of sharing blocks is through the **XBind** command. You can use a separate drawing as part of any other drawing by simply inserting it with the **Insert** command.

 Note: The **Insert** command is used to place blocks or other entire drawings into your drawings. After insertion, the block becomes a part of the drawing. In the next section, we look at the **Insert** command.

INSERTING BLOCKS

The Insert command inserts blocks and external drawings into the current drawing. AutoCAD displays the **Insert** dialog box.

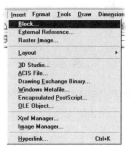

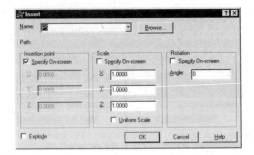

Figure 18.24 *Insert Dialog Box*

Inserting Blocks

A predefined block can be inserted in two ways. The first is to type the block name in the **Name** field.

The second is to click the down arrow in the **Name** field to display the list of block names. Select a block.

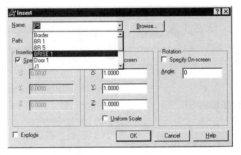

Figure 18.25 *Defined Blocks Dialog Box*

Inserting Drawing Files

The method of inserting entire drawing files into your drawing is similar to inserting blocks. You can either type a drawing name in the text box next to the **Browse** button, or choose the **Browse** button to display the **Select Drawing File** dialog box. You select a drawing file from the list box displayed in the dialog box.

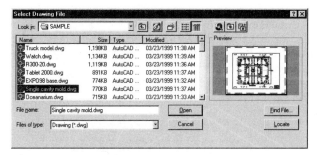

Figure 18.26 *Select Drawing File Dialog Box*

Insertion Point

The insertion point is the reference point for the block. When you identified the base point for the block, you chose the reference point from which the block will be inserted. The block is inserted into the drawing so that the insert point and the base point of the block are the same. To place a block accurately, use an object snap mode.

You can enter the coordinates in the dialog box or select the **Specify on screen** option to select a place for the block in the drawing.

Setting the Scale and Rotation

The scale factor and rotation can be preset or entered at the time of insertion. To preset the scale factors and rotation angle, deselect the **Specify on-Screen** check box.

When the box is selected, the factors are selected when the block is inserted and the scale factor fields are grayed out.

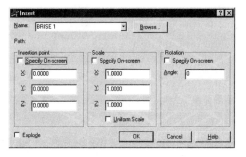

Figure 18.27 *Insertion Point, Scale, and Rotation Parameters*

Exploding the Inserted Item

You can designate that the block or drawing file be automatically exploded. To do this, select the **Explode** check box.

Note that you can only specify a single scale factor for an exploded block, which is applied equally to *X*, *Y*, and *Z*.

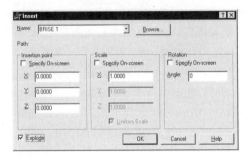

Figure 18.28 *Explode Check Box*

SCALE FACTORS

By default, AutoCAD uses a unit scale factor for the inserted block. There are times, however, when you might want the block to be inserted at a different scale. The scale factor is a multiplier at which the block was constructed. The *X* and *Y* scales can be the same or different, negative or positive. The following examples show the same drawing inserted at different *X* and *Y* scales.

Figure 18.29 *Using Insertion Scale Factors*

Negative Scale Factors

When you use negative scale factors, the drawing is mirrored around the axis to which the negative factor is applied. The following example shows illustrations of negative scale factor combinations for each axis:

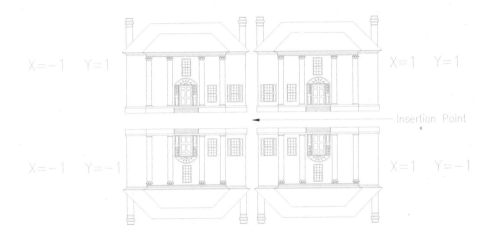

Figure 18.30 *Negative Insertion Scale Factors*

Note that the **Mirror** command is another way of producing images of blocks and other objects besides using negative scale factors. The **Mirror** command is covered in Chapter 19 "Applying Intermediate Edit Commands."

Rotation Angle

The angle of the inserted block is specified in the current AutoCAD angle format. This angle is in reference to the original orientation of the drawn block. If you wish, you can enter a point showing the desired angle. This point is shown immediately after you are prompted for the insertion point. If you move the crosshairs, the point rubber bands between the previously set insertion point and the crosshairs. Move the cursor until the rubber-banded line shows the desired angle and enter a point. The distance between the two points is irrelevant. The angle measured by the rubber-banded line between the insertion point and the angle point determines the angle of insertion.

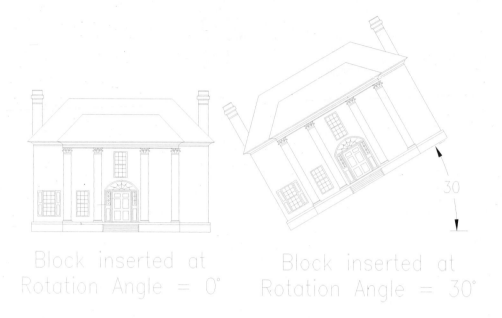

Block inserted at Rotation Angle = 0°

Block inserted at Rotation Angle = 30°

Figure 18.31 *Insertion Rotation Angle*

 Note: After the block has been inserted, use the insertion object snap mode to snap to the block's insertion point.

WHOLE DRAWINGS AS INSERTS

You can insert an entire drawing into another drawing by using the **Insert** command. When you are prompted for the block name, specify the desired drawing name, including the drive specifier.

The drawing will be inserted with the 0,0 point of the original drawing as the insertion base point.

If you wish to have greater control over the insertion of the drawing, use the **Base** command. The **Base** command is executed in the drawing to be inserted before you insert it into another drawing. In the drawing to be inserted, designate the **Base** command by entering:

Command: **base**

Specify base point <default>: *(Select the point you desire for the insertion base point.)*

Be sure to **Save** the drawing to save the base point location. The point you stipulate becomes the reference when you insert this drawing into another.

Redefining Inserts

If you have used many inserts of the same block in a drawing, redefining only one of the blocks changes all the blocks. This is an especially powerful feature. Imagine being able to change 100 drawing parts in a single operation!

There are two methods of redefining a block: (1) through the **Block** command and (2) through the **Xref** command.

If you inserted an entire drawing as a block, you can edit the original drawing. This alone will not redefine the block. You must then (while in the drawing in which the block was inserted) reissue the **-Block** command. When you are prompted for the block name, type the following:

> Block name=file name

This forces the regeneration of all instances of the inserted block and will result in your changes being incorporated in all of them!

If you defined the block "on the run" (that is, you defined a portion of your current drawing as a block), you can still redefine the block, but you need to use a different method.

Insert the block into the drawing using the **Explode** option. Make the desired changes. Then re-block the edited block using the same block name. AutoCAD will then inform you that you already have an existing block by this name and will be asked whether you wish to redefine it. Respond to this dialog box with **OK**. All instances of the block will be redefined (updated).

Another method of updating blocks is to use the "external reference" of specifying an insert. The **Xref** command allows you to insert a separate drawing in a procedure similar to a standard insert. Instead of storing the inserted drawing in the database of the new drawing, however, an external reference to the original drawing is stored. If the original inserted drawing is changed, the changes will be loaded into the "second" drawing when it is next loaded.

External references are covered in Chapter 25 "Attaching External Drawings."

ADVANTAGES OF BLOCKS

Using blocks in your drawings has several distinct advantages.

Libraries

Entire libraries of blocks can be built that can be used over and over again for repetitive details.

Time saving

Using blocks and nested blocks is an excellent method of building larger drawings from "pieces." (A *nested* block is a block that is placed within another block.)

Space saving

Several repetitive blocks require less space than copies of the same entities. AutoCAD must only store information on one set of entities instead of several sets. Each instance of the block can be referred to as one entity (a Block reference). The larger the block, the greater the space saving. This can be very valuable if there are many occurrences of the block.

Attributes

An attribute is a text record that is stored with a block. The text can be set to be displayed or invisible. Attributes must be attached to a block. These attributes can be loaded into database or spreadsheet programs or printed as a listing. This is useful in facilities management where there are multiple occurrences of items, such as desks, which can be stored as blocks and have attributes, such as a person's name or telephone number, associated with them.

Attributes are covered in Chapter 24 "Applying Attributes."

MULTIPLE BLOCK INSERTIONS

The **MInsert** (short for "multiple insert") command is actually a single command that combines the **Insert** and rectangular **Array** operations.

The sequence starts by issuing prompts in the same manner as the **Insert** command. You are then prompted for information to construct a rectangular **Array**. Let's explore the procedure step by step.

MInsert Operations

The **MInsert** command produces an array that has many of the same properties as blocks, with some exceptions. The following qualities apply to MInserts:

- The entire array reacts to editing commands as if it were one block. You cannot edit each individual item. If you select any one object to move or copy, for instance, the entire array is affected.

- You cannot MInsert a block as exploded.

- You cannot explode the block into individual objects. (The **Explode** command is covered in Chapter 19.) AutoCAD reports, "I was not able to be exploded."

- If the initial block is inserted with a rotation, the entire array is rotated around the insertion point of the initial block. This creates an array in which the original object appears to have been inserted at a standard zero angle, then the entire array rotated.

Figure 18.32 shows the block inserted at a 30-degree angle through the **MInsert** command, with four rows and six columns.

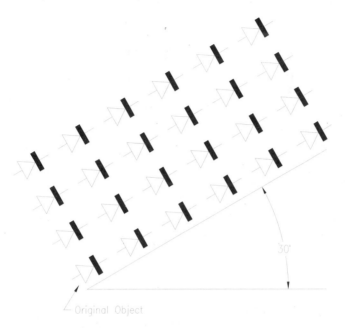

Figure 18.32 *MInsert at 30 Degrees*

Using MInsert

If you wish to create a true horizontal and vertical array in which the objects are themselves rotated, block the original object in its desired rotation, or use the **Insert** and **Array** commands separately.

The **MInsert** command can be used to save steps if you intend to array an inserted block by combining two command functions into one command.

Figure 18.33 *Original Diode Drawing Rotated*

Figure 18.34 *MInserted Diodes*

Create arrays using the **MInsert** command if you might need to edit the array as a whole later. For example, a seating arrangement consisting of several rows of chairs might need to be moved around in a space for design purposes. Creating the arrangement by using the **MInsert** command would allow you to move and rotate the seating as a whole.

EXERCISES

1. Let's use the **Ellipse** command to draw a can. Start by entering the **Ellipse** command. Use the following illustration to draw the ellipse. Don't worry about the size

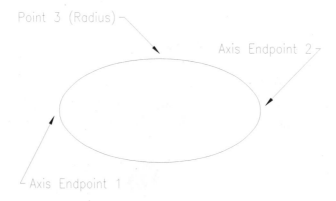

Figure 18.35 *Drawing the Ellipse*

Next, use the **Copy** command to copy the ellipse above the first to create the top of the can. Use F8 to turn on ortho mode so the copy will align perfectly with the original ellipse.

Now use the **Line** command to draw lines between the two ellipses as shown in the following illustration. Use object snap QUADrant to capture the outer quadrant of the ellipses.

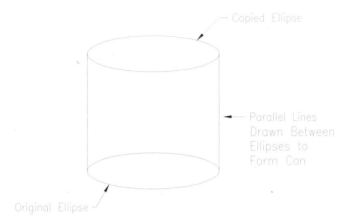

Copied Ellipse

Parallel Lines
Drawn Between
Ellipses to
Form Can

Original Ellipse

Figure 18.36 *Completing the Can*

2. Draw the bicycle as shown in Figure 18.37. Use polylines for the frame, lines for the spokes, and donuts for the tires.

Figure 18.37 *Bicycle*

3. Let's use the **Offset** command to help draw a site plan. Open the drawing named **OFFSET.Dwg** from the CD-ROM. This is a drawing of a cityscape. The edges of the streets are drawn with a different type of line called a poly-line (polylines are covered in Chapter 22 "Understanding Advanced Operations"). Polylines can be joined so that each segment is "glued" to the other segments.

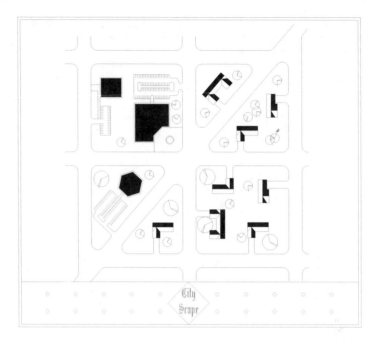

Figure 18.38 *OFFSET.DWG Work Disk Drawing*

Use the **Offset** command to offset a curb thickness (curbs are typically 6" thick). You can do this by using **Offset Through**. Start by selecting the curb to be offset, and then entering a relative coordinate to offset the 6". You may want to review relative coordinates in Chapter 4 "General AutoCAD Principles."

4. Draw the object shown on the left in Figure 18.39. Use the **Chamfer** command to edit the object so that it looks like the object on the right. Hint: You can set distances in AutoCAD by entering two points.

Figure 18.39 *Chamfer Exercise*

5. Use the **Polygon** command to draw the following object.

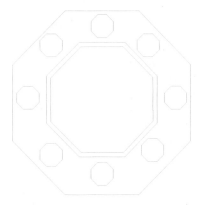

Figure 18.40 *Polygon Exercise*

6. Let's capture a block from a drawing. Open the drawing from the CD-ROM named **OFFICE.Dwg.** The following illustration shows the drawing. Execute the **Block** command and window the entire desk. Name the block "SDESK."

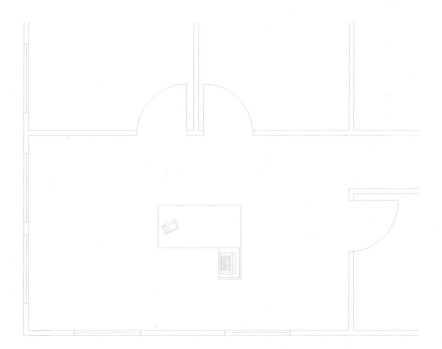

Figure 18.41 *OFFICE.DWG CD-ROM Drawing*

The block is now stored with the drawing and can be used as many times as you desire.

7. Use the **Insert** command to place the **SDESK** block into the drawing.

8. Open the drawing from the CD-ROM named INSERT.Dwg. This is the site plan you used in the **Copy** exercise in Chapter 8 "Editing the Drawing." The landscape items have already been drawn and blocked and are contained as resident blocks in the drawing.

   Use **Insert** to view the names of the blocks; then use **Insert** to insert the landscaping to create a design of your own.

9. In this exercise, you will draw a box and perform a multiple insert containing four rows and six columns. You use the "window" method of showing AutoCAD the column and row spacing. As you proceed, notice the similarities to the **Insert** and rectangular **Array** commands.

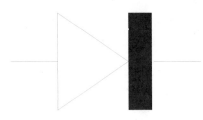

Figure 18.42 *Diode Drawing*

Construct a drawing as shown in the illustration above. Block the rectangle and name it "DIODE1." Use the left end of the diode as the insertion point. You have now created the block "DIODE1" in the drawing. (Notice, however, that you cannot see it.)

Use the **MInsert** command, as follows:

> Command: **minsert**
>
> Enter block name or ?: **diode1**
>
> Specify insertion point or
>   [Scale/X/Y/Z/Rotate/PScale/PX/PY/PZ/PRotate]: *(Pick point.)*
>
> Enter X scale factor, specify opposite corner, or [Corner/XYZ] <1>:
>   *(Press Enter.)*
>
> Enter Y scale factor <use X scale factor>: *(Press* ENTER.*)*
>
> Specify rotation angle <0>: *(Press* ENTER.*)*
>
> Enter number of rows (——) <1>: **4**
>
> Enter number of columns (||||) <1>: **6**

Enter distance between rows or specify unit cell (—-): *(Pick point A.)*

Specify opposite corner: *(Pick point B.)*

You may need to **Zoom All** to see the completed array.

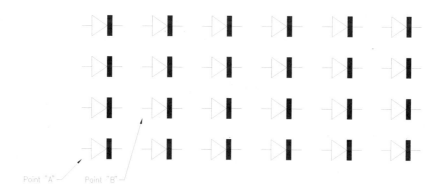

Point "A" Point "B"

Figure 18.43 *MInserted Diodes*

Notice the (—-) and (|||) notations in the rows and columns prompts that make it easy to remember which way rows and columns operate.

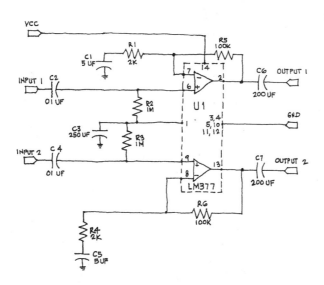

Figure 18.44 *Create a Symbol Library and Use It to Draw the Stereo Amplifier*

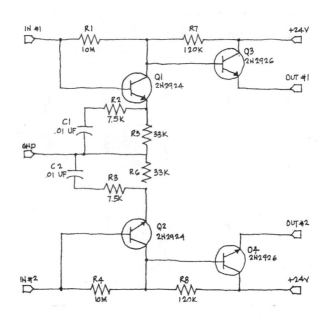

Figure 18.45 *Create a Symbol Library and Use It to Draw the Preamplifier*

CHAPTER REVIEW

1. What are the two axes of an ellipse?

2. How would you draw an isometric circle?

3. From what two types of objects can ellipses be constructed?

4. What types of objects are constructed with the **Donut** command?

5. How do you control whether the solid-filled areas of donuts are displayed and plotted?

6. How many objects can be offset at one time?

7. What does the **Through** option of the **Offset** command perform?

8. Draw an example of using the **Offset** option of **Through**.

9. What does the **Chamfer** command perform?

10. What is the procedure for setting chamfer distances?

11. The **Polygon** command constructs polygons of a specified number of sides. What is the minimum number of sides? The maximum?

12. What is a *block*?

13. What is the *insertion base point* of a block?

14. When placed in a drawing, how does a block handle its layer definitions?

15. What is a *nested* block?

16. How would you create a separate drawing file from an existing block?

17. How do you place a block in a drawing?

18. How can you place a block in another drawing?

19. When you place a block in the drawing, what is the insertion point?

20. Can you place one AutoCAD drawing in another AutoCAD drawing?

21. Name two advantages of using blocks.

22. What commands are combined to create the **MInsert** command?

CHAPTER 19

Applying Intermediate Edit Commands

After you have become proficient with the basic edit functions in AutoCAD, you can learn more powerful edit commands. This chapter covers the intermediate edit commands. After completing this chapter, you will be able to

- Change the properties of an existing object or objects

- Produce multiple copies of an object or objects with the array functions

- Edit existing objects by rotating, stretching, scaling, etc.

- Undo operations just performed with AutoCAD

INTRODUCTION

In AutoCAD, you can change the look (called the "property") of an object after you draw it. Properties include color, layer, linetype, linetype scale, elevation, thickness, and text style. You can use individual commands to change the related property: **Color**, **Layer**, **Linetype**, **Ltscale**, **Elev**, and **Style**. However, AutoCAD includes commands that let you change more than one property at a time. In this chapter, you learn about

Object Properties Toolbar

Lets you see and change properties of objects.

MatchProp

Lets you copy the properties from one object to a selection set.

Properties

Displays a dialog box that lets you change nearly every aspect of objects.

Changing properties is different from editing an object; editing involves changing the object's size, position, and shape. You have encountered some of AutoCAD's editing commands in earlier chapters. In this chapter, you learn about

Array

Creates rectangular and polar arrays.

Mirror

Mirrors objects.

Divide

Places a specified number of points or blocks along an object.

Measure

Places points or blocks a measured distance along an object.

Explode

Breaks apart complex objects, such as blocks and polylines.

Trim

Shortens an object to a trim line.

Extend

Lengthens an object to a trim line.

Lengthen

Changes the length of open objects.

Rotate

Rotates an object.

Scale

Changes the size of an object.

Stretch

Stretches or moves an object.

Undo

Undoes one or more previous operations.

DrawOrder

Changes the display order of overlapping objects.

DdGrips

Allows direct editing through grips.

EDITING WITH THE OBJECT PROPERTIES TOOLBAR

The **Object Properties** toolbar, located above the drawing area, lets you see and change several properties of objects: layer, color, linetype, and lineweight. It does not let you change an object's other properties—linetype scale, elevation, thickness, or style.

When you select a single object, AutoCAD changes the display of the **Object Properties** toolbar to reflect that object's properties. For example, when you select a red circle on layer Door with the hidden linetype, AutoCAD changes the **Object Properties** toolbar to display layer Door, color Red, and linetype Hidden.

When you select more than one object, AutoCAD only displays those object properties that are in common; the other properties are blanked out. For example, when you select the red circle and a black line (both are on layer Door and have linetype Hidden), then the Object Properties toolbar displays the Door layer name and Hidden linetype; however, the Color section is blank.

As you draw an object, all of the drop-down lists are grayed-out, indicating you cannot change them during the command.

From left to right, the controls on the **Object Properties** toolbar perform the following functions:

Layer: The **Layer Control** lists all layers in the current drawing. To set a current layer, simply select its name.

When this is a new drawing or you have not created any layers, **Layer Control** displays a single layer: 0. To create additional layers in the drawing, choose the adjacent **Layer** button.

The **Layer Control** lets you change the settings of layers by choosing the small icons next to each name:

> On or off.
>
> Freeze or thaw. When you freeze a layer, the layer name changes from black to gray.
>
> Unlock or lock.
>
> Plot or no plot.

Choosing the color square next to the layer name does nothing. Select the layer name to make it the current layer.

Color: The **Color Control** lists the first seven AutoCAD colors—red, yellow, green, cyan, blue, magenta, white—along with Bylayer and Byblock. To set a current color, simply select the color name (or number).

To list additional colors, select the **Other** item at the end of the list, which brings up the **Select Color** dialog box. The color you select is added to the list, to a maximum of four additional colors.

Linetype: The **Linetype Control** lists all linetypes currently loaded in the drawing. To set a current linetype, simply select the linetype name.

If this is a new drawing or you have not loaded any linetypes, **Linetype Control** displays just three linetypes: Continuous, Bylayer, and Byblock. To load additional linetypes into the drawing or to change the linetype scale, select the **Other** item at the end of the list.

Lineweight: The **Lineweight Control** provides a list of preset line widths, ranging from 0 mm (one pixel wide) to 2.11 mm. In addition, the standard Bylayer and Byblock are available, as well as Default.

Plot Style: the **Plot Style Control** lists the plot style names that are defined in this drawing. The default plot style is called Normal. The plot style is a collection of overrides for color, dithering, gray scale, pen assignments, screening, linetype, lineweight, end styles, join styles, and fill styles. Plot styles are assigned on an object basis, not by layer.

CHANGING OBJECT PROPERTIES

As described earlier, the **Object Properties** toolbar lets you see and set some of the properties of one or more objects. You can also use the **Object Properties** toolbar to change the properties of objects. The procedure is different from the way you may be used to operating AutoCAD, because you do not type any command; instead, you use your mouse to select, as follows:

Step 1

Select an object. AutoCAD highlights the object and displays its properties on the toolbar.

Step 2

Select the new properties you want for the objects. AutoCAD changes the properties. In Figure 19.2, I have selected layer TXT, color BYLAYER, and linetype BYLAYER. As I selected the properties, AutoCAD changed the look of the selected object.

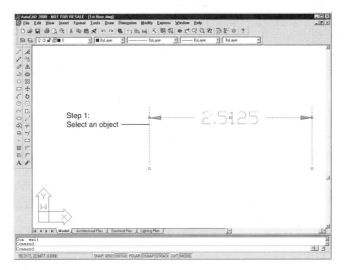

Figure 19.1 *Select An Object*

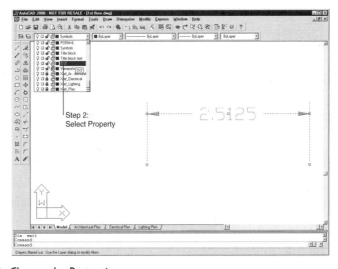

Figure 19.2 *Change the Properties*

Step 3

Press ESC twice to de-select the object. AutoCAD removes the highlighting.

MATCHING PROPERTIES

Sometimes you want to have a group of objects match the properties of some other object in your drawing. For instance, you accidentally placed several doors on the Landscape layer, instead of on the door layer. Or, you realize that some lines need to have a different linetype but you don't recall the name of the linetype.

You could use the procedure outlined above to change the layer and linetype, but it doesn't work well if you are not sure of the exact layer and linetype name. Complex drawings have hundreds of layer names, many of which look similar. One of the sample drawings provided with AutoCAD has the following layer names: ARCC, ARC-CLR, ARCDIMR, ARCDSHR, ARCG, ARCM, ARCR, ARCRMNG, ARCRMNR, and ARCTXTG. And that's just the first ten layer names! Similarly, linetypes can look confusingly similar.

The **MatchProp** command solves those problems. **MatchProp** lets you copy the properties from one object to a selection set of objects. (For compatibility with AutoCAD LT, you can also type **Painter** as the command name.)

> Command: **matchprop**
>
> Select source object: *(Pick object.)*

When you use the **MatchProp** command, you must select objects in the correct order: (1) first, you select the single object whose properties you want to copy and (2) then, you select the object(s) you want to take on those properties.

> Current active settings: Color Layer Ltype Ltscale Lineweight
> Thickness PlotStyle Text Dim Hatch
>
> Select destination object(s) or [Settings]: *(Pick object or type **S**.)*

When you select one (or more) destination objects, AutoCAD assumes you want to match *all* properties from the source object. You can use any form of object selection.

Type **S** when you want to match *some* properties. AutoCAD displays the **Property Settings** dialog box, which displays the settings of the source object.

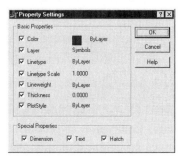

Figure 19.3 *Property Settings Dialog Box*

The properties you can match are more extensive than the properties that the **Object Properties** toolbar allows you to change. The full list includes: color, layer, linetype, linetype scale, lineweight, thickness, plot style, dimension properties, text properties, and hatch properties. (Lineweight is the width of an object, while thickness is its extrusion height in the *Z* direction.)

Property	Objects	Action
Color	All	Changes the color
Layer	All	Changes the layer
Linetype	All	Changes the linetype
Linetype Scale	All	Changes the linetype scale factor
Thickness	All except 3D faces, 3D polylines, 3D polygon meshes, dimensions, and viewports	Changes the thickness
Text	Text and MText only	Changes the text style (not attribute text)
Dimension	Dimension, leader, and tolerance objects only	Changes the dimension style
Hatch	Objects with hatching only	Changes the hatch pattern

Table 19.1

Not all properties, however, work with all objects. For example, it makes no sense to match the hatch properties of text objects, since text cannot be hatched. Table 19.1 gives a summary of objects that can (and cannot) be changed with the **MatchProp** command.

By default, all properties are turned on. You can turn off some properties but you cannot change any properties. When you turn off some properties, AutoCAD remembers those for the next time you use the **MatchProp** command. For example, if you turn off text, dimension, and hatch properties, you see a shorter list of active settings:

> Command: **matchprop**
>
> Select source object: *(Pick.)*
>
> Current active settings: Color Layer Ltype Text Dim Hatch
>
> Select destination object(s) or [Settings]: *(Pick.)*

THE PROPERTIES DIALOG BOX

The **Properties** command displays a dialog box that lets you change almost everything about an object.

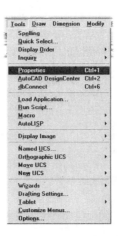

The **Properties** dialog box displays a different set of properties, depending on the object you select. AutoCAD has a varying set of properties box for every one of AutoCAD's objects: text, mtext, 3D face, 3D solid, multiline, arc, point, attribute definition, polyline, block insertion, ray, body, region, circle, shape, dimension, 2D solid, ellipse, spline, external reference, hatch, tolerance, image, trace, leader, viewport, line, and xline. That's because every object has a slightly different set of modifiable properties. For example, you can modify a line's endpoint X,Y,Z coordinates; for a circle, however, you can change the circle's center and radius.

However, there is a section common to all objects, called "General," in the upper part of the **Properties** dialog box, as shown below:

Figure 19.4 *Common Area of Properties Dialog Box*

Color: To change the color, select the color square. AutoCAD displays the standard color names; to select another color, select **Other**. AutoCAD displays the Select Color dialog box. Select a color, choose **OK**, and return to the **Properties** dialog box.

Figure 19.5 *Select Another Color*

Layer: To change the layer, select the Layer field. AutoCAD displays a list of the layer names defined in the drawing. Select a layer name.

Figure 19.6 *Select Another Layer*

Linetype: To change the linetype, select the Linetype field. AutoCAD displays a list of linetypes loaded into the drawing. Select another linetype name (notice that AutoCAD displays a sample).

Figure 19.7 *Select Another Linetype*

Linetype Scale: The current linetype scale factor is displayed in current units. To change the scale factor (if available), delete the number showing and type a new number.

Plot style: To change the plot style (if available), choose the field and select a name. Select **Other** (at the end of the list) to display the **Plot Style** dialog box.

Lineweight: To change the lineweight, select the field and select one of the preset lineweights.

Figure 19.8 *Select Another Lineweight*

Hyperlink: To change the hyperlink, select the ... button to display the Add Hyperlink button.

ARRAYING OBJECTS

There are times when multiple copies of an object or objects are desirable. Consider the number of seats in a movie theater. Or the number of parking spaces in a shopping center parking lot. If you were using traditional drafting techniques, you would draw each one separately. AutoCAD uses the **Array** command to draw repeated objects.

The **Array** command is used to make multiple copies of one or more objects in linear, rectangular, or circular patterns. After you have arrayed the object, each one can be edited separately.

CONSTRUCTING LINEAR AND RECTANGULAR ARRAYS

Let's first look at linear and rectangular arrays. A *linear array* copies an object in the horizontal direction (a row or in the *X* direction) or the vertical direction (a column or in the *Y* direction). A *rectangular array* copies the object in a rectangular pattern that is made up of rows and columns.

The first item of a rectangular array usually occupies the lower left position of an array. This is the object that will be identified as the selected object in the standard object selection process.

Figure 19.9 *Rectangular Array*

Before you invoke the **Array** command, this object must already be in existence. After you have this object in place, select **Modify | Array** from the menu bar.

> Command: **array**
>
> Select objects: *(Select one or more objects.)*
>
> Select objects: *(Press* ENTER.*)*
>
> Enter the type of array [Rectangular/Polar] <R>: **R**
>
> Enter the number of rows (———) <1>: *(Type a number.)*
>
> Enter the number of columns (||||) <1>: *(Type a number.)*
>
> Enter the distance between rows or specify unit cell (———): *(Indicate a distance.)*
>
> Specify the distance between columns (||||): *(Indicate a distance.)*

The number of objects in the rows and columns includes the selected item(s). In other words, an array of 1 column and 1 row creates no array at all: it is simply the selected object.

To create a horizontal linear array, enter "1" for rows; similarly, for a vertical linear array, enter "1" for columns.

The **unit cell** distance is "center to center." Many new CAD users make the mistake of thinking the unit cell distance is the distance *between* the items. It is not. The unit cell distance may be thought of as the distance from the center of one item to the center of the next item. The unit cell distance is defined in the current drawing units.

The distance between the items in the columns can be different from the row distance.

To create an array that "grows" to the left or downward, enter a negative distance.

 Note: Set your first point on the selected object so you can "see" the distances between the objects.

ROTATED RECTANGULAR ARRAY

You can construct a rotated rectangular array by changing the snap rotation angle. For example, if the snap angle is set to 30 degrees, the rectangular array is rotated as a whole to a 30-degree angle.

CONSTRUCTING A POLAR ARRAY

Polar arrays are used to array objects in a circular pattern. To construct a polar array, you must define the angle between the items (again, from center to center, not actually between) and either the number of items or degrees to fill. You have the option of rotating (or not rotating) each object as it is arrayed.

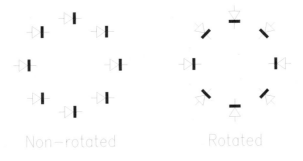

Figure 19.10 *Non-Rotated and Rotated Polar Arrays*

MIRRORING OBJECTS

The **Mirror** command is used to make a mirror image of objects. You can choose to retain or delete the original objects.

To use the **Mirror** command, you can also select **Modify | Mirror** from the menu bar.

> Command: **Mirror**
>
> Select objects: *(Select objects.)*
>
> Select objects: *(Press* ENTER.*)*
>
> Specify first point of mirror line: *(Pick point.)*
>
> Specify second point of mirror line: *(Pick point.)*
>
> Delete source objects? {Yes/No] <N>: *(Answer Y or N.)*

The *mirror line* is the line about which the objects are mirrored.

MIRRORED TEXT

You can mirror objects with text (and attribute-associated text) without producing "backward" text in the image. This is achieved through the **MirrText** system variable, as follows:

> Command: **mirrtext**
>
> New value for MIRRTEXT <1>: 0

Setting the variable to **0** produces non-mirrored text, while the default setting of **1** produces a true mirror image of text.

The following illustration shows examples of text mirrored with the **MirrText** variable set to each variable.

Figure 19.11 *Using the MirrText System Variable*

The **Mirror** command saves time when you are drawing symmetrical objects. Draw only half, or one-fourth, of an object, then construct the other parts by mirroring them.

DIVIDING AN OBJECT

The **Divide** command divides an object into an equal number of parts and places either a specified block or a point object at the division points on the object.

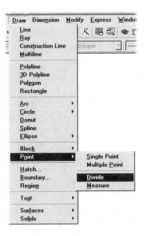

Use the cursor to select a single object; you cannot use Window, Crossing, or Last. From the menu bar, select **Modify | Divide**.

> Command: **divide**
>
> Select object to divide: *(Select one object.)*
>
> Enter the number of segments or [Block]: *(Enter a number.)*

Choosing an object other than a line, arc, circle, spline, or polyline results in the message:

Cannot divide that object. * Invalid*

The number of segments must be between 2 and 32,767. The current point setting (**Pdmode** and **Pdsize**) determines the type of point used at the divide points. See Chapter 7 "Creating the Drawing" for more about points.

Entering the **B option** (short for "block") causes AutoCAD to place the block along the object. The block must already exist in the drawing.

Enter the number of segments or [Block]: **b**

Enter name of block to insert: *(Type name.)*

Align block with object? [Yes/No] <Y>: *(Enter Yes or No.)*

Yes means the inserted block turns with the divided object, such as arc, circle, or spline. **No** means the block is always oriented in the same direction.

Enter the number of segments: *(Type a number.)*

If you are unsure whether a block reference exists, first use the **Insert** command to view the names of blocks in the drawing.

Note: You can use the Divide command to divide an object into an equal number of parts, then snap to these points using the Node object snap.

Create blocks, used with the **Divide** command, to place desired symbols at intervals along an object.

THE MEASURE COMMAND

The **Measure** command is similar to the **Divide** command, except that you can choose the length of segment by which the points are spaced.

Command: **measure**

Select object to measure: *(Pick a single object.)*

Specify length of segment or [Block]: *(Type a number or enter B.)*

The rules for execution are the same as for the **Divide** command. You can only measure a line, arc, circle, spline or polyline. You must use only the single point method of object selection. The current point setting is the one used to place the points.

If you use a block, the block reference must already exist in the drawing. You can choose to rotate (align) the block with the measured object.

> Enter name of block to insert: *(Enter name.)*
>
> Align block with object? [Yes/No] <Y>: *(Enter Y or N.)*
>
> Specify length of segment: *(Type a number.)*

 Note: The **Measure** command does not "measure" an object. To find the length of an object, use the **List** command.

EXPLODING OBJECTS

The **Explode** command breaks down the objects that make up a block or other complex objects, such as polyline. The block is replaced by the individual objects from which it was constructed. To explode a block into individual objects, you can also select **Modify | Explode** from the menu bar.

> Command: **explode**
>
> Select objects: *(Select objects.)*
>
> Select objects: *(Press* ENTER.*)*

Some objects must be "exploded" several times to reduce them to the graphic primitives of lines, arcs, and circles.

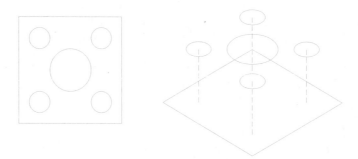

Figure 19.12 *Using the Explode Command on a Block*

Exploding Blocks

Explode breaks down blocks one level at a time. The **Explode** command breaks blocks into the simple objects that comprise the block. Nested blocks must be exploded after the initial block is exploded.

Block attributes are deleted, but the attribute definitions from which the attributes were created are redisplayed. The attribute values and changes made through **AttEdit** are discarded.

Blocks inserted with the **MInsert** command *cannot* be exploded.

Exploding Polylines

You can use **Explode** on polylines to break them down into simple lines and arcs. The width and tangent information of polylines is discarded, and the resulting lines and arcs follow the centerline of the old polyline. If the exploded polyline has segments of non-zero width, the following message appears:

> Exploding this polyline has lost width information.
>
> The UNDO command will restore it.

The new lines and arcs are placed on the same layer as the polyline and will inherit the same color.

Exploding 3D Meshes

3D polygon meshes are replaced with 3D faces and polyface meshes with 3D faces, lines, and points.

Exploding Dimensions

If you explode an associative dimension, the dimension entities are placed on layer "0", with the color and linetype of each "BYBLOCK."

Exploding 3D Solids

ACIS 3D solids explode into bodies. These bodies can be exploded into regions. Regions, in turn, can be further exploded to lines, arcs, and circles.

 Note: Explode inserted objects into individual objects so they can be edited to suit the specific purpose of that drawing.

TRIMMING OBJECTS

The **Trim** command allows you to trim objects in a drawing by defining other objects as cutting edges and then specifying the part of the object to be "cut" from between them.

Command: **trim**

Current settings: Projection=UCS Edge=None

Select cutting edges ...

Select objects: *(Pick object.)*

I found Select objects: *(Press* ENTER.*)*

Select object to trim or [Project/Edge/Undo]: *(Pick object.)*

Select object to trim or [Project/Edge/Undo]: *(Press* ENTER.*)*

To use **Trim**, you must define: (1) the object(s) to be used as the cutting edges and (2) the portion of the desired object to be removed.

 Note: A quick way to select all objects as cutting edges is to press ENTER at the "Select objects" prompt. AutoCAD will use all objects, including those not in view.

The *cutting edges* are defined through the standard object selection process. Cutting edges can be lines, arcs, circles, and/or polylines. If you use a polyline with a non-zero width as a cutting edge, the centerline of the polyline will be used as the point to trim to. Combinations of circles, arcs, and lines can be used as cutting edges. The following illustration shows how the **Trim** command can be used in a more complex manner.

Point to the part that you want trimmed.

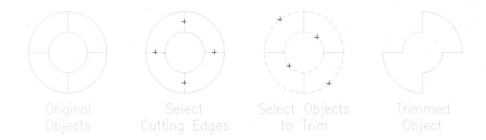

Figure 19.13 *Using the Trim Command on a Complex Object*

TRIMMING POLYLINES

Polylines are trimmed at the intersection of the centerline of the polyline and the cutting edge (polylines are covered in Chapter 22 "Understanding Advanced Operations"). The trim is performed with a square edge. Therefore, if the cutting edge intersects a polyline of non-zero width at an angle, the square-edged end may protrude beyond the cutting edge. Figure 19.14 shows an example of such a situation.

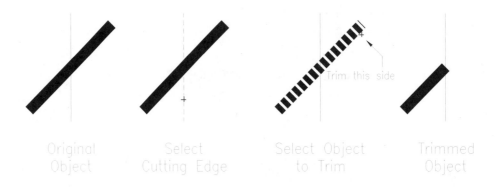

Figure 19.14 *Trimming a Wide Polyline*

If you select objects that cannot serve as cutting edges, AutoCAD displays the message:

> No edges selected.

If an object is chosen that cannot be trimmed, AutoCAD displays:

> Cannot TRIM this object

If the object to be trimmed does not intersect a cutting edge, the following message is displayed:

> Object does not intersect an edge.

TRIMMING CIRCLES

To trim a circle, you must either intersect it with two cutting edges, or with the same cutting edge twice (such as a line drawn through two points on the circumference). If the circle intersects only one cutting edge, AutoCAD displays the following message:

> Circle must intersect twice.

PROJECTION

The **Project** option is meant for use in 3D drafting. It allows you to specify the projection AutoCAD should use when it trims objects.

> Select object to trim or [Project/Edge/Undo]: **p**
>
> Enter a projection option [None/Ucs/View]: *(Enter an option.)*

Your options are

None

The **None** option specifies no projection. This means that AutoCAD trims objects that actually intersect with the cutting edge.

UCS

The **UCS** option projects the objects onto the *XY* plane of the current UCS. This means that AutoCAD trims objects, even if they do not intersect with the cutting edge in 3D space—in effect, flattening the 3D objects into a 2D plane.

View

The **View** option specifies a projection along the current view direction. This means that AutoCAD trims objects when it looks like they should be trimmed from your viewpoint, even if they don't physically intersect.

EDGES

The **Edges** option toggles whether the object is trimmed at an actual cutting edge or at an implied cutting edge.

Select object to trim or [Project/Edge/Undo]: **e**

Enter an implied edge extension mode [Extend/No extend]: *(Enter an option.)*

The options are:

Extend

When extend mode is turned on, AutoCAD projects the cutting edge so that it intersects the object in 3D space.

No Extend

When extend mode is off, AutoCAD trims objects only at cutting edges that physically intersect.

EXTENDING OBJECTS

The **Extend** command allows you to extend objects in a drawing to meet a boundary object. The **Extend** command functions very much like the **Trim** command.

Command: **extend**

Current settings: Projection=UCS Edge=None

Select boundary edges ...

Select objects: *(Select objects.)*

1 found Select objects: *(Press* ENTER.*)*

Select object to extend or [Project/Edge/Undo]: *(Pick a single object.)*

Select object to extend or [Project/Edge/Undo]: *(Press* ENTER.*)*

To use **Extend**, you must first specify the object(s) to be used as the boundary objects (the point to which a selected object will be extended), then the object to be extended.

You can use any type of object selection to choose the boundary objects. You must select the object to be extended, however, by pointing to the end (or close to the end) of the part to be extended. This is the process by which you tell AutoCAD the end of the object to extend.

If several boundary edges are selected, the object to be extended will be lengthened to meet the first boundary object encountered. If none of the selected boundary objects can be met, AutoCAD displays the message:

> No edges selected.

If an object that cannot be lengthened is chosen, the following message is displayed:

> Cannot EXTEND this object.

EXTENDING POLYLINES

Polylines are extended in much the same manner as lines and arcs. Polylines of non-zero width are extended until the centerline meets the boundary object. Objects extended to a polyline used as a boundary are extended to the centerline of the polyline.

Polylines of non-zero width are extended until the centerline intersects the boundary. Therefore, if the wide polyline and the boundary intersect at an angle, a portion of the square end of the polyline may protrude over the boundary.

Extending a tapered polyline adjusts the length of the segment that tapers—the taper extends over the longer length. Examples of using the **Extend** command on tapered polylines follow.

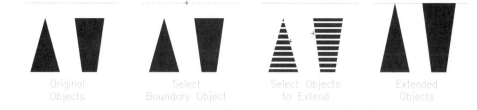

Figure 19.15 *Extending Tapered Polylines*

Only open polylines can be extended. If you attempt to extend a closed polyline, the following message will appear:

> Cannot extend a closed polyline.

Note: The **Extend** command's **Project, Edge,** and **Undo** options are identical in function to those of the **Trim** command.

LENGTHENING OPEN OBJECTS

The **Lengthen** command is a faster version of the **Trim** and **Extend** commands. **Lengthen** changes the length of open objects, such as lines, polylines, arcs, and splines; closed objects, such as circles and donuts, cannot be lengthened. **Lengthen** can change the length by making open objects either shorter or longer.

Lengthen is faster than **Trim** and **Extend** because you do not need to trim or extend to an existing object: you just point on the screen for the change to occur. Or, you can specify a percentage change numerically.

The **Lengthen** command has the following options:

> Command: **lengthen**
>
> Select an object or [DElta/Percent/Total/DYnamic]:

Reporting the Object

When you select an object at the initial "Select an object:" prompt, AutoCAD reports on its length, as follows:

> Current length: 10.2580

For an arc, AutoCAD also reports its angle, as follows:

> Current length: 8.4353, included angle: 192

When you select a closed object, such as a circle or closed polyline, AutoCAD complains, "This object has no length definition." You cannot lengthen dimensions or hatch patterns.

DELTA OPTION

The **DElta** option changes the length by adding the indicated amount to the object. You can indicate the amount by selecting two points anywhere in the drawing.

> Command: **lengthen**
>
> Select an object or [DElta/Percent/Total/DYnamic]: **de**
>
> Enter delta length or [Angle] <0.0000>: *(Enter a number or pick on screen.)*
>
> Select an object to change or [Undo]: *(Select object.)*
>
> Select an object to change or [Undo]: *(Press* ESC.*)*

The default response, "Enter delta length," prompts you to select an object, changes its length, then prompts you for another object until you press ESC. Enter a negative number to shorten the object.

Note that this command lengthens the end of the open object that you select. To lengthen both ends of an object, select it twice: once at one end, then again at the other end.

Lengthening Arc Angles

The **Angle** option changes the angle of a selected arc or polyarc. The command adds length to the *end* of the arc you select.

> Enter delta angle <0>: *(Type an angle.)*
>
> Select an object to change or [Undo]: *(Select an arc.)*
>
> Select an object to change or [Undo]: *(Press* ESC.*)*

The arc is lengthened by the angle you specify. When you select two points, AutoCAD uses the angle of the rubber-band line, not the length of the line.

The angle you specify must be small enough to not cause the arc to have 360 degrees; the angle must be large enough that the arc not be less than 0 degrees. If the angle is too large (or too small), AutoCAD curtly informs you, "Invalid angle."

PERCENT OPTION

The **Percent** option changes the length by the percentage of the object. For example, entering **25** shortens a line to 25 percent of its original length, while entering **200** doubles its length, as follows:

> Command: **lengthen**

Select an object or [DElta/Percent/Total/DYnamic]: **p**

Enter percentage length <100.0000>: **25**

Select an object to change or [Undo]: *(Select.)*

Select an object to change or [Undo]: *(Press ESC.)*

TOTAL OPTION

The **Total** option changes the length of a line to an absolute length. For example, a value of 5 changes the line to a length of 5.0000 units, no matter its existing length.

Command: **lengthen**

Select an object or [DElta/Percent/Total/DYnamic]: **t**

Specify total length or [Angle] <1.0000>: *(Enter a value.)*

Select an object to change or [Undo]: *(Select.)*

Select an object to change or [Undo]: *(Press ESC.)*

The **Total** option changes the length of an arc to an absolute circumference. The **Angle** option changes the length of an arc to the included angle specified.

DYNAMIC OPTION

The **DYnamic** option lets you visually change the length of open objects. Notice that this option reverses the order: first you select the object to lengthen, then you specify the length.

Command: **lengthen**

Select an object or [DElta/Percent/Total/DYnamic]: **dy**

Select an object to change or [Undo]: *(Select.)*

Specify new end point: *(Pick a point.)*

Select an object to change or [Undo]: *(Press ESC.)*

After you select the object, the length of the object changes as you move the cursor about. You can use object snap modes to make the dynamic lengthening more accurate.

ROTATING OBJECTS

The **Rotate** command allows you to rotate an object or group of objects around a chosen base point. The objects are not required to be part of a block.

You can choose the angle at which the chosen object rotates by specifying the angle to rotate from its existing angle, by dragging the angle, or by choosing the angle of rotation from a reference angle. Let's look at examples of each type of angle rotation.

ROTATING BY SPECIFYING ANGLE

Specifying an angle can rotate an object or a group of objects. You can specify a simple angle, or you may want to change one angle to another. For example, if an object is currently oriented at 58 degrees and you wish to rotate the object to 26 degrees, you could rotate the object the difference of −32 degrees.

> Command: **rotate**
>
> Current positive angle in UCS: ANGDIR=counterclockwise
> ANGBASE=0
>
> Select objects: *(Select objects.)*
>
> I found Select objects: *(Press* ENTER.*)*
>
> Specify base point: *(Pick point.)*
>
> Specify rotation angle or [Reference]: **-32**

Entering a positive angle results in a counterclockwise rotation of the object; a negative angle results in a clockwise rotation.

ROTATING FROM A REFERENCE ANGLE

It is sometimes necessary to obtain a particular rotation angle. The **Rotate** command allows you to achieve this by using the **Reference** option. To use the **Reference** option, you must first know the existing rotation angle of the object.

> **Note:** The **List** command is helpful in determining the existing angle of an entity when you use the Reference option.

> Specify rotation angle or [Reference]: **R**

If you know the "current" angle of the object, enter it at the next prompt.

Specify the reference angle <0>: *(Enter an angle.)*

You can rotate an object a specified number of degrees from its existing angle by entering zero for the "reference angle," regardless of the actual angle. (If you want to measure an angle in the drawing, select a point instead; When AutoCAD prompts "Specify second point:", select to show the angle.)

Specify the new angle: *(Enter an angle.)*

ROTATING AN OBJECT BY DRAGGING

If you wish to drag-rotate an object, just move the cursor in response to the "Specify rotation angle or [Reference]:" prompt. If the updating screen coordinates are set for relative distance and angle, you can read the angle from the screen. Select to fix the object at the location currently shown on the screen.

You can rotate part of an object by choosing only the parts to rotate in response to the "Select objects:" prompt.

SCALING OBJECTS

The **Scale** command allows you to change the scale of an object or objects. The *X, Y,* and *Z* scales of the object are changed equally.

The base point can be specified anywhere in the drawing. This point remains stationary, while the object is scaled from that point.

You can change the scale by a numerical factor, by dragging, or by referencing a known length, then entering a new length. Let's look at each method.

CHANGING SCALE BY NUMERICAL FACTOR

You can change the scale of an object by entering a numerical factor that serves as a "multiplier." Entering a positive number has the effect of multiplying the size of the object by that number. For example, entering **2.0** makes the object twice the original size in the direction of all three axes.

Entering a decimal factor results in an object smaller than the original. A factor of .25 results in an object 25 percent of the original size.

> Command: **scale**
>
> Select objects: *(Pick one or more objects.)*
>
> I found Select objects: *(Press* ENTER.*)*
>
> Specify base point: *(Pick.)*
>
> Specify scale factor or [Reference]: **2**

Original Objects   After 2x Scaling

Figure 19.16 *Changing the Scale Factor*

CHANGING SCALE BY REFERENCE

The **Reference** option is used to adjust an object to a "correct" size. This may occur when you scan a drawing, then bring it into AutoCAD with the **Image** command. To trace over the scanned image (a raster or bitmap image), you first must make the image the right size; you achieve this with the Reference option.

To do this, you need a line in the image that has dimension, or you must figure out its true length:

1. At the "Select objects:" prompt, select the raster image:

   Command: **scale**

   Select objects: *(Pick the raster image.)*

   I found Select objects: *(Press* ENTER.*)*

2. For the base point, select the lower left corner of the raster image. Then, specify the **Reference** option:

   Specify base point: *(Pick.)*

   Specify scale factor or [Reference]: **r**

3. Select the end of the known line in the raster image.

   Specify reference length <1>: *(Pick one end of the raster line.)*

4. Select at the other end—this allows AutoCAD to measure the current distance of the image.

   Specify second point: *(Pick other end of raster line.)*

5. For the new length, enter the value of the known dimension.

   Specify new length: *(Enter value.)*

You may find this technique useful for rescaling an odd drawing that did not follow good CAD practice, such as Rule #1: Always Draw Full Size.

STRETCHING OBJECTS

The **Stretch** command permits you to move selected objects while allowing their connections to other objects in the drawing to remain unchanged. As you will see, the **Stretch** command is one of the most useful edit commands in AutoCAD.

Command: **stretch**

Select objects to stretch by crossing-window or crossing-polygon...

Select objects: **C**

Specify first corner: *(Pick.)*

Specify opposite corner: *(Pick.)*

I found Select objects: *(Press* ENTER.*)*

Specify base point or displacement: *(Pick on screen; use an osnap mode if necessary.)*

Specify second point of displacement: *(Pick on screen.)*

When you select the objects with **Crossing** or **CPolygon** mode, the whole object is highlighted. It is, however, only the defining points that are inside the crossing window that are adjusted. To modify an object, make sure you have at least one part of the object *outside* the selection to anchor the object.

If the complete object is inside the selection, it will be moved rather than stretched.

STRETCH RULES

There are several rules associated with the **Stretch** command that must be understood in order for you to execute it properly. Let's look at some of these rules.

- Lines, arcs, elliptical arcs, solids, traces, rays splines, and polylines can be stretched.

- You can choose the objects to be edited by any combination of object selection options, but one of these options must be **Crossing** or **CPolygon**.

- Any objects contained entirely within a window are moved in the same manner as with the Move command.

- If a line, arc, trace, solid, or polyline segment is chosen with the "Crossing" method, the endpoints contained in the window are moved, but those outside the window remains "fixed."

- Arcs are stretched similarly to lines, except that the arc's center, start, and endpoints are adjusted so the distance from the midpoint of the chord to the arc is constant.

- Polylines are handled by their individual segments. The polyline width, tangent, and curve fitting are not affected.

- Some objects are just "moved" or left alone. The decision depends on the "definition point." If the definition point lies inside the window, the objects are moved. If it occurs outside the window, it is not affected. The definition point of certain objects is as follows:

Point: Center of the point

Circle: Center point of the circle

Block: Insertion point

Text: Leftmost point of the text line

- If more than one window specification is used in the object selection process, the last window used will be the one used for the Stretch.

- If you do not use a window selection, AutoCAD displays the message:

You must select a window to stretch.

UNDOING DRAWING STEPS

The **Undo** command is used to undo several command moves in a single operation. You can also identify blocks of commands for reference later. Earlier in this book, you learned about the **U** command. The **Undo** command is similar, but has additional options. To execute the **Undo** command, type the **Undo** command. (Selecting **Edit | Undo** from the menu bar executes the **U** command).

Command: **undo**

Enter the number of operations to undo or
[Auto/Control/BEgin/End/Mark/Back]:

Pressing ENTER is the same as using the **U** command.

Responding to the prompt with a number causes AutoCAD to undo the specified number of commands. This has the same effect as entering the **U** command the same number of times, except that it is done in a single operation, thus causing only one regeneration.

Let's look at the options for the **Undo** command.

Auto

The **Auto** option causes a macro (one or more commands in a row) from menus to be treated as a single command. Some menus combine several operations such as inserting a window in a wall of a floor plan by breaking the wall and Inserting the window. If **Auto** is on, the entire command string will be undone as if it were a single command. **Auto** prompts with: "ON/OFF."

Control

The **Control** option limits the Undo commands or disables them entirely. This is sometimes necessary because of the large amount of disk space required if the **Undo** edit is extensive.

The **Control** option prompts:

> Enter an UNDO control option [All/None/One] <All>:

Following is an explanation of the sub-options:

All

Enables all **Undo** commands and features.

None

Disables all **Undo** commands and features, such as Begin, End, Auto, and Mark. Using None removes all markers, groups, and other stored information. If **Undo** is entered when the **None** option is on, AutoCAD displays the **Control** prompt:

> Enter an UNDO control option [All/None/One] <All>:

You can then enter the level of **Undo** performance that you desire.

One

Allows the **Undo** commands to function for one undo at a time. All markers, groups, and other stored information are removed. This frees disk space, making it a preferred setting when your computer is low on disk space. The **Group, Mark,** and **Auto** options are not available. The following prompt is displayed when **Undo** is executed:

> Control/<1>:

BEgin/End

The **BEgin** and **End** options cause all operations—from the time **BEgin** is entered and the time **End** is entered—to be treated as a single command. This means that all operations between the **Begin** and **End** entries will be undone in a single step. For example, consider the following string of operations:

> Circle
>
> Arc
>
> **Undo BEgin**
>
> Line
>
> Line
>
> Fillet
>
> **Undo End**
>
> Arc
>
> Undo
>
> Undo
>
> Undo

The first **Undo** removes the arc. The second **Undo** removes all the operations between the Begin /End entries (Line, Line, Fillet). The third **Undo** removes the Arc. Only the first circle remains.

If you enter the **Begin** option while in a current group, the current group is ended and a new one is started.

If a **Begin** has been started, but not ended, and the **Undo** command is entered, it undoes one operation at a time, but cannot back up past the point where **Begin** was entered. To continue back, the current group must be ended, even if nothing remains in the current group.

When you use the **Begin** option, **Undo** will not display each operation. The message "GROUP" is displayed.

Mark

Think of your drawing as a list of functions that is added to the drawing file, one at a time. If this list were on paper, you could place a mark at a certain point for reference. The **Mark** option allows you to do this. Later, you can use the **Back** option to undo everything back to the mark.

You can have as many marks as you wish during the drawing session.

Back

The **Back** option will cause AutoCAD to undo all operations back to the preceding mark. Each operation will be listed and the message "Mark encountered" will be displayed. When the next marker (going back) is encountered, it is removed. The next **Undo Back** then undoes all operations back to the previous marker, if any. If there is no preceding mark, AutoCAD displays the message:

> This will undo everything. OK? <Y>:

Entering **Y** undoes every operation since last entering the drawing editor. Answering **N** causes the **Back** option to be ignored.

Multiple undoes are stopped by a mark. Entering a number in response to the undo prompt that is greater than the number of operations since the last mark will have no greater effect than the **Back** option. The undo will still stop at the mark.

GENERAL NOTES

Undo has no effect on the following commands:

ABOUT	GRAPHSCR	PSOUT	REINIT
AREA	HELP	QSAVE	RESUME
ATTEXT	HIDE	QUIT	SAVE

COMPILE	ID	RECOVER	SAVEAS
CVPORT	LIST	REDRAW	SHELL
DBLIST	NEW	REDRAWALL	STATUS
DELAY	OPEN	REGEN	TEXTSCR
DIST	PLOT	REGENALL	

Plotting clears the **Undo** information from the drawing.

The **Undo** command should not be confused with the **Undo** option contained in some commands such as the **Line** command. Using the **Undo** option from the **Line** command will remove one line segment at a time. If you use the **Undo** command, however, all line segments in the sequence will be undone in a single operation. For example, if you draw a box containing four lines, then exit the **Line** command and execute the **Undo 1** command, all four lines will be undone. If this does not produce the intended results, the **Redo** command will restore the lines.

DISPLAY ORDER

Normally, AutoCAD displays overlapping objects in the reverse of the order in which you draw them. For example, when you place some text, followed by an overlapping 2D solid, the text is obscured by the 2D solid, as shown in Figure 19.17.

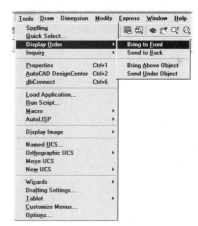

The **Draworder** command lets you change the display and plotting order of objects, as is commonly found in other Windows-based drawing programs. The command gives you "absolute" and "relative" control over display order. The absolute reordering brings the selected object to the front (visually) or back of all other overlapping objects. The relative reordering lets you bring the object above or below another selected object.

Figure 19.17 *Overlapping Objects*

Command: **draworder**

Select objects: *(Select one or more objects.)*

Select objects: *(Press* ENTER.*)*

Enter object ordering option [Above object/Under
   object/Front/Back] <Back>: *(Select an option.)*

Select reference object: *(Pick one object.)*

The options have the following meaning:

Select objects

When you select multiple objects, the relative display order of the selected objects is maintained. The order of selection has no impact on drawing order.

Above object

Moves the display order of the selected object above the reference object.

Under object

Moves the display order of the selected object below the reference object.

Front

Moves the display order of the selected object above all other objects.

Back

Moves the display order of the selected object below all other objects.

Note: The first time you use the **Draworder** command in a drawing, AutoCAD automatically turns on all of the object sort methods (**SortEnt** system variable). Autodesk warns that this can result in slower regeneration and redrawing times.

EDITING WITH GRIPS

In Chapter 18, you learned how to use the object selection process. This process is used to select objects for editing. AutoCAD provides a shortcut process to object selection and editing. You can set AutoCAD to display grips on the objects on the screen. *Grips* are small rectangles that are placed on objects and can be used as grip points when editing. The grip boxes are generally placed at the positions available with object snap. Figure 19.18 shows some AutoCAD drawing objects with their grip locations.

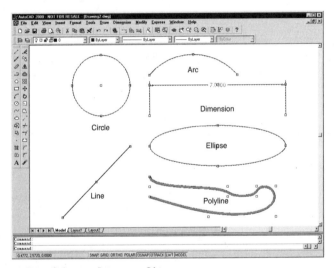

Figure 19.18 *Grips Selection Points on Objects*

ENABLING GRIP EDITING

Grip editing is enabled or disabled through the **DdGrips** command, which displays the **Options** dialog box's **Selection** tab. To enable grip editing, select the **Enable Grips** check box.

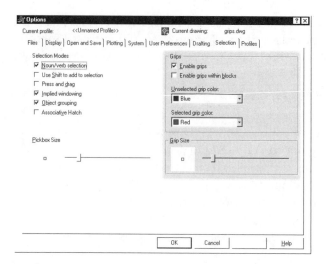

Figure 19.19 *Dialog Box for Setting Grips Preferences*

Normally, a grip box is displayed at the insertion point of an inserted block. If you wish to display grips on the objects within blocks, select the **Enable grips within blocks** check box.

Setting Grip Colors

As a visual aid, grip boxes are assigned a color. When a grip box is selected, the interior of the box is filled with a solid color. Both the grip box and solid fill are assigned a color through this dialog box.

To set the colors, choose either the **Unselected grip color** (the warm grip box is blue, by default) or the **Selected grip color** (the hot solid-filled grip box is red, by default) button. AutoCAD displays the **Select Color** dialog box. Select the desired color from the displayed palette, then choose the **OK** button to set the color.

Setting the Grip Size

The size of the grip boxes is set with the slider bar in the **Grips Size** section.

The preview to the right of the slider bar shows the current grip box size. As you move the slider bar, the grip box changes scale to indicate the actual grip box size. The grip box size ranges from 1 to 255 screen pixels.

USING GRIPS FOR EDITING

When grip editing is enabled, a small selection box (called the *pickbox*) is displayed at the intersection of the crosshairs.

To select an object for editing, place the pickbox at any point on an object in the drawing and click. The object is highlighted. Notice that grip boxes are placed on the object. You can continue to select as many objects as you wish by selecting several times.

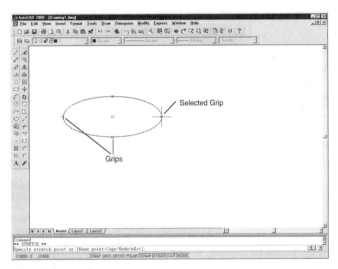

Figure 19.20 *Grip Selection Box*

After you select the object(s) to be edited, you must select one of the blue grips as a base point for editing. As you move the selection over the grip points, the crosshairs snap to the grip box similar to using object snap. To select a grip box, move to it and click. The grip box changes color from blue to red to denote selection.

(If you find you cannot select more than one object, hold down the SHIFT key on the keyboard when selecting the grips.)

Pressing ESC twice clears all selected objects and their grips. Using ESC once clears the selected objects, but not the grips. This facilitates using a grip on a non-selected object as a base grip.

When you select two objects, the display shows warm grips. Pressing ESC turns them into cold grips: these are blue boxes with unhighlighted objects. Press ESC a second time to remove the grip boxes.

Grip Mode.

When you select a single grip so that it turns red, AutoCAD displays editing commands on the command line. The selections cycle as you press the Spacebar or Enter.

The following command line listings show the options available as you "page" through with the Spacebar or Enter (right-click to see a cursor menu listing the same options).

> ** STRETCH **
>
> Specify stretch point or [Base point/Copy/Undo/eXit]: *(Press* SPACE-BAR.*)*
>
> ** MOVE **
>
> Specify move point or [Base point/Copy/Undo/eXit]: *(Press* SPACE-BAR.*)*
>
> ** ROTATE **
>
> Specify rotation angle or [Base point/Copy/Undo/Reference/eXit]: *(Press* SPACEBAR.*)*
>
> ** SCALE **
>
> Specify scale factor or [Base point/Copy/Undo/Reference/eXit]: *(Press* SPACEBAR.*)*
>
> ** MIRROR **
>
> Specify second point or [Base point/Copy/Undo/eXit]: *(Press* SPACE-BAR.*)*

Press the SPACEBAR until the desired edit command is listed, then proceed with the command. For example, to scale the object, press the spacebar until **SCALE** appears. Type a scale factor at the "Specify scale factor" prompt.

The selected grip point is used as a base point for the edit operations. Alternatively, you can select one of the options listed on the command line. The commands and their options are covered later in this section.

When selecting several grip points, select the base point grip last. The base grip can be either a selected or nonselected grip. It is only after the base grip is selected that the command options are displayed.

GRIP EDITING COMMANDS

The grip editing commands of **Stretch, Move, Rotate, Scale,** and **Mirror** are displayed on the command line when you enter grip mode. You can rotate through the edit command options by pressing the SPACEBAR. Let's look at each choice.

Stretch Mode

Stretch mode functions similarly to the **Stretch** command, except that AutoCAD uses the selected grip points to determine the stretch results.

> ** STRETCH **
>
> Specify stretch point or [Base point/Copy/Undo/eXit]:

If you select the midpoint grip of a line or arc or the center of a circle, the object is moved, but not stretched. Figure 19.21 shows an example of using the **Stretch** mode.

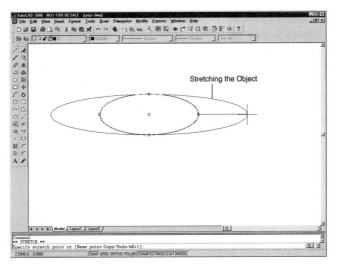

Figure 19.21 *Stretching with Grips*

Stretch options:

Base point

Entering **B** at the command line allows you to use a base point other than the base location.

Copy

Entering **C** (for **Copy**) makes copies from the stretched objects, leaving the original objects intact. The prompt changes to:

> ** STRETCH (multiple) **
>
> Specify stretch point or [Base point/Copy/Undo/eXit]:

Undo

Undoes the operation.

eXit

Exits grip editing.

Move Mode

When you press the SPACEBAR until the **Move** mode is displayed, grip editing moves the selected objects.

> ** MOVE **
>
> Specify move point or [Base point/Copy/Undo/eXit]:

The distance the objects move is determined by the distance from the base grip to a point entered in response to AutoCAD's "Move to point" prompt. You can enter either an absolute or a relative coordinate to specify the distance and direction of the move.

If you hold down the apostrophe (') after selecting the base grip and before placing the point designating the move distance and direction, AutoCAD places a copy of the object(s) at the new location, leaving the original object(s) unchanged. You can use this technique to place multiple copies. This functions similarly to the Copy option (explanation follows).

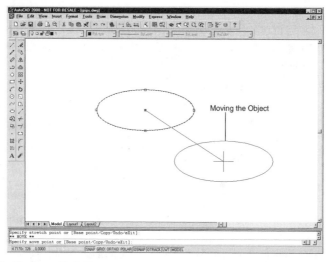

Figure 19.22 *Moving with Grips*

Move options:

Base point

Entering **B** at the command line allows you to use a base point other than a grip location.

Copy

Entering **C** (for **Copy**) makes copies of the objects, leaving the original objects intact.

Undo

Undoes the operation.

eXit

Exits grip editing.

Rotate Mode

The **Rotate** option allows you to rotate the selected objects around the base point.

> ** ROTATE **
>
> Specify rotation angle or [Base point/Copy/Undo/Reference/eXit]:

Unless you use the **Base** point option to position the base point at a location other than a grip, the rotation occurs around the base grip.

When rotating, you can dynamically set the rotation angle by moving the crosshairs, or you can specify a rotation in degrees.

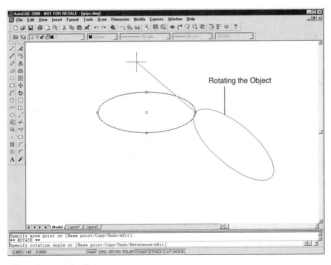

Figure 19.23 *Rotating with Grips*

Rotate options:

Base point

Entering **B** at the command line allows you to use a base point other than a grip location.

Copy

Entering **C** (for **Copy**) makes copies of the rotated objects, leaving the original objects intact.

Undo

Undoes the operation.

Reference

Allows you to set a reference angle. When you enter **R**, AutoCAD prompts for the "Reference angle." This is the angle at which the object is currently rotated. AutoCAD next prompts for the "New angle." The new angle is the actual angle at which you want the object to be rotated.

eXit

Exits grip editing.

Scale Mode

If you wish to scale the selected objects, cycle through the mode list until the Scale mode is listed on the command line.

```
** SCALE **
Specify scale factor or [Base point/Copy/Undo/Reference/eXit]:
```

The scale mode is used to rescale the selected object(s). The base grip serves as the base point for the scaling operation. You dynamically scale the objects by moving the crosshairs away from the base grip. You can also scale the selected objects by entering a scale factor. A scale factor greater than one enlarges the object by that multiple. For example, a scale factor of 2.0 results in an object twice the size. Entering a decimal scale factor results in an object that is smaller than the original. For example, entering a scale factor of 0.5 shrinks the object to one-half the size of the original.

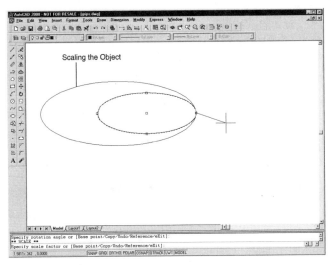

Figure 19.24 *Scaling with Grips*

Scale options:

Base point

Entering **B** at the command line allows you to use a base point other than a grip location.

Copy

Entering **C** (for **Copy**) makes copies of the scaled objects, leaving the original objects intact.

Undo

Undoes the operation.

Reference

Allows you to set a reference scale. Let's look at an example. Let's suppose you have an object 6 units in length and you wish to scale it to 24 units in length. Select the Reference option by entering R at the command line. AutoCAD prompts:

> Reference length <1.0000>:

Enter **6** and press ENTER. AutoCAD then prompts for the new length.

<New length>/Base point/Copy/Undo/Reference/eXit:

Enter **24** and press ENTER. AutoCAD calculates the scale factor and applies the scale.

eXit

Exits grip editing.

Mirror Mode

Mirror mode is used to mirror the selected object(s) in a similar manner to the Mirror command.

** MIRROR **

Specify second point or [Base point/Copy/Undo/eXit]:

When you use Mirror mode, the first point of the mirror line is stipulated as the point of the base grip. The second point of the mirror line is entered at any point on the drawing.

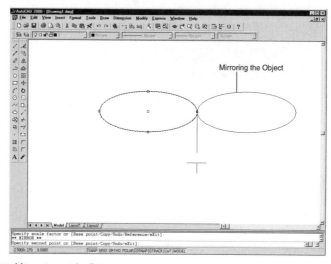

Figure 19.25 *Mirroring with Grips*

Mirror options:

Base point

Entering **B** at the command line allows you to use a base point other than a grip location.

Copy

Entering **C** makes copies of the mirrored objects, leaving the original objects intact.

Undo

Undoes the operation.

eXit

Exits grip editing.

EXERCISES

1. Let's try a rectangular array. Start a new drawing. Set limits of 0,0 and 12,8. As an aid, display a grid with a value of one.

   Draw a circle with the center point located at 1,1 and a radius of one.

   Execute the **Array** command:

   Command: **array**

   Select objects: **L**

   Select objects: *(Press Enter.)*

   Enter the type of array [Rectangular/Polar] <R>: **R**

   Enter the number of rows (—) <1>: **3**

   Enter the number of columns (||||) <1>: **5**

   Enter the distance between rows or specify unit cell (—): **2**

   Specify the distance between columns (||||): **2**

   Your array should look like the following illustration:

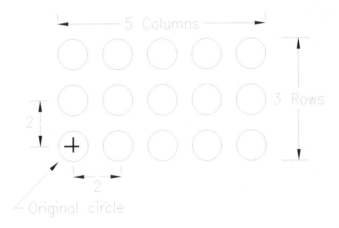

Figure 19.26 *Example of Using the Array Command*

You can use a window to specify the distances between the items. Simply enter a point when prompted for the "Unit cell distance" and enter a second point to define the window. The *X* and *Y* distances are measured by AutoCAD and used as the values. Figure 19.27 shows this method of distance definition:

Save the drawing as "ARRAY."

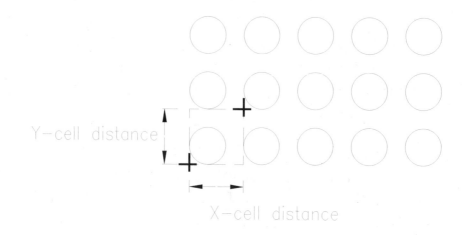

Y-cell distance

X-cell distance

Figure 19.27 *Using a Window for Array Spacing*

2. Start another drawing (which you'll turn into a polar array). Set limits of 0,0 and 12,8. Draw a square with sides that are one unit each.

   Execute the **Array** command:

   Command: **array**

   Select objects: *(Select square.)*

   Select objects: *(Press* ENTER.*)*

   Enter the type of array [Rectangular/Polar] <R>: **P**

   Specify center point of array: *(Pick point.)*

   Number of items: **7**

   Angle to fill (+=ccw, –=cw) <360>: **270**

   Rotate arrayed objects? [Yes/No] <Y>: **N**

Your polar array should look like the one in Figure 19.28. The item is arrayed around the center point you chose.

A positive angle causes the array to build in a counterclockwise direction; a negative angle causes the array to build in a clockwise direction from the selected object.

You are asked whether you wish the object to be rotated about the center point:

Rotate arrayed objects? [Yes/No] <Y>:

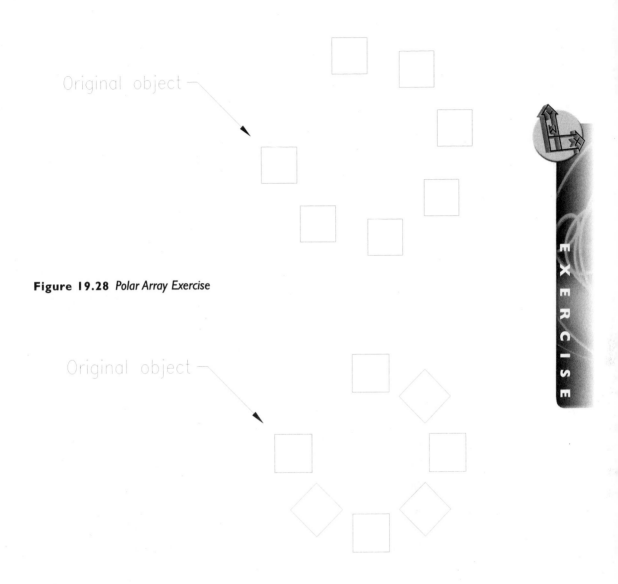

Figure 19.28 *Polar Array Exercise*

Figure 19.29 *Rotated Polar Array*

Respond with a **Y** or **N** (yes or no) to tell AutoCAD which you require. If the square you used in your circular array was a block and you responded with **Y**, the block would be rotated around its base point. Other objects are rotated about specific points. Table 19.2 lists these points.

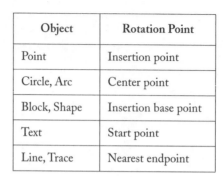

Object	Rotation Point
Point	Insertion point
Circle, Arc	Center point
Block, Shape	Insertion base point
Text	Start point
Line, Trace	Nearest endpoint

Table 19.2

Save your drawing as "PARRAY."

3. Figure 19.30 shows how the **Mirror** command can be a great time-saver. Open the CD-ROM drawing named "MIRROR1.Dwg." Let's suppose that you are designing a house and you want to reverse the layout of the bathroom. The following drawing shows the room to be reversed.

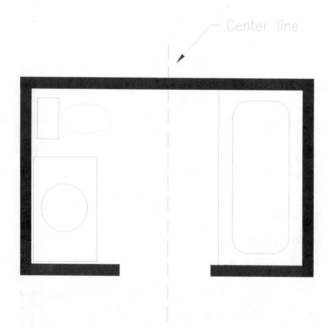

Center line

Figure 19.30 *MIRROR1.DWG CD-ROM Drawing*

Let's reverse the room. First, ensure ortho mode is turned on (press F8).

Command: **mirror**

Select objects: *(Select the fixtures.)*

Select objects: *(Press ENTER.)*

Specify first point of mirror line: *(Pick point 1.)*

Specify second point of mirror line: *(Pick point 2.)*

Delete source objects? [Yes/No] <N>: **Y**

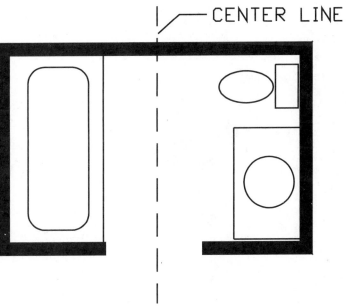

CENTER LINE

Figure 19.31 *Mirrored Room*

The mirror line can be placed at any angle in respect to the selected object.

4. Let's look at an example using a **PdMode** of 34. Start a new drawing. First, draw a circle on the screen: center the circle in the viewport; use a radius of 3. Now, enter the **Divide** command:

Command: **divide**

Select object to divide: *(Select the circle.)*

Enter the number of segments or [Block]: **8**

You won't notice any difference to the circle.

Select **Format | Point Style** from the menu bar. From the **Point Style** dialog box, choose the style shown in Figure 19.32. Choose **OK**.

Figure 19.32 *Point Style Dialog Box*

If necessary, use the **Regen** command to make the new point style visible. The circle should now be divided into eight equal segments as shown in Figure 19.33.

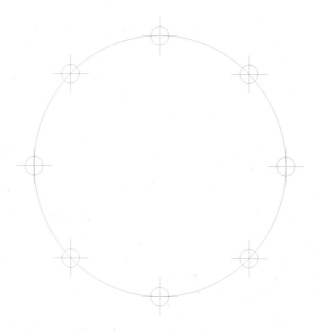

Figure 19.33 *Using the Divide Command*

You can choose the number of divisions to be between 2 and 32,767. The object is equally divided into the specified number of parts and a point object is placed at the division points.

5. Now, draw a symbol similar to the one in Figure 19.34.

Figure 19.34 *A Block*

Now, block the symbol and name it "SYMBOL1." Draw another circle on the screen. Enter the **Divide** command.

> Command: **divide**
>
> Select object to divide: *(Select the circle.)*
>
> Enter the number of segments or [Block]: **B**
>
> Enter name of block to insert: **SYMBOL1**
>
> Align block with object? [Yes/No] <Y>: **Y**
>
> Enter number of segments: **8**

When you respond to the "Align block with object?" prompt with Y, the **Divide** command rotates the block around its insertion point so that its horizontal lines are tangent to the object being divided.

Figure 19.35 *Dividing with the Rotated Block*

Figure 19.36 shows the same procedure, except the block is not rotated:

Figure 19.36 *Dividing with the Non-rotated Block*

6. Let's perform a measure on an object. Draw a horizontal line on the screen, 6 units in length as in Figure 19.37. (Keep the point setting of 34 to make the points visible.) Enter the **Measure** command:

Command: **measure**

Select object to measure: *(Select the line.)*

Specify length of segment or [Block]: **1**

Figure 19.37 *Using the Measure Command*

You can also show AutoCAD the segment distance by placing two points on the screen.

The measurements of lines and arcs start at the endpoint closest to the point used to select the object. The measurement of a circle starts at the angle of the center set by the current snap rotation. Measurements of polylines start at the first vertex drawn.

7. Let's use the **Trim** command to trim an intersection. Draw four intersecting lines about 6 units long, as shown in Figure 19.38. Now, enter the **Trim** command:

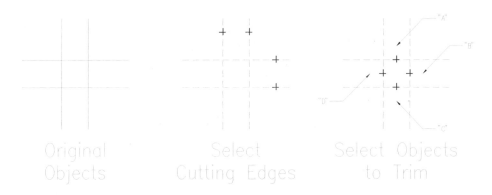

Original Objects   Select Cutting Edges   Select Objects to Trim

Figure 19.38 *Using the Trim Command*

Command: **trim**

Current settings: Projection=UCS Edge=None

Select cutting edges ...

Select objects: **W** *(Window the four lines.)*

Select objects: *(Press* ENTER.*)*

Select object to trim or [Project/Edge/Undo]: *(Select points A,B,C, and D.)*

Notice that the "Select objects to trim" prompt repeats to allow multiple trimming. Pressing ENTER returns you to the command prompt.

The intersection is trimmed and should appear as shown in Figure 19.39.

Figure 19.39 *Objects After the Trim*

8. Draw two vertical parallel lines 6 units long and one horizontal line as shown in Figure 19.40. You now extend the horizontal line to the vertical lines. Enter the **Extend** command:

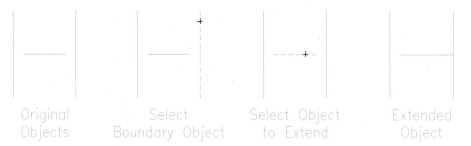

Figure 19.40 *Using the Trim Command*

Command: **extend**

Current settings: Projection=UCS Edge=None

Select boundary edges ...

Select objects: *(Select the vertical lines and *)*

Select object to extend or [Project/Edge/Undo]: *(Select.)*

If you wish to extend the other end (remember, you chose both vertical lines as boundaries), simply respond to the repeating "Select object to extend" prompt by selecting a point at the other end of the horizontal line. Pressing Enter returns you to the command line.

9. Draw the object as shown in Figure 19.41. Enter the **Rotate** command:

   Command: **rotate**

   Select objects: **W** *(Window the entire object.)*

   Select objects: *(Press *)*

   Specify base point: *(Pick point "A".)*

   Specify rotation angle or [Reference]: **45**

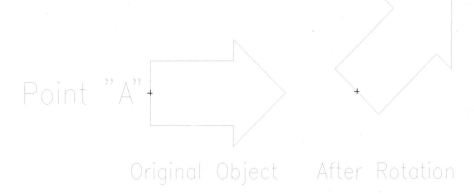

Point "A"

Original Object   After Rotation

Figure 19.41 *Rotating an Object*

10. Draw an object similar to the one shown in Figure 19.42. Enter the **Rotate** command.

    Command: **rotate**

    Select objects: **W** *(Window the object.)*

    Select objects: *(Press *)*

Specify base point: *(Pick point "A".)*

Specify rotation angle or [Reference]: **r**

Specify the reference angle <0>: **0**

Specify the new angle: **-45**

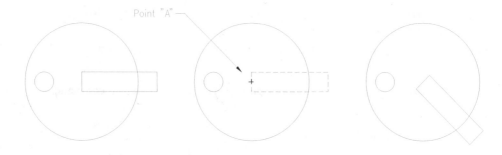

Figure 19.42 *Rotating in Reference to a New Angle*

11. Draw the object as in Figure 19.43. Enter the **Scale** command.

Command: **scale**

Select objects: *(Select object.)*

Select objects: *(Press* ENTER.*)*

Specify base point: *(Select bottom left corner.)*

Specify scale factor or [Reference]: **R**

Specify reference length <1>: **2**

Specify second point: **4**

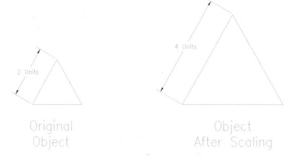

Figure 19.43 *Change the Scale by Reference*

Notice that AutoCAD rescales all the objects that make up the object, not just the length of the known object.

12. Let's use the **Stretch** command to move the location of a window in a wall. Figure 19.44 shows a wall containing a window that we wish to move to the right. This is a work disk drawing. Start the drawing named "STRETCH1" and use the following command sequence.

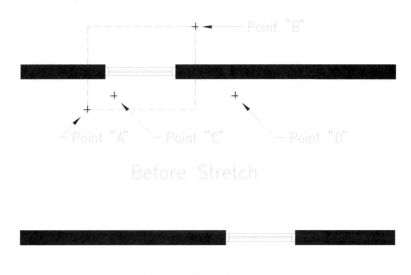

Figure 19.44 *Stretch1.Dwg File from CD-ROM*

Let's enter the **Stretch** command and move the window.

> Command: **stretch**
>
> Select objects to stretch by crossing-window or crossing-polygon...
>
> Select objects: **C**
>
> First corner: *(Pick point "A".)*
>
> Other corner: *(Pick point "B".)*
>
> Specify base point or displacement: *(Pick point "C".)*
>
> Specify second point of displacement: *(Pick point "D".)*

13. Start the drawing from the CD-ROM disk named "STRETCH2.Dwg." This is a pencil as shown in Figure 19.45.

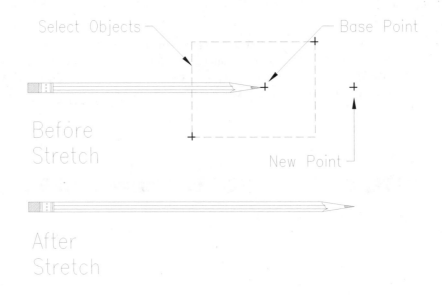

Figure 19.45 *Stretch2.Dwg File from CD-ROM*

14. Use the **Stretch** command to make the pencil longer and shorter.

CHAPTER REVIEW

1. What properties of an object can be altered with the Properties command?

2. What is meant when you change an object's color to "BYLAYER"?

3. Why would you use the **Properties** command instead of the **Object Properties** toolbar?

4. What are the three types of arrays?

5. Using simple circles, draw an example of an array using five columns and three rows. Using four-sided boxes, draw an example of a polar array with six objects that are rotated.

6. What does the **Mirror** command perform?

7. What variable controls whether text is mirrored?

8. What is the *mirror line*?

9. Use point type 34 to show a horizontal line divided with the **Divide** command into four segments.

10. Show a circle divided into six segments with the **Divide** command, using a small square as the block used with **Divide**. Show the same situation again, using the rotated block option.

11. What objects can you use with the **Measure** command?

12. What is the result when you explode a block?

13. What happens when you explode a polyline?

14. What is a trim cutting edge?

15. At what point is a polyline trimmed?

16. What are the requirements for trimming a circle?

17. What is the object called that is selected with the Extend command and acts as a borderline for the extended object?

18. What are the three ways to rotate an object?

19. Can an object be scaled differently in the *X* and *Y* axes with the **Scale** command?

20. What value would you enter to scale an object to one-fourth of its original size?

21. Can the **Object Properties** toolbar be used to change objects? If so, how?

22. When using the **Stretch** command, if you select objects with three window operations, which operation is used to determine the stretch?

23. What does the **Mark** option of the **Undo** command do?

24. What effect does display order have on plotting?

25. What is the difference between a hot and a cold grip?

26. What editing operations can you perform with grips?

27. How do you deselect a gripped object?

28. What is the **Matchprop** command used for?

29. Describe five properties that the **Properties** command lets you change.

CHAPTER 20

Working with Intermediate Operations

Intermediate use of AutoCAD requires the CAD operator to possess knowledge of several operations. After completing this chapter, you will be able to

- Store and retrieve drawing views and pictorial images

- Save and view slide images of a drawing screen

- Remove unwanted layers, blocks, styles, and other named objects from a drawing

- Rename objects in the drawing

- Import and export PostScript image files

- Set and control the color of objects independent of the layer color settings

INTRODUCTION

In this chapter, you learn about

View

Lets you save and restore views by name.

MSlide

Makes a slide of the current view.

VSlide

Displays an SLD slide file.

Script

Executes a number of commands in a row.

SlideLib

Collects slide files into a SLB slide library.

Rename

Renames named objects, such as blocks and views.

PsOut

Exports the drawing in EPS (encapsulated PostScript) format.

Psin

Imports a PostScript file into the drawing as a block.

PsDrag

Controls how the PostScript image looks during importation.

PsQuality

Controls the look of the PostScript image after importation.

Color

Lets you change an object's color independent of layer.

Purge

Removes unused named objects, such as layers and text styles.

STORING AND DISPLAYING DRAWING VIEWS

In AutoCAD, a *view* is a stored zoom identified by name. You can recall a view at any time in the drawing. This has the same effect as zooming and panning to that location.

To invoke the **View** command, select **View | Named Views** from the menu bar. The **View** dialog box lists the named views stored in the current drawing. Model means the view was saved in model space, while Layout means the view was saved in paper space.

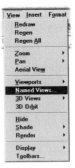

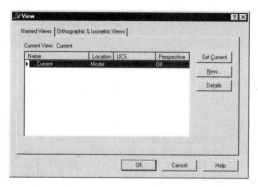

Figure 20.1 *View Dialog Box*

Set Current

To restore a view, select the name from the list, then choose **Set Current** and **OK**. Views labeled **Model** can only be restored when the drawing is in model space, while views labeled **Layout** can only be restored when the drawing is in layout mode (paper space). If you are in the "wrong" space, AutoCAD prompts, "Select Viewport for view:". You then need to select the **Model** or **Layout** tab, as appropriate.

New

To create a new named view, choose the **New** button, which displays the **New View** dialog box. Type a name in the **View Name** text box (maximum 255 characters). Select either the **Current display** or the **Define** window radio button. Choose the **Define View Window** button (looks like an arrow cursor) to select the coordinates of the windowed view. When done, choose the **OK** button.

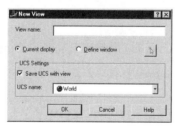

Figure 20.2 *Define New View Dialog Box*

Delete

To select a named view, select a view name and right-click. From the cursor menu, select **Delete** to remove the view from the list without warning.

Rename

To rename a named view, select a view name and right-click. From the cursor menu, select **Rename**.

Details

Displays a dialog box with information about the selected view.

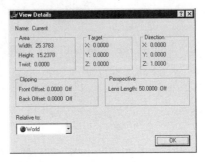

Figure 20.3 *View Details Dialog Box*

 Note: Each view can also save a UCS to restore with the view. (See Chapter 32 "The User Coordinate System.")

STORING AND DISPLAYING DRAWING SLIDES

There may be times when you want to show several views of a drawing or several drawings to someone. You could, of course, take the time to load each drawing, but this would be very laborious. The **MSlide** and **VSlide** commands provide much faster ways to load pictures.

The slide commands capture a view of a drawing like a camera shot, then show it later. Think of it as making a slide with a camera, then showing the shot with a projector.

You can use slides to view reference material.

Let's look at how to use each.

MAKING A SLIDE

The **MSlide** (short for "make slide") command is used to make a slide file. First, display the drawing or any part of the drawing you wish to make a slide of on the screen. Then enter:

    Command: **mslide**

AutoCAD displays the **Create Slide File** dialog box. Enter the name you want the slide file to have. AutoCAD adds a file extension of .sld to the slide name.

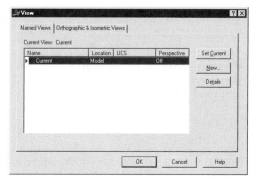

Figure 20.4 *Create Slide File Dialog Box*

Portions of the drawing not currently displayed are not captured. The **MSlide** command is a "what you see is what you get" operation. Blip marks, grid dots, and the UCS icon, however, are not captured.

VIEWING A SLIDE

The **VSlide** command is used to view a slide. The following command is used to view a single slide.

       Command: **Vslide**

AutoCAD displays the **Select Slide File** dialog box.

Figure 20.5 *Select Slide File Dialog Box*

It is not necessary to include the file extension. You can view the slide at any time and from any drawing. The slide file overwrites the current screen display—but does not destroy it. A slide cannot be edited or zoomed. It is a fixed snapshot of the drawing.

The current drawing is not affected by the **VSlide** command. To return to the original drawing, issue a **Redraw** command.

SLIDE SHOWS

You can write a self-running script that presents a series of slides on the screen.

To do so, utilize the **Script** command. You eliminate the required loading time by entering an "*" before the slide name, which makes AutoCAD "preload" the next slide while the current one is being displayed. The following is an example of a script for a self-running slide show:

Script	Meaning
VSLIDE A	Begins the slide show with the first slide
VSLIDE *B	Preloads slide B
DELAY 5000	Displays slide A for 5 seconds (5000 milliseconds)
VSLIDE	Displays slide B
VSLIDE *C	Preloads slide C
DELAY 1000	Displays slide B for 1 second (1000 milliseconds)
VSLIDE	Displays slide C
DELAY 2500	Displays slide C for 2.5 seconds (2500 milliseconds)
RSCRIPT	Repeats the script from the beginning

Create the three slides named A, B, and C.

Type the script, shown above, into a text editor, such as Notepad. Save the file with any name and the SCR extension. For example, name the file slideshow.scr. Load the slideshow.scr script file into AutoCAD with the **Script** command. The script execute automatically.

Refer to the **Script** command for more information.

SLIDE LIBRARIES

Slide libraries are a collection of slides stored in a special library format. Slide libraries are excellent for storing slides for use with icon menus.

To build a slide library, you must first prepare a text file (ASCII format) that contains the names of the slide files you wish to assemble into a library. Using the Notepad text editor, list one slide file per line in the document. Next, start DOS and use the **SlideLib** utility program included in the ACAD 2000\Support subdirectory. The correct usage for the **SlideLib** program is as follows.

C:\program files\acad 2000\support\>**slidelib** (library name) <(file list name)

Let's look at an example.

If the text file is named "FLIST" (for furniture list) and you want the library name to be "FURNLIB" (for furniture library), you would enter the following:

C:\program files\acad 2000\support\>**slidelib furnlib <flist**

Once a library is built, you cannot change a slide. If you wish to add or change a slide, you must rebuild the library.

If you wish to view a slide that is part of a slide library, turn off the **Filedia** system variable (set it to 0). Then use the **Vslide** command and enter the slide library name, followed by the slide file name. For example, if you wanted to view a slide named "sofa" from the "furnlib" library, the following format would be used.

Command: **vslide**

Slide file name: **furnlib(sofa)**

Remember to turn **Fildia** on again (set it to 1), so that you regain use of file dialog boxes.

RENAMING PARTS OF YOUR DRAWING

The **Rename** command is used to rename parts of the drawing known as "tables." The tables include:

Blocks	Text Styles
Dimension styles	Named user coordinate systems
Layers	Names views
Linetypes	Named viewports
Plot styles	

This command makes it easy to change the name of, say, layers or blocks. Note that layer 0 cannot be renamed.

To rename a part of the drawing, enter the **Rename** command or select **Format | Rename** from the menu bar. To rename an object, take these steps:

1. Select a table name in the **Named Objects** column.

2. Select a name from the **Items** column. Notice that the name appears in the **Old name** field.

3. Type the new name next to the **Rename to** button.

4. Choose the **Rename to** button, then choose **OK**.

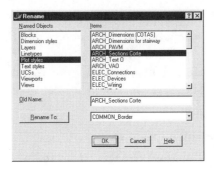

Figure 20.6 *Rename Dialog Box*

PRODUCING AND USING POSTSCRIPT IMAGES

Images using the PostScript drawing format can be imported or exported within AutoCAD. PostScript images are universally used by desktop publishing packages and other types of drawing programs. The use of PostScript images allows a direct interface with these types of programs and makes AutoCAD an invaluable illustrating program for these programs.

EXPORTING A POSTSCRIPT IMAGE

You can export a drawing in encapsulated PostScript format (ESP) by using the **PsOut** command. When you issue the **PsOut** command, the dialog box shown in Figure 20.7 is displayed.

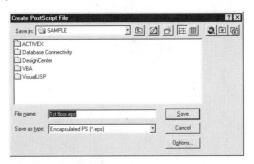

Figure 20.7 *Create PostScript File Dialog Box*

Choose the **Options** button. Let's look at how each prompt controls the PostScript file.

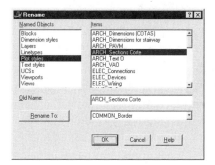

Figure 20.8 *Export Options*

Determining What Part of the Drawing to Plot

You can plot any part of the drawing to a PostScript file. The choices are similar to the **Plot** command where you select the part of the drawing you want to plot.

Adding a Screen Preview

An external program can use a screen preview image to preview the image before placement. If you want to preview the image, select one of the image format options: **ESPI** (for Macintosh computers) or **TIFF** (for PCs). Select none if you find the EPS won't load into another software package.

 Note: Check your desktop publishing or drawing program documentation to verify the use of either format with your program.

Setting the Preview Resolution

If you select either the ESPI or TIFF format, AutoCAD prompts for the pixel resolution to use. Select 128, 256, or 512.

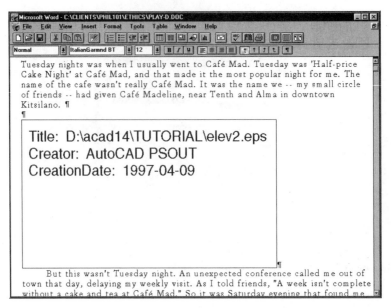

Figure 20.9 *EPS File with No Preview*

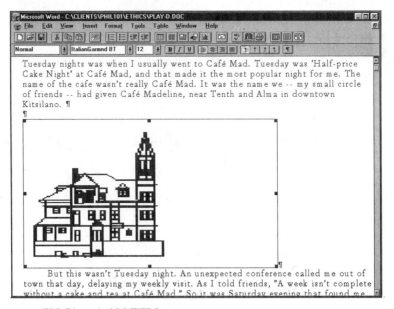

Figure 20.10 *EPS File with 128 TIFF Preview*

Note: Although higher resolution screen previews are clearer, they require larger file sizes and display more slowly in your desktop publishing or drawing program. The lowest acceptable resolution is recommended.

Setting the Output Units

The output units of either inches or millimeters can be selected.

 Note: Set the size units to the same setting as the external program to which you will import the PostScript file.

Setting the Output File Scale

The scale for the output file determines the final size of the drawing in the same manner as the plot scale function in the **Plot** command.

The scale is important if you will be printing the PostScript file to paper (see also the next section on output size). Or, you can have the plot fit the selected size of paper.

Setting the Paper Size

The output size is the size to which the plot will be drawn. You can enter a USER size or select one of the standard sizes listed.

Choose **OK,** provide a file name, and choose **Save.** AutoCAD writes the drawing to a file. Exported PostScript files contain a file extension of EPS (short for Encapsulated PostScript File).

IMPORTING A POSTSCRIPT IMAGE

PostScript images can be imported into AutoCAD with the **PsIn** (short for "PostScript in") command. When you enter the **PsIn** command, AutoCAD displays the following dialog box.

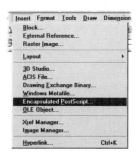

Figure 20.11 *Select PostScript File Dialog Box*

The PostScript file is imported into the drawing as a block. The block can be exploded and edited with AutoCAD edit commands.

When you select a PostScript file to import, depending on the setting of the **PsDrag** command, AutoCAD displays a box representing the size of the file. The box contains the name of the PostScript file. AutoCAD then prompts:

> Insertion point <0,0,0>:
>
> Scale factor:

The insertion point and scale factor function in the same manner as when inserting a block. You can also visually place the file by moving the cursor around the screen, moving the representative box into position.

You can also set the scale visually and dynamically by moving the cursor away from the box.

Setting the PsDrag

PsDrag controls how the PostScript image is displayed when it is imported into your drawing. The drag setting is controlled by the **PsDrag** command.

> Command: **psdrag**
>
> PSIN drag mode [0/1] <0>:

If **PsDrag** is set to 0, the PostScript image is displayed as a box when imported, similar to Figure 20.12 (left). If set to 1, the actual image is displayed.

 Note: Setting **PsDrag** to 1 will cause the image to handle slowly. If your system works slowly, set **PsDrag** to 0.

Setting PostScript Input Quality

The quality of a PostScript image during insertion is determined by the **PsQuality** command.

Command: **Psquality**

New value for PSQUALITY <current>:

If **PsQuality** is set to 0, the PostScript image is displayed only as a box with the file name enclosed within. The box is displayed even if the **PsDrag** setting is set to 1.

Any positive value sets the number of pixels per drawing unit. For example, a setting of 75 displays 75 pixels per drawing unit, as shown in Figure 20.12 (right).

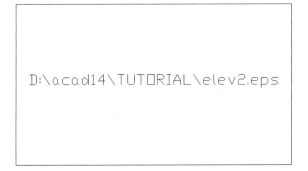

Figure 20.12 *Effect of PsQuality*

A negative value will display the drawing outline, but it will not include the fills. The absolute value of the negative number controls the quality.

DISPLAYING A POSTSCRIPT FILL

AutoCAD allows you to fill a closed polyline boundary with any of several PostScript fill patterns. The fill patterns will appear on the screen when you perform a PsOut and PsIn cycle, and the fill patterns will print when you output the drawing to a PostScript printer.

To place a PostScript fill, use the **PsFill** command.

Command: **psfill**

AutoCAD prompts you to select the polyline border and then asks for the PostScript fill pattern.

Select polyline:

Enter PostScript fill pattern name (.=none) or [?] <.>:

To see the selections, enter a question mark (?). AutoCAD displays the following listing of patterns:

Grayscale RGBcolor Allogo Lineargray Radialgray Square Waffle
Zigzag Stars Brick Specks

and then repeats the prompt.

Enter PostScript fill pattern name (.=none) or [?] <.>:

Figure 20.13 shows the available PostScript fills.

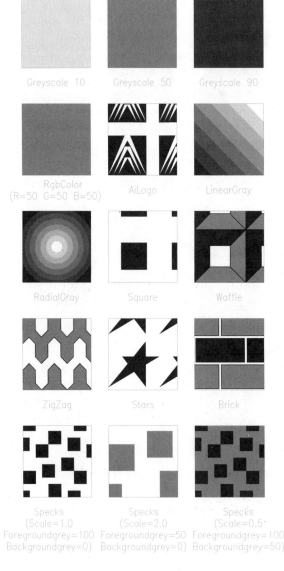

Figure 20.13 *PostScript Fill Patterns Supplied with AutoCAD*

Enter the name of the desired PostScript fill. AutoCAD continues.

>   Scale <default>:

The scale factor works in a similar manner to a hatch pattern scale.

The remaining prompts are dependent on the pattern you select. When you select gray scales, a value of 0 is white and 100 is black. Values between 0 and 100 represent the percent of black area of the gray scale.

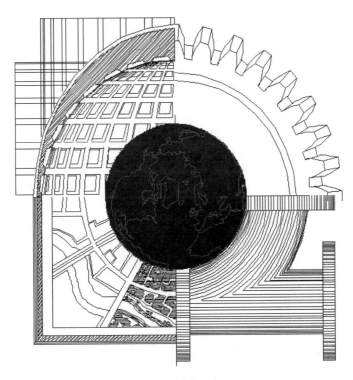

Figure 20.14 *Sample Drawing Filled with PostScript Patterns*

SETTING THE CURRENT COLOR

The **Color** command is used to set the color type for all subsequently drawn objects.

This differs from the **Color** option in the **Layer** command, which sets the color for objects drawn on that layer. The use of the **Color** command allows you to set colors that are contrary to the current color setting for the layer. Thus, it is possible that a layer will contain objects of different colors, regardless of the color set for that layer.

To set a new color for subsequently drawn objects, select **Format | Color** from the menu bar. AutoCAD displays the **Select Color** dialog box.

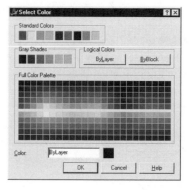

Figure 20.15 *Select Color Dialog Box*

You can set a new color by one of several methods. You can select a color sample; you can specify a number from 1 to 255; or you can enter a standard color name such as "BLUE". All new entities will be displayed in the designated color until a new color is selected by the **Color** command.

You can also select BYLAYER or BYBLOCK.

- Choosing BYLAYER causes all subsequently drawn objects to inherit the layer's color; thus you relinquish control of objects' colors to each layer's color setting.

- Choosing BYBLOCK causes all subsequent objects to be drawn in white until they are blocked. When the block is inserted, the objects inherit the color of the block insertion.

As an alternative to the **Select Color** dialog box, you can use the **Color** control in the **Object Properties** toolbar.

PURGING OBJECTS FROM A DRAWING

AutoCAD stores named objects (blocks, dimension styles, layers, linetypes, multiline styles, shapes, plot styles, and text styles) with the drawing. When the drawing is loaded with the **Open** command, AutoCAD determines whether other objects in the drawing reference each named object. You use the **Purge** command to delete any *unused* named objects.

The **Purge** command eliminates unused blocks, text styles, dimension styles, etc. This eliminates space-consuming parts of a drawing and simplifies drawing management.

For example, you can purge unused layers so the dialog box does not contain useless layers to list through. From the menu bar, select **File | Drawing Utilities | Purge**.

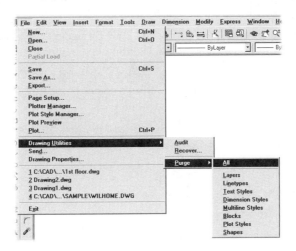

Command: **Purge**

Enter type of unused objects to purge
[Blocks/Dimstyles/LAyers/LTypes/Plotstyles/SHapes/textSTyles/
Mlinestyles/All]:

You use the capitalized letters to select any of the options, or "A" to purge all the objects. AutoCAD prompts you with the name of each unused object and asks if that object should be purged. Repeat the **Purge** command until it reports, "No unreferenced ... found."

The **Purge** command cannot be used to purge layer "0", the Continuous linetype, or the Standard text style.

EXERCISES

1. Start either an existing drawing you have completed or one of AutoCAD's sample drawings. Use **Zoom Window** to change the view of the drawing.

   Use the **MSlide** command to capture a slide named "SLIDE1."

   Now use **Zoom All** to redisplay the entire drawing. Next, issue the **VSlide** command and respond to the prompt with "SLIDE1."

   Note that you can view a slide from any drawing, not just the one from which it was captured.

2. While the slide is displayed, issue a **Redraw**. Did the slide go away?

3. Use **VSlide** to display the slide again. Select the **Line** command and try to draw a line on the slide. What happens? You cannot edit or zoom a slide. (Be sure to exit the drawing with **Quit** if you happen to edit the existing drawing and do not wish to save the edits.)

4. Set the current color to yellow and draw three circles. Next, set the current color to red. Draw three boxes. Did the objects appear in the correct color?

5. Use the **Layer** command to set the layer color to cyan. Draw some lines. Did the circles and boxes you drew change color? Notice that the **Color** command overrides the layer color setting.

6. Use the **Change** command and select all the objects on the screen. Select **Properties**, and then Color. Select blue as the color. Did all the objects change to blue?

CHAPTER REVIEW

1. When can you display a slide?

2. How do you "capture" and store a view?

3. Can a view be edited?

4. How do you make a slide?

5. Can you edit a slide?

6. What is the file extension of a slide file?

7. From where can you save a slide?

8. What can be purged with the Purge command?

9. What items cannot be purged?

10. What file extension is used with a PostScript file?

11. Name some uses for PostScript export files.

CHAPTER 21

Constructing Isometric Drawings

One of the standards in the drawing profession is the isometric drawing. Isometric drawings look like 3D but use special 2D construction techniques. This chapter covers the method of constructing isometric drawings with AutoCAD. After completing this chapter, you will be able to

- Distinguish the three basic types of pictorial drawings

- Manipulate the aspects of isometric drawing

- Use and understand the isometric drawing commands in AutoCAD

INTRODUCTION

AutoCAD provides the following commands for drawing in isometric mode:

DSettings

Enters isometric mode.

Isoplane

Switches the cursor between the three isoplanes.

Ellipse

Draws isometric circles.

Style

Creates isometric text style.

DimEdit

Create isometric dimensions through the Rotate and Oblique options.

Function key F5 (or CTRL+E) switches isoplane in mid-command.

ISOMETRIC DRAWINGS

There are three basic types of engineering pictorial drawings: axonometric, oblique, and perspective. Isometric drawings are one type of axonometric drawing. Figure 21.1 shows examples of each type of pictorial drawing.

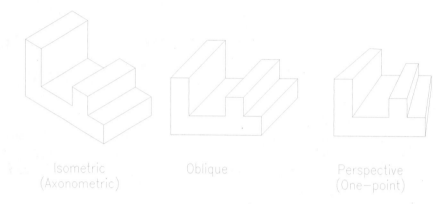

Isometric
(Axonometric)

Oblique

Perspective
(One–point)

Figure 21.1 *Engineering Pictorial Drawings*

Axonometric drawings are mostly used for engineering drawings. Each face of the object is shown in true length, resulting in the ability to measure each length.

Oblique drawings are primarily used as quick design drawings. The front face is shown in plane. The sides are drawn back at an angle. This makes an oblique drawing easy to construct from an elevation view of the front face.

Perspective drawings are the most realistic type. Perspective drawings show the object, as it would appear to the eye from a specified location and distance. This involves the use of vanishing points. (See Figure 21.2.) Although perspectives are the most realistic, most of the lengths in the drawing are not true lengths and cannot be accurately measured. Perspective mode is discussed in detail in Chapter 31 "Viewing 3D Drawings."

AutoCAD provides a mode for drawing isometric drawings. This mode displays three drawing planes which you utilize to construct your drawings.

PRINCIPLES OF ISOMETRIC DRAWINGS

As mentioned earlier, isometric drawings are a type of axonometric drawing. In isometric drawings, the lines that are used to construct a simple box are drawn at *30 degrees* from the horizontal. Figure 21.3 shows a box drawn in isometric.

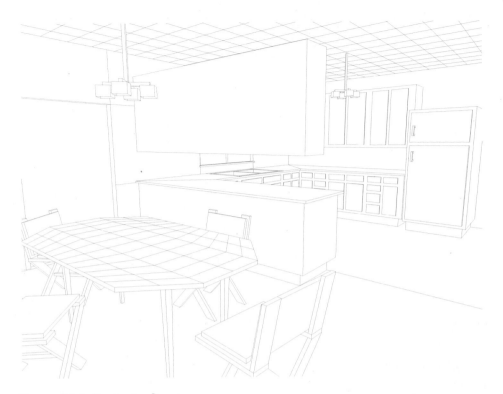

Figure 21.2 *Perspective Drawing*

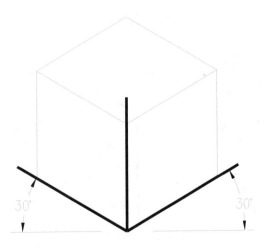

Figure 21.3 *Box in Isometric*

From this, we can derive an isometric axis as shown in Figure 21.4.

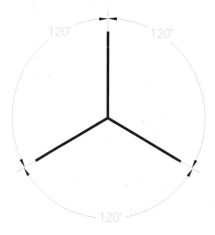

Figure 21.4 *Isometric Axes*

If we look at each of the isometric axes, we can identify a plane lying between each pair of axis. These are referred to as *isometric planes*. (The isometric planes are sometimes called "isoplanes" for short.)

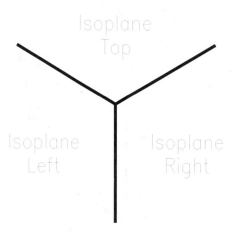

Figure 21.5 *Isometric Planes*

There is a top plane, left plane, and right plane as shown in Figure 21.5. As you draw isometrically with AutoCAD, you draw in one of these three planes.

A line parallel to one of the isometric axes is called an *isometric line*. A sloping surface requires lines that are not parallel to any axis; these lines are called *nonisometric lines*. The endpoints of these lines, however, are derived from points first determined by isometric lines.

Circles are drawn as ellipses. Text is slanted by 30 degrees, known as the obliquing angle in AutoCAD. The extension lines of dimensions are obliqued by 30 degrees; dimension text is also obliqued by 30 degrees.

Let's continue to learn how to create a drawing with AutoCAD's isometric capabilities.

ENTERING ISOMETRIC MODE

The isometric mode is actually a snap style. To enter isometric mode, use the **DSettings** command. From the menu bar, select **Tools | Drafting Settings**.

Command: **dsetting**

AutoCAD displays the **Drafting Settings** dialog box. In the **Snap & Grid** tab, turn on the following options:

- **Snap On**.

- **Grid On**.

- **Snap style & type: Isometric Snap** on. Notice that the **Snap X** and **Grid X** spacing fields are grayed out. This means that you cannot have an "aspect" ratio during isometric drawing.

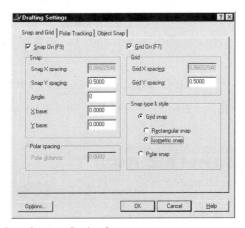

Figure 21.6 *The Drafting Settings Dialog Box*

Choose **OK**. You will notice that the crosshairs are displayed at a angle. The particular angle depends on the current axis.

SWITCHING THE ISOPLANE

To draw in each isoplane, you must change between the planes. This is accomplished by using either the **Isoplane** command or by using CTRL+E (or function key F5 as a toggle. Let's look at how to use each.

The **Isoplane** command is used to toggle between the three isometric planes. Type the command; there is no menu selection.

> Command: **isoplane**
>
> Current isoplane: **Left**
>
> Enter isometric plane setting [Left/Top/Right] <Top>: *(Press* ENTER.*)*
>
> Current isoplane: Top:

The following explains the options for the **Isoplane** command:

> **Left**
>
> Selects the left plane and uses axes defined by 90° and 150°.
>
> **Top**
>
> Selects the top plane and uses axes defined by 30° and 150°.
>
> **Right**
>
> Selects the right plane and uses axes defined by 90° and 30°.
>
> **Enter**
>
> Pressing ENTER toggles to the next plane. The planes are displayed in a rotating fashion: left, top, right, and back to left isoplane again.

You can switch between the isoplanes transparently by using CTRL+E or F5. Each time you press CTRL+E, the isoplane changes between the top, left, and right in a repeating order.

Figure 21.7 shows the crosshairs configuration for each of the isoplane settings.

If you display a grid while in Isometric mode, the grid display will reflect the axis markings.

DRAWING IN ISOMETRIC MODE

After you have set up the isometric mode, you can use AutoCAD's commands to construct and edit the drawing. Construct an isometric drawing by first boxing in the general shape of the object. The boxing-in process uses all isometric lines, creating the intersection points for nonisometric. Figure 21.8 shows the sequence for drawing an isometric object.

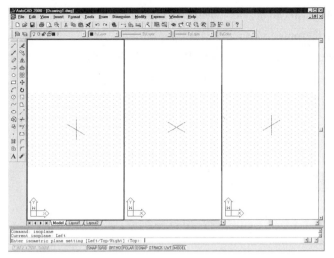

Figure 21.7 *The Three Isoplanes: Left, Top, and Right*

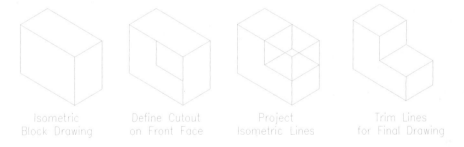

Isometric
Block Drawing

Define Cutout
on Front Face

Project
Isometric Lines

Trim Lines
for Final Drawing

Figure 21.8 *Isometric Drawing Sequence*

In general, you start by drawing an isometric box that describes the overall size of the object. You then add and trim lines and isocircles that create the shape.

Direct distance entry makes it very easy to draw lines in isometric mode. *Tracking* is an efficient way to move the starting point of a line, as shown in the tutorial.

ISOMETRIC CIRCLES

Isometric circles are constructed with the **Isocircle** option under the **Ellipse** command. This option appears only when isometric mode is turned on with the **DSettings** command.

Command: **Ellipse**

Specify axis endpoint of ellipse or [Arc/Center/Isocircle]: **i**

Specify center of isocircle: *(Pick point.)*

Specify radius of isocircle or [Diameter]: *(Pick point.)*

The isocircle is drawn correctly for the current isoplane.

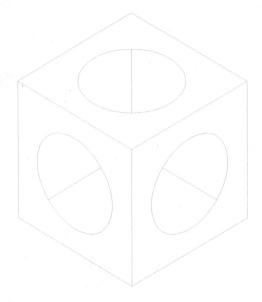

Figure 21.9 *Isometric Circles*

ORTHO MODE

When AutoCAD is in isometric mode, ortho mode switches to reflect the current isoplane. Choose the **ORTHO** button on the status line to turn on ortho mode. I recommend that you keep ortho mode turned on while creating an isometric drawing.

ISOMETRIC TEXT

Text must be aligned with the isometric planes to make it look "right." The best way is to create three text styles, one for each isoplane. AutoCAD does not ship with text styles appropriate for isometric drawing, so you'll need to create them yourself. (The support files that include the term "ISO" refer to the International Organization for Standardization, and not isometric drawing standards.)

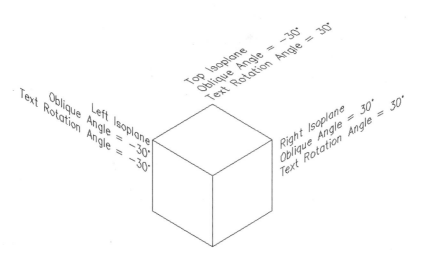

Figure 21.10 *Isometric Text Styles*

The three text styles need to have obliqued text rotation angles that align with the three axes, Then, when the text is placed, you need to specify a text rotation angle. The following table summarizes the combinations of positive and negative 30-degree angles required to place the text correctly:

Isoplane	Text Rotation	Style Oblique Angle
Left	−30°	−30°
Top	30°	−30°
Right	30°	30°

ISOMETRIC DIMENSIONING

Dimensioning an isometric object requires that the extension line and dimension text to align properly with the isometric plane. To dimension an isometric drawing, use the **DDim** command to create three dimension styles, one for each isoplane. AutoCAD does not ship with dimension styles appropriate for isometric drawing, so you'll need to create them yourself. (The support files that include the term "ISO" refer to the International Organization for Standardization, and not isometric drawing standards.)

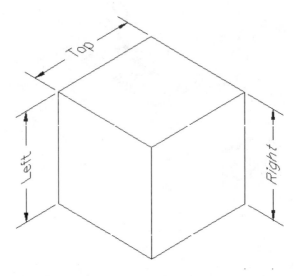

Figure 21.11 *Isometric Dimensioning*

The following table summarizes the combinations of positive and negative 30-degree angles required to place the dimensions correctly

Isoplane	Dimension Style	Text Oblique	DimEdit Oblique
Top	ISOTOP	−30°	−30°
Left	ISOLEFT	30°	30°
Right	ISORIGHT	30°	−30°

EXERCISES

1. Enter the isometric mode by selecting **Drafting** Settings from the **Tools** drop-down menu. In the dialog box, enter .25 in the **Y** spacing text box. Select the check boxes next to **Snap On**, **Grid On**, and **Isometric** snap. Choose **OK** to exit the dialog box.

2. Move the crosshairs around the drawing area. Use CTRL+E (or F5) to switch the crosshairs between the three isoplanes. Notice how the angle of the crosshairs cursor changes as you switch between isoplanes.

3. Start the **Ellipse** command. Choose the **Isocircle** option and then enter a point at the center of the screen. Move the crosshairs away from the center point and watch the isocircle dynamically drag into form. Use CTRL+E to switch between the isoplanes. Note how the isocircle changes form for each isoplane. Complete the isocircle.

4. Draw the following isometric objects and dimension each. After you have drawn the objects, you may want to use the **Color** command to set different colors for each face and use the **BHatch** command with solid hatching to make the faces solid.

You can make the isometric drawing process easier by setting the snap spacing to be an increment of the dimensions of the object. Also, using the ortho mode assists in drawing isometric lines perfectly along the isometric axes.

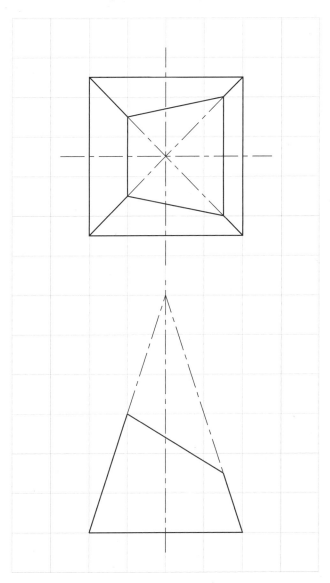

Figure 21.12

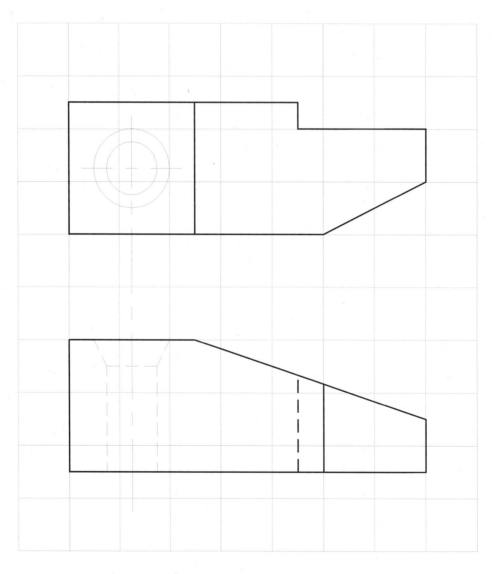

Figure 21.13

Figure 21.14

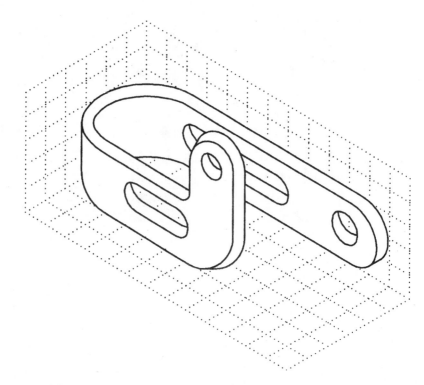

Figure 21.15

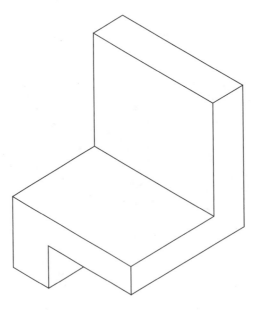

Figure 21.16

CHAPTER REVIEW

1. What are the three types of engineering pictorial drawings?

2. Which type is the most realistic in appearance?

3. What type of drawing is an isometric?

4. How many axes are used in isometric drawing?

5. How many degrees above horizontal are the isometric axes?

6. What are the three isometric planes?

7. How do you change between isometric planes in AutoCAD?

8. What commands are used to enter isometric mode?

9. How would you set isometric mode from a drop-down menu?

10. How would you draw a circle in isometric mode?

11. What command do you use to set up isometric text?

12. What dimension commands are used to convert standard dimension components to isometric dimension components?

TUTORIAL

In this tutorial, we create an isometric drawing of an L-bracket and then dimension it with isometric dimensions. Figure 21.17 shows the completed project.

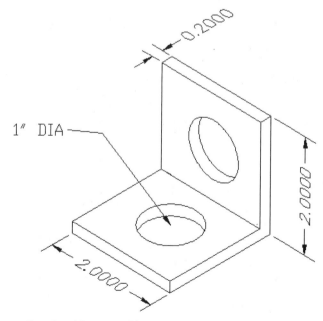

Figure 21.17 *Completed Isometric Project*

1. Start AutoCAD with a new drawing.

2. Select **Tools | Drafting Settings** from the menu bar. Turn on snap, grid, and isometric snap. Set the Y spacing to 0.1 for snap, and 0.5 for grid. Choose **OK**.

3. Turn ortho mode on by choosing the **ORTHO** button on the status line.

4. Set the isoplane to left with the **Isoplane** command. Notice how the cursor looks slanted.

5. Set the limits to 0,0 and 6,6. Perform the **Zoom All** command. Now you are ready to draw.

6. The easiest way to draw in isometric mode is to draw a cube first; the cube represents the bulk of the model. To draw the cube, you start with the left face. It is made of four lines, each 2 units long. You use direct distance entry to assist you.

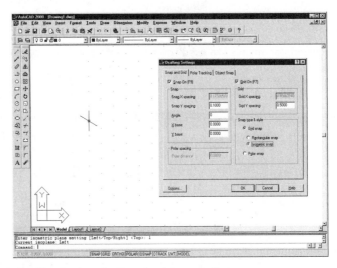

Figure 21.18 *Drafting Settings for Drawing Isometrically*

Command: **line**

Specify first point: *(Pick point "1" on top of a grid mark.)*

Specify next point or [Undo]: *(Move the cursor northwest to "2".)* 2

Specify next point or [Undo]: *(Move the cursor north to "3".)* 2

Specify next point or [Close/Undo]: *(Move the cursor southeast to "4".)* 2

Specify next point or [Close/Undo]: **c**

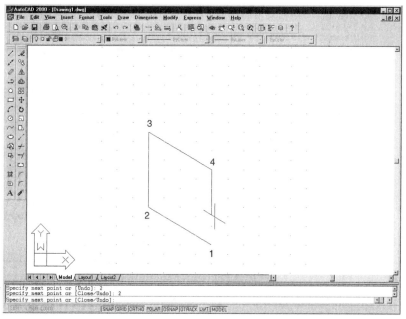

Figure 21.19 *Drawing the Left Face*

7. Press F5 to switch the isoplane to top. Notice how the cursor changes shape.

8. You now draw the top face, which is made of three lines, each also 2 units long. We will use INTersection object snap to start and end the lines accurately:

Command: **line**

Specify first point: **int**

of *(Move cursor over the upper-left corner of the left face at "5" and click).*

Specify next point or [Undo]: *(Move cursor to the northeast at "6".)* 2

Specify next point or [Undo]: *(Move cursor to the southeast at "7".)* 2

Specify next point or [Close/Undo]: *(Move cursor to the southwest at "8".)* 2

Specify next point or [Close/Undo]: *(Press* ENTER.*)*

Your drawing should now have the left and top face.

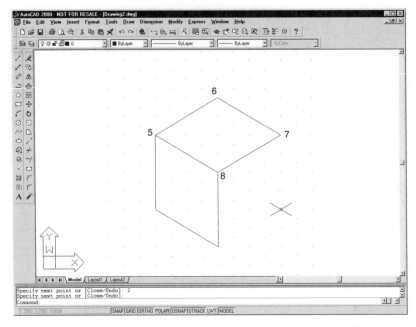

Figure 21.20 *Left and Top Faces of the Cube*

9. Press CTRL+E to switch the isoplane to the right. Again the cursor changes shape.

10. Draw the two lines that form the right face:

Command: *(press* ENTER.*)*

LINE Specify first point: **int**

of *(Move cursor over the rightmost corner of the top face at "9" and click.)*

Specify next point or [Undo]: *(Move cursor to the south at "10".)* 2

Specify next point or [Undo]: *(Move cursor to the southwest at "11".)* **2**

Specify next point or [Close/Undo]: *(Press* ENTER.*)*

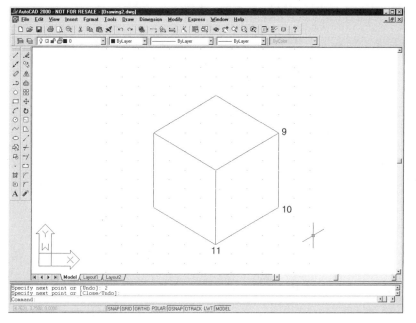

Figure 21.21 The Completed Cube

The cube is now complete.

11. With the cube in place, we now draw the "inner" parts that define the legs of the L- bracket. The legs of the bracket are 0.2 units thick. This time we employ tracking to help us start drawing the lines:

> Command: *(Press 3.)*
>
> <Isoplane left> Line
>
> Specify first point: **Tk**
>
> First tracking point: **Int**
>
> of *(Pick the bottom-most corner at "12".)*
>
> Next point (Press ENTER to end tracking): *(Move cursor up to "13".)* 0.2
>
> Next point (Press ENTER to end tracking): *(Press ENTER.)*
>
> Specify next point or [Undo]: *(Move cursor northwest to "14".)* 2
>
> Specify next point or [Undo]: *(Press ENTER.)*

You have now drawn the left face of the bracket's lower leg, as shown in Figure 21.22.

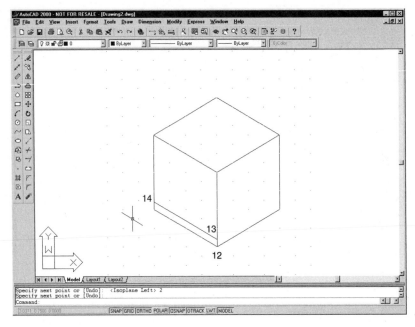

Figure 21.22 *Left Face of Bracket*

12. Let's continue on to draw the remainder of the bracket's lower leg. Press F5 to switch to the upper isoplane.

13. Draw the remainder of the lower leg, as follows:

    Command: **line**

    Specify first point: **int**

    of *(Pick the just completed line endpoint at "15".)*

    Specify next point or [Undo]: *(Move cursor northeast to "16".)* 1.8

    Specify next point or [Undo]: *(Move cursor southeast to "17".)* 2

    Specify next point or [Close/Undo]: *(Move cursor southwest to "18".)* 1.8

    Specify next point or [Close/Undo]: *(Press ENTER.)*

    That completes the lower leg.

14. While we are in the top isoplane, let's draw the top face of the upright leg. The procedure is similar to before:

    Command: *(Press ENTER.)*

    LINE Specify first point: **tk**

First tracking point: **Int**

of *(Pick uppermost corner of the cube at "19".)*

Next point (Press ENTER to end tracking): *(Move cursor southwest to "20".)* 0.2

Next point (Press ENTER to end tracking): *(Press ENTER.)*

Specify next point or [Undo]: *(Move cursor southeast to "21".)* 2

Specify next point or [Undo]: *(Press ENTER.)*

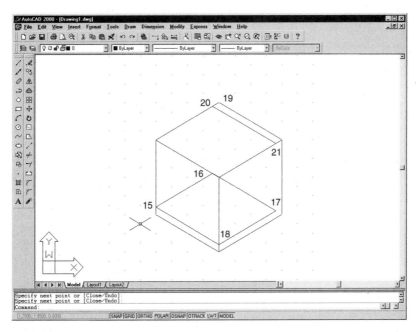

Figure 21.23 *Top Face of Upright Leg*

15. Let's finish drawing the leg by switching to the right isoplane. Press F5. You have two more lines to draw:

    Command: **line**

    Specify first point: **Int**

    of *(Pick end of the line you just drew at "22".)*

    Specify next point or [Undo]: *(Move cursor south to "23".)* 1.8

    Specify next point or [Undo]: *(Press ENTER.)*

16. Let's use the **Copy** command to draw the final line, using the **Last** option to select the line you just drew:

Command: **Copy**

Select objects: **l**

l found Select objects: *(Press* ENTER.*)*

<Base point or displacement>/Multiple: **Int**

of *(Pick end of line you just drew at "22".)*

Specify second point of displacement or <use first point as displacement>: *(Move cursor.)*

Oops! You find that you cannot move the line to where you want it to go. That's because you are in the wrong isoplane. Press CTRL+E to switch isoplanes and move the cursor. That's better!

Specify second point of displacement or <use first point as displacement>: *(Press* F5.*)*

<Isoplane Left> Near

of *(Pick at "24".)*

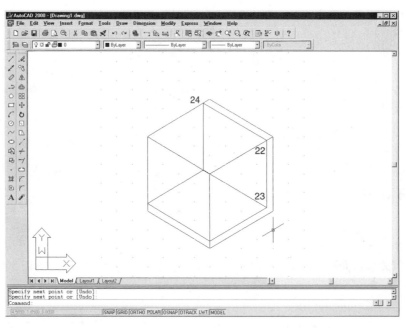

Figure 21.24 *Upright Leg Complete*

That completes the drawing of the two legs. Now you need to erase parts of the cube that we don't need. The best command for doing that is **Trim**:

Command: **Trim**

Current settings: Projection=UCS Edge=None

Select cutting edges ...

Select objects: *(Pick a line.)*

1 found Select objects: *(Pick another line.)*

Select the six lines shown dotted in Figure 21.25. These form the cutting edges.

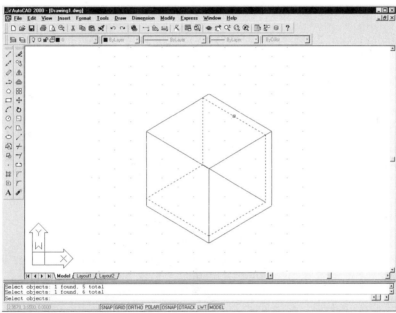

Figure 21.25 *Selecting the Cutting Edges*

Cutting Edges for Trim Command Select objects: *(Press* ENTER.*)*

Select object to trim or [Project/Edge/Undo]: *(Pick an unneeded line of the cube.)*

17. Select the five lines that are unneeded. The figure shows four lines trimmed.

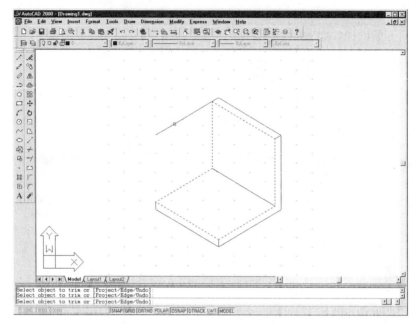

Figure 21.26 *Most Cube Lines Trimmed*

Select object to trim or [Project/Edge/Undo]: *(Press* ENTER.*)*

18. With the cube trimmed back to the bracket's legs, you now draw the holes in the bracket with the **Ellipse** command. It can be tricky finding the exact middle of an isometric object, so you employ tracking again. The center of the hole is 1,1 from the front corner of the lower leg. First, though, press F5 until you reach the top isoplane.

Command: *(Press* F5.*)*

<Isoplane top> Ellipse

Arc/Center/Isocircle/<Axis endpoint 1>: **i**

Specify center of isocircle: **tk**

First tracking point: **int**

of *(Pick corner at "25" on the lower leg.)*

Next point (Press ENTER to end tracking): *(Move northwest to "26".)* 1

Next point (Press ENTER to end tracking): *(Move northeast to "27".)* 1

Next point (Press ENTER to end tracking): *(Press* ENTER.*)*

Specify radius of isocircle or [Diameter]: **0.5**

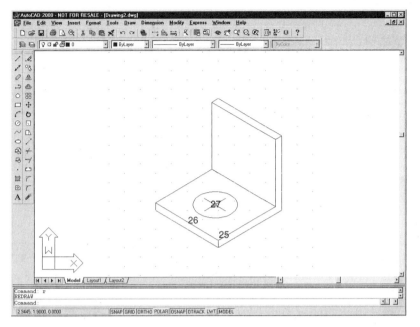

Figure 21.27 *Isometric Circle*

19. To draw the "bottom" of the hole, you use the **Copy** command. Since the copy takes place in the vertical direction, press F5 to get the right isoplane, which allows movement in the up-down direction.

    Command: *(Press F5.)*

    <Isoplane right> Copy

    Select objects: **I**

    I found Select objects: *(Press ENTER.)*

    <Base point or displacement>/Multiple: **Nea**

    to *(Pick the isocircle.)*

    Second point of displacement: *(Move cursor south.)* 0.2

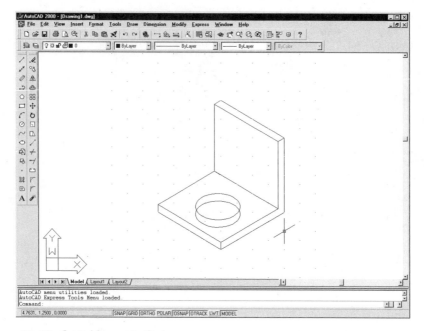

Figure 21.28 *Copied Isometric Circle*

20. You now use the **Trim** command to erase the parts of the lower isocircle that are "hidden."

   Command: **Trim**

   Current settings: Projection=UCS Edge=None

   Select cutting edges ...

   Select objects: *(Pick the upper circle.)*

   I found Select objects: *(Press* ENTER.*)*

   Select object to trim or [Project/Edge/Undo]: *(Pick unwanted portion of lower circle.)*

   Select object to trim or [Project/Edge/Undo]: *(Press* ENTER.*)*

We've now given the lower bracket a hole that looks quite realistic!

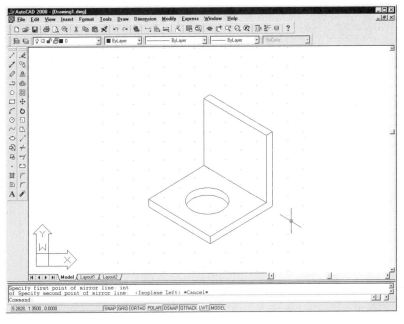

Figure 21.29 *Trimmed Isometric Circle*

21. You use the **Mirror** command to copy the "hole" to the upright leg. It will be a touch easier if you first switch to the left isoplane, first:

Command: *(Press* F5.*)*

<Isoplane left> Mirror

Select objects: *(Pick isocircle.)*

1 found Select objects: *(Pick the "isoarc".)*

1 found Select objects: *(Press* ENTER.*)*

Specify first point of mirror line: Int

of *(Pick one end of "fold" line.)*

Specify second point of mirror line: **Int**

of *(Pick other end of "fold" line, as shown in Figure 21.25.)*

Delete source objects? [Yes/No] <N>: *(Press* ENTER.*)*

Voila! Instant hole!

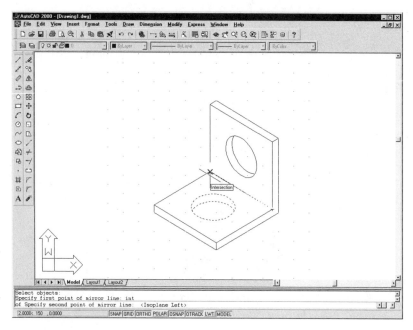

Figure 21.30 *Second Isometric Circle*

22. With the bracket drawn, you go on to dimension it. But it would be a good idea to save your work with the **Save** command. Call the drawing "Isobracket."

23. Before you can dimension the drawing, you need to change the dimensions so that the dimension text looks "correct" in isometric mode. That involves two steps: (1) setting up appropriate text styles and (2) changing dimension variables. You need to do this three times, once for each isoplane.

   From the menu bar, select **Format | Text Style**. When the **Text Style** dialog box appears, choose **New**. Type "ISOTOP" for the name of the new text style. Choose **OK**.

Figure 21.31 *New Text Style Dialog Box*

24. When the **Text Style** dialog box reappears, select **Simplex.Shx** for the Font Name. Change the **Oblique Angle** to **-30**. Choose **Apply**.

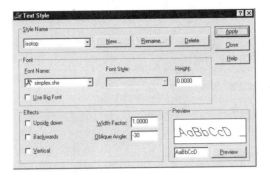

Figure 21.32 *Text Style Dialog Box*

25. Create the text styles for the isoplanes, as follows:

Style Name	Font Name	Oblique Angle	Action
ISOTOP	Simplex.Shx	-30	Choose Apply and New
ISORIGHT	Simplex.Shx	30	Choose Apply and New
ISOLEFT	Simpelx.Shx	30	Choose Apply and Close

Enter these values in the Text Style dialog box and choose Apply, then Close.

26. With the three text styles created, let's go on to create the dimension styles for the three isoplanes. Select **Format | Dimension Style**.

27. You need to set three variables:

Choose the **New** button and type "Isoleft" for the **New Style Name**. Choose **Continue**.

Choose **Text**. Force dimension text to align with the dimension line: choose **Aligned** with dimension line.

Specify the text style for the dimension text: select **ISOLEFT** from the text **Style** list box and choose **OK**.

The **Dimension Style Manager** does not, unfortunately, allow you to set the angle of the extension lines. They will have to be adjusted later. You have now created one of the three dimensions styles we need.

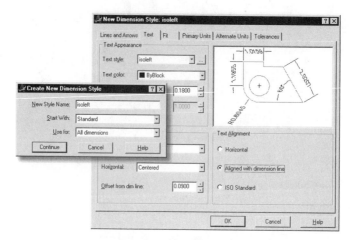

Figure 21.33 *Creating an Isometric Dimension Style*

28. Repeat for the other isoplanes using these parameters:

Dimstyle Name	Text Style
ISOLEFT	ISOLEFT
ISOTOP	ISOTOP
ISORIGHT	ISORIGHT

Remember to save each one.

29. Select the **ISOLEFT** dimension style, and choose **Set Current**. Choose **Close** to exit the **Dimension Style Manager** dialog box.

30. You can now place isometric dimensions. For linear dimensions, you always use the **DimAligned** command because it aligns the dimension along the iso-axes. You first do all dimensioning in one isoplane, and then switch to the next isoplane. Make sure AutoCAD is in the left isoplane by pressing F5 until \<Isoplane left\> shows up.

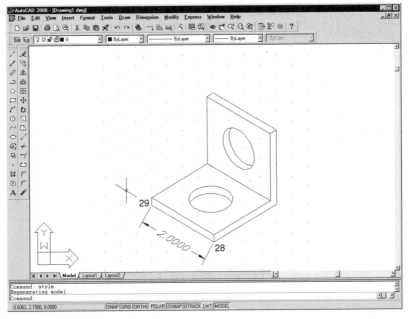

Figure 21.34 *Placing Dimension in Left Isoplane*

Command: **Dimaligned**

First extension line origin or press ENTER to select: **Int**

of *(Pick corner at "28".)*

Second extension line origin: **Int**

of *(Pick other corner at "29".)*

Dimension line location (Mtext/Text/Angle): *(Pick.)*

Dimension text=2.0000

31. Well, does the dimension look odd to you? To correct this, you need to apply the **Dimedit** command to skew the extension lines:

Command: **Dimedit**

Dimension Edit (Home/New/Rotate/Oblique) <Home>: **O**

Select objects: *(Pick dimension.)*

I found Select objects: *(Press ENTER.)*

Enter obliquing angle (press ENTER for none): **30**

Ahh, that's better: isometric dimension that looks good.

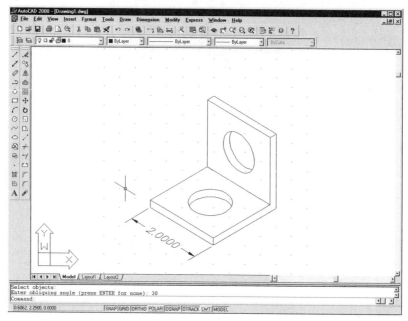

Figure 21.35 *Obliquing Dimension in Left Isoplane*

32. Let's now do the dimensioning in the top isoplane:

    a) Press F5 to switch to the top isoplane.

    b) Use the **DDim** command to select the **ISOTOP** dimension style.

    c) Place the dimension with the **DimAligned** command and INTersection object snaps.

    d) Use the **DimEdit** command's **Oblique** option to skew the dimension by −30 degrees.

33. Repeat for placing dimensions in the right isoplane, except that you use the **ISORIGHT** dimension style and skew the dimension by -30 degrees.

34. If you want to place a leader, use the Standard dimstyle and Standard text style.

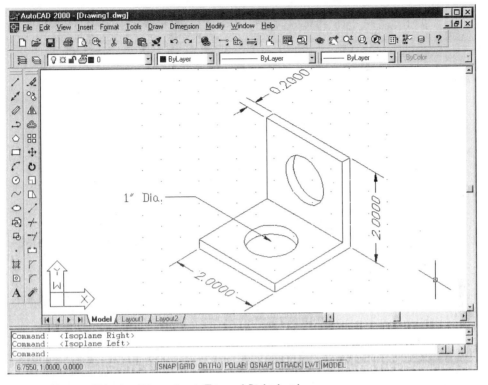

Figure 21.36 *Obliquing Dimension in Top and Right Isoplanes*

35. Save your work!

CHAPTER 22

Understanding Advanced Operations

Once you have become proficient with the basic draw, edit, and display commands of AutoCAD, you may want to try AutoCAD's advanced features. This chapter covers some of the advanced operations in AutoCAD. After completing this chapter, you will be able to

- Draw and edit polylines

- Create multilines

- Place spline curves

- Export and import DXF files for exchanging drawing with other software

- Recover damaged AutoCAD drawing files

INTRODUCTION

In this chapter, you learn about the following commands:

PLine

Draws polylines and arcs.

PEdit

Edits polylines and polyfaces.

MLine

Draw multiple, parallel lines.

MlEdit

Edits the vertices of mlines.

MlStyle

Creates multiline styles by name.

Spline

Draws a NURBS (non-uniform rational Bezier spline) curve.

SplineSegs

Controls the roundness of splines and splined polylines.

SplinEdit

Edits the position of a spline.

DxfOut

Exports the current drawing as a DXF (drawing interchange format) file.

DxfIn

Imports a DXF file into the current drawing.

Audit

Audits DWG and DXF drawing files for errors.

DRAWING POLYLINES

The polyline is perhaps the most unique of AutoCAD's objects. A polyline can have a tapered width, be made of arcs *and* straight lines, be splined, and have several other configurations.

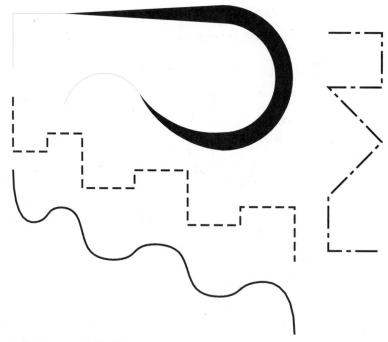

Figure 22.1 *Examples of Polylines*

While the polyline uses many editing commands you learned in earlier chapters, it has its own set of draw and edit commands. This chapter describes the draw command first, followed by the polyline editing command.

Polylines are a bit more difficult to learn than most AutoCAD objects. You should practice each option of the **PLine** and **PEdit** commands until you become familiar with the characteristics of each.

AutoCAD draws the following objects from polylines: donuts, polygon, rectangle, ellipse (when Pellipse = 1), and boundary.

USING THE PLINE COMMAND

To draw a polyline, issue the **PLine** (short for "polyline") command:

Command: **Pline**

Specify start point:

You always start a polyline by entering a "from" point. After you enter this point, AutoCAD displays a two-line prompt:

Current line-width is 0.0000

The stated width is the width for all your polylines unless you make a change. This is followed by a longer prompt:

Specify next point or [Arc/Close/Halfwidth/Length/Undo/Width]:

This prompt contains all the options from which the **PLine** command can branch. The following list explains each of these options:

A (Arc)

Switches to arc mode (more about this later in this section).

C (Close)

Works in the same way as the **Close** option for **Line**. A polyline segment is drawn back to the starting point and the command is terminated. The last width entered is used as the width for the closing polyline.

H (Halfwidth)

Functions like the **Width** option, except that you "show" AutoCAD the width by moving the crosshairs from the "Specify first point:" to one edge of the desired polyline width. This represents one-half the total width. This option is useful when you work with a centerline-to-edge distance.

L (Length)

Draws a line segment at the same angle as the previous segment. You must specify the length. This is similar to direct distance entry, except that AutoCAD uses the value stored in **Lastangle**. The default is 0 degrees. This option is useful when you want a polyline segment tangent to a polyarc.

U (Undo)

Works the same as the **Undo** command for line objects. The last polyline segment entered is "undone."

W (Width)

Selects the width for the succeeding polylines. A zero width produces a polyline one pixel wide, like the **Line** command. A width greater than zero produces a line of the specified width, similar to **Trace** lines. Polylines have both a starting and an ending width, which allows for tapered widths. You are prompted for both widths when choosing the **Width** option:

Specify starting width <0.0000>:

Specify ending width <0.0000>:

Setting both the starting and ending widths the same produces polylines of uniform width. If you specify different values, the polyline is tapered from the starting width to the ending width; the ending width becomes the default for the next segment.

Note: The display of wide polylines—on the screen and when plotted—is controlled by the **FillMode** system variable. When set to 0 (zero), only the outline of the polyline is displayed; when set to 1 (the default), the filled polyline is displayed.

DRAWING POLYLINE ARCS

To draw polyline arcs, you must enter the **PLine's** arc mode. Choosing the **Arc** option does this.

Specify next point or [Arc/Close/Halfwidth/Length/Undo/Width]: **A**

AutoCAD then switches to the arc mode and displays the following prompt:

Specify endpoint of arc or
[Angle/CEnter/CLose/Direction/Halfwidth/Line/Radius/Second
pt/Undo/Width]:

The default is "Specify endpoint of arc." Therefore, if you enter a point, it is used as the endpoint of the arc.

The **Halfwidth, Undo,** and **Width** options are the same as for polylines. The **Width** and **Halfwidth** options determine the width of the polyline used to draw the arc, and the **Undo** removes the most recent arc.

The arc starts at the previous point and is tangent to the previous polyline segment. If this is the first arc segment, the direction is the same as the direction of the last object drawn. The arc options allow you to modify the manner in which the arc is drawn. The following lists the options and their functions:

A (Angle)

Permits you to specify the included angle. The prompt is:

Specify included angle:

Just like regular arcs, a polyline arc is drawn counterclockwise. If you desire a clockwise rotation, use a negative angle, such as **–45**.

CE (Center)

Polyline arcs are normally drawn tangent to the previous polyline segment. The **CE** option allows you to specify the center point for the arc. The prompt is:

Specify center point of arc:

After the center point is entered, a second prompt is displayed:

Specify endpoint of arc or [Angle/Length]:

The **Angle** option refers to the included angle and Length refers to the chord length.

Notice that the **CEnter** option must be specified by the two letters **CE** to distinguish it from the **CLose** option. If you type "**C**", AutoCAD prompts: "Ambiguous response, please clarify...center or close?"

CL (Close)

Functions the same as the **Close** option for polylines, except that the close is performed using an arc. Closing a polyarc results in a "polycircle."

D (Direction)

The starting direction is determined by the last object's direction. The **Direction** option allows you to specify a new starting direction for the arc. You are prompted:

Specify the tangent direction for the start point of arc:

You show AutoCAD the starting direction by entering a second point in the desired direction. The next prompt is:

Specify endpoint of the arc: *(Pick.)*

H (Halfwidth)

Works in the same way as for straight polyline segments.

L (Line)

Exits Arc mode and returns to regular polyline mode.

R (Radius)

Allows you to specify the radius of the arc. After choosing the **Radius** option, you are prompted:

Specify radius of arc: *(type a number or show a distance)*

Then with:

Specify endpoint of arc or [Angle]:

S (Second pt)

Using the **Second pt** option allows you to switch to a three-point type of arc construction. You are prompted:

Specify second point on arc:

Specify end point of arc:

U (Undo)

Works in the same way as for straight polyline segments.

W (Width)

Works in the same way as for straight polyline segments.

POLYLINE EDITING

The polyline editor is used to modify your polylines (as well as 3D polylines and 3D polyfaces). The **PEdit** command is used to begin editing.

Command: **pedit**

Select polyline:

The desired polyline can be selected through any of the standard object selection processes. If the selected object is a line or arc, AutoCAD sometimes prompts you with:

Object selected is not a polyline.

Do you want to turn it into one? <Y>

If you respond with a **Y**, the object is converted to a polyline. If AutoCAD cannot convert the object to a polyline, it prompts, "Object selected is not a polyline."

After you select a polyline, the following prompt is displayed:

> Enter an option [Close/Join/Width/Edit
> vertex/Fit/Spline/Decurve/Ltype gen/Undo]:

Note: The **Open** option replaces the **Close** option, depending whether the polyline is open or closed.

The following list describes each option and its function:

C (Close)

Closes the open polyline by connecting the last point with the first point of the polyline. This command performs the same function as the **Close** option in the **PLine** command, except it allows you to close the polyline after you have exited the **PLine** command.

O (Open)

Opens a closed polyline by removing the segment created by the **Close** option. If the closing segment was not created by the **Close** option, the **Open** command has no effect.

J (Join)

The **Join** option converts and connects non-polyline objects to the current polyline. The string of connected objects becomes a polyline that is part of the original polyline.

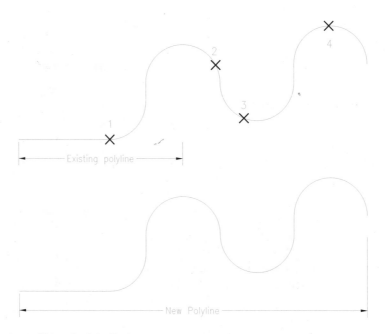

Figure 22.2 *Using the Join Option*

When you enter the **J** option, AutoCAD prompts:

Select objects:

You can then select the objects that you wish to join. Several objects can be selected in a continuous string. AutoCAD then determines which arcs, lines, and polylines share a common endpoint with the current polyline and merges them into that polyline.

To be successfully joined, the objects must have a perfect endpoint match (pretty close doesn't count). Use **Fillet** or **Change** and/or object snap to ensure a perfect match.

Figure 22.2 shows the use of the Join option:

Command: **pedit**

Select polyline: *(Enter point "1".)*

Enter an option [Close/Join/Width/Edit
      vertex/Fit/Spline/Decurve/Ltype gen/Undo]: **J**

Select objects: *(Enter points 2, 3, and 4.)*

The objects selected by points 2, 3, and 4 will be converted to polylines and merged into the polyline selected by point 1.

W (Width)

The **W** option allows you to choose a new width for the entire polyline. The width is applied uniformly to all segments of the polyline. When you choose the **Width** option, AutoCAD prompts:

Specify new width for all segments:

The new width can be entered as either a numerical value from the keyboard or by the selection of two points on the screen. Figure 22.3 shows a polyline of varying widths and the same polyline after use of the **Width** option:

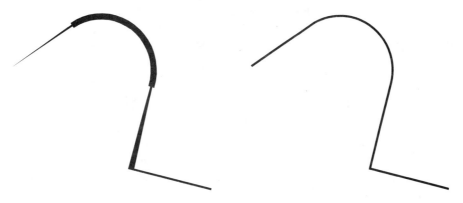

Figure 22.3 *Using PEdit with Width Option*

E (Edit vertex)

The connecting points of the segments of a polyline are each called a vertex. The **E** option allows you to edit a vertex and all the segments that follow it. Vertex editing is another mode and is covered later in this chapter.

F (Fit curve)

The fit curve option constructs smooth curves from the vertices in the polyline. AutoCAD inserts extra vertices where necessary.

S (Spline curve)

Creates a spline curve from the edited polyline. See the following section on spline curves.

D (Decurve)

Negates the effects of the two curve options.

L (Linetype generation)

Controls the generation of linetype through the vertices of the polyline. When the polyline linetype is set to other than continuous, the linetype dashes and dots are generated for each individual segment of the polyline. Depending on the linetype scale, some polyline segments will not show the linetype. If linetype generation is turned on, the polyline is considered as one segment when the linetype is applied, resulting in a uniform linetype pattern along the entire length of the polyline.

U (Undo)

Reverses the last **PEdit** operation.

CONSTRUCTING MULTIPLE PARALLEL LINES

The **MLine** command lets you draw up to 16 parallel lines at the same time. Each of the lines can have a different color, different linetype, and have a different offset distance. Unlike lines drawn parallel to each other with the Offset command, the objects drawn with the **MLine** command are treated as a single object.

The **MLine** command draws multilines; by default, a pair of parallel lines 1.0 units apart is drawn. The **MIEdit** command displays a dialog box that lets you perform some changes to the multiline. The **MIStyle** command lets you define custom multilines. Other editing commands, such as **Fillet**, **Chamfer**, and **Trim**, *cannot* be used on multilines.

USING THE MLINE COMMAND

You begin the **MLine** command by selecting **Draw | Multiline from the** drop-down menu, or by entering **MLine** at the 'Command:' prompt, as follows:

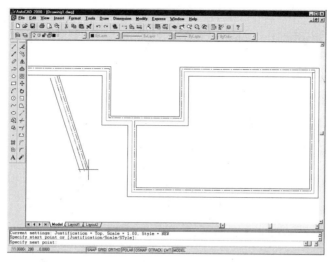

Figure 22.4 *Drawing Multilines with the MLine Command*

Command: **mline**

Current settings: Justification = Top, Scale = 1.00, Style = STAN-
   DARD

Specify start point or [Justification/Scale/STyle]:

Drawing a multiline is similar to drawing a line with the **Line** command: Keep
selecting points on the screen or supply coordinates, as follows:

Current settings: Justification = Top, Scale = 1.00, Style =
   STANDARD

Specify start point or [Justification/Scale/STyle]: *(Pick.)*

Specify next point: *(Pick.)*

Specify next point or [Undo]: *(Pick.)*

Specify next point or [Close/Undo]: *(Pick.)*

Specify next point or [Close/Undo]: **c**

The **Close** option joins the start point with the last end point. The **Undo** option lets
you back up, segment by segment.

In addition to those three options, the **MLine** command reports the current status of
three additional options: **Justification**, **Scale**, and **Style**:

Current settings: Justification = Top, Scale = 1.00, Style =
   STANDARD

Justification

The **Justification** option determines how the multiline is drawn relative to the cursor. Type **J** to access the justification options:

Specify start point or [Justification/Scale/STyle]: **j**

Enter justification type [Top/Zero/Bottom] <top>:

The meaning of the three options is:

Option	Meaning
Top	Draws multiline below the cursor (default)
Zero	Draws multiline centered on the cursor
Bottom	Draws multiline above the cursor

When a multiline is defined (discussed later in this chapter), you specify the offset of each parallel line. Thus, the Top justification mode draws the multiline so that the cursor selects the points that define the parallel line with the greatest positive offset.

Scale

The **Scale** option draws the multiline larger or smaller than as defined:

Specify start point or [Justification/Scale/STyle]: **s**

Enter mline scale <1.00>:

A value of 2 draws the multiline twice as wide; 0 draws a single line.

STyle

The **STyle** option lets you choose from a selection of predefined, named multiline styles. By default, AutoCAD comes with just one style, named Standard, which consists of a pair of lines 1 unit apart. You created custom multiline styles with the **MIStyle** command (discussed later in this chapter).

Specify start point or [Justification/Scale/STyle]: **st**

Enter mline style name or[?: **?**

Loaded mline styles:

Name	Description
STANDARD	

Enter mline style name or ?: **standard**

As with the **Text** command's **Style** option, the **?** option lists the names of available styles.

EDITING MULTILINES WITH MLEDIT

After the multiline is drawn, you edit its vertices with the **MlEdit** command. Select **Modify | Multiline** from the menu bar. A very few editing commands can be used with multilines, such as **Move, Copy,** and **Stretch. Trim, Extend, Lengthen, Fillet,** and other editing commands do *not* work with multilines.

Command: **mledit**

displays the **Multiline Edit Tools** dialog box (see Figure 22.5).

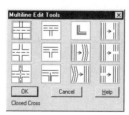

Figure 22.5 *Multiline Edit Tools Dialog Box*

As you select the images, their meaning is displayed at the bottom of the dialog box. From left to right, top to bottom, the symbols in the **Multiline Edit Tools** dialog box have the following meaning:

Closed Cross	Closed Tee	Corner Joint	Cut Single
Open Cross	Open Tee	Add Vertex	Cut All
Merged Cross	Merged Tee	Delete Vertex	Weld All

The **Closed** options do not cut the multiline; the **Open** options cut multilines; the **Merged** options cut just the outermost lines of the multiline, while interior lines remain. You cannot use the **Tee** options on closed multilines.

DEFINING MULTILINES WITH MLSTYLE

You can define many different kinds of multilines, each with its own style name. You might want to create one multiline for exterior walls, another for interior lines, and yet another for walls to be demolished.

You create new multiline styles by defining the color, linetype, and offset for each line. You can specify whether the multiline is filled with a color and whether it is capped with a line or arc at its ends. Here's how:

Command: **mlstyle**

displays the **Multiline Styles** dialog box (see Figure 22.6).

Figure 22.6 *Multiline Styles Dialog Box*

The top half of the dialog box lets you enter the name of a new style (maximum eight characters), along with a description up to 64 characters long. (After you finish defining the style, choose the **Save** button to save the style to the Acad.mln file.)

The **Save** button saves a multiline style to disk. The **Load** button loads a previously saved multiline style into the drawing. The **Rename** button allows you to rename a style.

Two buttons bring up the **Element Properties** and **Multiline Properties** dialog boxes for defining the properties of individual elements and of the multiline as a whole:

Element Properties

Choose the **Element Properties** button to bring up the dialog box, shown in Figure 22.7. Here you add and change elements from the multiline style by changing the offset distance, color, and linetype. Choose the **Add** button to add a new element; choose the **Delete** button to remove it. A multiline can have up to 16 parallel lines, called elements. The only element available is the straight line; a multiline cannot have arcs or splines.

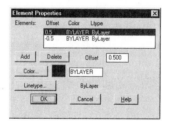

Figure 22.7 *Element Properties Dialog Box*

While the **Multiline Styles** dialog box displays the result of changes you make in the **Elements Properties** dialog box, the changes are not made in real time. Instead, you need to exit the dialog box periodically to check the effect of the changes you made.

Multiline Properties

Choosing the **Multiline Properties** button brings up a second dialog box, shown below.

Figure 22.8 *New Properties for a Multiline Style*

Here you have a number of options over the multiline's inside and endpoints:

Display Joints

When turned on, makes the multiline display a joint line at every vertex.

Caps

Either or both ends of an open multiline can have a cap consisting of a straight line or an arc. The line cap is drawn at right angles (90 degrees, the default) or at any angle down to 10 degrees (or up to 170 degrees). You can have a different angle of line cap at each end.

The alternative is to draw an arc as the cap. An outer cap draws an arc between the two outermost lines making up the multiline; an inner cap draws an arc between the innermost lines. Once again, you can have a different arc cap at each end of the multiline.

You can have both a straight line and arc cap at an end. For example, you can arc cap the innermost lines but straight line cap the outermost lines.

Fill

You can specify that the multiline be filled with any of AutoCAD's 255 colors. You cannot specify which multiline elements form the boundary of the fill: the entire multiline is filled. The preview image in the main dialog box does not display the look of the fill.

CONSTRUCTING SPLINE CURVES

AutoCAD has two methods for drawing spline curves: (1) use the **Spline** option of the **PEdit** command on an existing polyline and (2) use the **Spline** command to draw a NURBS-based spline curve.

Here we first look at polyline splines; later we'll look at NURBS (short for non-uniform rational Bezier spline).

SPLINE ANATOMY

To properly construct a spline curve, you must understand the manner in which a spline is defined.

The spline uses a *frame* of the spline curve. This frame can be open or closed. In the case of the open frame, the spline is "connected" to the first and last vertex points (beginning and endpoints of the polyline). Each vertex point between these exerts a pull on the curve.

Notice that a larger number of central points about an area exerts more pull on the curve in that area.

The spline curve frame (if constructed from a polyline) can be displayed or not, and consequently, plotted or not. The **Splframe** variable is used to control the visibility of the frame. If **Splframe** is set to 0, frames are not displayed. If set to 1, the frames will be displayed after a subsequent regeneration.

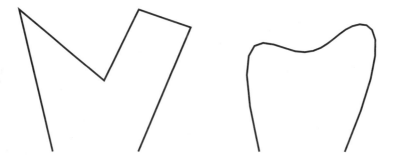

Figure 22.9 *Spline Frame (left); Spline-Curves Polyline (right)*

The **PEdit** command's **Decurve** option allows you to restore the polyline to its original configuration. **Decurve** can be applied to curves constructed with either the Fit or Spline options.

Polyarc segments are straightened for spline frame purposes. If the polyline contains differing line widths, a resulting spline curve will contain a smooth taper that begins with the width of the first endpoint and ends with the width of the last endpoint. This differs from a curve constructed with the Fit curve option, which maintains width information for each segment.

Spline curves construct curves in a different manner than the **Fit curve** method. Fit-curve constructs curves using a pair of arcs at each central point. More central points are required to obtain a reasonably accurate curve. The nature of the spline curve allows it to yield a more accurate curve with fewer central points. Figure 22.9 shows the relationship of applying both the **Fit curve** and **Spline** method of curve construction to the polyline shown on the left.

SPLINE SEGMENTS

Each polyline spline contains a specified number of straight-line segments used to construct the spline. The variable **Splinesegs** (short for spline segments) is used to control this number. The default is eight segments.

> Command: **splinesegs**
>
> New value for SPLINESEGS <8>:

Setting a larger **SplineSegs** value yields a smoother curve. This should be balanced against the increased drawing size and longer regeneration time created by the larger number of segments. If a negative value is used for **SplineSegs**, such as −8, the resulting curve will be constructed as a **Fit curve**, with the numerical value determining the number of segments between central points.

EFFECT OF EDIT COMMANDS ON SPLINES

Editing commands cause spline curves to react in different ways. The following describes the effects of each.

Move, Copy, Erase, Rotate, Scale, Mirror

Changes both the curve and the curve frame.

Explode, Break, Trim

Deletes the frame and creates a permanent change to the curve. **Decurve** cannot be applied after using.

Offset

Copies only the curve (without the frame) for the new offset object.

Stretch

The frame itself is stretched and the curve is refitted to the new frame. The frame can be stretched whether it is visible or not.

Divide, Measure, Hatch, Chamfer, Fillet

Applies only to the curve and not the frame.

Pedit (Join)

Decurves the polyline. After the **Join** process is completed, the **Spline** curve fit can be reconstructed.

Pedit (Vertex edit)

Moves the markers (denoted by "x") to the vertices on the frame (whether visible or invisible). When the spline curve is edited with **Insert**, **Move**, **Straighten**, or **Width**, the curve is automatically refit. When the **Break** option is applied, the polyline is decurved.

Object Snap

Object snap recognizes only the curve.

THE SPLINE COMMAND

Unlike the **Spline** option of the **PEdit** command, the **Spline** command creates a *true* spline object, constructed as a cubic or quadratic NURBS (short for non-uniform rational Bezier spline). Like **PEdit Spline**, the result is a smooth curve that fits between three or more control points; see Figure 22.10. (A two-point spline results in a straight line.)

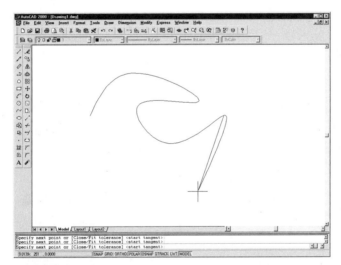

Figure 22.10 *Drawing a NURBS (Spline) Curve with the Spline Command*

Start the **Spline** command, as follows:

> Command: **Spline**
>
> Specify first point or [Object]: *(Pick.)*
>
> Specify next point: *(Pick.)*
>
> Specify next point or [Close/Fit tolerance] <start tangent>: *(Pick.)*
>
> Specify next point or [Close/Fit tolerance] <start tangent>: *(Press ENTER.)*
>
> Specify start tangent: *(Pick.)*
>
> Specify end tangent: *(Pick.)*

As you select points on the screen, AutoCAD automatically draws the NURBS between the points. When done, the **Spline** command asks you for the starting and ending tangencies, which determine the "tangent angle" for the start and end of the spline. You can use the TANgent and PERpendicular object snaps to force the spline tangent and perpendicular to an existing object.

Object Option

The **Object** option is an addition to the **Spline** command: it converts a 2D or 3D poly-line spline to a true NURBS curve.

Fit Tolerance Option

The **Fit tolerance** option prompts you, as follows:

> Specify fit tolerance:

The value of the fit tolerance determines how closely the curve fits to its control points. A tolerance of zero forces the spline to pass through the points.

Close Option

The **Close** option closes the spline: the end is joined with the start point.

EDITING SPLINES WITH SPLINEDIT

To edit a NURBS-based spline, you can use the **Splinedit** command. The **Property** and **Grip** commands also can be used to edit a spline.

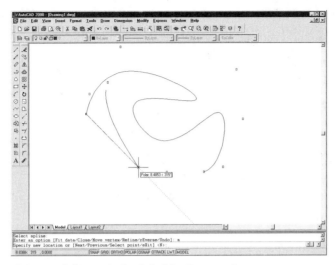

Figure 22.11 *Editing a Spline with the SplinEdit Command*

Command: **Splinedit**

Select spline: *(Pick.)*

Enter an option [Fit data/Close/Move vertex/Refine/rEverse/Undo]:

The **Splinedit** command has several options, as follows:

Fit Data Options

The **Fit data** option lets you edit the control points that define the position of the spline.

Enter a fit data option

[Add/Close/Delete/Move/Purge/Tangents/toLerance/eXit] <eXit>:

Add

Adds a fit point to the spline.

Close or Open

The **Close** option closes an open spline; if closed, the **Open** option opens up the spline.

Delete

Removes a fit point.

Move

Moves a fit point to another location.

Purge

Purges fit point data from the drawing.

Tangents

Changes the tangency of the start and end of the spline.

ToLerance

Changes the tolerance value of the spline.

After each change, AutoCAD reflows the spline to fit the new parameters.

Close or Open Options

The **Close** option closes an open spline. If the selected spline is already open, then the **Open** option appears instead.

Move Vertex Option

The **Move** vertex option moves the control vertices, one at a time.

Refine Options

The **Refine** option changes the primary spline parameters, as follows:

> Enter a refine option [Add control point/Elevate order/Weight/eXit] <eXit>:

Add control point

Adds a control point nearest to the cursor position between two existing control points.

Elevate order

Uniformly changes the number of control points along the spline; maximum value is 26.

Weight

Changes the distance between the spline and a control point. A larger number forces the spline closer to its control points.

Reverse Option

The **rEverse** option reverses the direction of the spline. The start point is now the ending point.

Undo Option

The **Undo** option undoes the most recent spline edit.

POLYLINE VERTEX EDITING

Vertex editing is used to modify the vertices of a polyline. The **PEdit** command edits polylines. Type **PEdit** and AutoCAD prompts:

Select polyline: *(Pick polyline.)*

Enter an option [Close/Join/Width/Edit
vertex/Fit/Spline/Decurve/Ltype gen/Undo]: **E**

If you wish to edit the vertices in a polyline, choose the **Edit vertex** option and you are presented with another option line.

Enter a vertex editing option

[Next/Previous/Break/Insert/Move/Regen/Straighten/Tangent/Width/
eXit] <N>:

The first vertex is marked by an X. If you have specified a tangent direction for that particular vertex, an arrow is displayed designating the direction.

The following list explains the options on the vertex editing option line and their functions:

N (Next) & P (Previous)

These options are used to move the identification marker X. If you enter **N**, you step through the vertices of the polyline by pressing ENTER each time a move is desired. Choosing the **P** (Previous) option allows you to "back up" in the same manner. When you reach the last vertex in the polyline, you must back up to reach previous vertices. You cannot wrap around the polyline, even if it is closed.

B (Break)

The **Break** option performs the same function as the normal **Break** command, except that the break can only occur at vertices. To perform a break, move the marker to the vertex where you wish the break to begin and enter **B**. The following prompt then appears:

Enter an option [Next/Previous/Go/eXit] <N>:

Use the **Next** and/or **Previous** options to move to the end of the desired break and enter **Go**. The segments between the points will be deleted.

Figure 22.12 *Vertex Ending*

If you wish to cancel the break while it is in progress, enter **X** (exit).

It is not possible to delete the entire polyline by entering break points at the first and last vertices.

If you break a closed polyline, it becomes "open" and AutoCAD removes the closing segment.

I (Insert)

A new vertex can be added to the polyline through the **Insert** option. The following prompt is displayed:

Specify location for new vertex:

The new vertex will be added at the location after the vertex that is currently marked by the X. Figure 22.13 shows a vertex insert.

M (Move)

Use the **Move** option to relocate a vertex. The following prompt appears:

Specify new location for marked vertex:

Enter the point that represents the new location for the current vertex.

R (Regen)

Regenerates the polyline, without your needing to exit the **PEdit** command.

S (Straighten)

The **Straighten** option is used in the same manner as the **Break** option described earlier, except that the segments between the selected points are

Figure 22.13 *Inserting a Vertex*

deleted and then replaced with a straight line segment. If you wish to straighten an arc segment, select the vertex immediately preceding the arc and enter both points on that vertex.

Figure 22.15 shows the effect of the **Straighten** option:

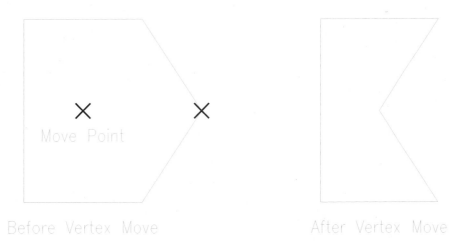

Figure 22.14 *Move Option*

Figure 22.15 *Straightening a Polyline*

T (Tangent)

Used to assign a tangent direction to the current vertex for use at a later time in curve fitting. You are prompted:

 Specify direction of vertex tangent:

You can specify the direction either by entering a numerically described angle or by showing AutoCAD the angle by entering a point in relation to the currently marked vertex.

W (Width)

The **Width** option in the vertex editing portion of **PEdit** differs from the normal **Width** option. Whereas the normal **Width** command determines the width for the entire polyline (and all its segments), the **Width** option in the vertex editing option line changes the width of just the segment following the vertex currently marked by the X. AutoCAD prompts:

 Specify starting width for next segment <0.0000>:

 Specify ending width for next segment <0.0000>:

The default starting width is equal to the current starting width for the segment being edited. As usual, the default ending width is shown equal to the starting width.

You must use the **Regen** option (see above) to redraw the screen and display the new segment width.

X (eXit)

Exits vertex editing and returns to the **PEdit** option line.

EXCHANGE FILE FORMATS

AutoCAD drawing files can be converted to a form that can be read and used by external programs. For example, a database program can read a DXF (drawing interchange format file) and extract blocks for compilation. The file format is available to programmers for development purposes and is generated in ASCII or binary format.

Other programs can also produce DXF files for AutoCAD to read. Let's look at AutoCAD's file exchange formats.

DRAWING INTERCHANGE FILE FORMAT

Drawing interchange files (DXF) are a popular way to exchange graphics files between software programs. AutoCAD uses the **Dxfin** and **Dxfout** commands to use and produce DXF files.

Producing a Dxfout file

The **Dxfout** command produces a drawing interchange format file from an AutoCAD drawing for use by external programs. To produce a DXF file, issue the **Dxfout** command while in the drawing from which you wish to produce the file.

> Command: **dxfout**

AutoCAD displays the **Save Drawing As** dialog box with the DXF file type selected. Enter the name of the file you wish to produce. Choose the **Options** button, then select the **DXF** tab.

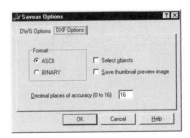

Figure 22.16 *The DXF Option Dialog Box*

Selecting **Binary** causes the file to be in binary format, as opposed to **ASCII** format. In most cases you should leave the default set to **ASCII**, unless you know the other program can read binary DXF files.

Selecting **Objects** causes AutoCAD to request the entities that you want extracted; block definitions are not included. When the **Objects** check box is turned off, the entire drawing is exported in DXF format.

The **Save thumbnail preview image** option adds a small image of the drawing to the DXF file. This is useful, provided the other application is able to display the image.

You can specify the precision of the output file, from 0 to 16 decimal places. In most cases, you should leave the default set to 16; a lower number results in a smaller but less accurate DXF file.

When done, choose the **Save** button. The file is written with a DXF file extension (do not specify the extension; AutoCAD automatically appends it). The output file will be a file separate from the drawing; it will not replace it.

Using an Existing DXF File

A drawing interchange file can be converted to an AutoCAD drawing file through the **Dxfin** command.

> Command: **Dxfin**

AutoCAD displays the **Select File** dialog box with file type set to DXF. Select the name of the DXF file and choose the **Open** button. It is not necessary to include the DXF file extension. While converting a DXF file, AutoCAD opens a new drawing.

DRAWING FILE DIAGNOSTICS

The **Audit** command is used to examine an existing drawing for damage. This diagnostic tool can also be used to correct damage to a file. The command sequence for the **Audit** command is as follows.

> Command: **audit**
>
> Fix any errors detected? [Yes/No] <N>:

If you answer "No" to the "Fix any errors detected?" question, AutoCAD displays a report, but does not fix the errors. If you answer "Yes," AutoCAD displays both the report and fixes the errors.

If a drawing contains no errors, AutoCAD will display a report showing the audit activity and the conclusion that no errors were found.

In addition to the screen report, the **Audit** command can print to file an audit report of the drawing file. This is controlled by the AuditCtl system variable:

- 0 means no file is produced (the default).

- I means AutoCAD writes the ADT file.

This file has the same name as the drawing file in the same folder with a file extension of ADT. You can read this file (with the Notepad text editor or a word processor) or print it on a printer.

DRAWING FILE RECOVERY

In a perfect world, everything works as planned. In our real world, however, we sometimes encounter difficulties. One day you will see a message on the screen that says:

INTERNAL ERROR (followed by a host of code numbers)

or

FATAL ERROR.

This means that AutoCAD has encountered a problem and cannot continue. You will usually be given a choice of whether or not you wish to save the changes you have made since the last time you saved your work. The following message is displayed:

AutoCAD cannot continue, but any changes to your drawing made
up to the start of the last command can be saved.

Do you want to save your changes? <Y>:

If you select "Yes," AutoCAD will attempt to write the changes to disk. If it is successful, the following message will be displayed:

DRAWING FILE SUCCESSFULLY SAVED

If the save is unsuccessful, one of the following messages is displayed:

INTERNAL ERROR

or

FATAL ERROR

If you see this, you can wave goodbye to all the changes made since you last saved your work. (Of course, as a good CAD operator, you save your work regularly.)

You can manually recover damaged drawing files by using the **Recover** command. When you use the **Recover** command, AutoCAD displays the **Open** Drawing File dialog box. Select the damaged drawing file and choose the **OK** button. AutoCAD will attempt to recover the damaged file. If successful, the drawing will be displayed.

If a drawing file is detected as damaged when you open it with the **Open** command, AutoCAD will perform an automatic audit of the drawing. If the audit is successful, the drawing is loaded for use. If it is not, the drawing will usually be unrecoverable.

If you exit the drawing without saving, the "repair" performed by AutoCAD will be discarded. If the recovery is successful and you save the drawing, you can load it normally the next time.

EXERCISES

1. Let's draw some basic polylines!

   We'll start by drawing a box. Start a new drawing. Set limits of 0,0 and 12,9.

   Command: **Pline**

   Specify start point: 1,2

   Current line-width is 0.0000

   Specify next point or [Arc/Close/Halfwidth/Length/Undo/Width]: **w**

   Specify starting width <0.0000>: **.2**

   Specify ending width <0.2000>: *(Press Enter.)*

   This sets a polyline with a width of .2. Now let's draw the box:

   Specify next point or [Arc/Close/Halfwidth/Length/Undo/Width]:
   1,5

   Specify next point or [Arc/Close/Halfwidth/Length/Undo/Width]:
   3,5

   Specify next point or [Arc/Close/Halfwidth/Length/Undo/Width]:
   3,2

   Specify next point or [Arc/Close/Halfwidth/Length/Undo/Width]: **C**

   This exits from the Polyline mode back into the command mode.

2. Let's draw and edit some multilines!

   Start the **MLine** command and draw the five-sided shape, as follows:

   Command: **Mline**

   Current settings: Justification = Top, Scale = 1.00, Style = STANDARD

   Specify start point or [Justification/Scale/STyle]: **2,2**

   Specify next point: **2,4**

   Specify next point or [Undo]: **8,8**

Specify next point or [Close/Undo]: **14,4**

Specify next point or [Close/Undo]: **14,2**

Specify next point or [Close/Undo]: **c**

Notice how the **Close** option automatically joins the start and end segments of the multiline.

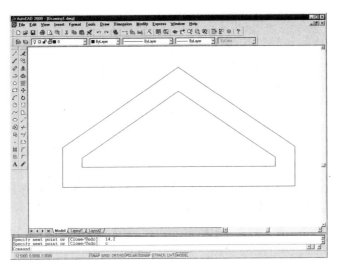

Figure 22.17 *Drawn with a Multiline*

You can use grips to edit a multiline, just like any other AutoCAD object. First, though, we'll use the **MlEdit** command to insert a vertex along the bottom of the multiline:

Command: **Mledit**

The **Multiline Edit Tools** dialog box appears. Select the **Add Vertex** icon (second row, third from the left). Choose **OK**. AutoCAD prompts you:

Select mline: *(Pick one of the multiline's bottom segments, in the middle.)*

Select mline (or Undo): *(Press* ENTER.*)*

AutoCAD adds a vertex (a joint) to the multiline where you made your selection, although you can't see it yet. To see the vertex, select the multiline. AutoCAD displays the blue grip boxes, including one where you added the vertex.

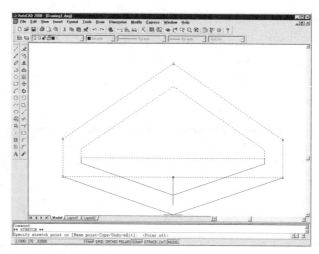

Figure 22.18 *Stretching the New Vertex in the Multiline*

Select the blue grip box where you added the vertex. It changes to red.

Stretch the multiline to a new shape.

Pick to place it.

Press ESC twice to dismiss the grips.

Now let's see how AutoCAD handles intersecting multilines. With the **MLine** command, draw a second multiline so that it crosses the first multiline.

Sart the **MIEdit** command again. This time select the **Open Cross** icon (second line, first icon). AutoCAD prompts you:

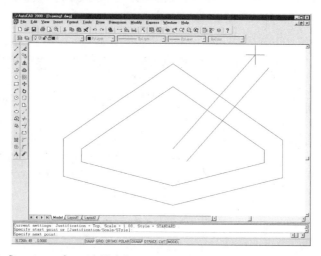

Figure 22.19 *Drawing a Second Multiline*

Command: **Mledit**

Select first mline: *(Pick the first multiline near the intersection.)*

Select second mline: *(Pick the second multiline.)*

Select first mline (or Undo): *(Press ENTER.)*

AutoCAD cleans up the intersection.

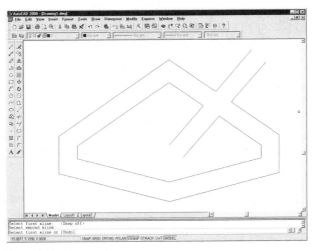

Figure 22.20 *Cleaned-Up Intersection*

CHAPTER REVIEW

1. What does the **PLine** command's **Close** option perform?

2. What option is used to create a tapered polyline?

3. What mode controls the display of solid areas in polylines?

4. What command is used explicitly to edit polylines?

5. What is a polyline vertex?

6. How do you edit a polyline vertex?

7. What types of vertex editing can you perform?

8. Can a spline curve be used on an object other than a polyline?

9. What variable setting controls the number of segments used in a spline curve?

10. How do you create a DXF file?

11. What is the **Audit** command used for?

12. How do a NURBS spline and a polyline spline differ?

13. How many lines can **MLine** draw at one time?

14. Which command is used to edit:

NURBS splines? _____

Multiline vertices? _____

Polylines? _____

15. What is the multiline good for drawing?

CHAPTER 23

Working with Viewports and Layouts

Two of the most interesting aspects of AutoCAD are the multiple viewports and paper space capabilities. Learning to use these is very helpful in many types of work, especially 3D design. After completing this chapter, you will be able to

- Construct and use multiple viewports

- Save and retrieve viewport configurations

- Understand the aspects of the two types of working spaces in AutoCAD

- Manipulate viewports and working spaces proficiently

INTRODUCTION

In this chapter, you learn about the following commands:

Viewports or Vports

Creates viewports in model mode.

Layout

Switches AutoCAD to layout mode.

Model

Switches AutoCAD to model tab.

MView

Creates viewports in layout mode.

Zoom XPThe paper space scaling option of the **Zoom** command.

USING VIEWPORTS IN AUTOCAD

The AutoCAD drawing screen can be divided into several "windows" called *viewports*. Each viewport can display a different view or zoom of the current drawing.

The following illustration shows a four-viewport screen, with different views and zooms of the same drawing.

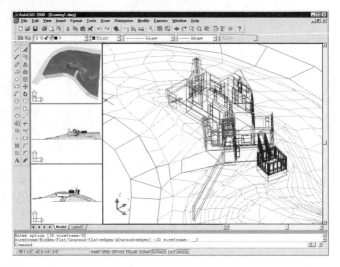

Figure 23.1 *Viewports*

THE CURRENT VIEWPORT

You draw in each viewport individually. After you finish drawing an object in one viewport, it appears in the other viewports. When you select and edit the object in one viewport, it shows as selected and being edited in the other viewports (where the object is visible). You can have the grid, snap, ortho, and UCS icon set independently in each viewport. Other modes cannot be set independently: osnap, polar, otrack, lwt, and model/paper.

Only one viewport, however, is active at one time. This viewport is called the *current viewport*. When drawing in the current viewport, the crosshairs cursor is displayed in that viewport, and all command activities are performed in the normal manner.

When the crosshairs cursor is moved into another view window, it becomes an arrow. To change another viewport to the current viewport, move the arrow into it and click the left button on the mouse. The crosshairs cursor is displayed, and normal drawing activities can be performed.

As an alternative, you can change the current viewport by pressing CTRL+R on the keyboard. As the active viewport is changed, the border around the viewport thickens.

The horizontal and vertical scrollbars pan the current viewport.

DRAWING BETWEEN VIEWPORTS

It is sometimes helpful to draw from viewport to another. For example, you may want to zoom in to a particular area of a drawing and connect a line to another area of the drawing not displayed in the current viewport. You can start the line in the current viewport and then move the cursor into the other viewport and click the mouse (or press CTRL+R). You have now changed the current viewport. The crosshairs are displayed in the new current viewport and you continue the Line command by connecting to the point displayed in the new current viewport.

 Note: Drawing between viewports is especially helpful when you draw in 3D. You can display the top view in one viewport and the plan view in another.

You cannot change viewports while some commands are active. These commands are as follows.

Dview	Vpoint
Grid	Vport
Pan	Zoom
Snap	

SETTING VIEWPORT WINDOWS

You can create dozens of viewport displays, up to a maximum of 64 viewports per drawing. The maximum number of viewports is controlled by the system variable **MaxActVp**, with a range of 2 to 64.

AutoCAD allows you to change the configuration (size and number) of viewports. In addition, you can save the configurations by name.

To set the number and design of viewports displayed on the screen, select **View | Viewports | New Viewport** from the menu bar, or type either **Viewports** or **Vports** (both versions of the command are valid). AutoCAD displays the **Viewports** dialog box with two tabs:

Figure 23.2 *Viewports Dialog Box: New Viewport Tab*

- **New Viewports** allows you to select from a dozen standard viewport configurations. Here is where you can, optionally, give a name to a viewport configuration.

- **Named Viewports** allows you to retrieve a named viewport configuration.

NEW VIEWPORTS

The purpose of the **New Viewports** tab is to change the viewport configuration. You can create two, three, four, or dozens of viewports—or collapse all the viewports back to a single viewport.

New Name

(Optional.) Type a name—up to 255 characters—for the viewport configuration. You do not need to enter a name. Note that, however, if you do not name the viewport configuration, you cannot use it in Layout mode (more later). Naming the viewport automatically saves it with the current drawing.

The following information is stored with the named view:

- ID number of viewport
- Viewport position
- Grid and snap mode, and spacings for each
- ViewRes mode
- UcsIcon setting and UCS name
- Views set by either the DView or VPoint command
- DView perspective mode, and clipping planes

Standard Viewports

The standard viewport configurations are listed here. This list changes, depending on the options selected below. The *Active Model Configuration* is the current viewport configuration.

Select a viewport name; a preview is displayed in the **Preview** area. When the word *Current* appears in a viewport, that means the current view will be displayed in the viewport. Other view names may appear—such as *Top*, *Front*, and SE Isometric—depending on the options you choose at the bottom of the dialog box.

Apply To

The **Apply** to list box lets you choose whether the viewport configuration is applied to the entire display or to the current viewport.

Display

The viewport configuration replaces the current configuration. For example, suppose you select **Three Vertical**. If the AutoCAD window has two horizontal viewports, those two are replaced by the three vertical viewports.

Current viewport

The viewport configuration is displayed within the current viewport. This allows you to further subdivide a viewport. For example, suppose you select **Three Vertical**. If the AutoCAD window has two horizontal viewports, the current viewport is subdivided into three vertical viewports. The AutoCAD window now has four viewports. You can keep subdividing viewports to a maximum of 64 viewports per drawing.

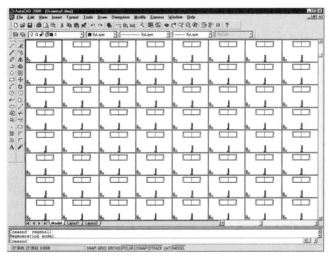

Figure 23.3 *64 Viewports*

Setup

The **Setup** list box allows you to choose whether you want 2D or 3D views automatically set up.

2D

Each new viewport contains the current view.

3D

Each new viewport contains a standard 3D view:

Number of Viewports	Engineering Views Created
Single	Current view
Two	Top, SE Isometric
Three	Top, Front, SE Isometric
Four	Top, Front, Right, SE Isometric

Change View To

When you select **3D** for the **Setup**, this list box lets you change the view displayed by each viewport. You do this in two steps:

1. Click on a viewport in the **Preview** window.

2. Select a view name from the **Change View To** list box. You can choose from these engineering views:

   Current

   Top

   Bottom

   Front

   Back

   Left

   Right

   SW Isometric

   SE Isometric

   NE Isometric

   NW Isometric

Notice that the view name changes in the viewport.

Note: The quick way to change the drawing back to a single viewport is to use the **-Vport** command with the **Single** option, as follows:

Command: **-vports**

Enter an option [Save/Restore/Delete/Join/SIngle/?/2/3/4] <3>: **si**

Regenerating model.

The new, single viewport displays the view that was in the current viewport, before you began the command.

NAMED VIEWPORT TAB

The purpose of the **Named Viewport** tab is to let you select one of the viewport configurations that you previously saved by name.

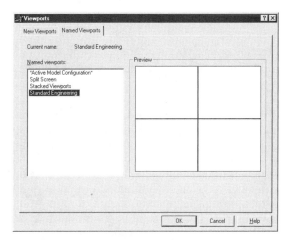

Figure 23.4 *Viewport Dialog Box: Named Viewports Tab*

If you haven't saved any configurations (by giving them names in the **New Viewport** tab), the only name in the list is *Active Model Configuration*, which is the current configuration.

If you have saved configurations in this drawing, their names are listed under **Named Viewports**. Choose a name to see its configuration in the **Preview** window.

Choose **OK** and AutoCAD reconfigures the entire display with the named viewport configuration.

REDRAWS AND REGENERATIONS IN VIEWPORTS

The **Redraw** and **Regen** (short for "regeneration") commands affect only the current viewport.

If you wish to perform either a redraw or regeneration on *all* viewports simultaneously, use **RedrawAll** or **RegenAll**, respectively.

LAYOUT MODES

In AutoCAD, you work in either model mode or layout mode (previously called "paper space").

Model mode is the mode you are used to working in. Most of your work in this book has been—and still will be—performed in model space. AutoCAD has one model mode per drawing. In model mode, you can only have tiled viewports.

Layout mode is used to arrange and detail views of your drawing in preparation for plotting. AutoCAD 2000 allows many layouts per drawing. You arrange views in layout space by moving and sizing viewports; in layout mode, you can have tiled or overlapping viewports.

In layout mode, viewports are handled as objects and can be edited. This allows you to place and plot different views of your work on the same drawing sheet (you can only plot the currently active viewport from model space). Any viewports you create in model space disappear in paper space.

SWITCHING BETWEEN MODEL AND LAYOUT MODE

To use layout mode, you simply choose the **Layout I** tab, located at the bottom of the drawing area. To switch back to model mode, choose the **Model** tab. As an alternative, you can type the **Model** command.

Figure 23.5 *Model and Layout Tabs*

When a drawing contains many layouts, the arrows next to **Model** allow you to scroll the tabs until you find the layout you are looking for.

PAPER SPACE ICON

When you are in layout mode, AutoCAD displays an icon at the lower left of the screen. **PAPER** is displayed on the status line at the bottom of the screen.

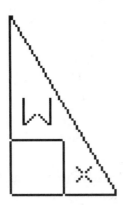

Figure 23.6 *Paper Space Icon*

You can turn off the paper space icon with the **UcsIcon Off** command.

CREATING LAYOUTS

The first time you choose the **Layout I** tab, AutoCAD switches to layout mode and displays the **Plot** dialog box. Choose **Cancel** to dismiss the dialog box (for more information about plotting, see Chapter 9 "Plotting Your Work").

AutoCAD displays a "sheet of paper," the print margins, and a single viewport.

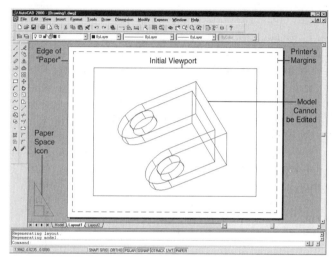

Figure 23.7 *Layout Mode: Paper Space*

It is crucial that you understand that your drawing now has two independent areas: (1) inside the viewport and (2) outside the viewport.

Inside the Viewport

The area inside the viewport is where the model resides. Try to edit the model, such as moving it with the **Move** command—you cannot. Notice the status bar: it reads **PAPER**. When you cannot edit the model, AutoCAD is in layout mode's *paper space*.

Think of looking at a vase though a window; when the window is closed, you cannot reach inside the window and touch the vase. This is like AutoCAD in layout mode: when you cannot edit the model.

To edit the model, choose **PAPER**; it changes to **MODEL**. Notice that the paper space icon changes to a model space icon inside the viewport. In addition, a heavy border surrounds the current viewport. You can now edit the model.

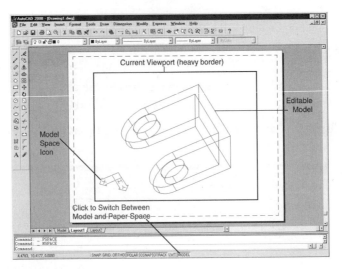

Figure 23.8 *Layout Mode: Model Space*

When the cursor is inside the viewport, you see the familiar crosshairs cursor. When you move the cursor outside of the viewport, it changes to an arrow cursor, indicating you cannot edit the drawing outside the viewport. To edit outside the viewport, choose the **MODEL** button on the status bar; it changes to **PAPER**.

Outside the Viewport

The area outside the viewport is called *paper space*. This is where you draw objects that relate to plotting the drawing. For example, you draw the title block, drawing border, and notes in paper space. In addition, you can manipulate viewports in paper space, as described below.

While in paper space, the cursor always has the crosshairs shape. This means you can draw and edit anywhere: outside the "page" border, across the model, inside and outside the viewport. Your drawing and editing, however, does not affect the model. As well, anything drawn outside the page margins (the dotted rectangle) is not plotted.

Model versus Paper Space

For the new AutoCAD user—and to some seasoned professionals—having two spaces can be confusing. In layout mode, it can be tricky knowing which space you are working in. Some commands do not operate as you expect them to in paper space. Other commands have different reactions, depending which space AutoCAD is in.

"Why does AutoCAD need to have two spaces?" you might be grumbling. The answer comes down to this:

- In model space, you draw the *model* full size.

- In paper space, you draw the *page* full size.

Layout mode combines the two, so that the model is plotted at its correct scale (usually a scale other than 1:1), while the paper space objects are plotted at their correct scale (usually 1:1).

LAYOUT VIEWPORTS

The viewports that are created in layout mode are different from the viewports in model mode. These differences include:

Model Viewports	Layout Viewports
Rectangular	Rectangular or any other shape
Tiled	Tiled, overlapping, or separated
Edited with the **VPorts** command	Edited with most AutoCAD commands
Visible border	Visible or invisible border

As the comparison table illustrates, layout viewports are much more flexible than model viewports. Figure 23.9 shows a drawing with several viewports. Can you count how many there are? (The figure has been enhanced to make it easier to see the viewport borders.)

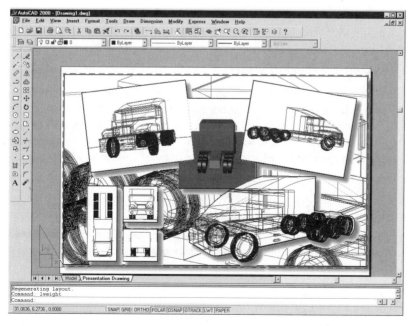

Figure 23.9 *Layout Viewports*

The drawing has eight viewports in total: five rectangular viewports, two rectangular viewports that have been rotated, and one truck-shaped viewport. Some viewports overlap others; one large rectangular viewport takes up the entire page. Most viewports show 3D wireframe drawings of the truck; one viewport shows a rendered image. The important point to remember is: WYSIWAP—what you see is what AutoCAD plots.

CREATING LAYOUT VIEWPORTS

To create viewports in paper space, you use the **MView** command. **MView** only works in paper space (and only if already in a layout tab). When you select **MView** while in model space, AutoCAD switches automatically to paper space and issues the message "Switching to paper space" to inform you of the change.

To create viewports, type the **MView** command. (You can select **View | Viewports | New Viewports** from the menu bar but the menu selection is sensitive to which mode you are in—model or paper—and displays a dialog box with fewer options than the **MView** command does.)

> Command: **mview**
>
> Specify corner of viewport or
>   [ON/OFF/Fit/Hideplot/Lock/Object/Polygonal/Restore/2/3/4]
>   <Fit>:

The **MView** command has many options.

Let's look at the **MView** command's many options, then look at an example of using paper space.

Specify Corner of Viewport

This is the default option. If you simply select a point on the screen, you set the first corner of a box that becomes a new viewport. AutoCAD prompts you for the other corner: "Specify opposite corner:". Move the cursor away from the first point and you see a box forming. Select the second point and you see the viewport form with your drawing contained within the viewport.

ON/OFF

The **On** and **Off** options are used to turn on and off the display of the contents of a viewport. This is helpful if you do not want to regenerate all the viewports as you edit. When AutoCAD prompts, "Select objects:", select the viewport border. If you turn off all the viewports, you will not be able to work in model space.

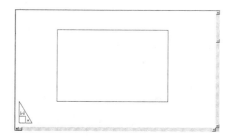

Figure 23.10 *Viewport Off (at left) and On (at right)*

Fit

The **Fit** option creates a new viewport that fills the entire "page." Note that this fills the entire limits of the drawing, even if you are zoomed in to a part of the drawing.

Hideplot

The **Hideplot** option allows you to specify that the viewport will be plotted with the hidden lines removed. You choose to have the hidden lines removed either in *all* viewports or in just in the ones you select.

This is especially useful if you are plotting a sheet that contains a 3D view of an object in one of the viewports. When you select **Hideplot**, you can turn the **Hideplot** function either on or off. If you turn it off, you can choose individual viewport(s) to be plotted with hidden lines.

Command: **mview**

Specify corner of viewport or
    [ON/OFF/Fit/Hideplot/Lock/Object/Polygonal/Restore/2/3/4]
    <Fit>: **h**

Hidden line removal for plotting [ON/OFF]: **on**

Select objects: *(Pick viewport.)*

I found Select objects: *(Press* ENTER.*)*

You select one or more viewports to be plotted with hidden lines removed: click the viewport's border. Note that when you select a viewport border, it does not create a hidden line view; it will plot with hidden lines removed. If you wish to display a hidden line view, use the **Hide** command.

Lock

This option locks the selected viewport. The effect is similar to locking a layer. You can view the objects in the viewport, but you cannot modify them. You can lock as many viewports as you wish.

Object

AutoCAD 2000 has two ways to create a non-rectangular viewport. You can use the **Object** or the **Polygonal** options. The **Object** option allows you to select a closed polyline (must have at least three vertices), an ellipse, a spline, a region, or a circle; AutoCAD converts the object to a viewport.

Polygonal

This option allows you to create an irregularly shaped viewport by selecting points in the drawing, much like drawing a polyline. AutoCAD prompts you:

Specify start point: *(Pick a point.)*

Specify next point or [Arc/Close/Length/Undo]: *(Pick another point.)*

The **Arc** option allows you to include arc segments to the viewport boundary. Its options are the same as for the **Arc** command.

The **Close** option closes the viewport boundary, if the boundary has at least three vertices.

The **Length** option draws a line segment tangent to the previous arc.

The **Undo** option removes the last-drawn boundary segment.

Restore

The **Restore** option permits you to create a new set of viewports based on a previously saved configuration. AutoCAD prompts:

?/Name of window configuration to insert <layout1>:

Entering a question mark (?) displays the names of saved viewport configurations. Enter the name of the saved configuration and press ENTER and AutoCAD prompts:

Fit/<First Point>:

As before, **Fit** fills the entire currently displayed area with the viewports, while "<First Point>" permits you to define an area for the viewports by creating a box with two corner points.

2, 3, 4

Creates two, three, or four viewports. The use of this option is similar to setting a specified number of viewports with the **VPorts** command, except the viewports are created to fill the currently displayed screen area or an area you specify with a box by placing two corner points. If you select **2**, **3**, or **4**, AutoCAD responds with:

Fit/<First Point>:

Fit fills the currently displayed screen area with the specified number of viewports, while "<First Point>" (the default) allows you to specify the area for the viewports by selecting two corner points that define the area.

Since the choices can conceivably build configurations of different arrangements (for example, side by side or over and under for two viewports), AutoCAD prompts for the arrangement.

The following is an overview of the prompts and responses used with each choice.

2

A prompt asks if you want a horizontal or vertical division. Vertical is the default.

3

The following prompt is displayed:

[Horizontal/Vertical/Above/Below/Left/Right] <Right>:

4

Creates four equally spaced viewports automatically.

MODIFYING LAYOUT VIEWPORTS

After the viewports are created, you can use the **Move** command to reposition the viewports. You can use grips editing, too: select a viewport, select a blue grip box, then drag viewport's corner to resize it, as shown in Figure 23.11.

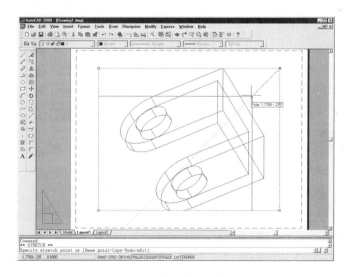

Figure 23.11 *Resizing a Viewport*

You can edit (move, stretch, copy, etc.) paper space viewports created with the **MView** command (you *cannot* edit viewports created with the **Viewports** command).

RELATIVE LAYOUT SCALES

The primary use of AutoCAD's layout mode is to create drawings with various views and details in a single, plottable drawing. Many times this requires that parts of the drawing be plotted at different scales. Let's look at an example.

Suppose that you have the floor plan of a house. The scale of the floor plan is 1/4"=1'0". You want to place an enlarged plan of a bath, with dimensions, on the same drawing sheet. The problem is that the bath area should be plotted at a different scale: 1/2"=1'0" (or a scale of 1:24).

AutoCAD can create this scale through the **XP** option of the **Zoom** command. Make a viewport active and then zoom at a factor of the paper scale, like this:

Command: **zoom**

Specify corner of window, enter a scale factor (nX or nXP), or
    [All/Center/Dynamic/Extents/Previous/Scale/Window] <real
    time>: **1/24xp**

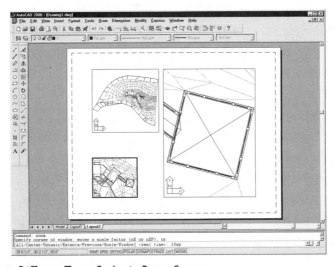

Figure 23.12 *Different Zoom Scales in Paper Space*

CREATING LAYOUTS

When you start a new drawing, AutoCAD automatically creates two tabs: **Model** and **Layout 1**. A drawing can have just one model view, but many layouts. To create more layouts, use the **Layout** command. From the menu bar, select **Insert | Layout | New Layout**.

Command: **layout**

Enter layout option
[Copy/Delete/New/Template/Rename/SAveas/Set/?] <set>:

CREATE ANOTHER LAYOUT

To create another layout, type **N** for the **New** option. AutoCAD prompts you:

Enter new Layout name <Layout1>: *(Type a name.)*

You can type a name up to 255 characters long; only the first 32 characters are displayed in the tab. A new tab appears with the name you gave it, as shown by Figure 23.13. Choose the tab to display the layout.

Figure 23.13 *Creating a New Layout*

CREATING A LAYOUT FROM A TEMPLATE

To create a new layout with a title block automatically installed, type **T** for the **Template** option. AutoCAD displays the **Select File** dialog box, listing DWT template files found in the \Acad 2000\Templates folder.

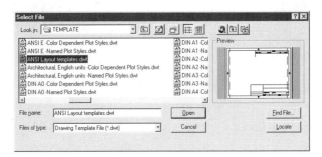

Figure 23.14 *Select a Template File*

Select a template file, and choose Open. AutoCAD displays the **Insert Layout(s)** dialog box.

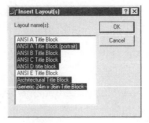

Figure 23.15 *Insert Layout(s) Dialog Box*

The dialog box lists the names of layouts stored in the template drawing. If the dialog box lists more than one layout name, you can select more than one named layout: hold down the CTRL key and choose each layout name you wish to insert.

Choose **OK**. AutoCAD creates a new layout tab for each. Notice that the tab is named after the template drawing. The layout contains the title block, drawing border, and an irregularly shaped viewport (notice the heavy line in Figure 23.16).

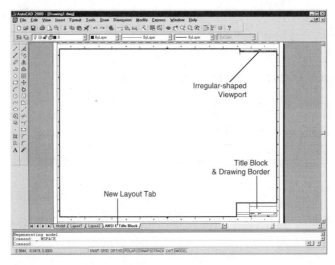

Figure 23.16 *Template Layout*

COPYING A LAYOUT

The **Layout** command's **Copy** option copies an existing layout:

> Enter layout to copy <layout 1>: *(Type the name of an existing layout.)*
>
> Enter layout name for copy <layout1(2)>: *(Type a name or press* ENTER.*)*

If you just press ENTER, AutoCAD creates a name based on the copied layout, such as layout1(2)—the name has an incremented number in parentheses added.

OTHER LAYOUT OPTIONS

The **Delete** option removes a layout. (You cannot delete the **Model** tab.) As an alternative, you can use the **Undo** command to remove a layout that has just been created.

The **Rename** option renames a layout.

The **Save** option saves a layout in a drawing template (DWT) file.

The **Set** option makes a layout current, although it is much easier to simply choose the layout's tab. This option is meant for macros and AutoLISP programming.

The **?** option lists the names of layouts in the current drawing. AutoCAD displays a list that looks something like this:

Active Layouts:

Layout: ANSI E Title Block	Block name: *Paper_Space.
Layout: Layout1	Block name: *Paper_Space0.
Layout: layout1 (2)	Block name: *Paper_Space10.
Layout: Layout2	Block name: *Paper_Space9.

EXERCISE

1. Start a new drawing. Let's set up four viewports. From the menu bar, select **View | Viewports | 4 Viewports**.

   You should now see four viewports on the screen as shown in the following illustration.

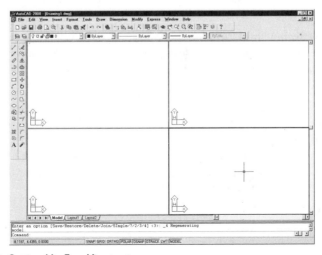

Figure 23.17 *Setting Up Four Viewports*

2. Move the crosshairs between the viewports. Notice how the crosshair cursor shows up in only one viewport. When you move the cursor into other viewports, it turns into an arrow.

3. Move the cursor into one of the viewports where it is displayed as an arrow. Click the left button of your mouse. You should now see crosshairs in that viewport. This is how you change active viewports.

4. If you are not there now, move into the upper left viewport, Click to make it active. With the **Circle** command, draw several circles in the drawing area. Notice how the circles appear in all viewports, after you finish drawing each circle.

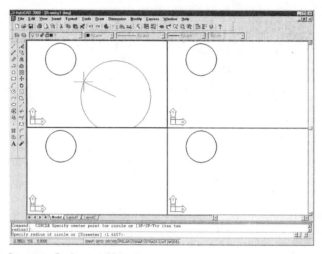

Figure 23.18 *Drawing Circles in a Viewport*

5. Move to the upper right viewport; make it active. Use the **Zoom** command to zoom in on one of the circles. Move to the lower left viewport, and zoom in on another circle. Notice how each viewport can contain a different view of the drawing.

6. Start the **Line** command and, with CENter object snap, select the center of the circle. Before placing the second endpoint of the line, move to the upper right viewport; click to make it active. With CENter object snap, place the line's second endpoint at the center of the other circle. Notice how you can monitor the activity of the entire drawing in the viewport at the upper left, since it is in a "zoom all" display.

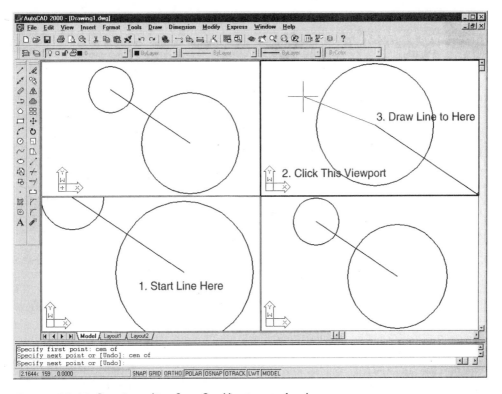

Figure 23.19 *Drawing a Line from One Viewport to Another*

CHAPTER REVIEW

1. How many viewports can be active at one time?

2. Can you switch between viewports while in a drawing command?

3. How would you save a viewport configuration that you wanted to use again?

4. How would you regenerate all the viewports in one operation?

5. When can the **MView** command be used?

6. How would you place a new viewport that would fill the entire screen in paper space?

7. How do you switch from paper space to model space?

8. What are the differences between viewports created by the **Viewports** and **MView** commands?

TUTORIAL

Open the drawing from the CD-ROM named **PsPlan.Dwg.** The drawing is of a floor plan, as shown in the following illustration.

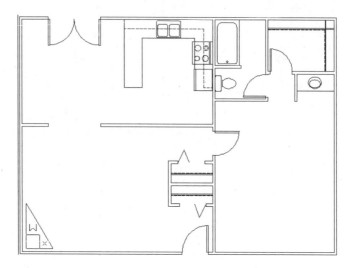

Figure 23.20 *PsPlan.Dwg Drawing*

The drawing is drawn in architectural units with limits of 0,0 and 96',72'. This is set up to be plotted at 1/4" =1'0" on a C-size (24"x18") sheet of paper.

You need to enlarge an area containing the bath, and show this area at a scale of 1/2"=1'0". Let's look at the steps for doing this.

1. First, switch to layout mode. Choose the **Layout1** tab.

2. AutoCAD displays the **Page Setup** dialog box; choose **OK**. Notice that AutoCAD displays the model inside a single viewport.

3. Let's now resize the viewport, using the grips editing method. Choose the viewport. Notice it changes to a dotted line, with one grip at each corner.

4. Select the grip in the upper right corner. Notice that it turns red. Drag the grip to the center of the drawing. As you do, notice how the viewport becomes smaller. Click to place it.

5. Choose **PAPER** on the status bar. Use the **Zoom E** command to make the drawing fit the viewport.

6. Let's create another viewport. Choose **MODEL** on the status bar.

7. Start the **MView** command and specify the **Polygon** option, as follows:

   Command: **mview**

Specify corner of viewport or
[ON/OFF/Fit/Hideplot/Lock/Object/Polygonal/Restore/2/3/4]
<Fit>: **p**

Specify start point: *(Pick a point.)*

Specify next point or [Arc/Close/Length/Undo]: *(Pick a point.)*

Specify next point or [Arc/Close/Length/Undo]: *(Pick a point.)*

8. Include an arc in the viewport boundary with the **Arc** option, as follows:

Specify next point or [Arc/Close/Length/Undo]: **a**

Enter an arc boundary option
[Angle/CEnter/CLose/Direction/Line/Radius/Second
pt/Undo/Endpoint of arc] <Endpoint>: *(Pick a point.)*

Enter an arc boundary option
[Angle/CEnter/CLose/Direction/Line/Radius/Second
pt/Undo/Endpoint of arc] <Endpoint>: *(Pick a point.)*

9. Switch back to line drawing mode with the **Line** option, as follows:

Enter an arc boundary option
[Angle/CEnter/CLose/Direction/Line/Radius/Second
pt/Undo/Endpoint of arc] <Endpoint>: **l**

Specify next point or [Arc/Close/Length/Undo]: *(Pick a point.)*

10. Select addition vertices and then close the viewport:

Specify next point or [Arc/Close/Length/Undo]: **c**

Your drawing should now have two viewports. One viewport is rectangular;
the other viewport has an irregular boundary.

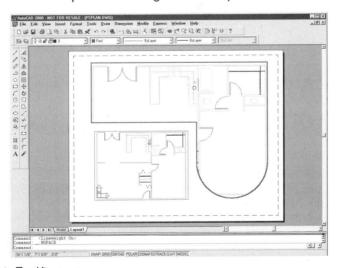

Figure 23.21 *Two Viewports*

11. Let's switch to the model. Choose **PAPER** on the status bar so that it changes to **MODEL**.

12. Move the cursor into the new viewport and click to make it the active viewport.

13. Let's now use AutoCAD's **Zoom** command to zoom and scale the contents of the new viewport in relation to the paper space units.

    Command: **zoom**

    All/Center/Dynamic/Extents/Previous/<Scale (X/XP) Window:<.01
        /<Real tine>: **0.01XP**

14. Now zoom window into the area of the bath. Notice how the scale of the viewport changes. The area of the plan you wanted to show may not be correctly centered.

15. While in model space, use the **Pan** command to reposition the drawing.

16. After you have finished, change to paper space: choose **MODEL** on the status bar. Use the **Text** command to create titles for each drawing. If this is performed in paper space, the text can be placed anywhere on the screen, even outside the viewports.

Attributes

Many types of CAD drawings can make use of attributes. Once you are familiar with the concept and learn the commands, attributes can help you extract data from the drawing. After completing this chapter, you will be able to

- Understand what attributes are and how they are used

- Comprehend the anatomy of attributes

- Manipulate attributes using AutoCAD commands

- Produce attribute output files for printing and use with other computer programs

INTRODUCTION

Attributes help you get text data out of AutoCAD. Some of the other ways to get the data include the **List** and **Dxfout** commands, as well as **dbConnect**, which uses SQL (structured query language), an advanced topic not covered by this book. Attributes, however, give you the best "price-performance" of all these methods.

Attributes are text data—any kind of text—attached to a block (symbol). The data could include a description, part number, and price. Each time you insert the block, you are prompted to fill in the attribute data. Then you can extract the attribute data and use another program—such as a spreadsheet—to process the data.

In this chapter, you learn about the following commands:

AttDef

Defines attributes and default values.

AttDisp

Controls the display of attributes in the drawing.

-AttEdit

Edits attributes at the command line.

AttEdit

Edits attribute values.

AttReq

Toggles whether attributes are filled in for you (using default values) or you enter the attribute text.

AttExt

Extracts attributes from drawing to text file.

In addition, you learn how the **Block** command binds attribute definitions to symbols, while the **Insert** command places attributes in the drawing.

ATTRIBUTES

AutoCAD allows you to add attributes to a block. An attribute can be considered a "label" that is attached to the block. This label contains any information you desire.

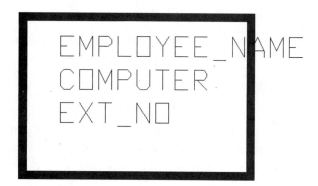

Figure 24.1 *Attributes*

Attributes are placed in the drawing with the block they are attached to. When the block is inserted, AutoCAD requests the values for the attributes. You determine what information is requested, the actual prompts, and the default values for the information requested.

The information from each block can be taken from AutoCAD's drawing file and used in other places, such as database programs.

For example, you can create a chair symbol with the attributes of Owner, Manufacturer, Model, Fabric, Date Purchased, and Purchase Price. These are the steps you go through:

Step 1
You draw the chair symbol with drawing commands, like **Line** and **Arc**.

Step 2
You create the attribute prompts with the **Attdef** command.

Step 3
You combine the symbol and the attribute information with the **Block** command.

Step 4
You insert the block with the **Insert** command, answering the prompts for the Owner, Manufacturer, and other data.

Step 5
If you change your mind, you edit the attributes with the **AttEdit** command.

Step 6
Finally, you extract the attribute data with the **AttExt** command from the drawing to a text file on disk.

Step 7
You import the text file into a spreadsheet or database program for further processing, such as adding up all occurrences of the chair block.

The type of information you can extract from the drawing through the **AttExt** command includes the names of blocks, the block's X,Y coordinates, the total number of each block, the layer the block is located on, and the attribute data itself.

EXERCISE
The best way to understand attributes is to use them in an example. You are now the manager of an engineering department that uses CAD (see how CAD has already helped your career?). The department contains several CAD workstations, each with an employee, a computer, and a telephone. You use attributes to keep information on each of these items contained in the floor plan of the department.

1. Start a new drawing.
2. Set limits of 0,0 and 24,18.
3. Draw a 20'x10' room with walls using the **PLine** command (with width=0.1), as shown in Figure 24.2:

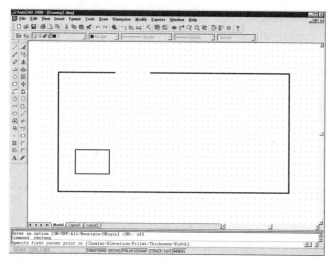

Figure 24.2 *Office Walls and Desk*

Creating the Symbol

4. First, you must draw the 2'x3' desk. (Do this now with the **Rectang** command.) It might help to zoom a bit. Now to set up the attributes.

Defining Attributes

The **AttDef** command is used to set up a template for the attribute (or label). Each attribute is made up of different bits of information. The attributes include: the name of employee; the type of computer used; and the telephone extension number. These items are called *tags*.

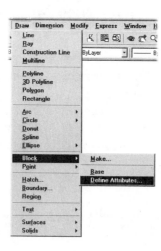

Attribute Tags

A tag is the name of a part of an attribute. The following tags will be used in your attribute:

Employee_name

Computer

Ext._no.:

Notice that underscores are used instead of spaces, because blank spaces are not allowed in tag names. You can also use a backslash (\) as a leading character in lieu of spaces.

5. Jump right in and execute the **AttDef** command:

Command: **AttDef**

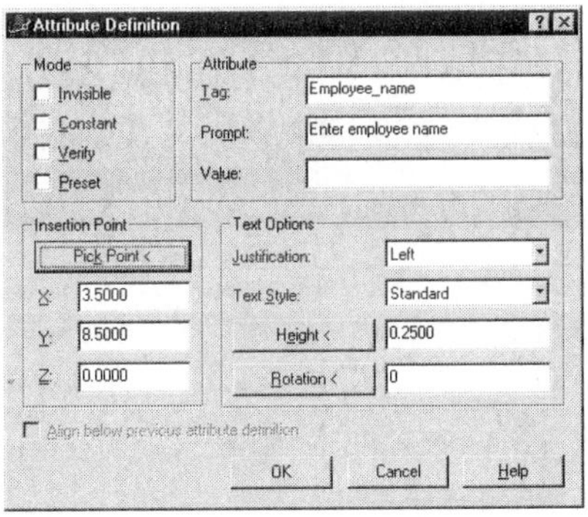

Figure 24.3 *Attribute Definition Dialog Box*

Let's look at the options for the attribute **Modes**:

Invisible:

The **Invisible** option determines whether the label is invisible when the block containing the attribute is in the drawing. If you later want to visibly display the attribute, you can use the **AttDisp** command.

Constant:

The **Constant** option gives every attribute the same value. This might be useful if every computer on every desk is the same. Beware! If you designate an attribute to contain a constant value, you cannot change it later.

Verify:

If you use the **Verify** option, you will be asked to verify that every value is correct.

Preset:

The **Preset** option allows you to preset values that are variable, but not be prompted when the block is inserted. The values are automatically set to their preset values.

Each option is either activated (check mark) or not activated (no check mark). To change the current setting for each, select it.

Let's leave the modes as they are for now.

6. Let's look at the **Attribute** section:

Tag:

The attribute *tag* is the name of the attribute. You enter the tag of "Employee_name."

7. You then provide the attribute *prompt*. The **Prompt** is the text that appears on the text line when the block containing the attribute is inserted. For the default prompt, you will use "Enter employee name."
   If you want the prompt to be the same as the tag name, press TAB (do not press ENTER, since that will dismiss the dialog box). If the **Constant** mode was specified for the attribute, the prompt field is grayed out.

8. Finally, you provide the default attribute *value*. You won't have a default employee name, so you leave **Value** blank. This value is displayed as the default later when you insert the block containing this attribute.

9. The remainder of the dialog box consists of a series of fields that are similar to text, except that a text string is not requested. The attribute information is used as the text string. The location and text size you specify will become the location of the information in the inserted block. Enter a text **Height** of **0.25** and a **Rotation** of **0**.

10. Choose the **Pick Point** button. AutoCAD clears the dialog box and prompts:

   Start point: *(Pick.)*

Select a point inside the desk. AutoCAD returns the dialog box. Choose the OK button.

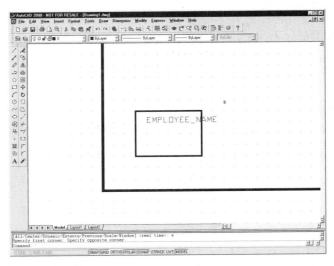

Figure 24.4 *First Attribute Tag*

11. Continue by repeating the **AttDef** command:

    Command: *(Press* SPACEBAR.*)*

12. This time we want to enter the tag for the computer:

    Tag: **Computer**

    Prompt: **Enter computer type**

    Value: **IBM-AT**

Notice that we used **IBM-AT** as the default attribute value. Every time the block is inserted, AutoCAD will show this as the default.

13. This time, choose the **Align** below previous attribute definition option to turn it on. Choose **OK**.

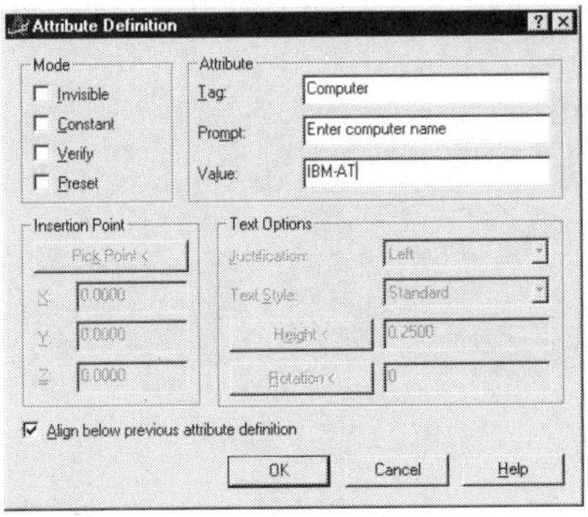

Figure 24.5 *Second Attribute Definition*

14. Now, for the telephone extension number. Press the SPACEBAR to repeat the **AttDef** command:

    Command: *(Press* SPACEBAR.*)*

Enter the following:

   Tag: **Ext_no.**

   Prompt: **Enter telephone extension number**

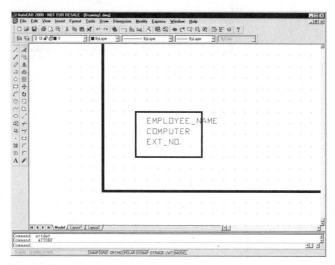

Figure 24.6 *The Attributes Defined*

This time, do not specify a default for **Value**.

Creating the Block

15. Now, to turn the desk and attributes into a block. From the menu, select **Draw | Block | Make**.

    Command: **block**

AutoCAD displays the **Block Definition** dialog box. Enter these values:

Figure 24.7 *The Block Definition Dialog Box*

Name:

For the name of the block, type "station" in the **Name** field.

Base point:

In the Base point section, choose the **Pick point** button. The dialog box disappears and AutoCAD prompts:

> Specify insertion base point: *(Select the lower left corner of the desk.)*

The **lock Definition** dialog box returns. Notice that AutoCAD has filled in the *X,Y,Z* coordinates of the pick point.

Objects:

In the **Objects** section, choose the **Select objects** button. Again, the dialog box disappears and AutoCAD prompts:

> Select objects: **c**
>
> Specify first corner: *(Pick point.)*
>
> Specify opposite corner: *(Pick point.)*
>
> 4 found
>
> Select objects: *(Press* ENTER.*)*

Use Crossing mode to select the desk and three attribute tags. When finished selecting the objects, press ENTER to return to the dialog box. Notice that the dialog box reports "4 objects selected."

For this tutorial, you want the objects deleted when they are turned into a block. Choose the **Delete** radio button.

Note: If you want the attributes to appear in a specific order (when you later insert the block), at the "Select objects" prompt do not use a Window or Crossing selection mode; instead, select the attribute definitions in the order you want them presented.

Preview Icon:

A preview icon is nice, since it helps remind you what the block looks like. In the **Preview** icon area, choose the **Create** icon from block geometry radio button.

Ensure that the **Insert** units is set to **Inches**.

For the **Description**, you can type anything, such as "Computer work station."

Choose the **OK** button. The block disappears. You are now ready to insert your block with its attributes!

Inserting the Block

16. Insert the block named Station. From the menu bar, select **Insert | Block**. The Insert dialog box appears.

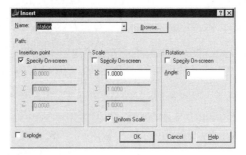

Figure 24.8 *The Insert Dialog Box*

17. Ensure the name of the block is Station. For the other options:

Insertion point: specify on-screen.

Scale: ensure that *X*, *Y*, and *Z* are set to 1.0 (the **Uniform scale** option may be turned on).

Rotation: ensure that angle is set to 0.

Explode: ensure that this option is off (no check mark showing).

Choose **OK**.

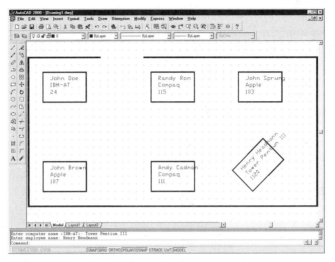

Figure 24.9 *Inserting the Blocks*

18. AutoCAD now switches to the command line, prompting you for the block's insertion point.

    Specify insertion point or
        [Scale/X/Y/Z/Rotate/PScale/PX/PY/PZ/PRotate]: *(Pick a point anywhere inside the office walls.)*

19. The final step is to provide values for the attributes. Notice that AutoCAD— curiously—asks for the attribute values in reverse order:

    Enter attribute values

    Enter telephone extension number: **24**

    Enter computer name <IBM-AT>: *(Press* ENTER.*)*

    Enter employee name: **John Doe**

20. Place several more desk symbols. You can use any values you wish, or follow along with these values:

Employee Name	Computer Name	Tel. Ex.
Randy Ron	Compaq	115
John Sprung	Apple	103
John Brown	Apple	107
Andy Cadman	Compaq	111
Henry Headmann	Tower Pentium III	1120

CONTROLLING THE DISPLAY OF ATTRIBUTES

You don't always want the attributes to show up on the display screen. You can use the **AttDisp** (short for "attribute display") command to determine the visibility of the attributes.

Command: **attdisp**

Enter attribute visibility setting [Normal/ON/OFF] <Normal>:

The options are listed below:

Normal

Attributes are visible, unless you specified them to be invisible as an attributes mode with the **AttDef** command when the attributes were formed. This option is useful if you want some attributes displayed and not others.

On

All attributes are visible, whether or not you set the Invisible toggle.

Off

All attributes are invisible, which helps reduce screen clutter.

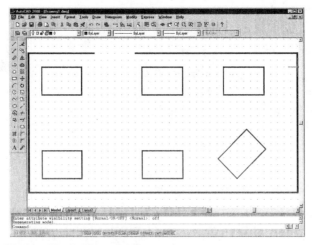

Figure 24.10 *Attribute Display Turned Off*

After the **AttDisp setting** is changed, the display is regenerated to show the new state (unless the **Regenauto** is off).

EDITING ATTRIBUTES

Now that everything is all set up properly, John Smith leaves, and Andy Cadman is hired to take his place. You have to make a change in your attribute database. The **-AttEdit** command is used to make changes in attributes. (Recall that prefixing a command name with a hyphen forces AutoCAD to display the prompts at the command line.) We cover the command-line version of the **AttEdit** command here because it is more powerful than the dialog box version. Even the **Properties** window cannot do what the **-AttEdit** command can!

    Command: **-attedit**

    Edit attributes one at a time? <Y>

The response determines the string of options that will follow:

> *Yes*
>
> Selects individual (one by one) editing. The attributes that are currently visible on the screen can be edited. The attributes to be edited can be further restricted by object selection or block names, tags, and values of the attributes to be edited.
>
> *No*
>
> Used for global editing of attributes (all). You can also restrict the editing to block names, tags, values, and on-screen visibility.

You are next asked to select the method of editing. You can choose the attributes to be edited by global wild card (*), or by using the ? symbol to replace common characters. AutoCAD prompts for the parts of the attributes to be edited:

Block name specification <*>:

Attribute tag specification <*>:

Attribute value specification <*>:

Your reply to each prompt determines the parts of the attribute that can be edited.

You choose to edit individual attributes or all attributes.

INDIVIDUAL EDITING

You can choose the individual attributes to be edited by using the object selection process. The prompt issued after you have selected the block, tag, and values that are possible to edit is:

Select Attributes:

Use the standard object selection process to choose the attributes to be edited. (The attribute set selection limits the attributes to be edited. When you press ENTER to each prompt at that time, all attributes are edited.)

After you have selected each attribute to be edited with the object selection process, an X marks the first item that can be edited. The X marks the current spot to be edited until you enter **Next** (or press ENTER) and a new spot is marked. You are then prompted:

Figure 24.11 *Attribute Selected for Editing*

Enter an option
[Value/Position/Height/Angle/Style/Layer/Color/Next] <N>:

The options are:

Value: attribute value.

Position: text position.

Height: text height.

Angle: text angle.

Style: text style name.

Layer: layer for the attribute.

Color: color of the attribute.

Next: next attribute value.

You can use the first letter to select the appropriate option, or press ESC to cancel.

When you enter **Value**, AutoCAD prompts with:

Enter type of value mofidication [Change/Replace] <R>

Change is used to change a few characters, as for a misspelling. If you choose **C**, the following prompt appears:

String to change:

New string:

You should respond to the first string with the string of characters to be changed and to the second string with the string you want it replaced with.

Replace is used to change the attribute value. You are prompted:

Enter new attribute value:

Responding with **Position, Height, Angle, Style, Layer,** or **Color** results in prompts that request the new text parameters and layer location.

GLOBAL EDITING

Global editing is used to edit all the attributes at one time. As usual, the limits you set for editing will be used. You can choose global editing by responding to the initial prompt:

Command: **-attedit**

Edit attributes one at a time? <Y>: **N**

AutoCAD then prompts:

Performing global editing of attribute values.

Edit only attributes visible on screen? [Yes/No] <Y>

An N response to this prompt will result in the comment:

Drawing must be regenerated afterwards.

All this means is that your changes will not immediately be shown on the screen. You will have to use the **Regen** command to see the changes.

You must then restrict the set of attributes to be edited to the specific tags, values, or blocks.

VISIBLE ATTRIBUTES

If you responded with a Y to the prompt, you will only edit visible attributes. You can then restrict the set of attributes to be edited to specific blocks, values, or tags. You are prompted:

Select Attributes:

Use the standard object selection process to choose the group of attributes to edit. You are then prompted:

String to change:

New string:

Respond to the prompts with the string you wish to change and the changes you wish to make.

SUPPRESSION OF ATTRIBUTE PROMPTS

Use the system variable **AttReq** (short for "attribute request") to suppress attribute requests. If **AttReq** is set to 0, no attribute values are requested, and all attributes are set to their default values. A setting of 1 causes AutoCAD to prompt for attribute values.

EDITING ATTRIBUTES WITH A DIALOG BOX

Attribute values can be edited by the use of a dialog box. The **AttEdit** command displays the dialog box.

Command: **attedit**

Select block:

Select the block with attributes to be edited and the dialog box is displayed. Notice that the dialog box allows you to change only the attribute text, and not the properties of the attribute. In addition, the dialog box is not capable of global attribute editing. The **Properties** dialog box, new to AutoCAD 2000, does not display attributes values and thus cannot edit them.

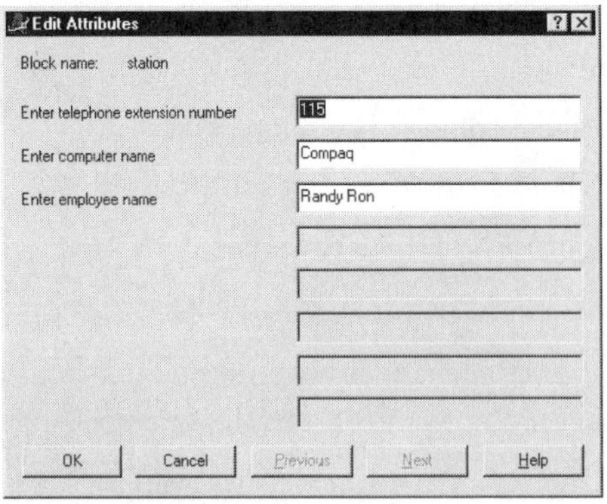

Figure 24.12 *Edit Attributes Dialog Box*

EXERCISE

1. Consider the workstations and corresponding attribute values, as shown in Figure 24.13.

   Let's assume that you want to change John Doe's name to Jane Doe. Start the **AttEdit** command by selecting **Modify | Attribute | Single**:

   Command: **attedit**

   Select block reference: *(Select the John Doe block.)*

2. AutoCAD displays the **Edit Attributes** dialog box. Change the name from John Doe to Jane Doe.

3. Choose **OK**. Notice that the name changes in the drawing.

4. Some of the workstations originally showed Apple as the type of computer. Change it to IBM 9000 by using global editing. For this, you need to use the command-line version of the **AttEdit** command. Prefix the command name with a dash, as follows:

   Command: **-attedit**

   Edit attributes one at a time? [Yes/No] <Y>: **n**

   Performing global editing of attribute values.

   Edit only attributes visible on screen? [Yes/No] <Y>: **y**

   Enter block name specification <*>: **station**

   Enter attribute tag specification <*>: **computer**

   Enter attribute value specification <*>: **Apple**

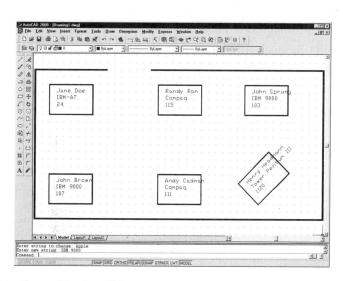

Figure 24.13 *Changed Attribute Values*

Select Attributes: **w**

Specify first corner: *(Pick one corner.)*

Specify opposite corner: *(Pick other corner.)*

2 found Select Attributes: *(Press* ENTER.*)*

2 attributes selected.

Enter string to change: **Apple**

Enter new string: **IBM 9000**

Each station now contains the attribute value of IBM 9000 in place of Apple.

ATTRIBUTE EXTRACTION

Attributes are a great feature for keeping records of your inserted blocks. You could maintain a database on furniture, model and cost figures on parts used in a design, or the number, type and cost of windows in a plan. It would be useful if you could print out all these items in a report. The **AttExt** command can!

The **AttExt** command (short for "attribute extract") is used to extract database information from the drawing in a specified form.

Command: **attext**

The **Attribute Extraction** dialog box is displayed. The attributes can be extracted in three formats, as listed in the **File Format** section:

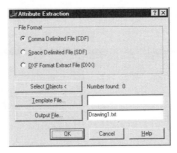

Figure 24.14 *Attribute Extraction Dialog Box*

CDF

Comma Delimited Format (CDF) produces an ASCII (plain text) file that uses delimiters (usually commas) to separate the data fields. Character fields (text) are enclosed in quotes. This format is read by spreadsheet and database programs. For example:

"STATION",1.00,1.00,"Henry Headmann","Tower Pentium III",120

SDF:

Space Delimited Format (SDF) produces an ASCII file that uses a fixed number of spaces for each field; it requires fixed field lengths. This format is read by spreadsheet and database programs. It is a format that is easier for us humans to read. For example:

STATION 1.00 1.00 Henry Headmann   Tower Pentium III   120

DXX:

DXF Format Extract File (DXX) is similar to the AutoCAD DXF (drawing interchange format) file. This format contains only the block reference, attribute, and end-of-sequence entities.

Template File

To extract the information, you must first prepare a *template file*. The template file guides AutoCAD in extracting attribute data from the drawing. The template file indicates the format of each *record* (line of data) to extract. You can use a text editor, such as Notepad, to prepare the template file.

Each line of the template represents one field to be listed in the extract file. You must specify the width of each field (in characters) and the number of decimal places to be displayed by numerical fields. Each field is listed in the order shown by the template file.

Table 24.1 shows the possibilities of field choices. Each field must be a character field (C) or a numerical field (N). The first character designates the type. The next three numbers represent the field width. The last three represent the number of decimal places in a numeric field. Thus, a character field representing a tag with a field spacing of 18 characters would be represented as "C018000".

Name	Format	Meaning
BL:Level	Nwww000	Block nesting level
BL:NAME	Cwww000	Block name
BL:X	Nwwwddd	X coordinate of block
BL:Y	Nwwwddd	Y coordinate of block
BL:NUMBER	Nwww000	Block counter
BL:HANDLE	Cwww000	Block's handle
BL:LAYER	Cwww000	Block insertion layer
NT	Nwwwddd	Block rotation angle
BL:XSCALE	Nwwwddd	X scale factor of block
BL:YSCALE	Nwwwddd	Y scale factor of block
BL:ZSCALE	Nwwwddd	Z scale factor of block
BL:XEXTRUDE	Nwwwddd	X component of extrusion
BL:YEXTRUDE	Nwwwddd	Y component of extrusion
BL:ZEXTRUDE	Nwwwddd	Z component of extrusion
Tag	Cwww000	Attribute tag (character)
Tag	Nwww000	Attribute tag (numeric)

Table 24.1

EXERCISE

Let's extract the attributes from the Office drawing. You want to obtain the block name, X coordinate, Y coordinate, employee name, computer type, and telephone extension number.

1. First, you must prepare a template file. Use a text editor, such as Notepad, to prepare a template like the one shown in Figure 24.15. Call the file "Template.Txt." The template file must have a file extension of **TXT**.

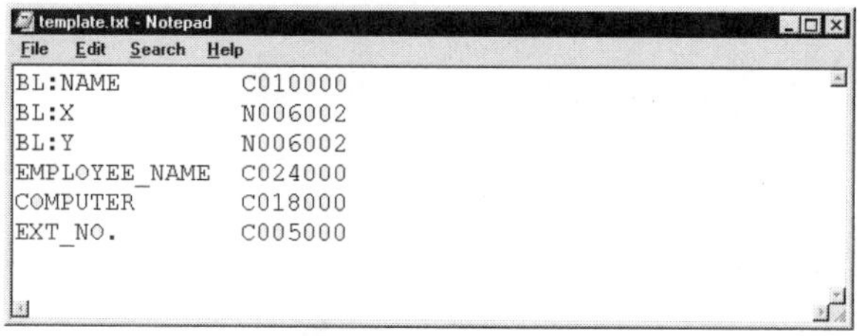

Figure 24.15 *Template File*

Field Name	Format	Meaning
BL:NAME	C010000	Name of the block
BL:X	N006002	X coordinate of the insertion point; a number with two decimal places
BL:Y	N006002	Y coordinate of the insertion point; a number with two decimal places
EMPLOYEE_NAME	C024000	Name of the employee
COMPUTER	C018000	Name of the computer
EXT_NO.	C005000	Telephone extension number

When typing in the field definition data, keep these important items in mind:

- The tag names in the template file must exactly match the tag names in the attribute.

- Do not use tabs; use spaces to separate the field name from its format.

- The file must have a blank line at the end.

2. Now, return to AutoCAD. Type the **AttExt** command (there is no menu choice):

   Command: **attext**

AutoCAD displays the **Attribute Extraction** dialog box.

3. You use the SDF format to prepare an attribute extraction of information from your drawing. Choose the **SDF** radio button.

4. Choose the **Select Objects** button to select the blocks whose attributes are to be exported. The quick way is to select **all** objects; AutoCAD will reject: (1) any objects that are not blocks and (2) any blocks that do not contain attributes.

   Select objects: **all**

7 found Select objects: *(Press Enter to return to the dialog box.)*

5. Choose the **Template File** button and select the Template.Txt file that you created earlier.

6. Ensure that there is a name in the **Output File** field. The default, Drawing1.txt, is fine.

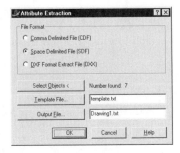

Figure 24.16 *Attribute Extraction Dialog Box*

7. Choose **OK**. If you made no errors writing the template file, AutoCAD reports:

   6 records in extract file.

You have now created an extract file with the name **Drawing1.Txt**. You should find it in the same folder as the Template.Txt file.

You can use the print command of your word processor to obtain a copy of the listing. Your listing should look roughly like Figure 24.17.

```
Drawing1.txt - Notepad
File  Edit  Search  Help
station    17.00   6.50Henry Headmann          Tower Pentium III 1120
station    10.00   6.50Andy Cadman             Compaq           111
station     2.50   6.50John Brown              IBM 9000         107
station    16.00  12.50John Sprung             IBM 9000         103
station    10.00  12.50Randy Ron               Compaq           115
station     2.50  12.50Jane Doe                IBM-AT           24
```

Figure 24.17 *Output from AttExt*

CHAPTER REVIEW

1. What is an *attribute*?

2. What is an attribute *tag*?

3. What command do you use to create attributes?

4. What is the attribute prompt?

5. How would you suppress attribute prompts?

6. What determines whether attributes are displayed in the drawing?

7. How are attributes edited?

8. What parts of an individual attribute can be changed with the attribute edit capabilities?

9. How would you globally edit all the attributes?

10. How would you obtain a file of all the attribute values in your drawing?

11. What is a *template file*?

12. What are the three types of attribute extract file formats?

CHAPTER 25

Attaching External Referenced Drawings

External reference drawings are extremely useful for certain types of CAD work. Those who assemble a large part of their work from library symbols and other drawings should consider the use of external reference drawings. After completing this chapter, you will be able to

- Develop the concept and use of external reference drawings
- Perform the AutoCAD commands used with the external reference functions
- Manage externally referenced files

INTRODUCTION

In this chapter, you learn about ways of bringing drawings and other files into the current AutoCAD drawing with the following commands:

XRef

Controls externally referenced drawings.

XAttach

Attaches another drawing as an externally referenced drawing.

XBind

Binds named objects from an externally referenced drawing.

RefEdit

Selects an xref or a block for editing.

RefSet

Adds or removes objects during in-place editing of an xref or a block.

RefClose

Saves (or discards) in-place editing changes made to an xref or a block.

EXTERNALLY REFERENCED DRAWINGS

One of the strengths of CAD is its ability to draw small components, then assemble them into a larger, more complex drawing. In Chapter 20, you learned how to do this with the **Block** and **Insert** commands. The Insert command allows you to merge a drawing into another drawing, controlling the placement and scale of the insertion.

When you insert a drawing, AutoCAD creates a block reference of that drawing in the destination drawing file. If you insert that drawing again, AutoCAD uses the block reference to obtain a copy of the previously inserted drawing. It is not an easy task to change the part of the drawing that was inserted. You can delete all copies of the block, then use the **Purge** command to delete the block reference and reinsert the drawing with the changes, or you can use the **Explode** command to break the inserted block into its simple objects, then edit the objects. This means that any changes made to a component drawing must also be changed in the drawing into which it was previously inserted.

AutoCAD's *external reference* capabilities can be useful when you have such a situation. The external reference feature allows you to insert a drawing into another as you would with the **Insert** command. If the drawing is inserted as an external reference drawing, however, AutoCAD does not load a block reference. Each time the master drawing is loaded, the component drawing is scanned, then loaded at that time. Thus, any changes made to the component drawing are updated in the master drawing automatically.

Figure 25.1 shows a sample drawing, 1st floor.dwg, which ships with AutoCAD 2000. It looks like a single drawing, but in fact it is a nearly-empty drawing (containing a title block, viewports, layouts, and some text) that references four other drawings:

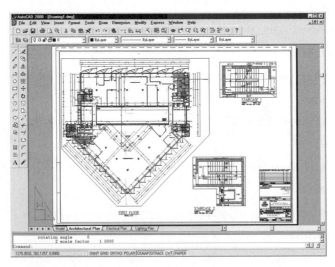

Figure 25.1 *Four Xref Drawings*

- 1st floor architectural.dwg

- 1st floor electrical.dwg

- 1st floor lighting.dwg

- 1st floor plan.dwg

Let's consider another example. Suppose that you have a design firm that designs machinery. Your machinery drawings are made up of many standard and nonstandard parts. You normally insert the parts (created and stored as component drawings) into a master drawing to create the finished machine drawings.

One day, you decide to make changes to one of the component parts. The problem is that you have 25 master drawings that refer to that part. If you use the **Insert** command to place the component drawings into the master drawings, you have 25 master drawings to correct. If you use external references, however, each master drawing is automatically updated when it is next started or plotted.

External references can be useful for any type of application that uses component drawings, or even for multi-station offices that have several people drawing parts of the work. There are several features of external references that you need to be aware of. Let's review these:

- The **XRef** command is used to insert a drawing as an external reference. The process is similar to the **Insert** command.

- An external reference drawing is *not* stored as a block reference in the master drawing. Because of this, the original component drawing must be available for AutoCAD to scan and load into the master drawing. The drive and path to the component must be either maintained or, if it is moved, redefined.

- Since a block reference isn't loaded, the master drawing can have many referenced drawings without the disadvantage of excessive single drawing file size. Note, however, that the component drawing must also reside on disk or network for the master drawing to load it.

- You can choose to *bind* the externally referenced drawing into the master drawing. This has the effect of turning the component drawing into a block, with all the standard aspects of a block.

- AutoCAD codes the referenced drawing's layers for identification. Let's look at an example: assume that you have a drawing named **Widget**. The **Widget** drawing has a layer named **Details**. When you reference the Widget drawing to a master drawing, a layer is created with the name **Widget|Details**. This represents the combination of the referenced drawing name and its layer name, separated by the vertical bar (|). This allows layers of the same name to coexist with other referenced drawings.

- External reference drawings may themselves contain other referenced drawings. For example, the **Widget** drawing may contain an external reference to another drawing named **Cog**. Thus, when you place the Widget drawing in a master drawing, AutoCAD includes the **Cog** drawing as an external reference.

- Changes made in externally referenced drawings will be reflected in the master drawing either the next time it is started or at the time it is updated if the master drawing is current. This will continue until you bind the externally referenced drawing.

- You can edit an external reference using the **RefEdit** command. The command is designed to allow you to make minor changes to xrefs and blocks, without opening the referenced drawing, or, in the case of the block, without exploding and redefining it.

Let's look at the commands used with external references.

XREF COMMAND

The **XRef** command is used to insert externally referenced drawings into a master drawing. It is also used to bind a referenced drawing, remove a referenced drawing, reset the path to a referenced drawing, and update a referenced drawing.

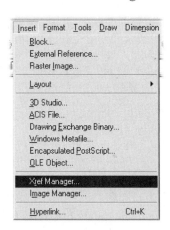

When you select **Insert | Xref Manager**, or type the **XRef** command, a dialog box is displayed. Let's look at each of the options and see how they work.

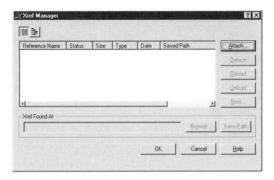

Figure 25.2 *External Reference Manager Dialog Box*

ATTACH (ADDING AN EXTERNAL REFERENCE)

Attach is the first button you choose when you use the **XRef** command. When you choose **Attach**, AutoCAD displays the file dialog box (as an alternative, you can go directly to this dialog box by selecting **Insert | External Reference** from the menu bar):

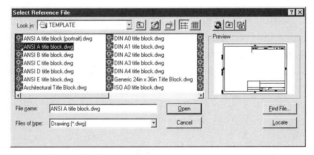

Figure 25.3 *Select Reference File Dialog Box*

Select any DWG file to attach. You cannot attach a DXF or DWT file; you can attach raster images using the **Image** command.

When AutoCAD detects a block name that matches the drawing, it issues an error message "Error: *bkname* is already a standard block in the current drawing. *Invalid*" and terminates the **XRef** command, since you cannot have a block reference and an external reference by the same name.

After you select a drawing file, AutoCAD displays the **External Reference** dialog box. (You can get directly to this dialog box with the **XAttach** command.)

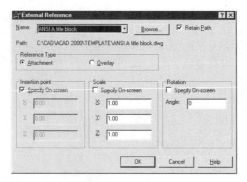

Figure 25.4 *External Reference Dialog Box*

Name

In the **Name** area, you select the externally referenced drawing file in two ways: (1) choose the **Browse** button to display the file dialog box or (2) select a previously loaded drawing from the list box.

Retain Path

The path specifies the location of the xref file, such as **D:\acad 2000\support**. Normally, this option is left on so that AutoCAD knows where the externally referenced drawing is located. If, however, the file has moved, turn this option off.

Reference Type

AutoCAD supports two kinds of externally referenced drawings:

Attachment

An attached drawing reflects changes each time it is loaded. This is the default and is the style you use most commonly.

Overlay

An overlaid drawing ignores nested external references.

Insertion Point, Scale, & Rotation

Xref files are placed in the drawing in the same way as a block is placed by the **Insert** command. You can control the insertion point, the *X,Y,Z* scales, and the rotation angle. As with the **Insert** command, you can specify the parameters here in the dialog box or during insertion.

Choose **OK** to exit the dialog box. AutoCAD loads the referenced drawing. The remaining prompts are identical to the **Insert** command:

Attach Xref "filename": C:\CAD\ACAD 2000\filename.dwg

"filename" loaded.

Specify insertion point or
    [Scale/X/Y/Z/Rotate/PScale/PX/PY/PZ/PRotate]: *(Pick an insertion point.)*

If AutoCAD detects that an external reference by that name is already present, it alerts you and proceeds. For example, the following command sequence shows an external drawing named "FLANGE" being reloaded as an external reference drawing.

Xref FILENAME has already been defined.

Using existing definition.

Insertion point:

LIST EXTERNAL REFERENCE INFORMATION

AutoCAD displays information about externally referenced drawings in two formats. Choose the **List View** button to view each file name with details:

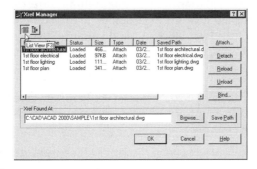

Figure 25.5 *List View*

Choose the **TreeView** button to display a "tree" that shows the connection between drawings.

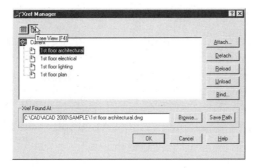

Figure 25.6 *Tree View*

BIND AN XREF TO THE DRAWING

Choosing the **Bind** button binds the external reference drawing to the master drawing, causing it to become a regular block. AutoCAD prompts with this dialog box:

Figure 25.7 *Bind Xrefs Dialog Box*

Actually, there are two ways to bind the xref to the current drawing:

Bind

Named objects (such as blocks, text styles, layer names, and linetype names) from the xref drawing have their names changed from **dwgname|objname** to **dwgname0objname**. In **dwgname|objname**, *dwgname* refers to the xrefed drawing filename, the | (vertical bar) separates the names, and *objname* is the name of the layer, block, etc. In **dwgname0objname**, notice that the vertical bar is replaced with 0.

Insert

Inserts the drawing as if you had used the **Insert** command. Named objects from the xref drawing have their names changed from **dwgname|objname** to just plain **objname**.

If you want to bind just one particular block or style, use the **XBind** command instead (more later in this chapter).

REMOVE AN XREF FROM THE DRAWING

The **Detach** button allows you to remove an external reference from the master drawing. This is equivalent to erasing all occurrences of a block, then purging its reference. When you select Detach, AutoCAD gives no warning.

CHANGE PATH TO AN XREF

The **Save Path** button is used to specify a new path for an xref. As you learned earlier, an attached xref must remain on the disk or network drive and directory where it was located when it was attached. The **Browse** button lets you change it.

This option can also be used to change the drawing brought in as an xref.

When you select the **Browse** option, AutoCAD displays the file dialog box.

UPDATE EXTERNAL REFERENCES

When you first start a drawing that contains external references, each xref is automatically reloaded. The **Reload** button is used to update one or more xrefs without exiting and reentering the drawing. When you select the **Reload** option, AutoCAD reloads the drawing.

XREF LOG FILE

Your xref operations can be recorded in a log that is kept by AutoCAD. The log file is written to disk as an ASCII file. You can view this log by using the Notepad text editor, or with a word processor. The log file is named the same as the drawing file, except that it has a file extension of XLG. Thus, if your drawing file is named "MASTER1," the log is named "MASTER1.XLG."

To keep a log file, change system variable **XrefCtl** from 0 (the default) to 1.

XBIND COMMAND

You have learned how to use **XRef Bind** to bind an xref drawing to a master drawing. There are times, however, when you may only want to bind a part of a drawing. For example, you may want to bind a linetype or layer to the drawing, without binding the rest of the externally referenced drawing. You can do this with the **XBind** command.

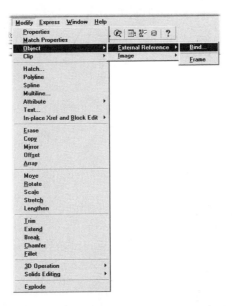

When you select **Modify | Object | External References | Bind** from the menu bar or type the **XBind** command, AutoCAD displays the **Xbind** dialog box:

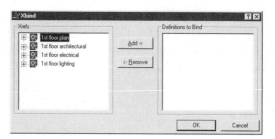

Figure 25.8 *Xbind Dialog Box*

The **Xbind** dialog box initially displays a list of externally referenced drawings on the left side. Notice the + symbol next to each drawing file name. Click it to display a list of named objects in that drawing.

In Figure 25.9, you can tell that the first drawing contains blocks, dimstyles, layers, and text styles because of the + sign next to each. When there is no + sign next to one, this indicates the drawing has none loaded.

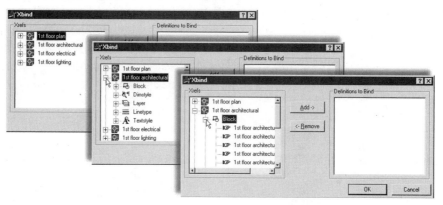

Figure 25.9 *Opening the Tree*

Click on the + next to **Block**. The tree opens further to show the names of blocks in the drawing.

To bind an object to the current drawing, select its name (such as a block name), then choose the **Add** button. AutoCAD displays the name in the **Definitions to Bind** list. Notice the **dwgname|objname** format. To remove a bound item, choose the **Remove** button.

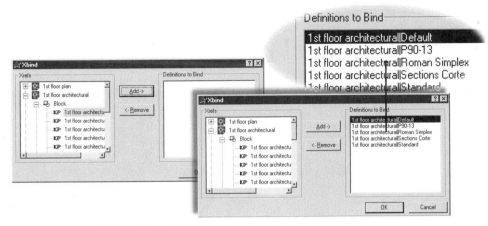

Figure 25.10 *Adding a Named Object*

The items are renamed when they are bound. You learned earlier how AutoCAD lists xref items, such as layers, with a vertical bar (|). AutoCAD removes the vertical bar and replaces it with two dollar signs ($) and a number that is usually 0.

For example, a layer named "PARTA|HEXAGONS" will be named "PART0HEXAGONS." If there is already a layer by that name, AutoCAD will try "PARTS1HEXAGONS", and so forth. The number of characters must be 255 or less, or AutoCAD terminates the command, and undoes the effects of the **XBind** command.

IN-PLACE XREF EDITING

AutoCAD 2000 allows you to edit blocks and xrefs *in-place*. This means you can edit a block without exploding it; you can edit an externally referenced drawing without explicitly loading it.

The process of in-place editing follows these steps:

1. Start the **RefEdit** command.

2. Select one block or xref; if it is nested, you must select one nested item.

3. Select one or more objects *within* the block or xref for editing; all objects in the drawing not selected are displayed as dimmed.

4. Edit the object(s); you can add (and remove) objects outside the block or xref with the **RefSet** command.

5. When finished editing, save or reject the changes with the **RefClose** command.

Note: Automatic save is disabled during reference editing. Between step 1 and 2 (listed above), all objects in the drawing are locked against editing.

Selecting an Xref for Editing

From the menu bar, select **Modify | In-place Xref and Block Edit | Edit Reference** or type **RefEdit**. AutoCAD prompts you:

Select reference: *(Select a single xref or a block.)*

AutoCAD expects you to select just one xref or block. It does not accept

- Associative dimension, which is a form of block

- Non-uniformly scaled blocks

- Blocks inserted with the **MInsert** command

You can select other forms of blocks, including associative hatch patterns. Once you select an xref or block, AutoCAD locks its layer so that no other editing can be performed on the object.

AutoCAD displays the **Reference Edit** dialog box.

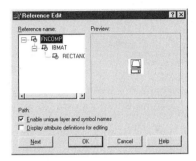

Figure 25.11 *Reference Edit Dialog Box*

The dialog box displays a preview of the xref or block if it has a preview image—you can give blocks a preview image with the **BlockIcon** command. You can in-place edit only one xref or block at a time. If you select an object that is part of one or more nested references, the nested references are displayed in the dialog box under **Reference Name**. You cycle through nested references by choosing the **Next** button.

The dialog box has two options:

Enable Unique Layer and Symbol Names

This check box determines whether the xref's symbol names are *altered* during reference editing (this setting does not apply to blocks). Symbol names include layer names, named views, UCSs, linetypes, text styles, and plot styles—in short, any name displayed by the **Rename** dialog box.

- When on (the default), symbol names are altered in a manner similar to bound xrefs; xref symbol names are prefixed with 1.

- When off, symbol names are unchanged. This setting does not apply to blocks.

Display Attribute Definitions for Editing

This check box determines whether attribute *definitions*— as opposed to attribute data— are displayed during reference editing (this setting applies only to blocks with attributes).

- When on, attribute data (except constant attributes) changes to the attribute definitions (see Figure 25.12). Note that when changes are saved, the attributes of the original block are unchanged; the edited attribute definitions come into effect with subsequent insertions of the block.

- When off (the default), then attributes data are displayed but cannot be edited.

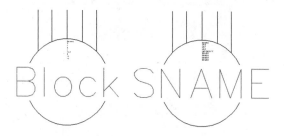

Figure 25.12 *Attribute Data (at left) and Attribute Definitions (at right)*

Choose **OK**. AutoCAD prompts you:

Select nested objects: *(Select one or more objects.)*

Select nested objects: *(Press* ENTER.*)*

n items selected

Use REFCLOSE or the Refedit toolbar to end reference editing session.

Select the object(s) in the block or xref you want to edit. Notice that when you select an object contained in the block or xref, you select the individual object— not the entire block. You cannot use the **All** option to select objects; you are limited to these selection options: pick a point, Window, Crossing, WPolygon, CPolygon, Add, Remove, and Undo.

After you press ENTER, AutoCAD displays the **Refedit** toolbar. Notice that a change happens to the drawing: all of the objects *not* in the selection of nested objects become dimmer.

You can now apply most of AutoCAD's editing commands to the selected objects. For example, you can change colors, lineweights, and linetypes; you can move, erase, and copy objects; you can lengthen, fillet, and chamfer objects.

Refedit Toolbar

The **Refedit** toolbar provides a shortcut to the five in-place editing commands, in addition to displaying the name of the block or xref currently being edited.

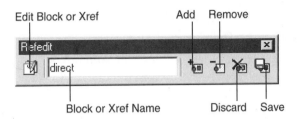

Figure 25.13 *The Refedit Toolbar*

Edit a block or xref: choose this button to select a block or xref for in-place editing. If one is already being edited, AutoCAD responds, "Command not allowed, *file name already checked out for editing.*"

Name: name of the block or xref currently being edited. You cannot change the name or use this text field to select another block/xref.

Add objects to working set: allows you to select other objects to add to the block/ref.

Remove objects from working set: allows you to select objects to remove from the block/ref.

Discard changes to reference: negates all editing changes you made to the block/ref, and exits in-place editing mode. AutoCAD displays a warning dialog box, to help save you from making a mistake.

Figure 25.14 *Discard Changes to Reference*

Save back changes to reference: saves all editing changes you made to the block/xref, and exits in-place editing mode. AutoCAD displays a warning box (similar to above), to make sure you make the right decision.

Changing the Edit Set

During the in-place editing of the block or xref, you can add and remove objects from the editing set. From the menu bar, select **Modify | In-place Xref and Block Edit | Add to Workset** or **Remove from Workset**, or type the **RefSet** command:

    Command: **refset**

Transfer objects between the Refedit working set and host drawing...

Enter an option [Add/Remove] <Add>:

Both options prompt you to "Select options"; the **Add** option adds objects to the editing set. This allows you to add objects from the drawing to the block (or xref). You can add any object, even if it appears dimmed. Later, when you save your work from this in-place editing session, the object becomes part of the block (or xref).

In the opposite manner, the **Remove** option removes objects from the set. When the in-place editing is saved, the removed objects are no longer part of the block. The removed objects remain in the drawing.

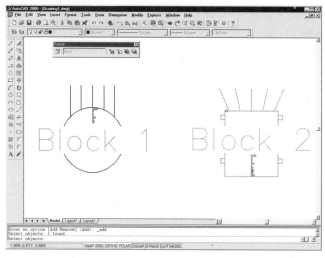

Figure 25.15 *Adding to the Editing Set*

Saving the Edits

The working set is a set of objects extracted from a reference that can be modified and then saved back to update the xref or block definition from within the current drawing.

Command: **refclose**

Enter option [Save/Discard reference changes] <Save>: **(Type s or d.)**

Regenerating model.

The **Save** option saves all your editing changes to the externally referenced drawing or block definition. If you have removed objects, they are moved from the block/xref to the drawing.

The **Discard** option returns the externally referenced drawing or block definition to its original state.

You can use the **Undo** command to reverse the effects of the **RefEdit** command, whether saved or discarded.

EXERCISE

Let's use an external reference. The CD-ROM contains a drawing named **PartA.Dwg**. Figure 25.16 shows the drawing.

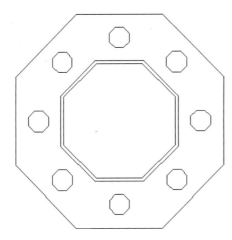

Figure 25.16 *PartA CD-ROM Drawing*

1. Start a new drawing. Set the units to decimal, and the limits to 24,18.

2. Let's now attach the "PARTA" drawing as an external reference. Use the **XAttach** command as follows.

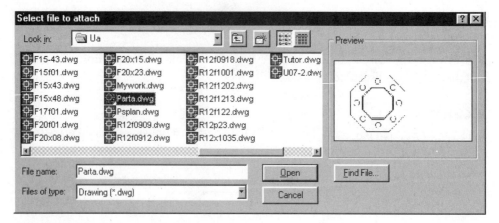

Figure 25.17 *Select Drawing*

Select PARTA.DWG from the file dialog box.

Choose **OK.**

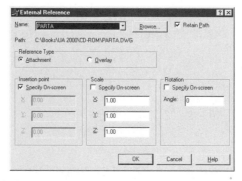

Figure 25.18 *Attach External Reference Dialog Box*

The externally referenced drawing is now attached to the drawing.

3. Now use the **Save** command to save the drawing as "Master1." Close Master1.Dwg.

4. Open the drawing named "PARTA." This is the externally referenced drawing.

5. Use the **Erase Fence** command to erase the hexagons so the drawing looks like Figure 25.19.

Figure 25.19 *Updated PartA Drawing*

6. Save and close the drawing.

7. Now open the MASTER1 drawing again. Notice the messages about resolving the xref PARTA. When the drawing appears on the screen, you will notice that the edits you performed in the PARTA drawing were scanned and incorporated into the MASTER1 drawing.

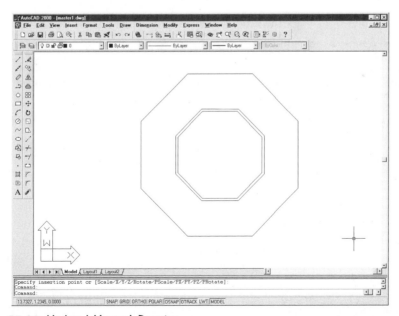

Figure 25.20 *Updated Master1 Drawing*

8. Continue and use the **XRef Bind** command to bind the xref drawing PARTA. Save the drawing and open the PARTA drawing. Edit the drawing again in any way you want. Save the drawing and open the MASTER1 drawing again. Were the last edits incorporated?

CHAPTER REVIEW

1. Explain the difference between an inserted drawing and one placed as an external reference.

2. Is an external reference stored as a block?

3. Can an external reference contain another external reference?

4. What are some advantages of using external references?

5. Which commands are used to place an external reference drawing?

6. How would you obtain a listing of the external references placed in a drawing?

7. How would you convert an external reference to a block?

8. Under what conditions would you want to convert the external references to blocks?

9. How would you remove an external reference from a drawing?

10. Why must an externally referenced drawing remain in its original drive and directory?

11. How would you change the drive and directory of an externally referenced drawing?

12. What option is used to update externally referenced drawings in the master drawing?

13. How would you write a log file that describes the external reference activity to disk?

14. Can you bind only part of an externally referenced drawing to the master drawing? Explain.

CHAPTER 26

Exploring the AutoCAD DesignCenter

The **XRef** and **XBind** commands allow you to see and access other drawings, as you learned in the previous chapter. AutoCAD 2000 adds a new interface to access information about any drawing called the AutoCAD DesignCenter. After completing this chapter, you will

- Understand the user interface of the AutoCAD DesignCenter
- Learn the commands related to the DesignCenter

INTRODUCTION

AutoCAD DesignCenter provides you with the following services:

- Views the drives on your computer, network drives, folders, shortcuts, and files
- Previews the content of DWG and some bitmap formats

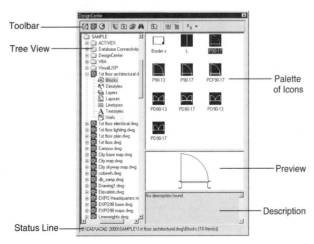

Figure 26.1 *AutoCAD DesignCenter*

- Views the named content of AutoCAD drawings

- Displays the layers, linetypes, text styles, blocks, dimension styles, external references, and layouts of every drawing

- Copies content from one drawing to another

In addition to pressing CTRL+2 to turn the DesignCenter on and off, you can use these commands:

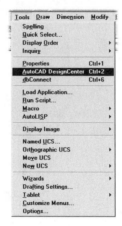

AdCenter

Opens AutoCAD DesignCenter.

AdcClose

Closes AutoCAD DesignCenter.

AdcNavigate

Specifies the initial path for DesignCenter:

Command: adcnavigate

Enter pathname <>: *(Enter a path, such as c:\program files\acad2000\sample.)*

DESIGNCENTER'S TOOLBAR

To open the AutoCAD DesignCenter, type the **AdCenter** command, or select **Tools | AutoCAD DesignCenter** from the menu bar. As an alternative, you can press CTRL+2 to toggle on and off the display of the dialog box.

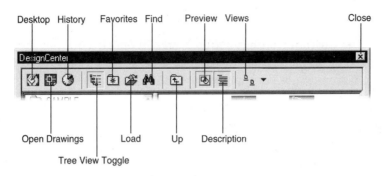

Figure 26.2 *DesignCenter's Toolbar*

Desktop

The **Desktop** button displays the names of folders (subdirectories) and files on the drives of your computer and on the network—similar to the Windows File Explorer. This acts as a toggle with three other buttons: **Open Drawings**, **Custom Content**, and **History. DesignCenter** displays the content of one of these buttons.

Open Drawings

The **Open Drawings** button displays a list of the drawings currently open in AutoCAD. The named content of each drawing is also displayed: blocks, xrefs, layouts, layers, linetypes, dimension styles, and text styles.

Figure 26.3 *DesignCenter with Open Drawings*

Custom Content

(*Optional.*) The **Custom Content** button is present only if a drawing contains proxy data generated by registered (third-party) applications. Either the third-party application must be running or the application must be registered with DesignCenter.

History

The **History** button displays the last 120 documents that the DesignCenter viewed. Double-click a file name to display its named content (if an AutoCAD drawing) or a thumbnail (if a raster image).

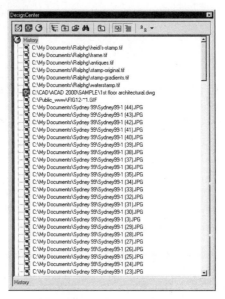

Figure 26.4 *DesignCenter's History List*

Tree View Toggle

The **Tree View Toggle** button hides and displays the tree views: **Desktop** and **Open Drawings**.

Favorites

The **Favorites** button displays the **\Windows\Favorites\Autodesk folder**. This folder contains a shortcut to the **\Acad 2000\Sample\Designcenter** folder, which contains 16 DWG files, each containing 20 of discipline-specific blocks.

Load

The **Load** button displays the **Load DesignCenter Palette** dialog box. This file dialog box is an alternative method to opening a drawing to view its content with DesignCenter. This dialog box also gives you access to content stored on the Internet.

Find

The **Find** button displays the **Find** dialog box. This lets you search for a drawing, provided you know its file name or some data stored in the drawing's **Drawing Properties** dialog box.

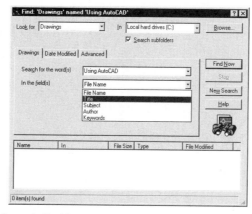

Figure 26.5 *DesignCenter's Find Dialog Box*

Up

The **Up** button moves your view up by one level. For example, if you are viewing blocks, choosing this button displays the drawing.

Preview

The **Preview** button toggles on and off the display of the preview image. The preview is typically a thumbnail of a block or a raster image, as shown in Figure 26.6. This is useful for previewing the TGA images (stored in the **\Acad 2000\Textures** folder) that AutoCAD uses for renderings (see Chapter 38 "Making a Realistic Rendering"). This version of DesignCenter displays preview images of DWG, TIFF, Targa, JPEG, GIF, and BMP files; it does not display other files you might expect, such as DXF, DWT, and WMF. The preview area is empty when the object does not contain an image.

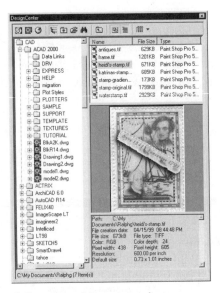

Figure 26.6 *DesignCenter's Preview and Description Areas*

Description

The **Description** button displays a description of the selected item at the bottom of the palette. Descriptions are typically found in DWG files, raster files, and blocks, as shown in Figure 26.6. The description area reads "No description found" if the user has not saved a description.

Views

The **Views** button allows you to choose a display format for the palette area. Select a view from the list, or choose the **View** button to cycle through the formats: Large Icon, Small Icon, List View, and Detail View. Large Icon and Detail View are probably the preferred formats; Large Icon shows thumbnails of blocks and drawings, while Detail View displays (and allows you to sort by) file name, file size, and type.

DESIGNCENTER CURSOR MENUS

The AutoCAD DesignCenter is loaded with cursor menus, some of which are crucial to the integration of DesignCenter with AutoCAD. It is unfortunate that DesignCenter's two most important functions—**Add** and **Copy**—are "hidden" in cursor menus.

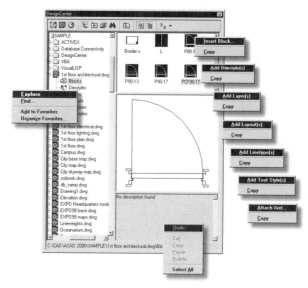

Figure 26.7 *DesignCenter's Cursor Menus*

Palette

Right-click the palette (the area displaying the icons) to display a cursor menu. All palette cursor menus have one item in common: select **Copy** to copy the content from one drawing to another. For example, to copy a block from one drawing to another:

1. Right-click the block you want to copy.

2. From the cursor menu, select **Copy**.

3. Click to select the drawing you want the block copied to. The block is stored in the second drawing but is not displayed.

As an alternative, you can drag content from DesignCenter into the drawing. For example, to drag all layers from one drawing to another drawing:

1. Select the first layer name.

2. Hold down the SHIFT key and select the last layer name. Notice that DesignCenter highlights all layers between your two selections. (To select files in a non-consecutive order, hold down the CTRL key and select files.)

3. Drag the group of highlighted layer names into the drawing.

Notice that AutoCAD warns, "Duplicate names ignored." This means that if a layer of that name already exists, AutoCAD does not duplicate it.

The second item in the cursor menu varies, depending on the content. Most cursor menus display **Add**. This option adds the content to the current drawing. For example, to add a linetype to the current drawing, right-click the linetype and select **Add** from the cursor menu. As before, AutoCAD warns that duplicate names are ignored.

The cursor menus for blocks and external references are different:

Blocks

The cursor menu displays **Insert Block.** Selecting this item displays the **Insert** dialog box, which is identical in function to AutoCAD's **Insert** dialog box. See Chapter 18 "Learning Intermediate Draw Commands."

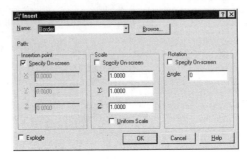

Figure 26.8 *DesignCenter's Insert Dialog Box*

External References

The cursor menu displays **Attach Xref**. Selecting this item displays the **External Reference** dialog box, which is identical in function to AutoCAD's **External Reference** dialog box. See Chapter 25 "Attaching External Drawings."

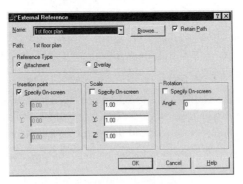

Figure 26.9 *DesignCenter's External Reference Dialog Box*

Tree View

Right-click the tree view to display a cursor menu that mimics the functions of some buttons on the toolbar: **Find** and **Favorites**. Selecting the **Explore** item has the same effect as clicking the small + symbol next to a folder or file name. Select **Add to Favorites** to add the drawing or folder to the list of "favorites," which acts like a bookmark.

Description

Right-click the description area to display a cursor menu with commands suitable for copying the text to the Clipboard.

Title Bar

Double-click the title bar to dock DesignCenter on the left side of the AutoCAD window.

CHAPTER REVIEW

1. Describe the purpose of AutoCAD DesignCenter.

2. What are two ways to insert a block from DesignCenter into a drawing?

    a.

    b.

3. What is the purpose of the **History** button?

4. Which key do you hold down to select more than one item?

5. List four types of DWG content displayed by DesignCenter:

    a.

    b.

    c.

    d.

6. Which shortcut keystroke toggles the display of the DesignCenter?

7. What is the purpose of the **AdcNavigate** command?

8. What is the meaning of the message "Duplicate names ignored"?

CHAPTER 27

Inserting Clipboard and OLE Objects

In additional to external reference drawings, AutoCAD can display objects from other Windows applications. These objects include spreadsheets, word processing documents, and animated images. In this chapter, you learn about ways to bring objects from other applications into the current AutoCAD drawing. After completing this chapter, you will be able to

- Copy and paste objects to and from the Windows Clipboard

- Use object linking and embedding with drawings

COPYING AND PASTING

Since AutoCAD is a Windows application, it can interact with other Windows applications. There are several ways AutoCAD can do this. In this chapter, we look at two methods: the Clipboard and OLE. The Clipboard allows the "cut and paste" operation that you are probably already familiar with from other Windows software, such as word processing and image editing. You can exchange a paragraph of text, a graphic, or a range of spreadsheet cells between Windows applications.

OLE allows more control over the cut-and-paste process. OLE is described later in this chapter.

It is rare to "cut" an object from a drawing, so in this book I refer primarily to "copying to the Clipboard." The process goes in both directions: you can copy objects *from* AutoCAD, and you can paste objects *into* AutoCAD. In addition, you can copy and paste objects from one AutoCAD drawing to another. To copy an object from one application to another takes just two steps:

1. In one application, select the object and press CTRL+C (the C is short for "copy").

2. In the other application, press CTRL+V (think of the V as shorthand for inserting something).

After you press CTRL+C, Windows stores the object in your computer's memory. This area of memory has been given the name of the "Clipboard." Just as a clipboard temporarily stores papers, so too the Clipboard temporarily stores objects. The next time you press CTRL+C, the new object replaces whatever was in the Clipboard.

Note: When you press CTRL+C but have nothing selected, AutoCAD's reaction is different from that of other applications. AutoCAD prompts you, "Select objects." Other applications do nothing.

Pressing CTRL+C and CTRL+V seems simple, but behind the scenes, a great deal of activity is taking place. AutoCAD actually translates the selected objects into no less than ten different formats, including Picture (also known as WMF, a vector format) and Bitmap (also known as BMP, a raster format). This allows the AutoCAD objects to be pasted into a variety of applications; more details later in the OLE section of this chapter. For this reason, AutoCAD may take a long time to copy or paste large drawings, particularly on slower computers and computers that are low on memory.

The opposite is true for pasting objects. You can press CTRL+V repeatedly, and Windows repeatedly pastes the same object—until a CTRL+C replaces the content of the Clipboard with a new object.

All of AutoCAD's Clipboard-related commands are found in the **Edit** menu:

COPYING TO THE CLIPBOARD

To copy one or more objects from the drawing to the Clipboard, press CTRL+C. As an alternative, you can select **Edit | Copy** from the menu bar or type the **CopyClip** command at the 'Command' prompt. AutoCAD prompts:

> Select objects: *(Pick one or more objects.)*
>
> Select objects: *(Press* ENTER.*)*

If you use the **All** selection mode, AutoCAD select all objects *visible in the current viewport*. This is different from using the **All** option with other commands (for example, **Erase All** erases all thawed and unlocked objects from the drawing, whether visible or not).

After you press ENTER, AutoCAD copies the objects to the Clipboard. As an alternative, you can select the objects first, then press CTRL+C. In this case, AutoCAD does not prompt you to select objects. Once the objects are in the Clipboard, you can paste them into another drawing or another document with the CTRL+V keystroke.

Copying All Objects

You may find the **CopyLink** command more useful than pressing CTRL+C. Despite its name, the **CopyLink** command has nothing to do with "linking." When you select **Edit | Copy Link** from the menu bar, AutoCAD copies *all* objects visible in the current viewport to the Clipboard.

You may be wondering, how is this different from responding with **All** to the **CopyClip** command's "Select objects" prompt? The **CopyLink** has the special property when AutoCAD displays a layout (paper space). In contrast, the **CopyClip** command copies either all paper space objects, or all objects in the current model space viewport. These following figures show you the difference:

The original drawing, as displayed by AutoCAD, is shown in Figure 27.1.

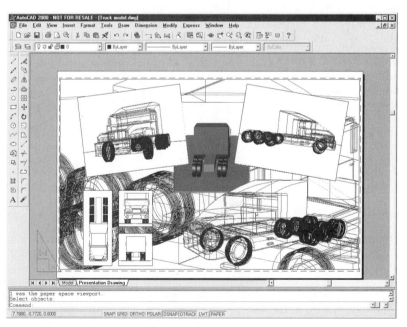

Figure 27.1 *Drawing Displayed by AutoCAD*

When the layout is in model space, the CopyClip command copies all objects in the current viewport, ignoring all other viewports.

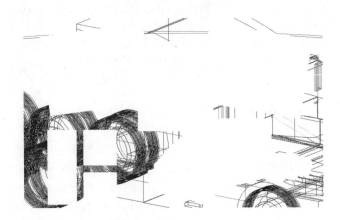

Figure 27.2 *Objects Copied by CopyClip in Model Sequence*

When the layout is in paper space, the **CopyClip** command copies all objects in paper space, ignoring the content of all model space viewports.

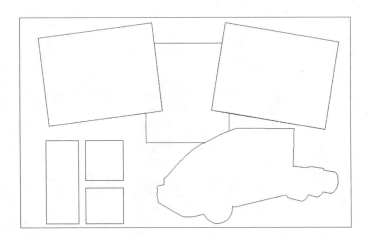

Figure 27.3 *Objects Copied by CopyClip in Paper Space*

When the layout is in either paper space or model space, the **CopyLink** command copies all objects, including all model space viewports.

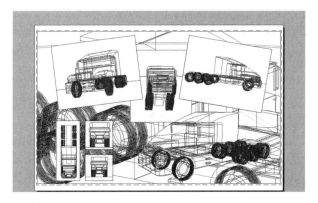

Figure 27.4 *Objects Copied by CopyLink*

Cutting to the Clipboard

Pressing CTRL+X is identical to CTRL+C, except that AutoCAD "cuts" objects from the drawing and then places them in the Clipboard. As an alternative, you can type **CutClip**, or select **Edit | Cut** from the menu bar.

The "cutting" operation consists of two steps: (1) AutoCAD copies the object to the Clipboard and (2) AutoCAD erases the object from the drawing. This command is rarely used, except for moving an object from one drawing to another.

Copying the Text Window

The **CopyHist** command copies all of the text in the Text window (called the command "history") to the Clipboard. As an alternative, you can select text (so that it is highlighted) and press CTRL+C. This copies the highlighted text to the Clipboard.

To see all your options, you can right-click in the Text window or the 'Command' prompt area to display a cursor menu.

Figure 27.5 *Text Copied by CopyHist*

PASTING INTO THE DRAWING

To paste an object from the Clipboard into the drawing, press CTRL+V. As an alternative, you can select **Edit | Paste** from the menu bar, or type the **PasteClip** command. Depending on the content of the Clipboard, AutoCAD reacts differently:

Unformatted Text

When the Clipboard contains unformatted text, AutoCAD inserts the text in the upper left corner of the current viewport; you have no control over the placement of the object. If you want the text in another area, move the text there after insertion.

Unformatted text comes from text editors, such as Notepad. The text is placed as an mtext object (paragraph text), with the current default settings of the **MText** command defaults. You can edit the text using the **DdEdit** command.

If the cursor is in the command line or the Text window, the text in the Clipboard is pasted at the 'Command' prompt. This can be useful for copying and pasting programming code, such as AutoLISP functions.

Formatted Text

When the Clipboard contains formatted text, AutoCAD inserts the text in the upper left corner of the current viewport, and displays the **OLE Properties** dialog box. Formatted text typically comes from word processing documents. Note that AutoCAD does not preserve some formatting, such as font and size differences. Other formatting is preserved, including colors, **boldface**, *italics*, and underlining.

Raster and Vector Graphics

When the Clipboard contains a raster image or a vector graphic (that came from a program other than AutoCAD), AutoCAD places the image in the upper left corner of the current viewport and then displays the **OLE Properties** dialog box (see Figure 27.7).

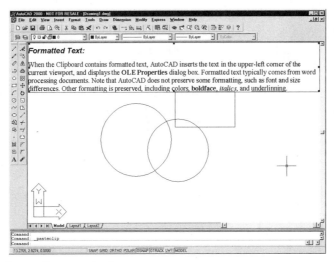

Figure 27.6 *Pasting Text into the Drawing*

AutoCAD Object

AutoCAD displays the graphic in the center of the viewport, and prompts you:

> Specify insertion point: *(Pick a point.)*

The object's base point is the lower left corner of the object's bounding box (an imaginary rectangle that delineates the extents of the object).

OLE Object

All objects, except unformatted text and AutoCAD objects, are inserted as embedded or linked objects. AutoCAD displays the **OLE Properties** dialog box. You edit this object by double-clicking it; this causes AutoCAD to open the originating application, create a new document, and insert the object. After you have finished editing the object, select **File | Close and Return to** *file name*.Dwg from the menu bar. AutoCAD displays the updated object.

For greater control over the format that is pasted into the drawing, use the **PasteSpec** command, as described in the OLE section of this chapter. When the Clipboard is empty, AutoCAD does not react to the CTRL+V keystroke.

OLE Properties Dialog Box

In most cases, when you use the **PasteClip** command to paste an object into the drawing, AutoCAD displays the **OLE Properties** dialog box, which allows you to control the placement of objects in the drawing.

Despite its name, the dialog box is not specific to OLE objects. The **OLE Properties** dialog box appears automatically any time you press CTRL+V (or select **Edit | Paste**), except when the Clipboard contains unformatted text. (The dialog box does not appear when you use the **PasteSpec** command.)

There is a toggle, called **Display Dialog When Pasting New OLE Object**, that you can turn off to prevent the dialog box from appearing. You can display the dialog box at any time with the **OleScale** command, or by right-clicking the pasted object and selecting **Properties** from the cursor menu.

Figure 27.7 *OLE Properties Dialog Box*

Size

The **Size** section specifies the size for the object. AutoCAD normally pastes the object at a somewhat arbitrary size. The **Height** and **Width** fields allow you to specify the size in current drawing units.

When **Lock Aspect Ratio** is on (a check mark shows), the **Width** is changed to maintain the proper aspect ratio between the height and the width of the object. The **Reset** button changes the object back to its size when it was originally inserted.

Scale

The **Scale** section allows you to size the object by scaling it by a percentage larger or smaller than its current size. You can scale the height and width independently; when **Lock Aspect Ratio** is on, the **Height** scale changes to maintain the ratio.

Text Size

The **Text Size** area allows you to change the font and point size of formatted text. You are, however, limited to the fonts and point sizes found in the text. Note that there are 72.72 points to the inch (one point = 0.01389" or 0.35278 mm).

For example, if the text contains a mixture of Arial, Times New Roman, and WingDings, then these three fonts are listed. If the text contains only Times New Roman font, then just that font is listed. There is a similar limitation to the point size. For example, if the text contains a mixture of 11- and 10-point text, then those are the only two choices. If the text is 10-point only, there is just that one choice.

 Warning. When you select a font name or point size, AutoCAD converts all text to that font and size—no matter the formatting of the original text. If you wish to keep the fonts and text sizes intact, choose **Cancel**. You *cannot* use the **DdEdit** command to change the text.

The = field allows you to specify the height of the text in drawing units. When you choose **OK**, the text in the OLE object of the selected font and point size changes to the height you entered.

OLE Plot Quality

The **OLE Plot Quality** list box determines the quality of the pasted object when plotted. I recommend the **Line Art** setting for text, unless the text contains shading and other graphical effects. Your choices are listed in Table 27.1

OLE Setting	Example	Plot Quality
Line Art	Spreadsheet	Text is plotted as text; no color or shading is preserved. Some graphical images are not plotted, while others are plotted as a monochrome image (black and white only, no shades of gray).
Text	Word Processing Document	All text formatting is preserved; text is plotted as graphics, which plots less cleanly than with the Line Art setting. Graphics are plotted less cleanly and at reduced colors compared to the Graphics and two Photograph settings.
Graphics	Pie Chart	Graphics are plotted at a reduced number of colors (fewer shades of gray, or "posterization"). All text formatting is preserved; text is plotted as graphics, but more cleanly than with the Text and Photograph settings.
Photograph	Raster Image	Graphics are plotted at reduced resolution and colors. All text formatting is preserved; text is plotted as graphics, but less cleanly than with the Graphics setting.
High Quality Photograph	Raster Image	Graphics are plotted at full resolution and colors. All text formatting is preserved; text is plotted as graphics, but more cleanly than with the Text and Photograph settings.

Table 27.1

Pasting Between Drawings

Since AutoCAD 2000 can open more than one drawing at a time, it now supports copying and pasting objects between drawings. Several commands have been added to help make this more accurate than the usual copy-and-paste operations described earlier.

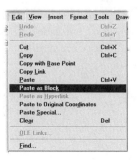

CopyBase

The **CopyBase** command (**Edit | Copy with Base Point**) copies object(s) from the drawing to the Clipboard, after prompting you for a base point. The *base point* is the same as the insertion point when you create a block.

> Command: **copybase**
>
> Specify base point: *(Pick point.)*
>
> Select objects: *(Select objects.)*
>
> Select objects: *(Press* ENTER.*)*

When you select the base point, I recommend you use an object snap mode to help you select the point accurately. Although this command is meant for use between AutoCAD drawings, you can paste the selected objects into any Windows application. When pasting the objects into the drawing, AutoCAD prompts:

> Command: *(Press Ctrl+V.)*
>
> _pasteclip Specify insertion point: *(Pick a point.)*

Like the **Insert** command, you specify where you insert the object. Unlike the **Insert** command, you cannot specify the scale factor nor rotation angle. If you need to change the size or rotation, you can do so after the object has been inserted.

PasteBlock

As an alternative, you can paste the object as a block. The **PasteBlock** command (**Edit | Paste as Block**) combines the objects in the Clipboard into a block. As with the **Paste** command, your only option is the insertion point:

> Command: **pasteblock**

Specify insertion point: *(Pick a point.)*

Upon insertion, the block is given an AutoCAD-generated name, such as A\$C3E8B2AFC. Even if you copy a block to the Clipboard, the block's name is "hidden" with the **PasteBlock** command, because a nested block is created. You can use the **Rename** command to change the name of the block. The scale factor is 1 for all three axes, and the rotation angle is 0 degrees.

PasteOrig

The **PasteOrig** command is a variation on the previous two. It pastes the object(s) in *another* drawing at the same insertion point and scale as in the original drawing. This makes it useful in creating an exact copy of drawing.

The **Edit | Paste to Original Coordinates** item appears only when two conditions are met: (1) the Clipboard contains AutoCAD objects *and* (2) you have switched to another drawing.

OBJECT LINKING AND EMBEDDING

As you saw in earlier chapters, the **Insert** and **XRef** commands work only with other AutoCAD drawings; the **Raster** command is limited to attaching raster images. The previous section showed how to paste any kind of object from the Clipboard and made reference to OLE. AutoCAD also lets you attach *any* kind of file using OLE (short for "object linking and embedding").

The file can be a text file, a spreadsheet, an image from any source (raster or vector), an animation file, or even a sound file. Figure 27.8 shows an AutoCAD drawing with four OLE objects: (clockwise from upper left) a multimedia object (double-click to run it), a floor plan from another CAD package, a WordArt object, and a raster image from a paint program.

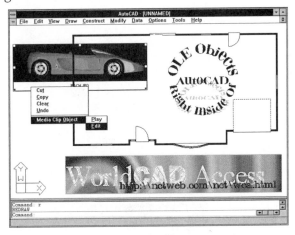

Figure 27.8 *OLE Objects Placed in AutoCAD Drawing*

Note: OLE objects in an AutoCAD drawing are only visible when loaded into Windows versions of AutoCAD. The objects disappear when loaded into other versions of AutoCAD, such as the DOS and UNIX versions.

Similarly, OLE allows you to place an AutoCAD drawing in just about any other Windows application. The drawing can be placed in a Word document (see Figure 27.9), an Excel spreadsheet, a Cardfile card, a paint program, or another CAD package. This feature is available in all Windows versions, with the exception of Release 11 (also known as AutoCAD Windows Extension).

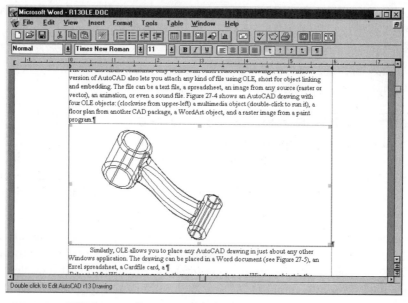

Figure 27.9 *AutoCAD Drawing Placed in a Word Document*

The commands for placing an OLE object in the drawing are found in the **Edit** and **Insert** menus. In summary, these are

Paste Special (PasteSpec)

Displays the **Paste Special** dialog box and gives you control over the paste process.

PasteAsHyperlink

Pastes object from the Windows Clipboard as a hyperlink into the drawing. This command is discussed in greater detail in Chapter 29 "AutoCAD on the Internet."

OLE Links (OleLinks)

Changes the settings of linked OLE objects.

Insert Object (InsertObj)

Selects a Windows application to open and then places its document in the AutoCAD drawing.

PLACING AN OLE OBJECT IN AUTOCAD

The concepts behind OLE can be confusing at times, so the best way to learn it is to walk through the steps. Let's place an object in a drawing. Here's how:

1. Start AutoCAD with a new drawing.

2. From the menu bar, select **Insert | OLE Object.**

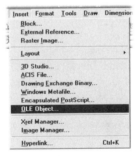

3. The **Insert Object** dialog box appears. You can select objects in one of two ways:

   • **Create New:** Start the application, then create the object, or load a file.

   • **Create from File:** The dialog box lets you select a file name. When you do, Windows launches the software application that it thinks is best suited for the file.

Select the **Create New** button. The dialog box lists the names of all applications that Windows knows can provide objects. Scroll through the list to find a suitable application, such as "Equation" for placing formula text in the drawing and "ClipArt Gallery" for selecting a piece of clip art, such as your firm's logo. (The applications listed by the dialog box depend on the applications loaded into your computer.)

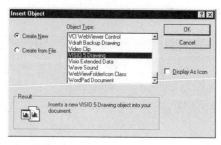

Figure 27.10 *Insert Object Dialog Box*

4. After selecting an application, choose **OK**.

5. The object appears in the AutoCAD drawing. AutoCAD always places OLE objects in the upper left corner.

 Note: For some older applications, Windows launches the application, letting you open and edit the object. When you are ready to leave the application, select **File | Update** from the menu bar; some applications have a **File | Exit and Return to AutoCAD** option.

WORKING WITH OLE OBJECTS IN AUTOCAD

Once the OLE object has been placed in the drawing, AutoCAD provides a few editing commands.

Resize the Object

Select the OLE object. Its border is surrounded by a rectangle (OLE objects can only be rectangular). Notice the eight small black (or white) squares that surround the object. These grips let you resize the object. Move the cursor over one of the squares, hold down the left mouse button, and drag the rectangle to a different size.

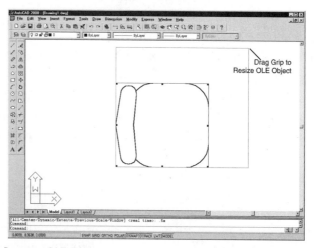

Figure 27.11 *Resizing OLE Object*

Move the Object

To move the object from the upper left corner to another location in the drawing, select the object, then drag it into place.

Additional Editing Commands

Move the cursor over the object, then click the right mouse button. AutoCAD displays a cursor menu with several primary options:

- **Cut** the object out of the drawing to the Windows Clipboard.

- **Copy** the object to the Clipboard.

- **Clear** the object (erase it).

- **Undo** the previous cursor menu operation.

- **Selectable** toggles whether it is included in selection sets.

- **Bring to Front** and **Send to Back** change the draw order relative to other objects in the drawing.

- **Properties** displays the **OLE Properties** dialog box, as discussed earlier.

- The final option on the cursor menu is specific to the object. In Figure 27.11, the object is a drawing created by the Visio Technical software. There are three options: **Edit** or **Open** the drawing (start Visio and load the file) or **Convert** the drawing to AutoCAD format. Because the Visio drawing is linked, any changes you make to the drawing while in Visio are automatically updated back in AutoCAD.

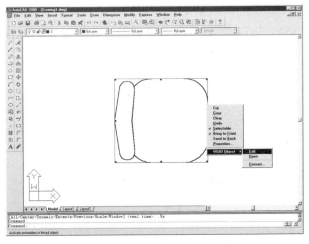

Figure 27.12 *Editing OLE Object*

If the object is a multimedia object, it can be either played (view the animation) or edited (launch the source application, a morphing program in this case).

CHANGING OLE LINKS

To change the nature of the link between the OLE object in AutoCAD and the originating software, select the **OLE Links** option from the **Edit** menu. AutoCAD displays the **Links** dialog box and the names of all linked OLE objects.

Just as the **XRef** command lets you change the links to externally referenced drawings, the **Links** dialog box lets you change the links to OLE objects. The dialog box has these options:

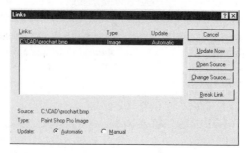

Figure 27.13 *Links Dialog Box*

Update Now

Automatic updates occur whenever the source document is changed; manual updates require you to choose the **Update Now** button.

Open Source

Starts the application that provided the OLE object.

Change Source

Lets you change where the object comes from, such as a new file or application or subdirectory.

Break Link

Removes the information that links the object back to its source.

Converting OLE Objects to AutoCAD Format

You must break the link if you want to convert the object to AutoCAD format. To do this, first break the link through the **Links** dialog box. Then, right-click on the object and select "Convert Picture Object." AutoCAD displays the Convert dialog box. This dialog box typically gives you one or more choices for handling the conversion.

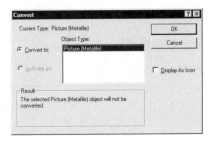

Figure 27.14 *Convert Dialog Box*

PLACING AN AUTOCAD DRAWING AS AN OLE OBJECT

OLE can be used in reverse; an AutoCAD drawing can be placed in the document of another Windows application. For example, you can use Cardfile to create a simple drawing manager by placing each AutoCAD drawing on a card, along with the drawing's file name. Here's how:

1. Start Cardfile.

2. From the menu bar, select **Edit | Index** to give the card a name on the index line, such as the drawing's file name, "Office.Dwg."

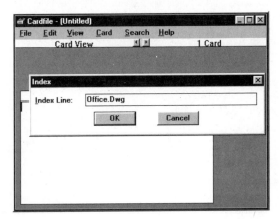

Figure 27.15 *Creating a New Card*

3. Switch Cardfile from text to graphics mode with **Edit | Picture**. The cardfile won't look any different at this point.

4. Place the AutoCAD drawing on the card by selecting **Edit | Insert Object**.

5. When the **Insert New Object** dialog box appears, select "AutoCAD Drawing," and choose **OK**.

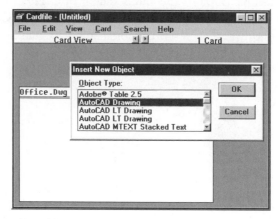

Figure 27.16 *Insert New Object*

6. Windows starts AutoCAD. Insert the drawing **Office.Dwg** provided on the CD-ROM.

7. After AutoCAD finishes loading the drawing, select **Update Cardfile** in the **File** menu. The **Update** command only appears when AutoCAD is being operated by "remote control" from another Windows application through OLE. And through OLE, Windows places an image of the drawing in Cardfile.

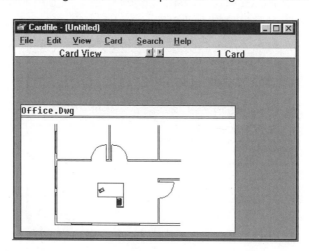

Figure 27.17 *AutoCAD Drawing in Cardfile*

8. Now that Cardfile is set up, you launch AutoCAD (together with the drawing) by simply double-clicking on the drawing's picture in Cardfile. The OLE link feature saves you from the manual updating process required when you use traditional cut and paste.

CHAPTER REVIEW

1. Explain the difference between an inserted drawing, one placed as an external reference, and one inserted as an OLE object.

2. What does OLE stand for?

3. What purposes can OLE be used for?

4. Can OLE objects be viewed in any version of AutoCAD?

5. What command places OLE objects in an AutoCAD drawing?

CHAPTER 28

CAD-Related Web Sites

The question is not *if* the Internet is in your future, but how. In this chapter, you tour half a dozen Web sites that show off the potential of the Internet for AutoCAD users. After completing this chapter, you will have

- Visited six Web sites that relate to CAD

- Stepped through a Web site that uses a database program to serve up CAD drawings

Web sites are viewed with a Web browser. There are several brands of Web browser available on the market, including Netscape Navigator®, Operasoft Opera, and Microsoft® Internet Explorer®. To view these Web sites, make sure your browser has the following options enabled: Automatic Image Load, Java™, and JavaScript®.

CONSTRUCTION JOBCAM

A construction company in Florida, USA has set up a video camera at its current job site. The camera is connected to a video capture board in a 486 computer. Every five minutes, a video image is digitized into JPEG format and the 41 KB file is transmitted to Winter Park Construction's Web server. At http://www.wpc.com/ anyone—the client, suppliers, subcontractors, and curious Web surfers—can see the progress of the current project.

At the end of the day, the 9:00 a.m. and 3:00 p.m. photos are archived with a file name consisting of the date and time, such as 97-01-31_0900.Jpg. The collection goes back to December 6, 1995. The archive, an automated photographic record, can serve for legal purposes. WPC estimates it cost them about $1,500 to set up their digital photograph facility (not counting the cost of the Web server) plus some in-house custom programming.

Figure 28.1 *Winter Park Construction Web Site*

PARTS ONLINE

Thomas Register (formerly Autodesk Data Publishing) has placed their entire PartSpec, PlantSpec, and CAD Blocks CD-ROM libraries on their Web site at http://data.autodesk.com. The Web site is useful, since, as a designer, you can get a lot of instant product information. The data includes performance, sizes, photographs, and 2D vector drawings.

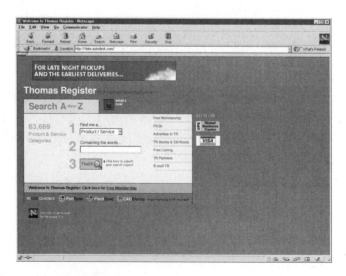

Figure 28.2 *Thomas Register Web Site*

MODEL PUBLISHING

Bentley Systems has taken a gamble in thinking about how to distribute drawings in a CAD world. They see the CAD world in three phases: (1) first, CAD software and data were stored on mainframe computers and was displayed by dumb terminals, (2) today, CAD software and data are on desktop computers, possibly networked, and (3) in the future, CAD software will continue to run on desktop computers, but the data will stored on a central server, with file viewing done by a Web browser-like program.

To get to the third stage, Bentley Systems is working on a suite of programs. The first product is ModelServer™ Publisher. Its purpose is to provide fast translation. When the CAD operator saves the drawing (AutoCAD DWG, MicroStation® DGN, or STEP format), it is saved to ModelServer, not the desktop PC. When someone wants to see the drawing over the Internet or intranet, Publisher translates the vector drawings into any of the popular Internet file formats your Web browser supports: SVF (simple vector format), CGM (computer graphics metafile), PNG (portable network graphics), WRL (VRML), and others.

Bentley demonstrates Publisher at http://www.bentley.com/modelserver/demo/, where you view the translated drawing in SVF, CGM, JPEG, and Java format.

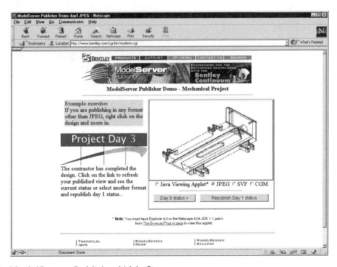

Figure 28.3 *ModelServer Publisher Web Site*

CAD DETAILS LIBRARY

The Tri-Service CADD/GIS Technology Center maintains a Web site for the Corps of Engineers to make their construction drawings available at http://cadlib.wes.army.mil. At that site, you can view and download complete drawings of three projects, a details library, and a standard symbols library in three different formats: raster GIF (meant for preview purposes), AutoCAD DWG and MicroStation DGN.

To narrow your search, the Web site asks you to select a discipline—architectural, mechanical, electrical, or HVAC. Then you choose a type of detail, such as exterior walls, stair construction, or fire protection. Finally, you select a format: AutoCAD or MicroStation. At this point, the Web site presents a list of drawings. You can preview the drawing in GIF format (about 10 KB) or download the drawing file to your hard drive.

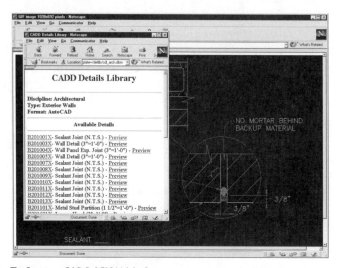

Figure 28.4 *Tri-Service CADD/GIS Web Site*

STEP ASSEMBLY TREE

STEP is an international file standard being created to document a 3D engineering drawing through all stages of its life cycle, from design, through construction, to disposal. STEP is short for Standard for The Exchange of Product model data. STEP has been under development for over ten years, but it is only now that you will start hearing more about it as Autodesk and other CAD vendors begin implementing STEP translators.

One provider of STEP management tools, STEP Tools®, has set up examples at its Web site, http://www.steptools.com, of how the Web can be used to disseminate STEP data. One example automatically converts STEP data to Web pages. The Web page displays the assembly tree of a control arm. Thumbnails in GIF format (a mere 1 KB in size) are next to hot links that display the part in either 3D wireframe (using Java) or fully shaded 3D (using VRML). You can rotate and zoom both the Java and

VRML 3D images, although the download takes a long time because the files are not optimized. (Netscape Navigator has the ability to view Java and VRML graphics.)

Along with the assembly tree, the Web site includes a costing calculator, also written in Java. When the price of a material changes (such as the cost-per-pound of steel), you enter its new unit cost and the total price is updated. The calculator knows the amount of steel in each STEP part. Color-coding warns when pricing information is missing are incomplete.

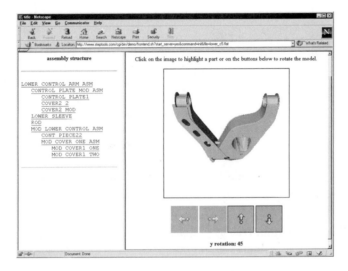

Figure 28.5 *STEP Tools Web Site*

CUSTOM-MADE ORDERING

MAN is a German manufacturer of heavy trucks. It maintains drawings of its trucks on its Web site. Their Web site is a good example of how to provide: (1) security (through an account number), (2) on-the-fly database access, and (3) CAD drawings in a variety of formats.

1. Point your Web browser to http://manted.man-nutzfahrzeuge.de/man/owa/man_ted.startup

2. Choose **English**, if you prefer that language.

Figure 28.6 *MAN Web Site*

3. When the English Web page appears, enter **342341** for the account number, and choose the **Sign-On** button.

4. Select a group of vehicles. Notice that you can specify a variety of parameters. For demo purposes, you best choice is to select from the **Vehicle Type** list.

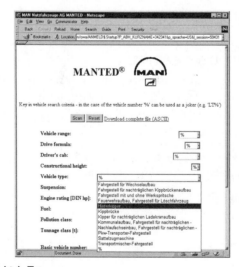

Figure 28.7 *Select Vehicle Type*

5. Choose **Scan**. An Oracle database retrieves the truck model numbers that correspond to your selection.

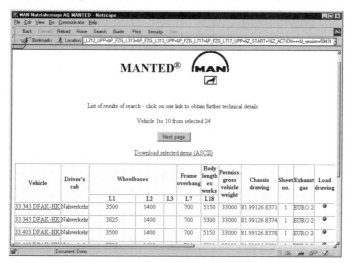

Figure 28.8 *Database-generated List of Vehicles*

6. Choose a green dot under the column, **Load Drawing**.

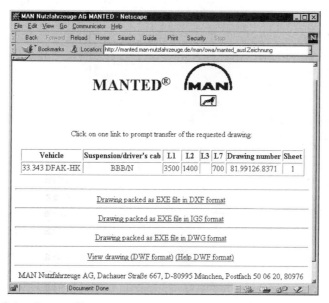

Figure 28.9 *Select Drawing Type*

7. Here you have a choice of drawing file formats. Three of them—DWG, IGES, DXF—are downloaded to your computer's hard disk as a self-extracting PkZip file. To view a drawing online, choose **View drawing (DWF format)**. For the truck model shown in the figures, the download size and time for each file format is as follows:

File Format	DWF	DXF	DWG	IGES
File Size (Compressed)	185 KB	245 KB	302 KB	552 KB
Download time, minutes (28.8Kbps)	0.75	1.0	1.25	2.25

8. Wait a few moments for the DWF plug-in to load. Then wait a few minutes longer for the DWF file to download. The drawing shown in Figure 28.10 is 185 KB. Right-click the drawing to display the DWF plug-in's menu of choices. For example, you can toggle layers, zoom in, and pan about, and save the DWF drawing to file. You learn more about DWF files in Chapter 29 "AutoCAD on the Internet."

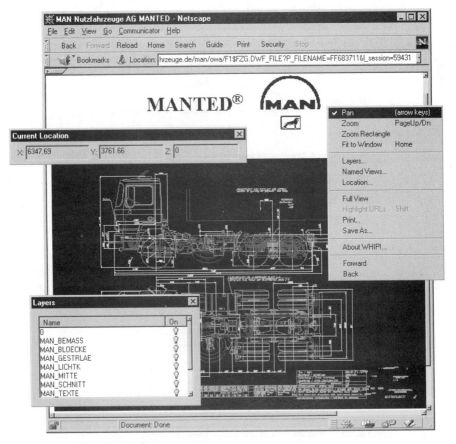

Figure 28.10 *Select Vehicle Type*

SUMMARY

Summarizes one third-party developer:

"I expect there will be a great demand for CAD drawings on the Web, for example, in technical documentation. Vector graphics can be transferred faster than bitmaps (because they are smaller), and they can be scaled without losing detail.

"In the future, it will be more and more important that information is up to date, and easy to access. So, there will be a strong trend toward Web manuals, because it's much easier to update a Web manual than a CD-ROM or a printed manual."

In this chapter, you took a tour of several CAD-related Web sites. They illustrate the potential of the Internet for AutoCAD users. In the next chapter, you learn how to use AutoCAD with the Internet.

CHAPTER REVIEW

1. Describe three kinds of data that you can get from a Web site:

   a.

   b.

   c.

2. Is it useful to archive photos of a construction site? If so, why?

3. You can obtain data through several sources. Name two:

   a.

   b.

4. Some file formats are meant for use on the Internet. What are these abbreviations short for?

   a. SVF

   b. PNG

   c. WRL

   d. CGM

5. Briefly describe the purpose of the STEP initiative.

6. What is an advantage of the DWF file format over other CAD file formats?

CHAPTER 29

AutoCAD on the Internet

The Internet has become the most important way in the world to exchange information. AutoCAD allows you to interact with the Internet in several ways. It can open and save files that are located on the Internet; it can launch a Web browser (AutoCAD 2000 includes a simple Web browser); and it can create drawing Web format (DWF) files for viewing as drawings on Web pages. After completing this chapter, you will be able to

- Launch a Web browser from AutoCAD

- Understand the importance of the URL (uniform resource locator) on the Internet

- Open and save drawings to and from the Internet

- Place hyperlinks in a drawing

- Convert drawings to DWF file format

- View DWF files with a Web browser

- Learn about the *WHIP!*® plug-in and the DWF file format

INTRODUCTION

Before you can use the Internet features in AutoCAD 2000, some components of Microsoft® Internet Explorer® must be present in your computer. If Internet Explorer v4.0 (or later) is already set up on your computer, then you already have these required components. If not, then the components are installed automatically during AutoCAD 2000's setup when you choose: (1) the **Full Install**; or (2) the Internet Tools option in the **Custom** installation.

This chapter introduces you to the following Web-related commands:

Browser:

Launches a Web browser from within AutoCAD.

Hyperlink:

Attaches and removes a URL to or from an object or an area in the drawing.

HyperlinkFwd:

Move to the next hyperlink (an undocumented command).

HyperlinkBack:

Moves back to the previous hyperlink (an undocumented command).

HyperlinkStop:

Stops the hyperlink access action (an undocumented command).

PasteAsHyperlink:

Attaches a URL to an object in the drawing from text stored in the Microsoft® Windows Clipboard (an undocumented command).

HyperLinkBase:

A system variable for setting the path used for relative hyperlinks in the drawing.

In addition, all of AutoCAD 2000's file-related dialog boxes are "Web enabled."

CHANGES FROM AUTOCAD RELEASE 14

If you used the Internet features in AutoCAD Release 14, then you should note the following differences in AutoCAD 2000.

Attached URLs from R14

In AutoCAD R14, attached URLs were not active until the drawing was converted to a DWF file. In AutoCAD 2000, URLs are active in the drawing.

If you attached URLs to a drawing in R14 (as well as LT 97 and LT 98), they are converted to AutoCAD 2000-style hyperlinks the first time you save the drawing in AutoCAD 2000 DWG format.

When you use the **SaveAs** command to save an AutoCAD 2000 drawing in R14 format, hyperlinks are converted back to R14-style URLs. Hyperlink descriptions are stored as proxy data and not displayed in R14/LT. (This ensures that hyperlink descriptions are available again when the drawing is brought back to AutoCAD 2000 format.)

WHIP! v3.1 Limitations

AutoCAD 2000 can create DWF files compatible with the *WHIP!* v3.1 plug-in. I recommend, however, that you download *WHIP!* v4.0 from the Autodesk Web site. *WHIP!* v3.1 does not display the following AutoCAD 2000-specific features:

- Hyperlink descriptions are not saved in v3.1 DWF format.

- All objects, except polylines and traces, are displayed at minimum lineweight.

- Linetypes defined by a plot style table are displayed as continuous lines.

- Nonrectangular viewports are displayed by rectangular viewports.

Changed Internet Commands

Many of R14's Internet-related commands have been discontinued and replaced. The following summary describes the status of all R14 Internet commands:

Browser (launches Web browser from within AutoCAD): continues to work in AutoCAD 2000; the default is now Home.Htm, a Web page located on your computer.

AttachUrl (attaches a URL to an object or an area in the drawing): continues to work in AutoCAD 2000, but has been superceded by the **-Hyperlink** command's **Insert** option.

SelectUrl (selects all objects with attached URLs): continues to work in AutoCAD 2000, but has been superceded by the **QSelect** dialog box's **Hyperlink** option.

ListUrl (lists URLs embedded in the drawing): removed from AutoCAD 2000. As a replacement, use the **Properties** dialog box's **Hyperlink** option.

DetachUrl (removes the URL from an object): continues to work in AutoCAD 2000, but has been superceded by the **-Hyperlink** command's **Remove** option.

DwfOut (exports the drawing and embedded URLs as a DWF file): continues to work in AutoCAD 2000, but has been superceded by the **Plot** dialog box's **ePlot** option.

DwfOutD (exports drawing in DWF format without a dialog box): removed from AutoCAD 2000. As a replacement, use the **-Plot** command's **ePlot** option.

INetCfg (configures AutoCAD for Internet access): removed from AutoCAD 2000; It has been replaced by the Internet applet of the Windows Control Panel (click on the **Connection** tab).

InsertUrl (inserts a block from the Internet into the drawing): automatically executes the **Insert** command and displays the **Insert** dialog box. You can type a URL for the block name.

OpenUrl (opens a drawing from the Internet): automatically executes the **Open** command and displays the **Select File** dialog box. You can type a URL for the file name.

> **SaveUrl** (saves the drawing to the Internet): automatically executes the **Save** command and displays the **Save Drawing As** dialog box. You can type a URL for the file name.

You are probably already familiar with the best-known uses for the Internet: email (electronic mail) and the Web (short for "World Wide Web"). Email lets users exchange messages and data at very low cost. The Web brings together text, graphics, audio, and movies in an easy-to-use format. Other uses of the Internet include FTP (file transfer protocol) for effortless binary file transfer, Gopher (presents data in a structured, subdirectory-like format), and Usenet, a collection of more than 10,000 news groups.

AutoCAD allows you to interact with the Internet in several ways. AutoCAD is able to launch a Web browser from within AutoCAD with the **Browser** command. Hyperlinks can be inserted in drawings with the **Hyperlink** command, which lets you link the drawing with other documents on your computer and the Internet. With the **Plot** command's **ePlot** option (short for "electronic plot"), AutoCAD creates DWF (short for "drawing Web format") files for viewing drawings in 2D format on Web pages. AutoCAD can open, insert, and save drawings to and from the Internet via the **Open**, **Insert**, and **SaveAs** commands.

UNDERSTANDING URLS

The uniform resource locator, known as the URL, is the file naming system of the Internet. The URL system allows you to find any resource (a file) on the Internet. Examples of resources include a text file, a Web page, a program file, an audio or movie clip—in short, anything you might also find on your own computer. The primary difference is that these resources are located on somebody else's computer. A typical URL looks like the following examples:

Example URL	Meaning
http://www.autodesk.com	Autodesk Primary Web Site
news://adesknews.autodesk.com	Autodesk News Server
ftp://ftp.autodesk.com	Autodesk FTP Server
http://www.autodeskpress.com	Autodesk Press Web Site
http://users.uniserve.com/~ralphg	Editor Ralph Grabowski's Web site

Note that the **http://** prefix is not required. Most of today's Web browsers automatically add in the *routing* prefix, which saves you a few keystrokes.

URLs can access several different kinds of resources, such as Web sites, email, news groups, but this always take on the same general format, as follows:

scheme://netloc

The scheme accesses the specific resource on the Internet, including these:

Scheme	Meaning
file://	File located on your computer's hard drive or local network
ftp://	File Transfer Protocol (used for uploading and downloading files)
http://	Hypertext Transfer Protocol (the basis of Web sites)
mailto://	Electronic mail (email)
news://	Usenet news (news groups)
telnet://	Telnet protocol
gopher://	Gopher protocol

The **://** characters indicate a network address. Autodesk recommends the following format for specifying URL-style file names with AutoCAD:

Resource	URL Format
Web Site	**http://***servername/pathname/filename*
FTP Site	**ftp://***servername/pathname/filename*
Local File	**file:///***drive:/pathname/filename*
or	**drive:***pathname\filename*
or	**file:///***drive|/pathname/filename*
or	**file://**/*localPC\pathname\filename*
or	**file:////***localPC/pathname/filename*
Network File	**file://***localhost/drive:/pathname/filename*
or	*\\localhost\drive:\pathname\filename*
or	**file://***localhost/drive|/pathname/filename*

The terminology can be confusing. The following definitions will help to clarify these terms.

Term	Meaning
servername	The name or location of a computer on the Internet, for example: www.autodesk.com.
pathname	The same as a subdirectory or folder name.

drive	The driver letter, such as C: or D:.
localpc	A file located on your computer.
localhost	The name of the network host computer.

If you are not sure of the name of the network host computer, use Windows Explorer to check the Network Neighborhood for the network names of computers.

LAUNCHING A WEB BROWSER

The **Browser** command lets you start a Web browser from within AutoCAD. Commonly used Web browsers include Netscape Navigator®/Netscape Communicator™, Microsoft Internet Explorer®, and Operasoft Opera.

By default, the **Browser** command uses whatever brand of Web browser program is registered as your default browser in your computer's Windows operating system. AutoCAD prompts you for URL, such as **http://www.autodeskpress.com**. The **Browser** command can be used in scripts, toolbar or menu macros, and AutoLISP routines to automatically access the Internet.

Command: **browser**

Browse <C:\CAD\ACAD 2000\Home.htm>: *(Enter the URL.)*

The default URL is an HTML file added to your computer during AutoCAD's installation. After you type the URL and press Enter, AutoCAD launches the Web browser and contacts the Web site. Figure 29.1 shows the popular Netscape Communicator/Navigator with the Autodesk Web site at **http://www.autodesk.com**.

Figure 29.1 *Netscape Communicator/Navigator Displaying the Autodesk Web Site*

CHANGING DEFAULT WEB SITE

To change the default Web page that your browser starts with from within AutoCAD, change the setting in the **INetLocation** system variable. The variable stores the URL used by the **Browser** command and the **Browse the Web** dialog box. Make the change as follows:

> Command: **inetlocation**
>
> Enter new value for INETLOCATION <"C:\CAD\ACAD 2000\Home.htm">: *(Type URL.)*

DRAWINGS ON THE INTERNET

When a drawing is stored on the Internet, you access it from within AutoCAD 2000 using the standard **Open**, **Insert**, and **Save** commands. (In Release 14, these commands were known as **OpenUrl**, **InsertUrl**, and **SaveUrl**.) Instead of specifying the file's location with the usual drive-subdirectory-file name format, such as c:\acad14\filename.dwg, use the URL format. (Recall that the URL is the universal file naming system used by the Internet to access any file located on any computer hooked up to the Internet.)

OPENING DRAWINGS FROM THE INTERNET

To open a drawing from the Internet (or your firm's intranet), use the **Open** command (choose **File | Open**). Notice the three buttons at the upper right of the **Select File** dialog box (see Figure 29.2). These are specifically for use with the Internet. From left to right, they are:

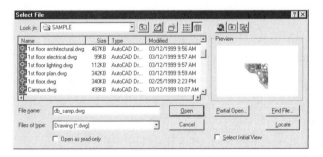

Figure 29.2 *Select File Dialog Box*

> **Search the Web:** Displays the **Browse the Web** "dialog box," a simplified Web browser.
>
> **Look in Favorites:** Opens the Favorites folder, equivalent to "bookmarks" that store Web addresses.
>
> **Add to Favorites:** Saves the current URL (hyperlink) to the Favorites folder.

When you choose the **Search the Web** button, AutoCAD opens the **Browse the Web** dialog box. This dialog box is a simplified version of the Microsoft's Web browser. The purpose of this dialog box is to allow you to browse files at a Web site.

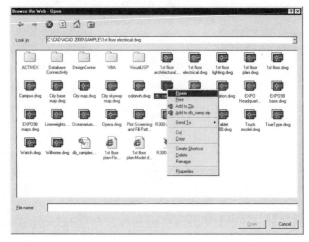

Figure 29.3 *Select File Dialog Box*

By default, the **Browse the Web** dialog box displays the contents of the URL stored in the **INetLocation** system variable. You can easily change this to another folder or Web site, as noted earlier.

Along the top, the dialog box has six buttons:

> **Back:** Go back to the previous URL.
>
> **Forward:** Go forward to the next URL.
>
> **Stop:** Halt displaying the Web page (useful if the connection is slow or the page is very large.)
>
> **Refresh:** Redisplay the current Web page.
>
> **Home:** Return to the location specified by the INetLocation system variable.
>
> **Favorites:** List stored URLs (hyperlinks) or bookmarks. If you have previously used Internet Explorer, you will find all your favorites listed here. Favorites are stored in the **\Windows\Favorites** folder on your computer.

The **Look in** field allows you to type the URL. Alternatively, click the down arrow to select a previous destination. If you have stored Web site addresses in the **Favorites** folder, then select a URL from that list.

You can either double-click a file name in the window, or type a URL in the **File name** field. The following table gives templates for typing the URL to open a drawing file:

Drawing Location	Template URL
Web or HTTP Site	**http:**//servername/pathname/filename.dwg
	http://practicewrench.autodeskpress.com/wrench.dwg
FTP Site	**ftp:**//servername/pathname/filename.dwg
	ftp://ftp.autodesk.com
Local File	**drive:**\pathname\filename.dwg
	c:\acad 2000\sample\tablet2000.dwg
Network File	\\localhost\drive:\pathname\filename.dwg
	\\upstairs\e:\install\sample.dwg

When you open a drawing over the Internet, it will probably take much longer than opening a file found on your computer. During the file transfer, AutoCAD displays a dialog box to report the progress. If your computer uses a 28.8 Kbps modem, you should allow about five to ten minutes per megabyte of drawing file size. If your computer has access to a faster T1 connection to the Internet, you should expect a transfer speed of about one minute per megabyte.

Figure 29.4 *File Download Dialog Box*

It may be helpful to understand that the **Open** command does not copy the file from the Internet location directly into AutoCAD. Instead, it copies the file from the Internet to your computer's designated **Temporary** subdirectory, such as C:\Windows\Temp (and then loads the drawing from the hard drive into AutoCAD). This is known as *caching*. It helps to speed up the processing of the drawing, since the drawing file is now located on your computer's fast hard drive, instead of the relatively slow Internet.

Note that the **Locate** and **Find File** buttons (in the **Select File** dialog box) do not locate files on the Internet. They are only for locating files on your local system.

TUTORIAL

The Autodesk Press Web site has an area that allows you to practice using the Internet with AutoCAD. In this tutorial, you open a drawing file located at that Web site.

1. Start AutoCAD.

2. Ensure that you have a live connection to the Internet. If you normally access the Internet via a telephone (modem) connection, dial your Internet service provider now.

3. From the menu bar, choose **File | Open**. Or, choose the **Open** icon on the toolbar. Notice that AutoCAD displays the **Select File** dialog box.

4. Choose the **Search the Web** button. Notice that AutoCAD displays the **Browse the Web** dialog box.

5. In the **Look in** field, type:

   practicewrench.autodeskpress.com

   and press ENTER. After a few seconds, the **Browse the Web** dialog box displays the Web site.

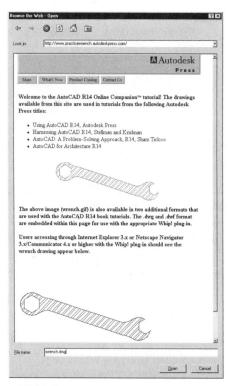

Figure 29.5 *Practice Wrench Web Site*

6. In the **Filename field**, type:

wrench.dwg

and press ENTER. AutoCAD begins transferring the file. Depending on the speed of your Internet connection, this will take between a couple of seconds and a half-minute. Notice the drawing of the wrench in AutoCAD.

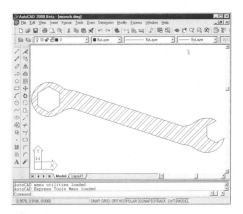

Figure 29.6 *Wrench Drawing in AutoCAD*

INSERTING A BLOCK FROM THE INTERNET

When a block (symbol) is stored on the Internet, you can access it from within AutoCAD using the **Insert** command. When the **Insert** dialog box appears, choose the **Browse** button to display the **Select Drawing File** dialog box. This is identical to the dialog box discussed above.

After you click the file name to select the file, AutoCAD downloads the file and continues with the **Insert** command's familiar prompts.

The process is identical for accessing external reference (xref) and raster image files. Other files that AutoCAD can access over the Internet include: 3D Studio, SAT (ACIS solid modeling), DXB (drawing exchange binary), WMF (Windows metafile), and EPS (encapsulated PostScript™). All of these options are found in the **Insert** menu on the menu bar.

ACCESSING OTHER FILES ON THE INTERNET

Most other file-related dialog boxes allow you to access files from the Internet or intranet. This allows your firm or agency to have a central location that stores drawing standards. When you need to use a linetype or hatch pattern, for example, you access the LIN or PAT file over the Internet. More than likely, you would have the location of those files stored in the Favorites list.

Some examples include:

> **Linetypes:** From the menu bar, choose **Format | Linetype**. In the **Linetype Manager** dialog box, choose the **Load**, **File**, and **Look in Favorites** buttons.
>
> **Hatch Patterns:** Use the Web browser to copy PAT files from a remote location to your computer.
>
> **Multiline Styles:** From the menu bar, choose **Format | Multiline Style**. In the **Multiline Styles** dialog box, choose the **Load**, **File**, and **Look in Favorites** buttons.
>
> **Layer Name:** From the menu bar, choose **Express | Layers | Layer Manager**. In the **Layer Manager** dialog box, choose the **Import** and **Look in Favorites** buttons.
>
> **LISP and ARX Applications:** From the menu bar, choose **Tools | Load Applications**.
>
> **Scripts:** From the menu bar, choose **Tools | Run Scripts**.
>
> **Menus:** From the menu bar, choose **Tools | Customize Menus**. In the **Menu Customization** dialog box, choose the **Browse** and **Look in Favorites** buttons.
>
> **Images:** From the menu bar, choose **Tools | Displays Image | View**.

You cannot access text files, text fonts (SHX and TTF), color settings, lineweights, dimension styles, plot styles, OLE objects, or named UCSs over the Internet.

SAVING THE DRAWING TO THE INTERNET

When you are finished editing a drawing in AutoCAD, you can save it to a file server on the Internet with the **Save** command. If you inserted the drawing from the Internet (using **Insert**) into the default Drawing.Dwg drawing, AutoCAD insists you first save the drawing to your computer's hard drive.

When a drawing of the same name already exists at that URL, AutoCAD warns you, just as it does when you use the **SaveAs** command. Recall from the **Open** command that AutoCAD uses your computer system's Temporary subdirectory, hence the reference to it in the dialog box.

USING HYPERLINKS WITH AUTOCAD

AutoCAD 2000 allows you to employ URLs in two ways: (1) directly within an AutoCAD drawing; and (2) indirectly in DWF files displayed by a Web browser. (URLs are also known as *hyperlinks*, the term we use from now on.)

Hyperlinks are created, edited, and removed with the **Hyperlink** command (displays a dialog box) and the **-Hyperlinks** command (for prompts at the command line).

The **Hyperlink** command (**Insert | Hyperlink**) prompts you to "Select objects" and then displays the **Insert Hyperlink** dialog box. (As a shortcut, you can press CTRL+K or choose the **Insert Hyperlink** button on the toolbar.)

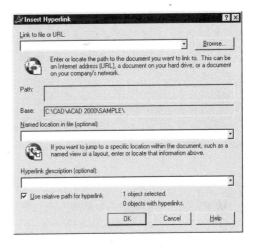

Figure 29.7 *The Insert Hyperlink Dialog Box*

As an alternative, you can use the **-Hyperlink** command. This command displays its prompts at the command line, and it is useful for scripts and AutoLISP routines. The **-Hyperlink** command has the following syntax:

> Command: **-hyperlink**
>
> Enter an option [Remove/Insert] <Insert>: *(Press ENTER.)*
>
> Enter hyperlink insert option [Area/Object] <Object>: *(Press ENTER.)*
>
> Select objects: *(Pick an object.)*
>
> I found Select objects: *(Press ENTER.)*
>
> Enter hyperlink <current drawing>: *(Enter the name of the document or Web site.)*
>
> Enter named location <none>: *(Enter the name of a bookmark or AutoCAD view.)*
>
> Enter description <none>: *(Enter a description of the hyperlink.)*

Notice that the command also allows you to remove a hyperlink. It does not, however, allow you to edit a hyperlink. To do this, use the **Insert** option, and respecify the hyperlink data.

In addition, this command allows you to create a hyperlink *area*, which is a rectangular area that can be thought of as a 2D hyperlink (the dialog box-based **Hyperlink** command does not create hyperlink areas). When you choose the **Area** option, the rectangle is placed automatically on layer URLLAYER and colored red.

Figure 29.8 *A Rectangular Hyperlink Area*

In the following sections, you learn how to apply and use hyperlinks in an AutoCAD drawing and in a Web browser via the dialog box–based **Hyperlink** command.

HYPERLINKS INSIDE AUTOCAD

AutoCAD allows you to add a hyperlink to any object in the drawing. An object is permitted just a single hyperlink; but a single hyperlink can be applied to a selected set of objects.

You can tell an object has a hyperlink by passing the cursor over it. The cursor displays the "linked Earth" icon, as well as a tooltip describing the link. See Figure 29.9.

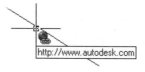

Figure 29.9 *The Cursor Reveals a Hyperlink*

If, for some reason, you do not want to see the hyperlink cursor, you can turn it off. From the menu, choose **Tools | Options**, and then click the **User Preferences** tab. The **Display hyperlink cursor and shortcut menu** item toggles the display of the hyperlink cursor, as well as the **Hyperlink tooltip** option on the cursor menu.

Tutorial: Attaching Hyperlinks

Let's see how that works by working through an example. In this example, we have the drawing of a floorplan. To this drawing we will add hyperlinks to another AutoCAD drawing, a Microsoft Word document, and a Web site. Hyperlinks must be attached to objects. For this reason, we will place some text in the drawing, then attach the hyperlinks to the text.

1. Start AutoCAD 2000.

2. Open the **1st floor plan.dwg** file found in the AutoCAD 2000**Sample** folder. (If necessary, click the **Model** tab to display the drawing in model space.)

3. Start the **Text** command, and place the following text in the drawing:

   Command: **text**

   Current text style: "Standard" Text height: **0.20000**

   Specify start point of text or [Justify/Style]: *(Select a point in the drawing.)*

   Specify height <0.2000>: **2**

   Specify rotation angle of text <0>: *(Press* ENTER.*)*

   Enter text: **Site Plan**

   Enter text: Lighting Specs

   Enter text: Electrical Bylaw

   Enter text: *(Press* ENTER.*)*

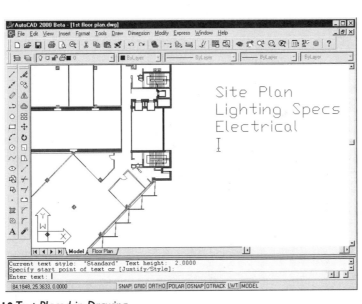

Figure 29.10 *Text Placed in Drawing*

4. Click the "Site Plan" text to select it.

5. Choose the **Insert Hyperlink** button on the toolbar. Notice the **Insert Hyperlinks** dialog box.

6. Next to the **Link to File or URL** field, choose the **Browse** button. Notice the **Browse the Web – Select Hyperlink** dialog box.

7. Go to AutoCAD 2000's Sample folder and select the **City base map.dwg file**. Choose **Open**.

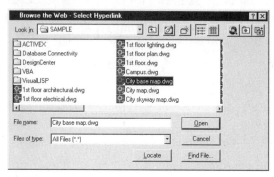

Figure 29.11 *Select a Drawing File*

AutoCAD does not open the drawing; rather, it copies the file's name to the **Insert Hyperlinks** dialog box. Notice the name of the drawing you selected in the **Link to File or URL** field.

Figure 29.12 *The Insert Hyperlinks Dialog Box*

You can fill in two other fields:

Named Location in File: When the hyperlinked file is opened, it goes to the named location, called a "Bookmark." In AutoCAD, the bookmark is a named view (created with the **View** command).

Hyperlink Description: You can type a description for the hyperlink, which is displayed by the tooltip. If you leave this blank, the URL is displayed by the tooltip.

8. Choose **OK** to dismiss the dialog box. Move the cursor over the Site Plan text. Notice the display of the "linked Earth" icon; a moment later, the tooltip displays "City base map.dwg".

9. Repeat the **Hyperlink** command twice more, attaching these files to the drawing text:

Figure 29.13 *The Hyperlink Cursor and Tooltip*

Text	URL
Lighting Specs	License.Rtf (found in the AutoCAD 2000 folder)
Electrical Bylaw	Home.Htm (also found in the AutoCAD 2000 folder)

You have now attached a drawing, a text document, and a Web document to objects in the drawing. Let's try out the hyperlinks.

10. Click "Site Plan" to select it. Right-click and choose **Hyperlink | Open "City base map.dwg"** from the cursor menu.

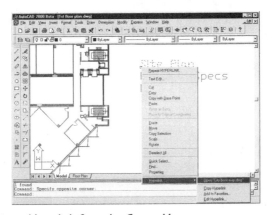

Figure 29.14 *Selecting a Hyperlink from the Cursor Menu*

Notice that AutoCAD opens "City base map.dwg". To see both drawings, choose **Window | Tile Vertically** from the menu bar.

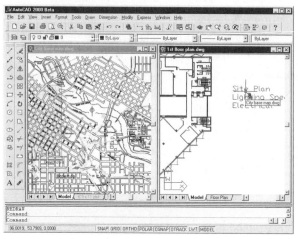

Figure 29.15 *Viewing Two Drawings*

11. Select, then right-click "Lighting Specs" hyperlink. Choose **Hyperlink | Open**. Notice that Windows starts a word processor and opens the License.Rtf file.

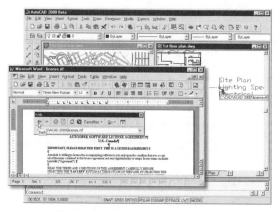

Figure 29.16 *Viewing a Word Document*

12. If your word processor has a Web toolbar, open it. Choose the **Back** button (the back arrow), as shown in Figure 29.16. Notice how that action sends you back to AutoCAD.

13. Select, then right-click "Electrical Bylaw" hyperlink. Choose **Hyperlink | Open**. Notice that Windows starts your Web browser and opens the Home.Htm file.

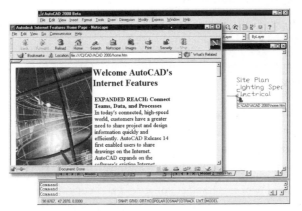

Figure 29.17 *Viewing a Web Site*

14. Back in AutoCAD, right-click any toolbar and choose **Web**. Notice that the Web toolbar has four buttons. From left to right, these are:

Figure 29.18 *AutoCAD's Web Toolbar*

Button	Command	Meaning
Go Back	HyperlinkBack	Move back to the previous hyperlink.
Go Forward	HyperlinkFwd	Move to the next hyperlink.
Stop Navigation	HyperlinkStop	Stops the hyperlink access action.
Browse the Web	Browser	Launches the Web browser.

15. Try clicking the **Go Forward** and **Go Back** buttons. Notice how these allow you to navigate between the two drawings, the RTF document, and the Web page.

When you work with hyperlinks in AutoCAD, you might come across these limitations:

- AutoCAD does not check that the URL you type is valid.
- If you attach a hyperlink to a block, be aware that the hyperlink data are lost when you scale the block unevenly, stretch the block, or explode it.
- Wide polylines and rectangular hyperlink areas are only "sensitive" on their outline.

Pasting As Hyperlink

AutoCAD 2000 has a shortcut method for creating hyperlinks in the drawing. The undocumented **PasteAsHyperlink** command pastes any text in the Windows Clipboard as a hyperlink to any object in the drawing. Here is how it works:

1. In a word processor, select some text and copy it to the Clipboard (via CTRL+C or the **Edit | Copy** command). The text can be a URL (such as http://www.autodeskpress.com) or any other text.

2. Switch to AutoCAD and choose **Edit | Paste As Hyperlink** from the menu bar. Note that this command does not work (is grayed out) if anything else is in the Clipboard, such as a picture.

3. Select one or more objects, as prompted:

   Command: **_pasteashyperlink**

   Select objects: *(Pick an object.)*

   I found Select objects: *(Press* ENTER.*)*

4. Pass the cursor over the object and note the hyperlink cursor and tooltip. The tooltip displays the same text that you copied from the document.

If the text you copy to the Clipboard is very long, AutoCAD displays only portions of it in the tooltip, using ellipses (…) to shorten the text. You cannot select text in the AutoCAD drawing to paste as a hyperlink. You can, however, copy the hyperlink from one object to another. Select the object, right-click, and choose **Hyperlink | Copy Hyperlink** from the cursor menu. The hyperlink is copied to the Clipboard. You can now paste the hyperlink into another document, or use AutoCAD's **Edit | Paste as Hyperlink** command to attach the hyperlink to another object in the drawing. (Note that the **MatchProp** command does not work with hyperlinks.)

Highlighting Objects with URLs

Although you can see the rectangle of area URLs, the hyperlinks themselves are invisible. For this reason, AutoCAD has the **QSelect** command, which highlights all objects that match specifications. From the menu bar, choose **Tools | Quick Select**. AutoCAD displays the **Quick Select** dialog box.

Figure 29.19 *The Quick Select Dialog Box*

In the fields, enter these specifications:

Apply to: **Entire drawing**

Object type: **Multiple**

Properties: **Hyperlink**

Operator: * **Wildcard Match**

Value: *

Choose **OK**, and AutoCAD highlights all objects that have a hyperlink. Depending on your computer's display system, the highlighting shows up as dashed lines or as another color.

Editing Hyperlinks

Now that you know where the objects with hyperlinks are located, you can use the **Hyperlink** command to edit the hyperlinks and related data. Select the hyperlinked object and start the **Hyperlink** command. When the **Edit Hyperlink** dialog box appears (it looks identical to the **Insert Hyperlink** dialog box), make the changes and choose OK.

Removing Hyperlinks from Objects

To remove a URL from an object, use the **Hyperlink** command on the object. When the **Edit Hyperlink** dialog box appears, choose the **Remove Hyperlink** button.

To remove a rectangular area hyperlink, you can simply use the **Erase** command; select the rectangle and AutoCAD erases the rectangle. (Unlike in Release 14, AutoCAD no longer purges the URLLAYER layer.)

As an alternative, you can use the **-Hyperlink** command's **Remove** option, as follows:

> Command: **-hyperlink**
>
> Enter an option [Remove/Insert] <Insert>: **r**
>
> Select objects: **All**
>
> I found Select objects: (Press ENTER.)
>
> I. www.autodesk.com
>
> 2. www.autodeskpress.com
>
> Enter number, hyperlink, or * for all: **I**
>
> Remove, deleting the Area.
>
> I hyperlink deleted.

HYPERLINKS OUTSIDE AUTOCAD

The hyperlinks you place in the drawing are also available for use outside of AutoCAD. The Web browser makes use of the hyperlink(s) when the drawing is exported in DWF format. To help make the process clearer, here are the steps that you need to go through:

> *Step 1*
>
> Open a drawing in AutoCAD.
>
> *Step 2*
>
> Attach hyperlinks to objects in the drawing with the **Hyperlinks** command. To attach hyperlinks to areas, use the **-Hyperlink** command's **Area** option.
>
> *Step 3*
>
> Export the drawing in DWF format using the **Plot** command's "DWF ePlot PC2" plotter configuration.
>
> *Step 4*
>
> Copy the DWF file to your Web site.
>
> *Step 5*
>
> Start your Web browser with the **Browser** command.
>
> *Step 6*
>
> View the DWF file and click on a hyperlink spot.

THE DRAWING WEB FORMAT

To display AutoCAD drawings on the Internet, Autodesk invented a new file format called drawing Web format (DWF). The DWF file has several benefits and some drawbacks over DWG files. The DWF file is compressed to make it as much as eight times smaller than the original DWG drawing file so that it takes less time to transmit over the Internet, particularly with relatively slow telephone modem connections. The DWF format is more secure; because the original drawing is not being displayed; another user cannot tamper with the original DWG file.

However, the DWF format has some drawbacks:

- You must go through the extra step of translating from DWG to DWF.

- DWF files cannot display rendered or shaded drawings.

- DWF is a flat 2D-file format; therefore, it does not preserve 3D data, although you can export a 3D view.

- AutoCAD itself cannot display DWF files.

- DWF files cannot be converted back to DWG format without using file translation software from a third-party vendor.

- Earlier versions of DWF did not handle paper space objects (version 2.x and earlier), or linewidths and non-rectangular viewports (version 3.x and earlier).

To view a DWF file on the Internet, your Web browser needs a *plug-in*—a software extension that lets a Web browser handle a variety of file formats. Autodesk makes the DWF plug-in freely available from its Web site at http://www.autodesk.com/whip. It's a good idea to regularly check for updates to the DWF plug-in, which is updated about twice a year.

Other DWF Viewing Options

In addition to *WHIP!* with a Web browser, Autodesk provides two other options for viewing DWF files.

CADViewer Light is designed to be a DWF viewer that works on all operating systems and computer hardware because it is written in Java™. As long as your Windows, Macintosh®, or UNIX computer has access to Java (included with most Web browsers), you can view DWF files.

Volo View Express is a stand-alone viewer that views and prints DWG, DWF, and DXF files. Both products can be downloaded free from the Autodesk Web site.

Viewing DWG Files

To view AutoCAD DWG and DXF files on the Internet, your Web browser needs a DWG-DXF plug-in. At the time of publication, the plug-ins were available from the following vendors:

Autodesk: http://www.autodesk.com/whip

SoftSource: http://www.softsource.com/

California Software Labs: http://www.cswl.com

Cimmetry systems: http://www.cimmetry.com

CREATING A DWF FILE

To create a DWF file from AutoCAD 2000, follow these steps:

1. Type the **Plot** command or choose **File | Plot** from the menu bar. Notice that AutoCAD displays the **Plot** dialog box.

2. In the **Name** list box (found in the Plotter Configuration area), choose "DWF ePlot pc3".

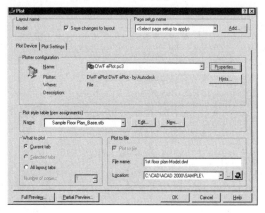

Figure 29.20 *Plot Dialog Box*

3. Choose the **Properties** button. Notice the **Plotter Configuration Editor** dialog box.

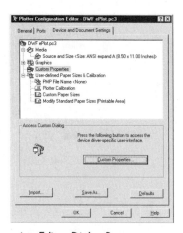

Figure 29.21 *Plotter Configuration Editor Dialog Box*

4. Choose Custom Properties in the tree view. Notice the **Custom Properties** button, which appears in the lower half of the dialog box.

5. Choose the **Custom Properties** button. Notice the **DWF Properties** dialog box.

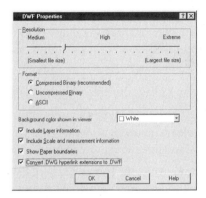

Figure 29.22 *DWF Properties Dialog Box*

Resolution: Unlike AutoCAD DWG files, which are based on real numbers, DWF files are saved using integer numbers. The **Medium** precision setting saves the drawing using 16-bit integers, which is adequate for all but the most complex drawings. **High** resolution saves the DWF file using 20-bit integers, while **Extreme** resolution saves using 32-bit integers. Figure 29.23 shows an extreme close-up of a drawing exported in DWF format. The portion at left was saved at medium resolution and shows some bumpiness in the arcs. The portion at right was saved at extreme resolution.

Figure 29.23 *DWF Output in Medium Resolution (at left) and Extreme Resolution (at right)*

More significant is the difference in file size. When I created a DWF file from the "1st floor plan.dwg" file (342 KB), the medium resolution DWF file was just 27 KB, while the extreme resolution file became 1,100 KB—about 40 times larger. That means that the medium resolution DWF file transmits over the Internet 40 times faster, a significant saving in time.

Format: Compression further reduces the size of the DWF file. You should always use compression, unless you know that another application cannot decompress the DWF file. Compressed binary format is seven times smaller than ASCII format (7,700 KB). Again, that means the compressed DWF file transmits over the Internet seven times faster than an ASCII DWF file.

Other Options: The following other options are available. In most cases, you would turn on all options.

> **Background Color Shown in Viewer:** while white is probably the best background color, you can choose any of AutoCAD's 255 colors.

> **Include Layer Information:** This option includes layers, which allows you to toggle layers off and on when the drawing is viewed in the Web browser.

> **Include Scale and Measurement Information:** This allows you to use the **Location** option in the Web browser *WHIP!* plug-in to show scaled coordinate data.

> **Show Paper Boundaries:** Includes a rectangular boundary at the drawing's extents.

> **Convert .DWG Hyperlink Extensions to .DWF:** Includes hyperlinks in the DWF file.

6. Choose OK to exit the dialog boxes back to the **Plot** dialog box.

7. Accept the DWF file name listed in the File name text box, or type a new name. If necessary, change the location that the file will be stored in. Note the two buttons: one has an ellipsis (…), which displays the **Browse for Folder** dialog box. The second button brings up AutoCAD's internal Web browser, a simplified version of the Microsoft's product.

8. Choose OK to save the drawing in DWF format.

VIEWING DWF FILES

To view a DWF file, you need to use a Web browser with a special plug-in that allows the browser to correctly interpret the file. (Remember: you cannot view a DWF file in AutoCAD.) Autodesk has named their DWF plug-in *WHIP!*, short for "Windows HIgh Performance."

Autodesk updates the DWF plug-in approximately twice a year. Each update includes some new features. The following is a summary of functions all versions of the DWF plug-in allow you to perform:

- View in a browser DWF files created by AutoCAD.

- Right-click the DWF image to display a cursor menu with commands.

- Use the real-time pan and zoom features to change the view of the DWF file as quickly as a drawing file in AutoCAD.

- Use embedded hyperlinks to display other documents and files.

- Compress files—the DWF file appears in your Web browser much faster than the equivalent DWG drawing file would.

- Print the DWF file alone or along with the entire Web page.

- View drawings using either Netscape Communicator/Navigator or Microsoft Internet Explorer. A separate browser-specific plug-in is required, depending on which browser you use.

- "Drag and drop" a DWG file from a Web site into AutoCAD as a new drawing or as a block.

- View a named view stored in the DWF file.

- Specify a view using *X,Y* coordinates.

- Toggle layers off and on.

If you don't know whether the DWF plug-in is installed in your Web browser, choose **Help | About Plug-ins** from the browser's menu bar. You may need to scroll through the list of plug-ins to find something like this:

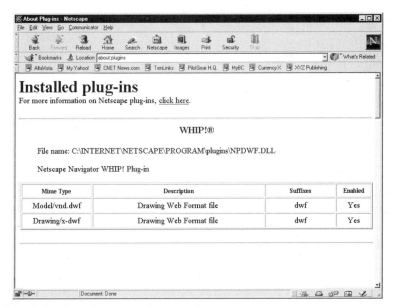

Figure 29.24 *Checking the Status of Browser Plug-ins*

If the plug-in is not installed, or it is an older version, then you need to download it from Autodesk's Web site at: http://www.autodesk.com/whip

For Netscape users, the file is quite large at 3.5 MB and takes about a half-hour to download using a typical 28.8 Kbaud modem. The file you download from the Autodesk Web site is either a self-installing file (has a .JAR extension) or a self-extracting installation file with a name such as Whip4.exe. After the download is complete, follow the instructions on the screen.

For Internet Explorer users, the DWF plug-in is an ActiveX control. It is automatically installed into Explorer when you set up AutoCAD 2000.

DWF PLUG-IN COMMANDS

To display the DWF plug-in's commands, position the cursor over the DWF image and click the right mouse button. This displays a cursor menu with commands, such as Pan, Zoom, and Named Views. To choose a command, place the cursor over the command name and click the left mouse button.

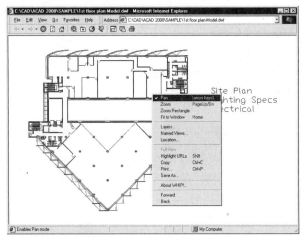

Figure 29.25 *WHIP!'s Cursor Menu*

Pan is the default command. Click the left mouse button and move the mouse. The cursor changes to an open hand, signaling that you can pan the view around the drawing. This is exactly the same as real-time panning in AutoCAD. Naturally, panning only works when you are zoomed in; it does not work in Full View mode.

Zoom is like the Zoom command in AutoCAD. The cursor changes to a magnifying glass. Hold down the left mouse button and move the cursor up (to zoom in) and down (to zoom out).

Zoom Rectangle is the same as Zoom Window in AutoCAD. The cursor changes to a plus sign. Click the left mouse button at one corner, then drag the rectangle to specify the size of the new view.

Fit to Window is the same as AutoCAD's Zoom Extents command. You see the entire drawing.

Layers displays a non-modal dialog box, which lists all layers in the drawing. (A *non-modal* dialog box remains on the screen; unlike AutoCAD's modal dialog boxes, you do not need to dismiss a non-modal dialog box to continue working.) Click a layer name to toggle its visibility between on (yellow light bulb icon) and off (blue light bulb).

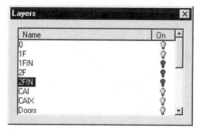

Figure 29.26 *WHIP!'s Layer Dialog Box*

The **Named Views** works only when the original DWG drawing file contained named views created with the **View** command. Choosing **Named Views** displays a non-modal dialog box that allows you to select a named view. Click a named view to see it; click the small x in the upper-right corner to dismiss the dialog box.

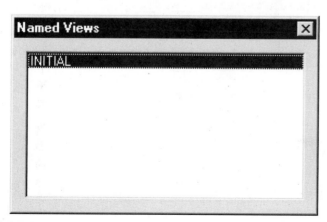

Figure 29.27 *WHIP!'s Named Views Dialog Box*

Location displays a non-modal dialog box that shows the X,Y,Z coordinates of the cursor location. Because DWF files are only 2D, the Z coordinate is always 0.

Figure 29.28 *WHIP!'s Location Dialog Box*

Full View causes the Web browser to display the DWF image as large as possible all by itself. This is useful for increasing the physical size of a small DWF image. When you have finished viewing the large image, right-click and select the **Back** command or click the browser's **Back** button to return to the previous screen.

Highlight URLs displays a flashing gray effect on all objects and areas containing a hyperlink. This helps you see where the hyperlinks are. To read the associated URL, pass the cursor over a highlighted area and look at the URL on the browser's status line. (A shortcut is to hold down the Shift key, which causes the URLs to highlight until you release the key). Clicking the URL opens the document with its associated application.

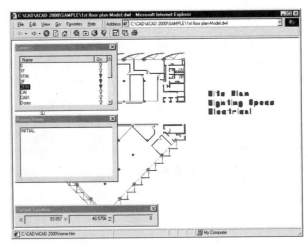

Figure 29.29 *Highlighting Hyperlinks*

Copy copies the DWF image to the Windows Clipboard in EMF (enhanced Metafile or "Picture") format.

Print prints the DWF image alone. To print the entire Web page (including the DWF image), use the browser's **Print** button.

Save As saves the DWF file in three formats to your computer's hard drive:

- DWF

- BMP (Windows bitmap)

- DWG (AutoCAD drawing file) works only when a copy of the DWG file is available at the same subdirectory as the DWF file.

You cannot use the **Save As** command until the entire DWF file has been transmitted to your computer.

About WHIP displays information about the DWF file, including DWF file revision number, description, author, creator, source file name, creation time, modification time, source creation time, source modification time, current view left, current view right, current view bottom, and current view top.

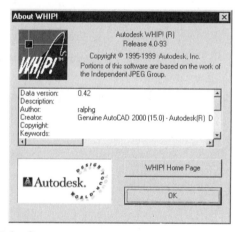

Figure 29.30 *About Dialog Box*

Forward is almost the same as clicking the browser's **Forward** button. When the DWF image is in a frame, only that frame goes forward.

Back is almost the same as clicking the browser's **Back** button. It works differently when the DWF image is displayed in a frame.

Drag and Drop from Browser to Drawing

The DWF plug-in allows you to perform several "drag and drop" functions. *Drag and drop* is when you use the mouse to drag an object from one application to another.

Hold down the Ctrl key to drag a DWF file from the browser into AutoCAD. Recall that AutoCAD cannot translate a DWF file into DWG drawing format. For this reason, this form of drag and drop only works when the originating DWG file exists in the same subdirectory as the DWF file.

AutoCAD executes the **-Insert** command, as follows:

> Command: _.-insert Enter block name or[?: **C:\CAD\ACAD 2000\SAMPLE\1st floor plan.dwg**
>
> Specify insertion point or [Scale/X/Y/Z/Rotate/PScale/PX/PY/PZ/PRotate]: **0,0**
>
> Enter X scale factor, specify opposite corner, or [Corner/XYZ] <1>: *(Press* ENTER.*)*
>
> Enter Y scale factor <use X scale factor>: *(Press* ENTER.*)*
>
> Specify rotation angle <0>: *(Press* ENTER.*)*

To see the inserted drawing, you may need to use the **Zoom Extents** command.

Another drag and drop function is to drag a DWF file from the Windows Explorer (or File Manager) into the Web browser. This causes the Web browser to load the DWF plug-in and then display the DWF file. Once displayed, you can execute all of the commands listed in the previous section.

Finally, you can also drag and drop a DWF file from Windows Explorer (in Windows 95/98) or File Manager (in Windows NT) into AutoCAD. This causes AutoCAD to launch another program that is able to view the DWF file. This does not work if you do not have other software on your computer system capable of viewing DWF files.

EMBEDDING A DWF FILE

To let others view your DWF file over the Internet, you need to embed the DWF file in a Web page. Here are the steps to embedding a DWF file in a Web page.

> **Step 1.** This hypertext markup language (HTML) code is the most basic method of placing a DWF file in your Web page:
>
> <embed src="filename.dwf">
>
> The **<embed>** tag embeds an object in a Web page. The **src** option is short for "source." Replace filename.dwf with the URL of the DWF file. Remember to keep the quotation marks in place.
>
> **Step 2.** HTML normally displays an image as large as possible. To control the size of the DWF file, add the Width and Height options:
>
> <embed width=800 height=600 src="filename.dwf">

The **Width** and **Height** values are measured in pixels. Replace 800 and 600 with any appropriate numbers, such as 100 and 75 for a "thumbnail" image, or 300 and 200 for a small image.

Step 3. To speed up a Web page's display speed, some users turn off the display of images. For this reason, it is useful to include a description, which is displayed in place of the image:

```
<embed width=800 height=600 name=description
    src="filename.dwf">
```

The **Name** option displays a textual description of the image when the browser does not load images. You might replace description with the DWF file name.

Step 4. When the original drawing contains named views created by the **View** command, these are transferred to the DWF file. Specify the initial view for the DWF file:

```
<embed width=800 height=600 name=description
    namedview="viewname" src="filename.dwf">
```

The **namedview** option specifies the name of the view to display upon loading. Replace **viewname** with the name of a valid view name. When the drawing contains named views, the user can right-click on the DWF image to get a list of all named views.

As an alternative, you can specify the 2D coordinates of the initial view:

```
<embed width=800 height=600 name=description view="0,0 9,12"
    src="filename.dwf">
```

The **View** option specifies the *X,Y* coordinates of the lower-left and upper-right corners of the initial view. Replace 0,0 9,12 with other coordinates. Since DWF is 2D only, you cannot specify a 3D viewpoint. You can use **View** or **Namedview**, but not both.

Step 5. Before a DWF image can be displayed, the Web browser must have the DWF plug-in called *WHIP!*. For users of Netscape Communicator/Navigator, you must include a description of where to get the *WHIP!* plug-in when the Web browser is lacking it.

```
<embed pluginspage=http://www.autodesk.com/products/autocad/
    whip/whip.htm width=800 height=600 name=description
    view="0,0 9,12" src="filename.dwf">
```

The **pluginspage** option describes the page on the Autodesk Web site where the *WHIP!* DWF plug-in can be downloaded.

The code listed above works for Netscape Communicator/Navigator. To provide for users of Internet Explorer, the following HTML code must be added:

```
<object classid ="clsid:B2BE75F3-9197-11CF-ABF4-08000996E931"
        codebase ="ftp://ftp.autodesk.com/pub/autocad/plugin/
        whip.cab#version=2,0,0,0" width=800 height=600>

<param name="Filename" value="filename.dwf">

<param name="View" value="0,0 9,12">

<param name="Namedview" value="viewname">

<embed pluginspage=http://www.autodesk.com/products/autocad/
        whip/whip.htm width=800 height=600 name=description
        view="0,0 9,12" src="filename.dwf">

</object>
```

The two **<object>** and three **<param>** tags are ignored by Netscape Communicator/Navigator; they are required for compatibility with Internet Explorer. The **classid** and **codebase** options tell Explorer where to find the plug-in. Remember that you can use **View** or **Namedview**, but not both.

Step 6. Save the HTML file.

CHAPTER REVIEW

1. Can you launch a Web browser from within AutoCAD?

2. What does DWF mean?

3. What is the purpose of DWF files?

4. What is URL is short for?

5. Which of the following URLs are valid:

   a. @   www.autodesk.com

   b. @   http://www.autodesk.com

   c. @   Both of the above

   d. @   None of the above

6. FTP is short for: _____ _____ _____

7. What is a "local host"?

8. Are hyperlinks active in an AutoCAD 2000 drawing?

9. The purpose of URLs is to let you create _____ between files.

10. When you attach a hyperlink to a block, the hyperlink data is _____ if you scale the block unevenly, stretch the block, or explode it.

11. Can you can attach a URL to any object?

12. The **-Hyperlink** command allows you to attach a hyperlink to
    _____ and _____.

13. To see the location of hyperlinks in a drawing, use the _____
    command.

14. Rectangular (area) hyperlinks are stored on
    layer: _____.

15. Compression in the DWF file causes it to take (less, more, the same) time to trans-
    mit over the Internet.

16. A DWF is created from a _____ file through the
    _____ command.

17. What does a "plug-in" let a Web browser do?

18. Can a Web browser view DWG drawing files over the Internet?

19. _____ is an HTML tag for embedding graphics in a Web page.

20. A file being transmitted over the Internet via a 28.8 Kbps modem takes about
    _____ minutes per megabyte.

Introduction to AutoCAD 3D

One of the most exciting parts of AutoCAD is 3D (three-dimensional) drafting. Before you start constructing 3D drawings, you must become familiar with concepts of three-dimensional drawing construction. After completing this chapter, you will be able to

- Use the three-dimensional coordinate system

- Distinguish 3D from perspective

- Apply the concepts used with clipping planes

INTRODUCTION

AutoCAD has the capability to produce 3D drawings in true and perspective views. You can use these capabilities interactively with 2D drawings.

The following chapters explain the use of AutoCAD 3D:

Chapter 30, this chapter, introduces you to 3D drawing.

Chapter 31 "Viewing 3D Drawings" covers the methods of creating 3D views, which is important while constructing the drawing and for modeling the final product.

Chapter 32 "The User Coordinate System" covers user-defined coordinate systems, an important tool in creating 3D drawings.

Chapters 33 through 37 explore the many methods of drawing in 3D.

Chapter 38 "Making a Realistic Rendering" shows how to make a realistic picture of your 3D model.

These chapters are organized in a learning order. You can achieve the best result by reviewing the information in the order presented. You need to become skillful in creating 3D views before learning to draw in 3D, since much of the drawing construction process is performed while in 3D views.

HOW TO APPROACH 3D

I assume that you have a good working knowledge of 2D drawing with AutoCAD. Using AutoCAD 3D requires expertise in many of the commands you learned in the previous chapters. If you are not familiar with commands or concepts in certain areas, review those in the previous chapters.

Although AutoCAD 3D uses many of the 2D commands you are familiar with, the approach to 3D drawing requires knowledge of some new concepts. The primary difference is the X,Y,Z coordinate system. Other concepts—such as camera position, target point, clipping plane, and User Coordinate System—are used to create and view 3D drawings. In this chapter, and in the next several chapters, you will learn about these concepts.

Take time to read and study these concepts, and to understand them. Practice each concept with simple shapes before proceeding to more advanced problems.

Three-dimensional drawing takes time and practice to learn, but in time, the results will be well worth the effort.

3D THEORY

Let's look at some of the concepts associated with three-dimensional drawing.

COORDINATE SYSTEM

To comprehend the construction of 3D drawings, you need to understand the concept of the X,Y,Z coordinate system. This is the same X,Y (Cartesian) coordinate system you work with when constructing a 2D drawing, with the Z axis representing the "height" of the objects.

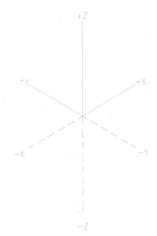

Figure 30.1 X,Y,Z Axes

Before you proceed to learn about the *Z* axis, it is necessary to understand that the 2D drawings you have constructed have used the *X,Y,Z* coordinate system all along. You have drawn only in the *X,Y* plane of this system. We now expand to also draw in the other planes that include the *Z* axis. Figure 30.1 shows the different axes. You may, at this point, wish to review the section on coordinates in Chapter 7 "Creating the Drawing."

Here is one way to think of the *X,Y,Z* coordinate system. Think of the "plan" view of your drawing as lying in the *X,Y* plane. Think of a room. Look into the corner: the floor is the *X,Y* plane, while the walls go up the *Z* coordinate. One of the two walls is the *X,Z* plane; that would be the wall over the *X* axis. The other wall is the *Y,Z* plane.

When working with a 3D drawing, you must draw in a manner that can be thought of as "up from the page." Let's take an example of a box with dimensions of 4x3x2 (in current units). Figure 30.2 shows the box sitting in the positive quadrants of an *X,Y,Z* coordinate system, with four units along the *X* axis, three units along the *Y* axis and two units along the *Z* axis.

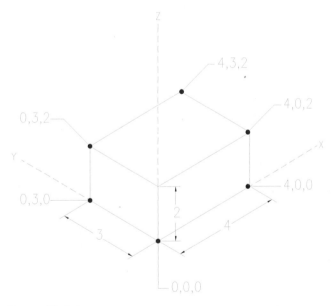

Figure 30.2 *Box on 3D Axis*

As in a 2D drawing, each intersection has coordinates. In the case of a 3D drawing, the coordinates are represented in an *X,Y,Z* format, such as 4,0,2. Notice the 3D coordinates listed in the figure.

3D VERSUS PERSPECTIVE

A 3D view can be perspective or non-perspective. AutoCAD is capable of displaying your drawings in either true 3D or 3D perspective views.

A *perspective view* shows the drawing in 3D with faces diminishing to a vanishing point and all forms shortened in distance. This is similar to the way a human eye perceives the environment. Figure 30.3 shows 3D perspective views.

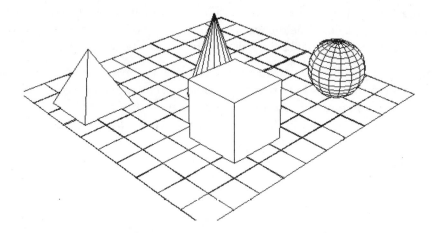

Figure 30.3 *A 3D Perspective View*

A *non-perspective view* displays the drawing with "true lengths" for all lines in the drawing, similar to an isometric drawing. When you create a 3D drawing, you mostly work in non-perspective mode.

CLIPPING PLANES

Some applications require that a "cutaway" type of view be displayed for the purpose of showing parts of the drawing that are otherwise concealed. This can be accomplished by a *clipping plane*.

You can think of a clipping plane as an imaginary plane, or surface, that "cuts" through the drawing and eliminates every part of the drawing either in front of or behind the plane. Figure 30.4 shows two clipping planes placed in a perspective drawing. Figure 30.5 illustrates the effect of the clipping plane cutting the objects in the view.

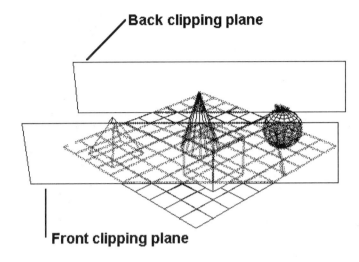

Figure 30.4 *Setting Up Front and Back Clipping Planes*

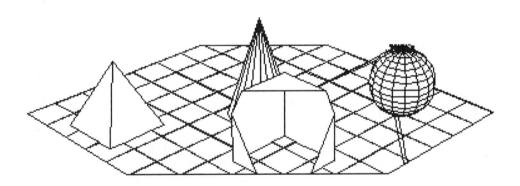

Figure 30.5 *Result of Clipping Planes*

EXERCISE

Refer to Figure 30.6 and write the coordinates for each corner in the space provided.

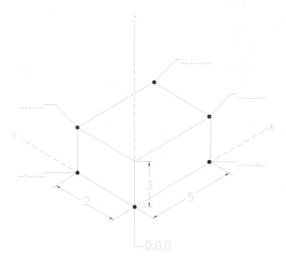

Figure 30.6 *3D Coordinate Exercise*

CHAPTER REVIEW

1. What are the axes in the Cartesian coordinate system?

2. From which direction from a page does the Z axis normally project?

3. What is the difference between normal 3D and *perspective*?

4. What is a *clipping plane*?

5. How many clipping planes are in AutoCAD? Name each.

6. What form of 3D most closely approximates the view of the human eye?

7. In AutoCAD, how would you show a coordinate that has values of X=3, Y=5, Z=9?

8. Which axes are two-dimensional drawings constructed in?

9. Which axis would normally represent the height of an object?

10. What does *true length* mean when you refer to a 3D drawing?

CHAPTER 31

Viewing 3D Drawings

To construct a 3D drawing, you must be capable of manipulating the drawing in the three planes so you can work with the different parts. After the drawing is completed, you can use the viewing commands to observe the drawing from different viewpoints. After completing this chapter, you will be able to

- Use commands that change the view of 3D drawings

- Place the imaginary camera and target to obtain the view you desire

- Manipulate the **3dOrbit** command to control the look of the 3D view

- Produce shaded views of the 3D objects, or with hidden lines removed.

INTRODUCTION

In this chapter, you learn about the following commands:

VPoint

Allows you to change the 3D viewpoint by variety of methods.

Plan

Displays the drawing in plan view.

DdVpoint

The dialog box version of the **VPoint** command.

3dOrbit

Dynamically changes the 3D view and creates perspective views.

Shade

Creates a quick, shaded view of the drawing.

Learning the methods for modeling your drawing in 3D is important not only for presentation purposes but also during the drawing process. In this chapter, we review the methods of displaying a drawing in 3D.

METHODS OF VIEWING 3D DRAWINGS

There are two primary methods of creating a 3D view of your drawings: through the **VPoint** or the **3dOrbit** command. The **VPoint** command can be referred to as a "static" method, while the **3dOrbit** command is a "dynamic" method of viewing. The **VPoint**'s static method is a simple shortcut method for creating quick viewpoints. The **3dOrbit**'s dynamic method allows you to see the drawing as it is being rotated into view.

SETTING THE 3D VIEWPOINT

The **VPoint** command is used to select a viewpoint for the current viewport by setting a direction and elevation of view.

When you select **View | 3D Views | VPOINT** from the menu bar (or type the **VPoint** command), the following prompt appears:

> Command: **vpoint**
>
> Current view direction: VIEWDIR=0.0000,0.0000,1.0000
>
> Specify a view point or [Rotate] <display compass and tripod>:

The *X, Y, Z* coordinates show the current 3D view coordinates. As the prompt indicates, they are stored in the **ViewDir** system variable.

The options allow you to define the viewpoint in three ways: *X,Y,Z* coordinates, rotation angles, or through interactive compass. Let's look at each.

DEFINING A VIEW BY COORDINATES

You can enter an *X,Y,Z* coordinate to specify a point from which to view the drawing. The coordinate is the point to look from, with the coordinate of 0,0,0 always the point to look at.

A viewpoint of 0,0,1 looks at the drawing in plan view (directly down along the *Z* axis). A negative coordinate value places the viewpoint at the negative end of the axis. Thus, a viewpoint of 0,0,-1 would look up at the drawing directly from below.

To view a 3D drawing isometric mode, use coordinates 1,–1,1 like this:

> Command: **vpoint**
>
> Current view direction: VIEWDIR=0.0000,0.0000,1.0000
>
> Specify a view point or [Rotate] <display compass and tripod>:
> **1,–1,1**

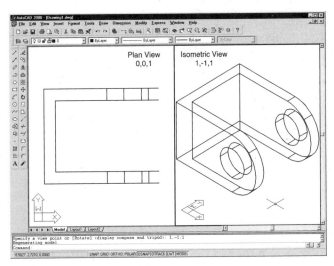

Figure 31.1 *Viewpoint Through 3D Coordinates*

DEFINING A VIEW BY AXES

When you respond to the prompt by pressing ENTER on the keyboard, the drawing screen temporarily displays a special axes diagram. The visual diagram consists of a tripod, representing the *X*, *Y*, and *Z* axes and a flattened globe. Figure 31.2 shows the tripod and globe.

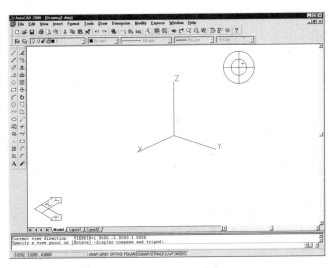

Figure 31.2 *Vpoint Axes (at center) and Globe (at right)*

To set the view, move the mouse. This causes the tripod to rotate, representing the rotation of each axis. It helps to think of your drawing as lying in the X,Y plane of the tripod when visualizing the desired view.

The compass-appearing icon to the right is a 2D representation of a globe. The center point is the North Pole, the middle circle is the equator, and the outer circle is the South Pole. The four quadrants of the globe can be thought of as the direction of the view. For example, the lower right quadrant would produce a view that is represented from the lower left of the plan view. Moving above the equator produces an "above-ground" view; below the equator, a "below-ground" view.

 Note: When you select a point outside the globe, unexpected results may occur. There are a few selection locations known to crash AutoCAD. Always click on or inside the globe.

DEFINING A VIEW BY ROTATION

The **Rotate** option allows you to specify the 3D view in terms of angles.

    Command: **vpoint**

    Current view direction: VIEWDIR=1.0000,-1.0000,1.0000

    Specify a view point or [Rotate] <display compass and tripod>: **R**

    Enter angle in XY plane from X axis <0>: *(Enter an angle.)*

    Enter angle from XY plane <90>: *(Enter an angle.)*

The first prompt is for an angle *in* the X,Y plane, starting from the X axis. This rotates the view about the Z axis.

The second prompt is for an angle *from* the X,Y plane. This rotates the view above or below the plane.

If you select points in response to the two prompts, AutoCAD draws a drag line showing the view direction.

RETURNING TO PLAN VIEW

When you wish to return to the plan view, type **Plan** and press ENTER.

As an alternative, use the **VPoint 0,0,0** command; for the **VPoint** command's **Rotate** option, specify 0 and 90 degrees.

Don't be afraid to experiment with different views. Try to associate the globe and tripod with the results of the 3D view. With a little practice, you will be able to obtain the results you want.

SETTING VIEW BY DIALOG BOX

You can use a dialog box to select the view angle of your 3D drawing. Select **View | 3D Views | Viewpoint Presets** from the menu bar, or type the **DdVpoint** command. Select the degree settings shown in the dialog box illustration to change to the preset degrees. You can also click inside the circle/arc for a fine-tuned selection.

As an alternative, you can type the degrees in the text box. Choose the **Set to Plan View** button for returning to the plan view.

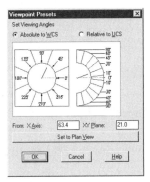

Figure 31.3 *Viewpoint Presets Dialog Box*

Another method is to use the Viewports dialog box, as described in Chapter 23 "Working with Viewports and Layouts." From the menu bar, select **View | Viewports | New Viewports**.

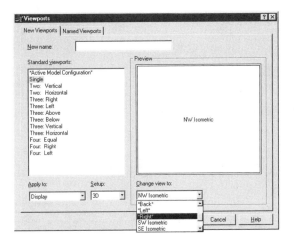

Figure 31.4 *Viewports Dialog Box*

When the dialog box appears, do the following:

1. In the **Standard Viewports** list, select **Single**.

2. In the **Apply** to list box, select Display.

3. In the **Setup** list box, select 3D.

4. In the **Change view to** list box, select one of the predefined views, such as **Right** or **SW Isometric**.

5. Choose **OK**. Notice that AutoCAD displays the 3D viewpoint.

DYNAMIC 3D VIEWING

The **3dOrbit** command is a powerful tool for viewing a 3D drawing dynamically. This means you can rotate the drawing and see the results as you go.

The **3dOrbit** command is also used to set *clipping planes* and set the distance from which the drawing is to be viewed, and other **3dOrbit** functions allow you to precisely control the appearance of the 3D view.

Select **View | 3D Orbit** from the menu bar or type **3dOrbit**. AutoCAD switches to dynamic viewing mode, and prompts you, "Press ESC or ENTER to exit, or right-click to display shortcut menu."

Notice the changes to the drawing area:

- The UCS icon changes from a flat, black 2-axis representation to a colorful, 3D-looking, 3-axis icon. The X axis is red; Y axis is green, and the Z axis is blue.

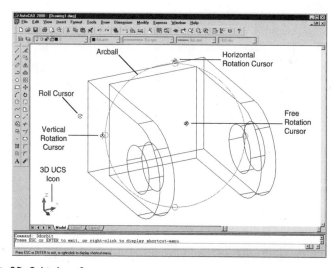

Figure 31.5 *3D Orbit Interface*

- A large green circle appears, with four smaller green circles. This is called the *arcball*.

- The cursor changes to circular arrow. This indicates that AutoCAD is in interactive rotation mode.

- The drawing loses its lineweight. That's because you are now viewing in "3D wireframe" view.

In the following sections, you learn about these and other aspects to 3D orbiting.

 Note: You cannot edit objects while the **3dOrbit** command is active.

THE ARCBALL

The *arcball* consists of one large circle and four smaller circles located at its quadrant points. The purpose of the arcball is to help orient you in 3D space.

The center of the arcball is the target point. The *target point* is the point at which you are looking; the point from which you are looking (that is, your eyeball) is called the *camera point*. (The center of the objects you're viewing is *not* the target point.) In AutoCAD, the target of the view stays stationary; the camera location moves around the target.

Move the cursor around the arcball to see it change shape. These cursor shapes indicate the direction in which the view will rotate. To make the view rotate, *drag* the cursor across the drawing. (To drag means to hold down the left mouse button and move the mouse.)

Cursor Location	Cursor Shape	Meaning
Outside arcball	Circular arrow	Rolls the view about the axis perpendicular to the view
Inside arcball	Double-circle	Rotate object horizontally, vertically, or diagonally
Left or right circle	Horizontal ellipse	Rotates the view about the vertical axis
Top or bottom circle	Vertical ellipse	Rotates the view about the horizontal axis

THE CURSOR MENU

Right-click in 3D Orbit mode to display a cursor menu. This menu lists all the options you have available in this mode: **Exit, Pan, Zoom, Orbit, More, Projection, Shading Modes, Visual Aids, Reset View, Preset Views**, and **Saved Views**. The

Orbit option allows you to rotate the 3D model in real time, as described in the previous section.

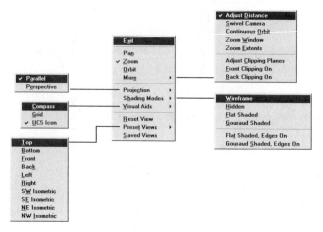

Figure 31.6 *3D Orbit Cursor Menu*

Projection Modes

Parallel projection is the mode you are most used to drawing in. Two parallel lines never converge at a single point; the drawing does not appear distorted.

Perspective projection causes parallel lines converge at one point, called the *vanishing point*. Objects recede into the distance; objects can become very distorted, as shown by Figure 31.7.

More Options

Adjust Distance moves the camera (your viewpoint) closer or farther away from the target point.

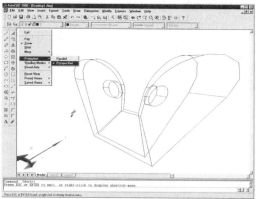

Figure 31.7 *Perspective Mode*

Swivel Camera turns the camera.

Continuous Orbit is pretty cool: drag the cursor across the screen, then let go: the view rotates continuously.

Zoom Window lets you drag a window around the area you want to zoom in to, like the **Zoom** command's **Window** option.

Zoom Extents sizes the view to display all objects; it may not work correctly in perspective mode.

Adjust Clipping Planes

Clipping planes are used to cut away a part of the 3D drawing. For example, you may want to cut away part of a building to see the interior. (Note that clipping planes are also used to control the fog effect of the **Render** command; see Chapter 38.)

You can think of clipping planes as transparent walls that slice through part of the 3D drawing and remove everything that is either in front of or behind them. The cutting plane is parallel to the viewing plane.

There is a *front* and a *back* clipping plane. Either or both planes can be positioned in the drawing. Positioning is accomplished with slider bars. The clipping planes can be positioned in either perspective or non-perspective mode.

The **Adjust Clipping Planes** option opens the **Adjust Clipping Planes** window. This independent window lets you move the front and back clipping planes. As you can see from Figure 31.8, clipping planes cut away the front and/or back of the object.

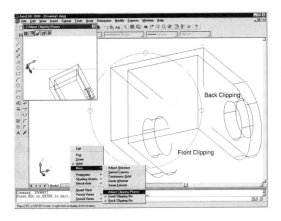

Figure 31.8 *Clipping Planes*

The toolbar of the **Adjust Clipping Planes** window has several buttons that assist you in changing the clipping plane:

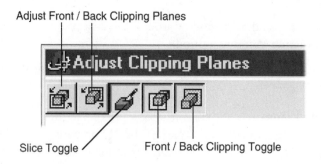

Figure 31.9 *Adjust Clipping Planes Toolbar*

Adjust Front Clipping

Choose the **Adjust Front Clipping** button to turn on the feature. With your cursor, move the black horizontal line to adjust the front clipping plane. As you move the line, you see the front of the object being clipped off (disappearing from view).

Adjust Back Clipping

Choose the **Adjust Back Clipping** button to turn it on; you can adjust only the front or the back, not both at the same time. With your cursor, move the green horizontal line to adjust the back clipping plane. As you move the line, you see the back of the object clipped off.

Create Slice

The **Create Slice** button locks both front and back clipping panes, causing them to move together. This allows you to "slice" through the objects. First adjust the front and back clipping planes independently, then do the slicing.

Front Clipping On

This button toggles the front clipping plane. When off, the object is not clipped; when on, the object can be clipped.

Back Clipping On

This button toggles the back clipping plane. When off, the object is not clipped; when on, the object can be clipped.

To close the **Adjust Clipping Planes** window, select the small x button in the upper right corner of the window. The effect of the clipping planes remains, even after the window is closed; you can rotate the object to see inside the object. As you rotate the object, it may disappear entirely, if it comes wholly in front of (or behind) a clipping plane.

Shading Modes

You can display the drawing in one of several shading modes, ranging from the traditional wireframe to Gouraud shading. You can turn on a shading mode, exit the **3dOrbit** command, and then continue to edit the drawing while it is fully shaded.

Shading modes are also controlled by the **View | Shade** menu selection and the **Shade** toolbar.

When you draw, the objects are displayed in shaded mode, as well—if appropriate. For example, a 2D line is drawn "unshaded." A 2D line with thickness is drawn shaded.

You cannot edit a drawing if perspective mode is on. AutoCAD complains, "You cannot point within a Perspective view."

The more advanced the shading, the slower AutoCAD is able to update the screen. Thus, while Gouraud shading is more realistic, it may slow down your operation of AutoCAD, especially with large drawings and/or slower computers.

Wireframe

This option displays objects in traditional "see through" wireframe mode. Note that lineweight is not displayed.

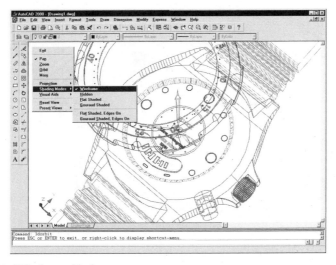

Figure 31.10 *Wireframe Mode*

Hidden

This option removes hidden lines from the drawing.

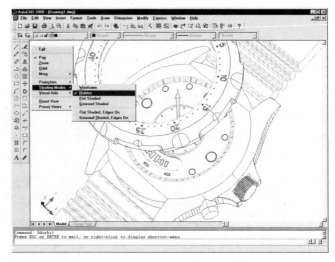

Figure 31.11 *Hidden Mode*

Flat Shaded

This option shades the faces of the 3D object in a uniform manner, giving a faceted appearance.

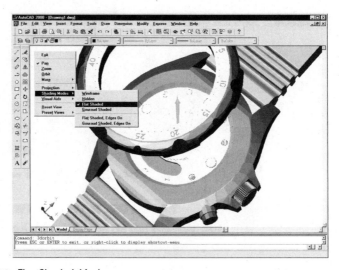

Figure 31.12 *Flat Shaded Mode*

Gouraud Shaded

This option applies smooth shading to the 3D object, giving a somewhat more realistic appearance. This option is named after Gouraud, the inventor of the shading algorithm.

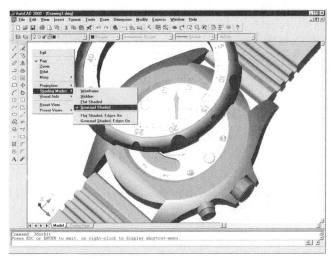

Figure 31.13 *Gouraud Shading Mode*

Flat Shaded, Edges On

This option flat shades the 3D object, and includes the hidden-line wireframe view. This allows 2D objects to appear.

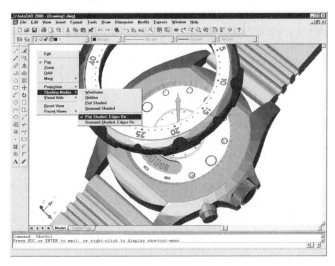

Figure 31.14 *Flat Shading Mode with Edges Turned On*

Gouraud Shaded, Edges On

This option smooth shades the 3D object and includes the hidden-line wireframe view. This allows 2D objects to appear.

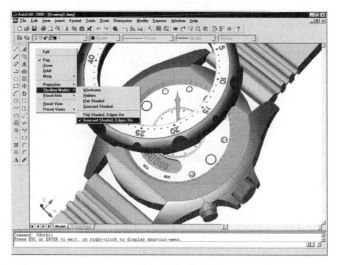

Figure 31.15 *Gouraud Shading Mode with Edges Turned On*

Visual Aids

While in the 3D Orbit mode, AutoCAD displays several visual aids to help you find your place in three-dimensional space—because it is easy to get "lost in space." In 3D Orbit mode, AutoCAD displays the arcball, described earlier. In addition, you can toggle the display of three other visual aids:

Compass

The *compass* is like a gimbal—three intersecting circles that represent the *X, Y,* and *Z* axes.

Grid

The *grid* is similar to the grid you have used in 2D drafting. Instead of an array of dots, the 3dOrbit Grid is a 2D array of lines (like graph paper) laying in the X,Y plane, constrained by the drawing limits. To change the parameters of the grid, use the **Snap and Grid** tab of the **DSettings** command. The spacing you specify indicates the number of *major* grid lines; ten *minor* grid lines are drawn horizontally and vertically between major grid lines.

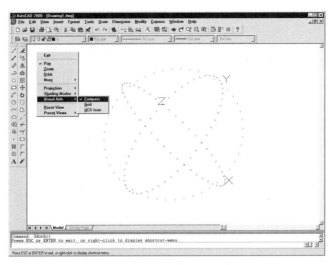

Figure 31.16 *The Compass*

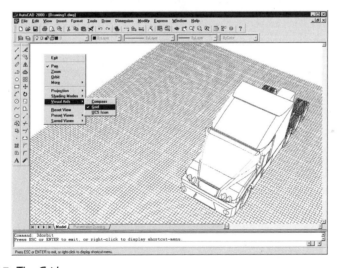

Figure 31.17 *The Grid*

UCS Icon

The UCS icon in 3D Orbit mode is a shaded, 3D UCS icon. The axes are labeled *X* (red), *Y* (green), and *Z* (blue). In perspective mode, the UCS icon takes on a perspective look. You can use the **UcsIcon** command (outside 3D Orbit mode) to control the position of the icon, either at the origin (0,0,0) or not at the origin of the drawing.

Figure 31.18 *The UCS Icon*

Other Options

The **Pan** option invokes real-time panning, just like the **Pan** command. The cursor changes to a hand; move the cursor to move the view.

The **Zoom** option switches to real-time zooming, as with the **Zoom** command. The cursor changes to a magnifying glass; move the cursor vertically to zoom in and out.

The **Reset View** option is perhaps the most important, because it is easy to get "lost" when viewing in 3D. It resets the view to what it was when you began the **3dOrbit** command.

Select **Preset Views** to display a list of standard views, such as Top, Bottom, and SE Isometric.

The **Saved Views** option displays a list of named views. You must have previously used the **View** command to create one or more named views. If no views are saved in the drawing, this option is not shown.

Select **Exit** to exit 3D Orbit viewing mode. As an alternative, you can press ESC. You must exit 3D Orbit viewing mode in order to continue editing the drawing. No other command operates while AutoCAD is in 3D Orbit mode.

CHAPTER REVIEW

1. What commands are used to generate a three-dimensional view of a drawing?
2. What is the **VPoint** command meant for?
3. What does the **3dOrbit** command do?
4. What is the *target point*?
5. Where is the *camera* located?
6. How do you turn on perspective mode? Turn off perspective mode?
7. Can you edit in perspective mode?
8. Can you edit in shaded mode?
9. What is meant by wireframe?
10. What does Gouraud shading mode do?

CHAPTER 32

The User Coordinate System

To effectively construct a 3D drawing, you must become proficient with the user coordinate system. After completing this chapter, you will be able to

- Comprehend the concept and use of the user coordinate system (UCS)
- Manipulate the UCS
- Identify the effect of the UCS on AutoCAD commands
- Save, restore, and manage UCS systems you create

INTRODUCTION

In this chapter, you learn about the following commands:

Ucs

Relocates the user-defined coordinate system in 3D space.

UcsIcon

Controls the display of UCS icon.

UcsFollow

System variable for setting UCS properties.

DdUcs

Dialog box lists defined UCSs in the drawing.

THE USER COORDINATE SYSTEM

When you draw in 2D plan view, you are working in the X,Y plane. Drawing is simple, since you are drafting in a single plane. Drawing in 3D, however, is more complicated. There may be many planes in which you wish to work. The *user coordinate system* is designed to make this process simpler. We'll refer to the user coordinate system as the "UCS."

To effectively draw in AutoCAD 3D, it is essential to understand the UCS. The primary purpose of the UCS is simplification of the 3D process. Mastering this system allows you to construct 3D drawings efficiently.

Let's consider the example of a sloped barn roof that has a graphic painted on it. It would be simple to draw the graphic in plan view, but placing it on the slope of the roof is quite different. Being able to draw on the slope of the roof as if it were in plan would be quite efficient—and that is the secret behind UCS.

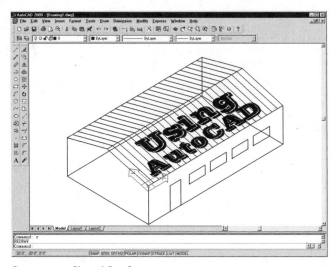

Figure 32.1 *Drawing on Sloped Roof*

The UCS allows you to do this. You can slope and rotate the view to change the plan view to match the slope of the roof. Once the view is rotated into place, you can draw on it (in this case, the roof) as if it were the 2D *X,Y* plane that you are used to.

It may help to think of the UCS as the *X,Y* plane that you are used to seeing when you draw in 2D plan view. Then imagine the ability to move, turn, and/or rotate it to any position on the 3D object you are drawing. Now you can redisplay the drawing with the new plan view "flat on the paper" and draw on it as though you were in plan.

Actually, this is exactly what you are doing. You can always return to "true" plan. The true plan is called the World Coordinate System or WCS, for short. The user coordinate system is a temporary, user-defined drawing plane to make drawing on the sides, slopes, etc. of the 3D object simpler. Let's continue and see how to manipulate the UCS to make 3D drawing simple.

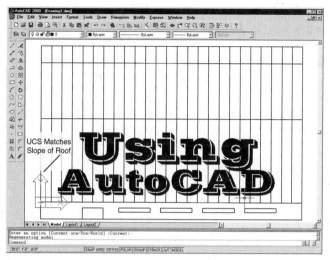

Figure 32.2 *Drawing in the new UCS*

THE UCS ICON

The *UCS icon* is used to denote the orientation of the current UCS. It shows the direction of the positive *X* and *Y* axes. The icon is displayed at the lower left of the drawing screen. The **UcsIcon** command is used to control certain display characteristics of the icon. We learn about the command later in this chapter.

Figure 32.3 *UCS Icon*

The icon provides you with three pieces of information: (1) the current *X,Y* axes orientation, (2) the origin of the UCS, and (3) the view direction.

When the icon contains a "W," the World Coordinate System is in effect.

When the icon contains a square, this indicates the drawing is being viewed from above. If the square is absent, the drawing is being viewed from below. (The "square" looks more like the intersection of the two arrows; see figure 32-3.)

A plus (+) in the center of the square indicates that the icon is located at the origin of the current UCS.

THE EDGE-OF-PLANE INDICATOR

It is possible to rotate the UCS (the view relative to the current UCS) to a position that looks head-on into the edge of the drawing. Such a position is almost useless as a drawing view. When such a view is current, AutoCAD displays a "broken pencil" icon. The indicator warns you that you cannot draw or edit in this view.

Figure 32.4 *Broken Pencil Icon*

CHANGING THE UCS

The UCS command is used to create and change named user coordinate systems. When you issue the **UCS** command, the following prompt is displayed.

> Command: **ucs**
>
> Current ucs name: ***WORLD***
>
> Enter an option
>   [New/Move/orthoGraphic/Prev/Restore/Save/Del/Apply/?/
>   World] <World>:

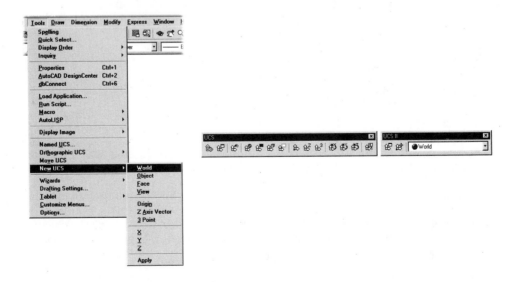

Let's look at each option of the **UCS** command. It may be helpful to use a 3D drawing from the CD-ROM, such as **Model1.Dwg**, to practice each option as you study it.

Note: The use of object snap is very helpful when you use many of the UCS options.

NEW

The **New** option allows you to create a new UCS. Typing **N** produces this prompt:

> Specify origin of new UCS or
> [ZAxis/3point/OBject/Face/View/X/Y/Z] <0,0,0>:

Specify Origin

The default option is to change the origin of the UCS, which is normally located at 0,0,0. The directions of the *X,Y,Z* axes are left unchanged. (The location of the UCS icon does not change unless the **Origin** option of the **UCSicon** command is on; the **UCSicon** command is covered later in this chapter.)

Enter an *X,Y,Z* coordinate point to describe a new origin point. If you enter only an *X,Y* point, the *Z* coordinate will be equal to the current elevation.

Alternatively, you can select a point on the screen to designate the new origin.

Object snap is especially useful in placing the new origin at a 3D point that can be described by a point on an object.

ZAxis

The **ZAxis** option defines a new UCS defined by two points: (1) the origin point and (2) a point anywhere along the *Z* axis. AutoCAD prompts:

> Specify new origin point <0,0,0>:

This prompt has the same effect as the **Specify Origin** option, discussed earlier.

> Specify point on positive portion of the Z axis <0,0,1>:

Responding with ENTER to the second prompt causes the *Z* axis of the new origin to be parallel to the UCS system.

3point

The **3point** option aligns a new UCS by three points that define: (1) the origin and (2) rotation of the *X,Y* plane of the new UCS. This is particularly helpful when you align the UCS with existing objects using object snap. Selecting the **3point** option displays the following prompts.

> Specify new origin point <0,0,0>:

Specify point on positive portion of the X-axis <1,0,0>:

Specify point on positive-Y portion of the UCS XY plane <0,1,0>:

The point given for each can be either a numerical coordinate entered from the keyboard or a point entered on the screen.

The point entered for the origin point prompt designates the new origin point. The second point defines the direction of the *X* axis. The third point will lie in the *X,Y* plane and define the direction of the positive *Y* axis. Simply pressing ENTER in response to any of the prompts will designate a value equal to the existing origin or direction.

Figure 32.5 shows the points entered to align the UCS with a barn roof.

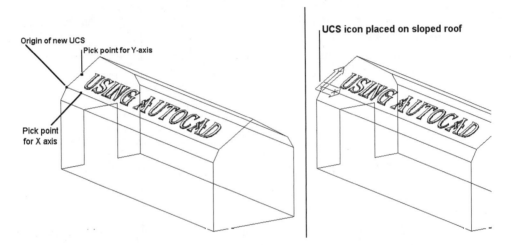

Figure 32.5 *Aligning the UCS*

OBject

The **OBject** option places the UCS relative to an existing object. Selecting **OBject** results in the following prompt.

Select object to align UCS: *(Pick an object.)*

You must use object pointing to select the object. No other object selection method is allowed. You cannot select the following objects for UCS alignment:

3D solid

3D polyline

3D mesh

Viewport

Mline

Region

Spline (as created by **Spline** command)

Ellipse (when **PEllipse** = 0)

Ray

Xline

Leader

Mtext

Arc	The center of the arc becomes the new UCS origin with the X axis passing through the point on the arc visually closest to the selection point.
Circle	The circle's center becomes the new UCS origin, with the X axis passing through the point on the arc visually closest to the selection point.
Dimension	The new UCS origin is the middle point of the dimension text. The direction of the new Y axis is parallel to the X axis of the UCS in effect when the dimension was drawn.
Line	The endpoint visually nearest the selection point becomes the new UCS origin. The X axis is chosen such that the line lies in the X,Z plane of the UCS (that is, its second endpoint has a Y coordinate of zero in the new system).
Point	The new UCS origin is the point's location. The X axis is derived by an arbitrary but consistent algorithm.
Polyline (2D only)	The polyline's start point is the new UCS origin, with the X axis extending from the start point to the next vertex.
Solid (2D only)	The first point of the solid determines the new UCS origin. The new X axis lies along the line between the first two points.
Trace	The "from" point of the trace becomes the UCS origin, with the X axis lying along its centerline.
3D Face	The UCS origin is taken from the first point, the X axis from the first two points, and the Y positive side from the first and fourth points. The Z axis follows by application of the right-hand rule. If the first, second, and fourth points are colinear, no new UCS is generated.
Shape, Text, Block Reference, Attribute Definition	The new UCS origin is the insertion point of the entity, while the new X axis is defined by the rotation of the entity around its extrusion direction. Thus, the entity you select to establish a new UCS will have a rotation angle of 0 degrees in the new UCS.

Table 32.1

Using this option will position the *X,Y* plane of the UCS parallel to the object selected. The direction of the axes, however, depends on the object selected. Table 32.1 explains the effect of each.

Face

The **Face** option aligns the UCS to the selected face of a 3D ACIS solid object:

> Select face of solid object: *(Click the face.)*
>
> Enter an option [Next/Xflip/Yflip] <accept>: *(Press ENTER to accept the face.)*

The **Next** option moves the UCS to an adjacent face or to the back face of the selected edge.

The **Xflip** option rotates the UCS by 180 degrees about the *X* axis.

The **Yflip** option rotates the UCS by 180 degrees about the *Y* axis.

View

It can be convenient to orient the UCS to the current view. Using either the **VPoint** or **3dOrbit** command may have set this view. Selecting the **View** option sets the UCS perpendicular to the current viewing direction. That is, the current view becomes the *plan view* under the new UCS created by the option. This results in a new UCS that is parallel to the computer screen.

There is no prompt. AutoCAD simply matches the UCS to the view; you will notice the UCS icon "straighten out."

X, Y, and Z

The **X**, **Y**, and **Z** options rotate the UCS around one of the three axes. You specify the angle that the UCS should rotate about one of the axes, as follows:

> Rotation angle about n axis <0.0>: *(Specify an angle.)*

(The *n* in the prompt is either X, Y, or Z.) Entering an angle will cause the UCS to rotate about the specified axis the designated number of degrees. You can also enter a point on the screen to show the rotation. AutoCAD provides a rubber-band line from the current origin point to facilitate this.

The direction of the angle is determined by the standard engineering "right-hand rule" method of determining positive and negative rotation angles. Figure 32.6 shows the method of determining angle directions.

Imagine curling your right hand around the axis; the thumb points in the positive direction. The direction of your curled fingers is the direction of positive angle of rotation of that axis. (Note that the **Angle Clockwise** setting in the **Units** command can reverse the direction.)

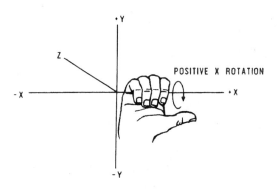

POSITIVE X ROTATION

Figure 32.6 *Right-hand Rule*

When **Tilemode** is off, the last 10 coordinate systems in both paper space and model space are saved. The "previous" coordinate system derived depends on the space in which you use the **Prev** option (discussed later in this chapter).

MOVE

The **Move** option moves the origin point of the current UCS. This option can also be used to change the Z coordinate of the current UCS. Note that this option does not add the UCS to the list of previous UCS.

>Specify new origin point or [Zdepth] <0,0,0>:

When you respond by typing an X,Y,Z coordinate (or selecting a point in the drawing), AutoCAD shifts the origin point of the UCS to that location.

When you type Z, AutoCAD prompts you to specify the distance to move the UCS's origin along the Z axis:

>Specify Z depth <0>:

Type a positive number to move the UCS origin up the Z axis, or a negative number to move the UCS down the Z axis.

ORTHOGRAPHIC

The orthoGraphic option selects one of six preset UCS orientations—top, bottom, front, back, left, and right—essentially the six sides of a cube. (For the dialog box version of this command, select **Tools | Orthographic UCS | Preset** from the menu bar.) AutoCAD prompts you at the command line:

>Enter an option [Top/Bottom/Front/BAck/Left/Right]<Top>:

Type an option, such as F for Front view, or press ENTER to keep the default (Top, in this case).

PREV

The **Prev** option returns the last UCS setting. You can use this option repeatedly to back up through as many as 10 previous UCS settings. The operation of this option is similar to the **Zoom Previous** command.

Note: The previously saved UCS becomes current, but the previous view does not. You may want to use the **Plan** command (explained later in this chapter) to restore the plan view of the now current UCS.

RESTORE

The **Restore** option restores a UCS that has been previously saved with the **Save option** (explained later). The following prompt is displayed:

Enter the name of UCS to restore or [?]:

Enter the name of the UCS to be restored. The named UCS then becomes current. When you enter a question mark (?), AutoCAD prompts:

Enter name of UCS name(s) to list <*>:

Press [Enter] to display all the saved coordinate systems. AutoCAD displays a list of named UCSs in a format similar to the following:

Current ucs name: "end_wall"

Saved coordinate systems:

"end_wall"

Origin = <0.0000,0.0000,0.0000>, X Axis = <1.0000,0.0000,0.0000>

Y Axis = <0.0000,1.0000,0.0000>, Z Axis = <0.0000,0.0000,1.0000>

SAVE

The **Save** option allows you to name and, hence, save the current UCS. By naming the UCS, you can later return to the UCS using the **Restore** option. (AutoCAD does *not* allow you to share the UCS with other drawings.) The name can be up to 255 characters in length and can contain letters, numbers, dollar signs ($), hyphens (-), and underscores (_).

Note: Use names that describe the location of the UCS, such as "ROOF- TOP" or "FRONT-HOUSE."

Enter name to save current UCS or ?:

DEL

The **Del** option erases a previously saved UCS name from the drawing. (You *cannot* use the **Purge** command to remove "unused" UCSs.) The following prompt is issued:

Enter UCS name(s) to delete <none>:

The wild-card characters of "?" and "*" can be used to delete several UCS names at a time. Entering the name of a single UCS name will delete only that system. You can enter several names by separating them with commas.

?

The **?** option lists the saved coordinate systems. This performs the same function as the **?** option under **Restore**. A listing of each saved UCS and the coordinates of the origin and *X,Y,Z* axes are displayed. AutoCAD switches to the Text window to display the listing. (Press F2 to return to the graphics window.)

Enter UCS name(s) to list <*>:

WORLD

The **World** option is the default; it resets the UCS to the world coordinate system.

PRESET UCS ORIENTATIONS

As described under the **orthoGraphic** option of the **UCS** command, AutoCAD provides several preset UCS orientations: top, bottom, front, back, left, and right—that are similar to the six sides of a cube. Type the **DdUcsP** command (short for "dialog UCS preset") or select **Tools | Orthographic Ucs | Preset** from the menu bar. AutoCAD displays the **UCS** dialog box with the **Orthographic UCSs** tab.

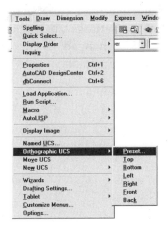

Select a preset name, such as **Top** or **Bottom**, to change the UCS to one of the six views. You have the option of displaying the preset UCS relative to the WCS or to any named UCS. Choose **Set Current**, then choose **OK**. If necessary, use the **Plan** command to view the new UCS in plan.

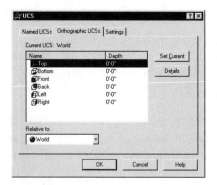

Figure 32.7 *Orthographic UCSs Dialog Box*

UCSFOLLOW SYSTEM VARIABLE

The **UcsFollow** system variable allows you to set the UCS plan view to follow a newly set user coordinate system.

Normally, when you set a new UCS, you select the **Plan** command to reset the display to align in true plan view with the new coordinate system. Turning on **UcsFollow** (set it to 1) causes the plan view to be set automatically each time the UCS is changed.

To change this setting in a dialog box, type **UcsMan** or select **Tools | Named UCS**, from the menu bar) then select the **Settings** tab. In the **UCS Settings** section, select **Update view to Plan when UCS is changed**. By default, the option is turned off; a check mark indicates the option is turned on.

CONTROLLING THE UCS ICON

The **UcsIcon** command controls the display of the UCS icons. The icon is located, by default, at the lower left corner of each viewport. Note that there are different UCS icons for model mode and layout mode; Figure 32.8 shows the UCS icon for each.

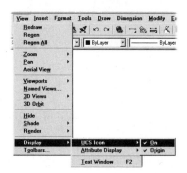

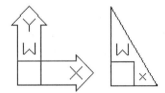

Figure 32.8 *UCS Icons*

Type the **UcsIcon** command, or select **View | Display | UCS Icon** from the menu bar. AutoCAD issues the following prompt.

Command: **ucsicon**

Enter an option [ON/OFF/All/Noorigin/ORigin] <ON>:

The following is an explanation of the options:

OFF

The **Off** option turns off the display of the icon. This is useful when you find the icon getting in the way or for 2D drafting, where the UCS icon is less useful.

ON

The **On** option turns on the icon display, if it is off.

All

The **All** option is like the **RedrawAll** command: it affects the UCS icon setting in all viewports. This option is meant for use when the current drawing has two or more viewports. Normally, the **UcsIcon** command affects only the icon in the current viewport.

To use the **All** option, first select **All**. The **UcsIcon** command's prompt repeats, allowing you to choose the option that will affect the icons in all viewports.

Noorigin

The **Noorigin** option forces the UCS icon to be displayed at the lower left of the viewport screen, without respect to the current UCS origin.

ORigin

The **ORigin** option displays the UCS icon at the UCS origin point of 0,0,0.

When the complete icon cannot be displayed at the origin of the viewport, the icon is displayed at the lower left corner of the viewport. For this reason, you may see the UCS icon bounce around the viewport (from origin to lower left corner, and back again) as you zoom and pan around the drawing.

UCS MANAGER

The **UcsMan** command displays a dialog box that assists in the management of the UCS system. Select **Tools | Named UCSs** from the menu bar. The **Named UCSs** tab of the **UCS** dialog box contains a listing of the coordinate systems that are defined in the active drawing.

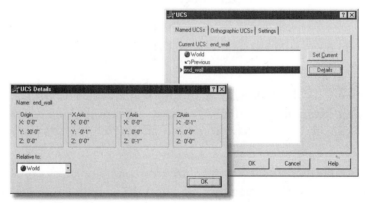

Figure 32.9 *UCS Control Dialog Box*

The first entry in the dialog box is always the World Coordinate System and is listed as "World." If you have changed the UCS, a listing named "Previous" is shown. A current UCS that has not been named is shown as "Unnamed." The list also shows the name of each saved UCS, if any.

CHANGING THE CURRENT UCS

The current UCS is prefixed with a small arrowhead. To select a new UCS, select another name, then choose **Set Current**. Choose **OK** to dismiss the dialog box.

LISTING UCS INFORMATION

You can obtain UCS information by choosing the **Details** button in the dialog box. The **UCS Details** dialog box displays the X,Y,Z vectors of the origin and each axis.

RENAMING AND DELETING A UCS

To rename a UCS, select the UCS name. Right-click to display the cursor menu, and select **Rename**.

You can delete a UCS by selecting the UCS name, then right-clicking to display the cursor menu. Select **Delete**. You cannot delete or rename the World and Previous UCS listings.

Note: Deleting a UCS that is no longer in use makes UCS management easier.

CHAPTER REVIEW

1. In what plane does the User Coordinate System exist?

2. What is the World Coordinate System?

3. What does the broken pencil icon mean?

4. What commands are used to manipulate the UCS?

5. What command is used to display the UCS icon?

6. What UCS option would you use to align the UCS with the current drawing rotation, placing the UCS parallel with the view?

7. What rule is used to determine the rotation of axes?

8. How would you store and recall a UCS?

9. What is the function of the **UcsFollow** system variable?

10. What are the two possible positions for the UCS icon?

TUTORIAL

Let's draw a small shop building so we can become familiar with the techniques of drawing in AutoCAD 3D.

Starting the Drawing

1. Start a new drawing.

2. Create the following setup parameters.

Units: Architectural

Limits: 0,0 / 80 ft.,50 ft. (remember to Zoom All)

Snap: 1 ft.

Grid: 10 ft. (initially)

Drawing the Basic Shape

3. Select **Draw | Line** and draw the outline of the floor. The dimensions are 50 feet in length and 30 feet in width. Figure 32.10 shows the plan. Do not dimension the plan.

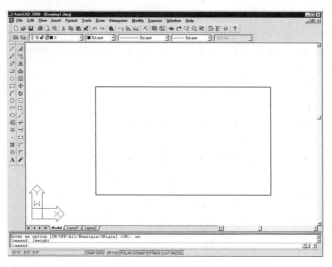

Figure 32.10 *Floor Outline*

When entering coordinates, remember to type 50' and 30' (feet) and not 50 (inches).

4. Use the **VPoint -1,1,.3** command to create a view similar to the one shown in Figure 32.11.

5. Let's save this view for recall later. Use the **View** command. Enter the name **view1** for this view.

6. You now set the UCS icon so that it moves to each new origin as we set it. Select **View | Display | UCS Icon | Origin** from the menu bar.

7. Let's set the first UCS. Select **Tools | New UCS | Object** from the menu bar. Select the line going right at the nearest corner. Notice how the UCS icon moves to the corner of the building. Observe the orientation of the axes.

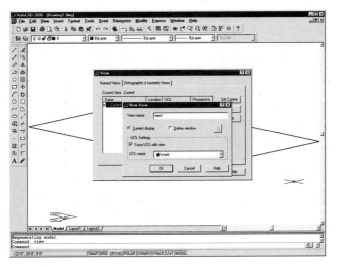

Figure 32.11 *Point View*

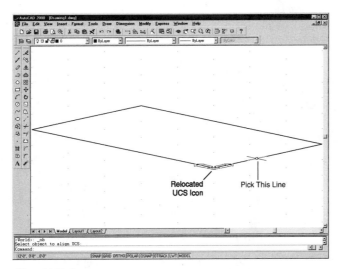

Figure 32.12 *New UCS Origin*

8. Let's now draw the first vertical corner line. Type **Line**, then type **.XY** filter. Now select INTersection snap. Snap to the intersection of the front corner. When the "need Z" prompt appears, enter a Z (height) value of **12** feet. When AutoCAD prompts "Specify next point:" use INT to snap to the same corner. Press ENTER to end the **Line** command.

9. Use the **Copy** command and INT object snap to copy the vertical line to the other end wall corner.

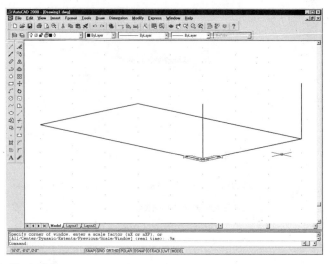

Figure 32.13 *Vertical Lines*

10. Let's now rotate the UCS so we can draw directly on the end wall.

    Type **UCS X**. The **X** option is used to rotate the UCS around the X axis. By the right-hand rule, you want to rotate the UCS 90 degrees. Enter **90** in response to the prompt. Notice how the UCS icon rotates.

11. You may want to save this UCS for recall later. Type **UCS Save**. Enter the name **END_WALL**.

12. Let's now change to the plan view for the current UCS. Type **Plan**. Press ENTER to default to the plan view for the current UCS.

13. You want to construct the roof angles as shown in Figure 32.14.

    You may want to zoom to a comfortable working size.

    Type **Line** and use ENDpoint object snap to connect to the top of the left vertical line. Enter the polar coordinate of **@16'<15** to define the endpoint. Connect a line to the other vertical line and enter the polar coordinate of **@16'<165**. Finish by filleting (ensure radius = 0) the ridge point to form a perfect intersection at the peak of the roof.

14. Restore the previously stored view through the **View** dialog box's **Set Current** button, selecting **VIEW1**.

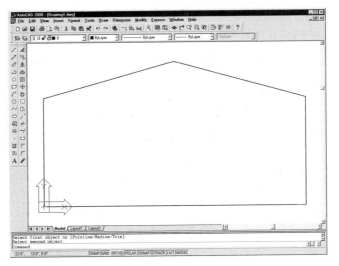

Figure 32.14 *End wall*

15. Copy the end wall "panel" to the opposite end (using INT object snap aids in exact placement).

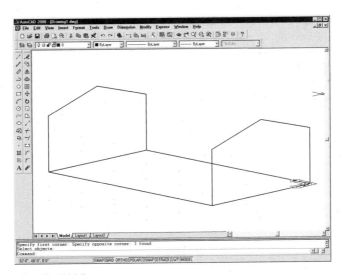

Figure 32.15 *Two End Walls*

16. Enter **VPoint -2,-2,.75** to obtain the view like Figure 32.16.

17. Now use the **Line** command and INT object snap to connect the ridge points and the edges of the roof as shown in Figure 32.16

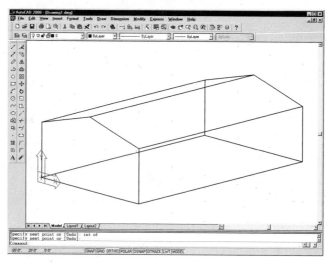

Figure 32.16 *Main Building Lines*

Drawing on the Roof

18. Let's now draw some lines on the slope of the roof to represent rafters. Type **UCS 3point**. Use INT and NEA object snap to snap to the origin, X-axis line, and Y-axis line as shown in Figure 32.17. Notice how the UCS icon relocates to the edge of the roof and rotates to match the slope.

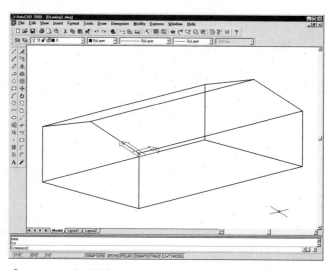

Figure 32.17 *Reorienting the UCS*

19. Array the edge lines of the roof. Type **Array Rectangular**, 1 row, **26** columns, and 24 inches between columns.

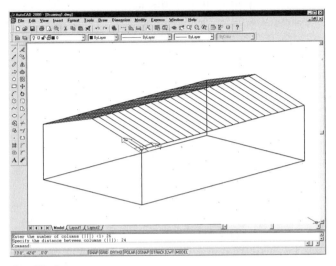

Figure 32.18 *Roof Rafters*

Drawing Doors on the Walls

20. Let's now proceed and draw some doors on one wall. Type **UCS 3point**. Select points as shown in the following illustration to set a new UCS. Type **Plan**, then press ENTER to set the plan view for the new UCS. The new plan view should look like Figure 32.19.

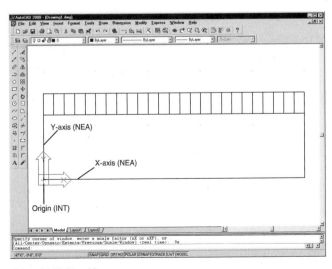

Figure 32.19 *New UCS Plan View*

21. Use the **Line** command to draw doors and windows on the side wall of the shop. You can draw any type of doors and windows you wish. Figure 32.20 shows as you may want to draw them.

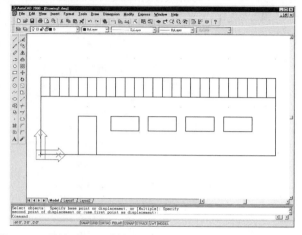

Figure 32.20 *Doors and Windows in Wall*

22. Let's now set the UCS system back to the World system before we model the drawing. Select **UCS World**, then **Plan** and press ENTER.

Viewing the Drawing in 3D

23. Use the **Dview** command to model the drawing. This is a good time to also practice using such **Dview** options as **Camera**, **Target**, and **Distance**.

24. Save the drawing as "Shop.Dwg."

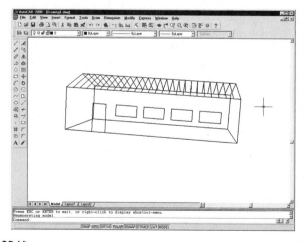

Figure 32.21 *3D View*

CHAPTER 33

Drawing in 3D

After you have become proficient with the use of 3D viewing and UCS commands, you must learn the methodology of constructing 3D drawings in AutoCAD. After completing this chapter, you will be able to

- Comprehend the concepts of elevation and thickness
- Draw in different UCS systems
- Create solid faces for objects
- Construct various types of 3D surface meshes
- Use AutoCAD's functions to create basic 3D objects

INTRODUCTION

In this chapter, you learn about many commands for creating 3D surface models.

Elev

Changes the default Z coordinate from 0.

3dFace

Draws 3D faces.

SplFrame

Toggles the display of 3D face edges.

SurfTab1 and **SurfTab2**

Specifies the number of rules in the M and N direction.

3dMesh

Draws a 3D mesh.

PFace

Draws a generalized 3D mesh.

RuleSurf

Draws a ruled 3D surface.

TabSurf

Draws a tabulated 3D surface.

RevSurf

Draws a revolved 3D surface.

EdgeSurf

Draws a Coons patch 3D surface.

AutoCAD includes the following commands to draw 3D surface primitives:

Ai_Box

Draws a rectangular box or cube.

Ai_Cone

Draws a cone.

Ai_Dome

Draws a hemisphere (top half of a sphere).

Ai_Dish

Draws the bottom half of a sphere.

Ai_Sphere

Draws a complete sphere.

Ai_Torus

Draws a donut shape.

Ai_Wedge

Draws a wedge shape.

WORKING WITH SURFACE MODELS

Now that you have learned about the user coordinate system and how to set up views of 3D drawings, it is time to learn the components of 3D drawing construction. In this chapter, you learn about working with surface models.

There are several methods for constructing the components of a 3D drawing. Some are used to draw a unique shape, while some are basic shapes that can be placed in the drawing. Many drawings are composed of a combination of constructed objects and basic shapes.

The first step in constructing a new 3D drawing is studying the shapes to be drawn. Determine the best method of approach. It is often easier to construct each individual shape than to insert it into a master drawing composed of many objects.

The ways to construct a 3D drawing that we study are as follows:

- Applying elevation and thickness to create extruded objects

- Standard drawing methods in different coordinate systems

- 3D polygon meshes

- 3D objects

Let's look at how we can use each of these to create 3D drawings.

EXTRUDED OBJECTS

This method of placing 3D shapes forms the components by applying extrusions to objects. An *extrusion* can be thought of as the application of a thickness to an object. Think of the object as "growing" up or down from the flat drawing plane (*X,Y* drawing plane).

Thus, a line appears as a sheet of paper on edge. A circle appears as a tube, etc. Figure 33.1 shows several drawing objects on the left and the same objects in extruded form on the right.

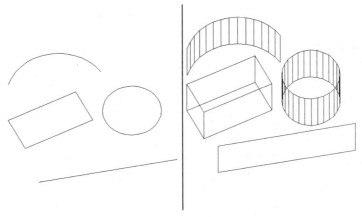

Figure 33.1 *2D Objects (at left) and Extruded Objects (at right)*

ELEVATION

The base, or bottom, of the extrusion can be set at different elevations. Consider an extruded tube (made from a circle) sitting on a table top. The top of the table could be considered as zero elevation. If you placed the bottom of the tube above the zero elevation, the tube would appear to be hovering above the table top. When the elevation is negative, the tube appears to be shoved downward through the table top.

Of course, the floor on which the table sits could be placed at the zero elevation. If the table were 30 inches to the top, the tabletop elevation would be 30 inches. If you wanted the tube to sit on the table, the elevation of the tube would also be 30 inches.

THICKNESS

The thickness of an extruded object is the distance from the base elevation of the object. For example, a thickness of 6 inches would make our tube 6 inches high. Note that its base elevation is not necessarily zero, as in the previous example of the tube sitting on a table top at a 30-inch base elevation.

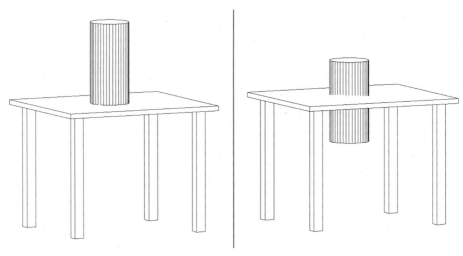

Figure 33.2 *3D Elevation*

SETTING ELEVATION AND THICKNESS

To draw an object with a specified elevation and/or thickness, the **Elev** command (short for "elevation") is used. This command presets both values, and every object that is subsequently drawn has these values. The default value in the ACAD prototype drawing is zero for each, resulting in a flat entity in 3D.

When you issue the **Elev** command, the following prompts are displayed:

> Command: **elev**
>
> Specify new default elevation <0.0000>:
>
> Specify new default thickness <0.0000>:

Setting new values only affects the entities drawn after the change. It is not retroactive, and changing either setting has no immediate visual effect on the drawing at the time the change is made.

CHANGING EXISTING OBJECTS

To create several objects of different elevations and/or thicknesses, you use the **Elev** command to change each value before drawing the new objects.

Sometimes, objects are drawn with an incorrect thickness. If this happens, it is not necessary to erase the objects and redraw them with a new thickness. The **Change** command is more convenient than the **Properties** dialog box for changing the elevation.

Simply issue the **Change** command.

> Command: **change**
>
> Select objects: *(Select.)*
>
> Select objects: *(Press* ENTER.*)*
>
> Specify change point or [Properties]: **p**
>
> Enter property to change
>   [Color/Elev/LAyer/LType/ltScale/LWeight/Thickness]: **e**
>
> Specify new elevation <0.0000>:

Enter either **E** or **T** (for Elevation or Thickness) and you are prompted for the new thickness for the objects you selected. (The **ChProp** command does not let you change the elevation.)

Note: It is often easier to construct all or part of a drawing at a single thickness and then use the **Change** command to reset one object or groups of objects when you are through drawing.

DRAWING WITH UCS

If you wish to draw more complex objects, you must use the User Coordinate System (or UCS, for short) to relocate the current system, as described in Chapter 32 "The User Coordinate System." This allows you to draw in a stipulated plane, as defined by the current UCS, as though you were in plan view.

This is especially useful, since most 3D drawings require detailing on the different faces (planes) of the objects in the drawing. In addition to drawing in different planes, you can use the thickness and elevation settings. Note that these are relevant to the current UCS when the object is drawn.

Let's look at an example of drawing in 3D using different coordinate systems.

CREATING 3D FACES

It is often desirable to create faces on objects that obscure others behind them when the **Hide** command is used.

As you have learned, many extruded objects naturally create solid faces. However, when you are drawing single objects (such as lines, circles, arcs, etc.), solid faces are not automatically formed.

3DFACE COMMAND

The **3dFace** command is used to create solid faces. The **3DFace** command is similar to the 2D solid in that it creates a "face plane." It is dissimilar, however, because it can be defined with different Z coordinates for each corner of the face, creating the possibility of a nonplanar or warped plane. It is void of solid fill.

Placing 3D Faces

The method of placing 3D faces is similar to the **Solid** command. When you stipulate the corner points, however, the points are entered in a consecutive clockwise or counterclockwise fashion. This differs from a 2D solid, where such a sequence creates a "bow tie" effect.

The placement of 3D faces creates a visual "edge frame." This can be undesirable in some situations. Take, for example, an area of irregular shape. Placing a 3D face creates several planes that fill in the irregular area. Normally, each frame would contain an edge that is visible in the drawing. Figure 33.3 shows the same drawing with and without edge frames on the 3D faces.

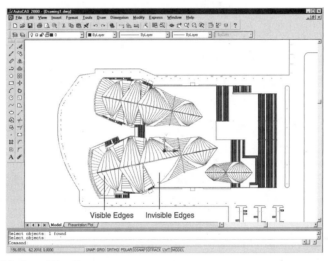

Figure 33.3 *3D Face Edge Frames*

SPLFRAME System Variable

Notice how the **I** (invisible) option is entered prior to the segment of the frame you wish to be invisible. Of course, all frame segments can be stipulated as invisible.

You can change the frame segments that are constructed with the **I** option to visible with the use of the **SplFrame** system variable. A zero setting maintains invisibility, while a nonzero (such as 1) setting makes them visible. This is convenient when you wish to edit sections that were constructed as invisible. The change is not apparent until the next regeneration is performed.

3D POLYGON MESHES

A 3D polygon mesh can be thought of as a general type of 3D object that can be curved and warped into shapes that cannot be described by other objects.

Mesh Density

There are several methods of constructing polygon meshes. All the methods of construction create a mesh that is defined by a density. A flat surface doesn't need a mesh, but a curved surface does. The density is the number of vertices used to describe the surface of the mesh. Figure 33.4 shows an object described by a mesh in two different densities.

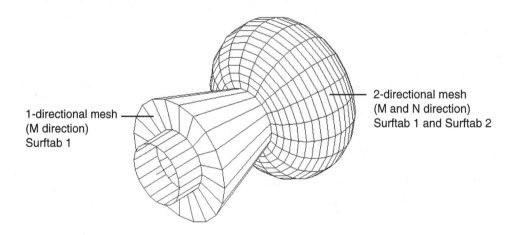

Figure 33.4 *3D Polygon Meshes*

The density is described by the system variables named **Surftab1** and **Surftab2**. One-directional meshes are controlled by **Surftab1**, while two-directional meshes are controlled by both variables; one for each direction. The directions are defined in AutoCAD as "M" and "N."

The commands for mesh construction are as follows:

3dMesh

PFace

Rulesurf

Tabsurf

Revsurf

Edgesurf

Let's look at each and study the construction method of each.

3DMESH COMMAND

Construct a 3D mesh by stipulating the number of vertices in each (M and N) direction, then specifying the X, Y, Z coordinate of each of the vertices.

It should be noted that construction of 3D meshes is very tedious. In most situations, it is more efficient to construct a mesh of the same type with one of the other construction methods. The **3dMesh** command is best utilized in LISP routines. With this in mind, let's proceed to briefly study how it works.

CONSTRUCTING A 3D MESH

To construct a 3D mesh, first issue the **3dMesh** command, then the number of vertices in the M and N directions, and finally the vertex coordinate of each. Let's look at an example.

Command: **3dmesh**

Enter size of mesh in M direction: **3**

Enter size of mesh in N direction: **3**

Specify location for vertex (0,0): **10,10,-1**

Specify location for vertex (0,1): **10,20,1**

Specify location for vertex (0,2): **10,30,3**

Specify location for vertex (1,0): **20,10,1**

Specify location for vertex (1,2): **20,20,0**

Specify location for vertex (1,3): **20,30,-1**

Specify location for vertex (2,0): **30,10,0**

Specify location for vertex (2,1): **30,20,1**

Specify location for vertex (2,2): **30,30,2**

The order of entry is one column at a time, then to the next column.

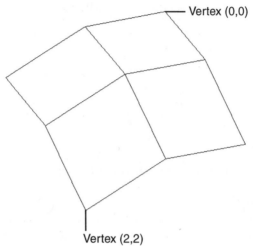

Figure 33.5 *3D Mesh*

3D meshes are displayed as wireframes; when they are coplanar, they are considered opaque when the **Hide** command is used. 3D meshes cannot be extruded.

PFACE COMMAND (POLYFACE MESH)

The **PFace** command (short for "polygon mesh") draws a polygon mesh that is independent of a continuous surface. Define these meshes by entering individual vertex values in *X,Y,Z* format.

Selecting the **PFace** command results in the prompt:

Command: **pface**

Specify location for vertex 1:

Specify location for vertex 2 or <define faces>:

and so forth, until you press ENTER on a blank line to close the mesh. Pressing ENTER again will terminate the construction of the mesh.

Polyface meshes can be used for simple mesh construction or they can be incorporated in a program to automatically enter many data points for a specialized application.

RULED SURFACES

A mesh created through the **RuleSurf** command represents a ruled surface between two objects. The objects can be lines, points, arcs, circles, 2D and 3D polylines.

You can achieve a similar effect on a drawing board by dividing two objects into the same number of segments and drawing a line between the corresponding segments. Figure 33.6 shows ruled surfaces between objects.

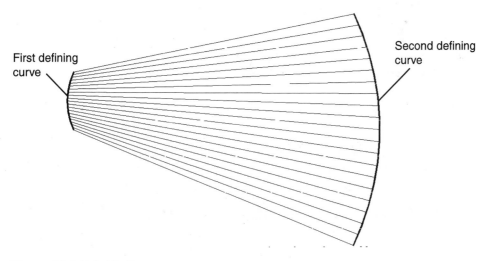

Figure 33.6 *Ruled Surface*

CONSTRUCTING RULED SURFACES

To construct a ruled surface, select the **RuleSurf** command, and then select the two objects you want the surface drawn between.

> Command: **rulesurf**
>
> Current wire frame density: SURFTAB1=6
>
> Select first defining curve: *(Pick object.)*
>
> Select second defining curve: *(Pick object.)*

AutoCAD draws the ruled surface from the endpoint nearest to the selection point on the object. Thus, selecting opposite ends of the objects creates a ruled surface that is crossed.

If an object is a circle, the selection point has no special effect. The ruled surface will begin at the 0-degree quadrant point. If a point is selected, the mesh lines are drawn from the single location of the point. Thus, a circle and a point would create a cone-type object as in Figure 33.7.

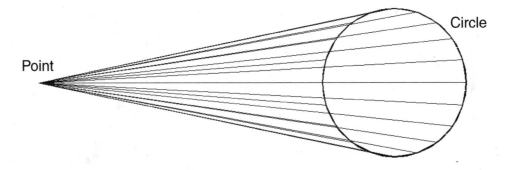

Figure 33.7 *Cone Constructed with RuleSurf*

TABULATED SURFACES

The **TabSurf** command constructs a mesh surface that is defined by a path and a direction vector. The effect is similar to an extrusion, except for the mesh surface.

CONSTRUCTING TABULATED MESH SURFACES

To construct a tabulated surface, select a *path curve* (from which the surface will be calculated) and a *direction vector* (describing the direction and length of the mesh from the path curve). Both objects must exist before you start the **TabSurf** command.

The following command sequence is displayed.

> Command: **tabsurf**
>
> Select object for path curve:
>
> Select object for direction vector:

The *path curve* can be a line, arc, circle, 2D or 3D polyline. The *direction vector* can be either a line or open polyline (2D or 3D).

The end closest to the selection point on the direction vector defines the direction in which the mesh is projected. Selecting the direction vector at one end causes the mesh to project in the direction of the opposite endpoint. Figure 33.8 shows the curve path, direction vector, and the selection point on the direction vector for the result shown.

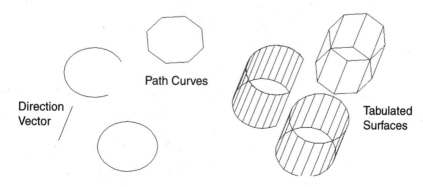

Figure 33.8 *Tabulated Surfaces*

REVOLVED SURFACES

The RevSurf command is used to create meshes from objects revolved around an axis.

CREATING REVOLVED SURFACES

To create a revolved surface, you must first have a path curve (object to be revolved) and an axis of revolution (defines an axis around which the object will revolve). The path curve can be a line, circle, arc, 2D or 3D polyline. The axis of revolution can be a line or open polyline (2D or 3D).

Figure 33.9 *Revolved Surface*

The displayed command prompts are:

> Command: **revsurf**
>
> Current wire frame density: SURFTAB1=6 SURFTAB2=6
>
> Select object to revolve:
>
> Select object that defines the axis of revolution:
>
> Specify start angle <0>:
>
> Specify included angle (+=ccw, -=cw) <360>:

The start angle point of zero is defined by the location of the path curve. If you specify a start angle other than zero, the starting point of the revolution will offset the specified number of degrees from the path curve.

The included angle designates the degrees of rotation of the path curve around the axis of revolution and from the start point.

The direction to which the revolution emanates is dependent on the selection point when the axis of revolution is selected. The right-hand rule is used to determine the direction. The revolution is in a positive rotation direction around the axis of revolution; the positive direction of the axis is defined as being from the end nearest the selection point to the following end. Thus, if your right hand is curved around the axis of revolution, with the thumb pointing toward the end of the axis furthest from the selection point, the direction of your fingers designates the positive rotation angle.

Figure 33.10 shows the effect of the location of the selection point on the axis of revolution in determining the direction of revolution. The revolution has a starting angle of 0 degrees and an included angle of 90 degrees.

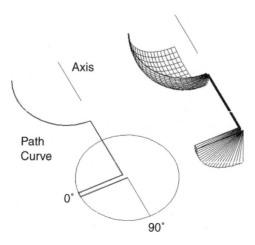

Figure 33.10 *Direction of Revolution*

EDGE-DEFINED SURFACE

The **EdgeSurf** command is used to construct a Coons surface patch from four edge surfaces. Figure 33.11 shows a Coons surface patch. The patch follows the contour of each edge surface, forming a mesh that approximates a surface attached to all points of each edge.

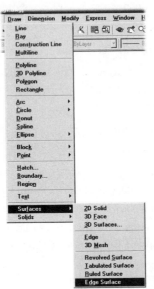

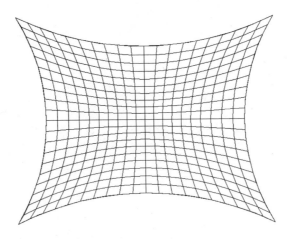

Figure 33.11 *Coons Surface Patch*

CONSTRUCTING EDGE-DEFINED SURFACES

To construct a Coons patch, you must first have four edges. These edges can be lines, arcs, or open 2D or 3D polylines. They must connect at their endpoints. The command sequence is as follows:

> Command: **edgesurf**
>
> Current wire frame density: SURFTAB1=6 SURFTAB2=6
>
> Select object 1 for surface edge:
>
> Select object 2 for surface edge:
>
> Select object 3 for surface edge:
>
> Select object 4 for surface edge:

If one edge is not connected to the adjacent edge, the message

> Edge x does not touch another edge

is displayed, where x defines the number of the edge that does not touch.

3D SURFACE OBJECTS

Several basic 3D shapes have been included for convenient use in drawings. These objects are actually constructed from the tools you have already learned to use.

The sequence to construct them has been preprogrammed. Prompts question you for pertinent information for proper construction to your specifications. The commands are

Ai_Box

Ai_Wedge

Ai_Pyramid

Ai_Cone

Ai_Sphere

Ai_Dome

Ai_Dish

Ai_Torus

Let's look at each of these and the information you will need to provide.

BOX

The **Ai_Box** command constructs a box of given length, width, and height, or a cube of a given edge dimension. The command sequence is as follows:

> Command: **ai_box**
>
> Specify corner point of box:
>
> Specify length of box:
>
> Specify width of box or [Cube]:
>
> Specify height of box:
>
> Specify rotation angle of box about the Z axis or [Reference]:

Figure 33.12 shows the components of box construction.

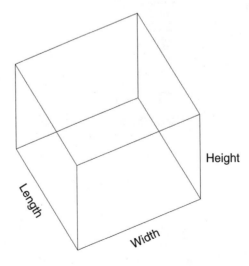

Figure 33.12 *Components of a 3D Box*

If you select the **Cube** option (on the fourth prompt line), a cube is constructed with the edge length equal to the length you previously defined.

Note: The length is in the X direction; the width is in the Y direction; and the height is in the Z direction.

CONE

The **Ai_Cone** command is used to construct cones. You define a cone by specifying the radius or diameter of the top and bottom and the height. The command prompts are:

Command: **ai_cone**

Specify center point for base of cone:

Specify radius for base of cone or [Diameter]:

Specify radius for top of cone or [Diameter] <0>:

Specify height of cone:

Enter number of segments for surface of cone <16>:

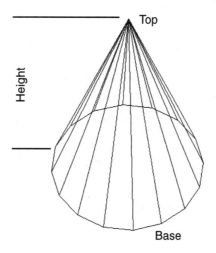

Figure 33.13 *Components of a 3D Cone*

DOME

Use the **Ai_Dome** command to create domes for your drawing.

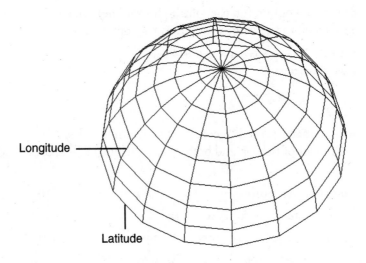

Longitude

Latitude

Figure 33.14 *Dome*

The command sequence is as follows.

> Command: **ai_dome**
>
> Specify center point of dome:
>
> Specify radius of dome or [Diameter]:
>
> Enter number of longitudinal segments for surface of dome <16>:
>
> Enter number of latitudinal segments for surface of dome <8>:

DISH

A dish is simply an inverted dome. The construction of a dish is similar to that of a dome.

The following command sequence is displayed.

> Command: **ai_dish**
>
> Specify center point of dome:
>
> Specify radius of dome or [Diameter]:
>
> Enter number of longitudinal segments for surface of dome <16>:
>
> Enter number of latitudinal segments for surface of dome <8>:

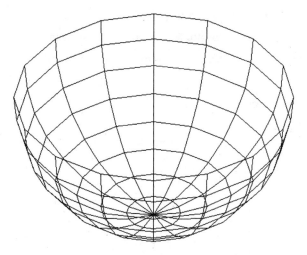

Figure 33.15 *Dish*

SPHERE

Spheres are created with the **AI_Sphere** command.

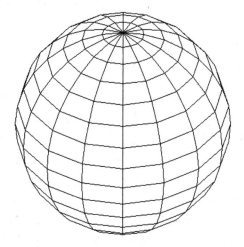

Figure 33.16 *Sphere*

The command sequence is as follows:

> Command: **ai_sphere**
>
> Specify center point of sphere:

Specify radius of sphere or [Diameter]:

Enter number of longitudinal segments for surface of sphere <16>:

Enter number of latitudinal segments for surface of sphere <16>:

TORUS

A torus is a closed tube, rotated around an axis.

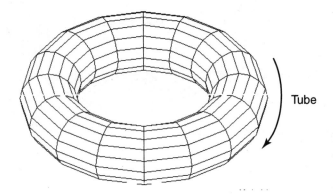

Figure 33.17 *Torus*

The command sequence for creating a torus is as follows:

Command: **ai_torus**

Specify center point of torus:

Specify radius of torus or [Diameter]:

Specify radius of tube or [Diameter]:

Enter number of segments around tube circumference <16>:

If you enter a tube radius or diameter that is greater than the torus radius or diameter, the following message is displayed:

Tube radius cannot exceed torus radius

You will then be given the opportunity to reenter the data.

WEDGE

A wedge could be thought of as a block, cut diagonally along its width.

The command sequence when constructing a wedge is as follows:

Command: **ai_wedge**

Specify corner point of wedge:

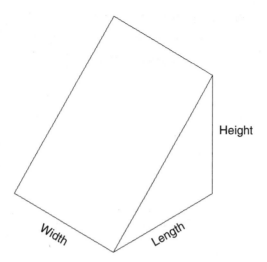

Figure 33.18 *Wedge*

Specify length of wedge:

Specify width of wedge:

Specify height of wedge:

Specify rotation angle of wedge about the Z axis:

Note: The wedge is drawn sloping down from the initial point, along the X-axis. You can then rotate it into position with the **Last** option.

PYRAMID

The **Ai_Pyramid** command constructs a variety of pyramid shapes. Depending on the options you select, you can construct

- A tetrahedron (four-sided solid made of four triangles)

- The base with 3 or 4 corners

- The top with a point or a ridge or a truncated top

The command sequence for drawing the "true" pyramid shown in Figure 33.19 is as follows:

Command: **ai_pyramid**

Specify first corner point for base of pyramid:

Specify second corner point for base of pyramid:

Specify third corner point for base of pyramid:

Specify fourth corner point for base of pyramid or [Tetrahedron]:

Specify apex point of pyramid or [Ridge/Top]:

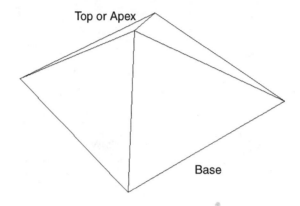

Top or Apex

Base

Figure 33.19 *Pyramid*

To draw the tetrahedron (left in Figure 33.20), follow this command sequence:

Command: **ai_pyramid**

Specify first corner point for base of pyramid:

Specify second corner point for base of pyramid:

Specify third corner point for base of pyramid:

Specify fourth corner point for base of pyramid or [Tetrahedron]: **t**

Specify apex point of tetrahedron or [Top]:

To draw the pyramid with a ridge (right in Figure 33.20, useful for the roof on a house), use these command options:

Command: **ai_pyramid**

Specify first corner point for base of pyramid:

Specify second corner point for base of pyramid:

Specify third corner point for base of pyramid:

Specify fourth corner point for base of pyramid or [Tetrahedron]:

Specify apex point of pyramid or [Ridge/Top]: **r**

Specify first ridge end point of pyramid:

Specify second ridge end point of pyramid:

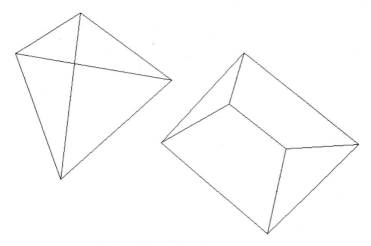

Figure 33.20 *Tetrahedron (left) and Ridge (right)*

3D COMMAND

The **3D** command prompts you to draw one of the objects described above:

> Command: **3d**
>
> Enter an option
>
> [Box/Cone/DIsh/DOme/Mesh/Pyramid/Sphere/Torus/Wedge]:

Alternatively, select **Draw | Surfaces | 3D Surfaces** from the menu bar. AutoCAD displays the **3D Objects** dialog box.

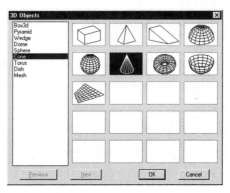

Figure 33.21 *3D Objects Dialog Box*

SUMMARY

Mastering the principles of 3D drawing takes time and effort. Try to simulate the examples in the text and experiment on projects of your own interest to build expertise in each area of 3D drawing.

Understand the principles of 3D, as discussed in Chapter 30 "Introduction to 3D."

Master the principles of viewing 3D drawings covered in Chapter 31 "Viewing 3D Drawings." This is not only important for viewing, but for displaying the drawing while it is in progress for effective 3D construction.

The user coordinate system explained in Chapter 32 "The User Coordinate System" is extremely important, since it is one of the primary methods of construction used in 3D drawings. You cannot obtain the maximum results from AutoCAD 3D without a complete understanding of this system.

The drawing methods described in this chapter should be considered as a toolbox to be used for 3D construction. Study each drawing to determine the best method of creating the objects in your drawing.

And most of all, have fun!

CHAPTER REVIEW

1. What is an *extrusion*?

2. Which axis direction is an extrusion normally projected into?

3. What is meant by the *elevation* and *thickness* of an object?

4. How could you change the thickness of an existing object?

5. How do you return to the plan view after rotating your drawing?

6. What does the **3dFace** command perform?

7. Compare the placement of a 3D face with a solid's face that would be placed with the **Solid** command.

8. What procedure is used if you want to place a 3D face without showing the edges?

9. Describe a use for the **RuleSurf** method of constructing a surface.

10. The density of a two-direction 3D surface is described by the values of M and N. Which system variable controls the values of each?

TUTORIAL

Let's try constructing a simple drawing using the **Elevation** command to set the base elevation and the thickness of the extrusion.

1. Begin a new drawing.

2. Set limits of 0,0 and 12,8. It is helpful to turn on snap and grid.

3. Now let's set the first elevation and thickness.

   Command: **elev**

   New current elevation <0>: **0**

   New current thickness <0>: **4**

4. Draw a box (see Figure 33.3) using the PLine command.

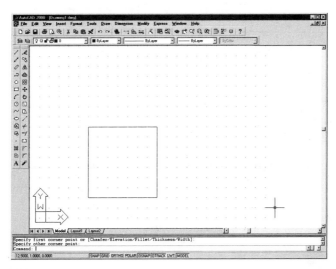

Figure 33.22 *Box*

5. Now change the elevation and thickness:

   Command: **elev**

   New current elevation <0>: **3**

   New current thickness <4.00>: **10**

   Notice how the currently set values show up as defaults. You set the elevation to 3 and changed the thickness to 10.

6. Now use the **Circle** command to draw a circle like the one shown in Figure 33.23.

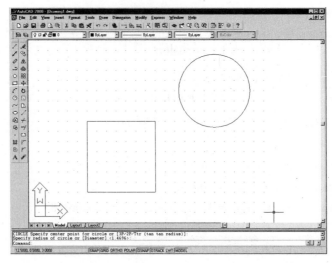

Figure 33.23 *Circle and Box*

7. Use the **VPoint -1,2,5** command to view the 3D shapes. Your drawing (depending on the viewpoint) should appear similar to the one in Figure 33.24.

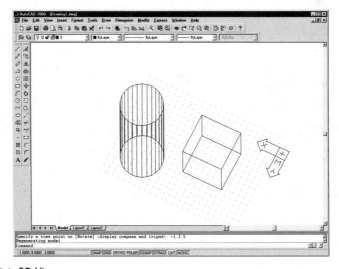

Figure 33.24 *3D View*

8. Try changing the elevation and thickness with the **Change** command.

9. Save the drawing as "3D.Dwg."

TUTORIAL

1. Use the **Polygon** command to draw the polygon shown in Figure 33.25.

2. With the **3dFace** command, enter the following sequence.

   Specify first point or [Invisible]: *(Enter point 1.)*

   Specify second point or [Invisible]: *(Enter point 2.)*

   Specify third point or [Invisible] <exit>: I (*) *(Enter point 3.)*

   Specify fourth point or [Invisible] <create three-sided face>: *(Enter point 4.)*

   Specify third point or [Invisible] <exit>: *(Enter point 5.)*

   Specify fourth point or [Invisible] <create three-sided face>: *(Enter point 6.)*

   Specify third point or [Invisible] <exit>: *(Press* ENTER.*)*

After the final ENTER, the second 3D face plane is closed.

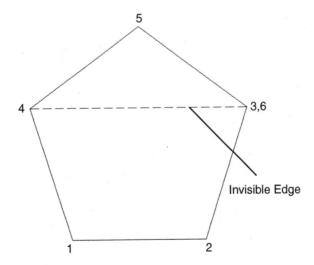

Figure 33.25 *3D Face Sequence*

CHAPTER 34

Constructing Solid Primitives

Solid modeling can be easy, fun, and useful. This chapter introduces you to the fundamentals of 3D solid modeling.

Solid primitives are used to create solid models from building block shapes. Many solid models can be created from these basic shapes. In order to become proficient in the solid modeling program, you should master the construction of each of these primitive shapes. After completing this chapter, you will be able to

- Understand the purpose and use of the AutoCAD ACIS modeler
- Differentiate the differences between solid models and 3D surface models
- Comprehend the Solids menu organization
- Utilize each of the solid primitive construction commands

INTRODUCTION

AutoCAD's ACIS modeler is another way to create solid three-dimensional (3D) objects. The solid objects are similar in appearance to wireframe or surface models created in AutoCAD 3D. Solid models, however, contain more information than a surface model. Let's look at some of the differences.

Surface models created in AutoCAD 3D are composed of solid faces. Imagine a cube that is constructed from thin pieces of cardboard. The cube has six faces, but no solid interior (see Figure 34.1). A solid model, however, is not constructed of faces; it is a solid object. The same cube constructed with a solid modeling program would have a solid interior.

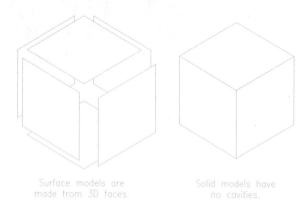

Surface models are
made from 3D faces.

Solid models have
no cavities.

Figure 34.1 *3D Versus Solid Object*

Creating objects as solids allows you to analyze the objects. Object properties such as center of gravity, surface area, and moments of inertia can be calculated.

WHO USES SOLID MODELING?

Solid modeling is useful to many disciplines. The mechanical engineer who needs to calculate properties of objects will find the ability to scientifically analyze an object of different materials invaluable. Architects can use solid modeling to visually represent complex intersections of building roofs. After the model is constructed, the drawing can be converted to a "standard" 3D form and embellished for presentation purposes.

Those who model the objects they create find that solid modeling offers the very useful possibilities of intersection construction, scientific analysis, and the simplicity of "building block" construction.

DRAWING WITH SOLIDS

Solid models are created from solid three-dimensional shapes. This "building block" approach is different from the method used with surface models. Objects created in 3D can be converted to solids. This allows the flexibility to create an object in the most efficient manner. Before you begin your work with solid modeling, you should be proficient in 3D.

Solid Primitives

AutoCAD provides commands that create basic building block shapes referred to as solid primitives. These simple 3D solid shapes are the box, wedge, cone, cylinder, sphere, and torus. These shapes can be "added" or "subtracted" from each other to create more complex shapes. In addition, you can edit with commands that revolve, extrude, chamfer, or fillet the shapes.

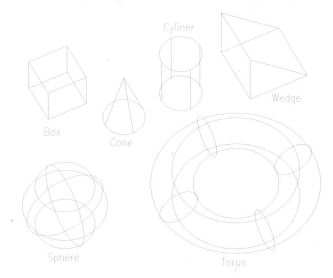

Figure 34.2 *Solid Primitives*

Composite Solids

When you create a solid object from several primitives, it is referred to as a composite solid. The primitives that combine to make the object can be consolidated to create a single object, enclosing the entire volume as one solid.

You can also add or subtract solids from each other. For example, a drill hole can be created from a cylinder that has been placed in a solid plate. Subtracting the cylinder from the plate creates the "drill hole."

Figure 34.3 *Solid with Drill Holes "Subtracted"*

Solid Modeling Commands

AutoCAD's solid modeling commands are indistinguishable from other commands. For example, to draw a box, you use the **Box** command.

SOLID PRIMITIVES

Solid primitives are the "building blocks" of solid modeling. The solid primitives available are as follows:

Box	Cylinder	Torus
Cone	Sphere	Wedge

You create solid primitives by providing the dimensions of the object. Let's draw each of the primitives. Start a new drawing of any name and follow the short tutorial for each.

You can use the **Draw | Solids** menu. You use these menus to select the commands to construct our solid primitives. Let's use the solid commands to construct some solid objects. Let's start with the **Box** command.

DRAWING A SOLID BOX

The **Box** command is used to draw a solid 3D box.

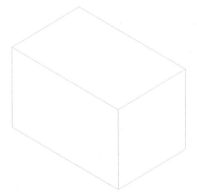

Figure 34.4 *Solid Box*

There are three methods of creating a box with this command. Let's look at each separately.

Specifying Opposite Corners and Height

This is the default method of creating a box. Start by selecting the **Box** command. Enter the first point, then the opposite corner. Refer to the following command sequence and Figure 34.5.

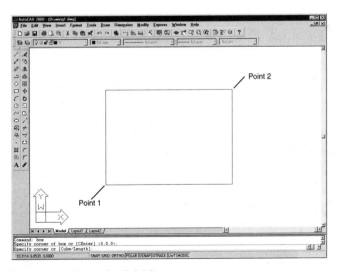

Figure 34.5 *Drawing the Base of a Solid Box*

Command: **box**

Specify corner of box or [CEnter] <0,0,0>: *(Enter point "1".)*

Specify corner or [Cube/Length]: *(Enter point "2".)*

The two points you have just entered specify the length and width of the box. You must now specify the height. You could also enter a relative coordinate to locate the opposite corner of the box after entering the first corner. The command line now prompts you for that height. In this example, let's make our box 3 units high.

Specify height: **3**

You may want to stop at this point and use AutoCAD's **VPoint** command to display a 3D representation of the box. Return to the Plan view and use a **Zoom All** after you are finished.

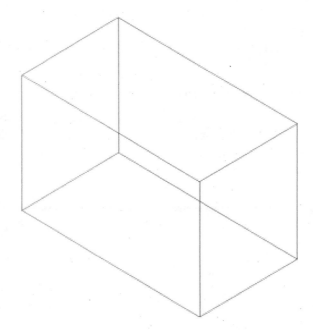

Figure 34.6 *3D View of Box*

Note that the length and width of the box are constructed relative to the current UCS and the height is perpendicular to the current UCS.

Let's continue and look at the other two ways to construct a solid box.

Length Option

The **Length** option allows you to construct a box by designating the actual length, width, and height of a box in numeric values. Let's construct a box that is 4 units long, 3 units wide, and 2 units high. The following command sequence will guide you through the process.

> Command: **box**
>
> Specify corner of box or [CEnter] <0,0,0>: *(Enter a point on the screen.)*
>
> Specify corner or [Cube/Length]: **L**
>
> Specify length: **4**
>
> Width: **3**
>
> Height: **2**

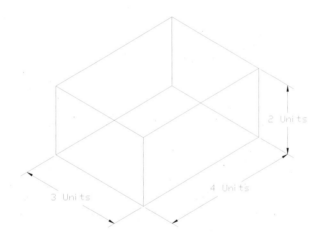

Figure 34.7 *Constructing a Box with Length Option*

Creating a Solid Cube

The **Cube** option is used to create a solid cube. Let's draw a cube.

> Command: **box**
>
> Specify corner of box or [CEnter] <0,0,0>: *(Enter a point showing the corner of the box.)*
>
> Specify corner or [Cube/Length]: **C**
>
> Specify length: **3**

AutoCAD constructs the cube, using the value entered for the length (in this case, the value of 3 you entered) to construct a cube with all sides equal to the entered value.

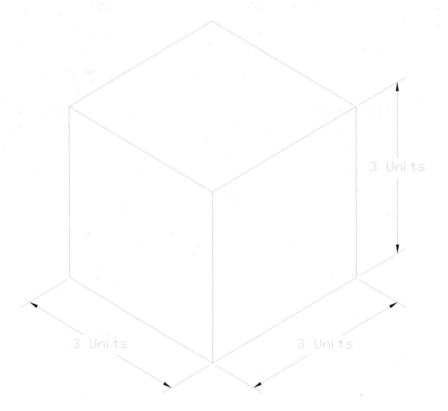

Figure 34.8 *Solid Cube*

CREATING A SOLID CONE

The **Cone** command is used to create a solid cone.

A cone is constructed with either a circular or elliptical base. The cone's base lies in the *X,Y* plane of the current UCS. The height describes the distance between the base and the point and is perpendicular to the current UCS.

Let's construct a cone with a circular base. Select the **Cone** command and use the following command sequence.

> Command: **cone**
>
> Current wire frame density: **ISOLINES=4**
>
> Specify center point for base of cone or [Elliptical] <0,0,0>: *(Enter a point on the screen.)*
>
> Specify radius for base of cone or [Diameter]: **2**

Note that you can construct the base by specifying either the diameter or the radius of the base. In our case, we entered a value of 2 units for the radius, since the radius is the default specification. If you wish to enter the diameter, enter **D** at this prompt and you will be prompted for the diameter.

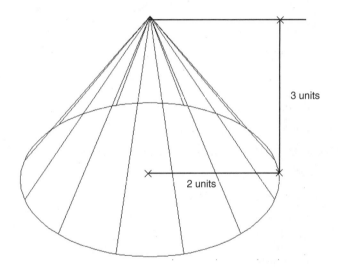

3 units

2 units

Figure 34.9 *Completed Cone in 3D*

Let's continue with the command sequence:

> Specify height of cone or [Apex]: **3**

Figure 34.9 shows the completed cone in 3D.

 Note: By default, AutoCAD draws solids with just four isolines, the lines that define the surface of the solid. You will find solids look better when you increase the value of system variable isolines to 12, as follows:

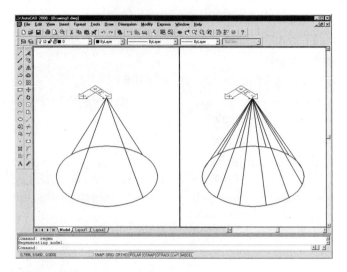

Figure 34.10 *Cone with 4 Isolines (at left) and 12 Isolines (at right)*

Command: **isolines**

New value for ISOLINES <4>: **12**

Command: **regen**

Regenerating drawing.

CONSTRUCTING A CONE WITH AN ELLIPTICAL BASE

The **Elliptical** option allows you to construct a cone with an elliptical base. Let's look at an example. Refer to Figure 34.11.

Command: **cone**

Current wire frame density: ISOLINES=4

Specify center point for base of cone or [Elliptical] <0,0,0>: **E**

Specify axis endpoint of ellipse for base of cone or [Center]: *(Enter a point for the first axis endpoint—shown as "Point 1".)*

Specify second axis endpoint of ellipse for base of cone: *(Enter "Point 2".)*

Specify length of other axis for base of cone: *(Enter "Point 3".)*

Specify height of cone or [Apex]: **3**

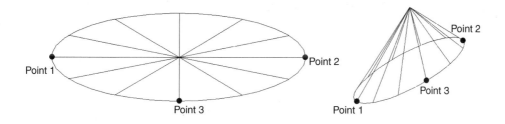

Figure 34.11 *Drawing a Cone with an Elliptical Base*

The base ellipse is constructed in the same manner as a standard ellipse through the AutoCAD **Ellipse** command. You can also construct an ellipse using the **Center** option (see the third line of the previous command sequence). The following command sequence shows the prompts for this type of solid cone construction:

> Command: **cone**
>
> Specify center point for base of cone or [Elliptical] <0,0,0>: **E**
>
> Specify axis endpoint of ellipse for base of cone or [Center]: **C**
>
> Specify center point of ellipse for base of cone <0,0,0>:
>
> Specify axis endpoint of ellipse for base of cone:
>
> Specify length of other axis for base of cone:
>
> Specify height of cone or [Apex]:

Notice, again, the similarity to the AutoCAD **Ellipse** command.

CREATING A SOLID CYLINDER

The **Cylinder** command is used to construct solid cylinders.

Figure 34.12 *Solid Cylinder (with hidden lines removed)*

Solid cylinders are constructed in a manner similar to solid cones. The difference, of course, is that the cylinders are not tapered to a point. Like cones, cylinders can be constructed with either circular or elliptical bases. Elliptical base cylinders are constructed in the same manner as cones with elliptical bases.

Let's construct a solid cylinder with a circular base. Refer to Figure 34.13 for the points to enter.

> Command: **cylinder**
>
> Current wire frame density: ISOLINES=4
>
> Specify center point for base of cylinder or [Elliptical] <0,0,0>:
>   *(Enter "Point 1".)*
>
> Specify radius for base of cylinder or [Diameter]: *(Enter "Point 2".)*
>
> Specify height of cylinder or [Center of other end]: **3**

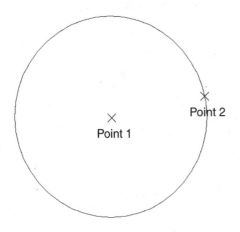

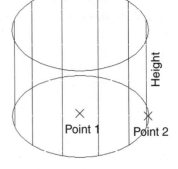

Figure 34.13 *Constructing a Solid Cylinder*

CREATING A SOLID SPHERE

The **Sphere** command is used to create a solid 3D sphere (ball).

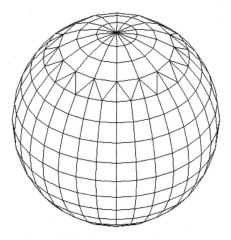

Figure 34.14 *Solid Sphere (with hidden lines removed)*

Constructing a solid sphere is very simple. You designate first the center point of the sphere and then either the radius or diameter. Let's construct a sphere. Use Figure 34. 15 for the points to enter.

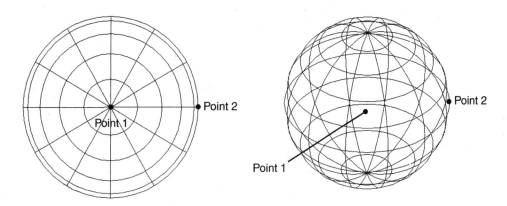

Figure 34.15 *Constructing a Solid Sphere*

Command: **sphere**

Current wire frame density: ISOLINES=4

Specify center of sphere <0,0,0>: *(Enter "Point 1".)*

Specify radius of sphere or [Diameter]: *(Enter "Point 2".)*

The sphere is constructed with the center point you entered on the *X,Y* axis of the current UCS. The circle described by the two entered points is then rotated about the center point to construct the sphere. The vertical axis of the sphere is perpendicul~~~ *X,Y* plane of the UCS.

CONSTRUCTING A SOLID TORUS

The **Torus** command is used to create a solid torus. A torus is a circle rotated about a point to create a tube, like a donut. The **Torus** command allows some variations of the traditional torus object. Figure 34.16 shows a solid torus.

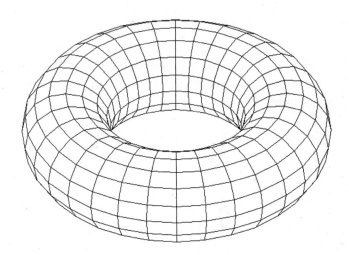

Figure 34.16 *Solid Torus (with hidden lines removed)*

Before we construct a solid torus, let's look at the components of a torus. Figure 34. 17 shows plan views of a torus with the components listed.

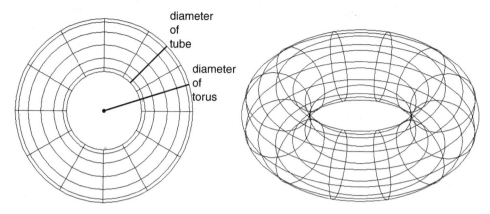

Figure 34.17 *Components of a Torus*

Let's construct a standard torus. Use the following command sequence and Figure 34.18.

> Command: **torus**
>
> Current wire frame density: ISOLINES=4
>
> Specify center of torus <0,0,0>: *(Enter "Point 1".)*
>
> Specify radius of torus or [Diameter]: *(Enter "Point 2".)*
>
> Specify radius of tube or [Diameter]: *(Enter "Point 3".)*

It is possible to construct a torus with a tube radius that exceeds the radius of the torus. Take a moment to create a torus that has a radius of 2 and a tube radius of 5. View the torus in 3D with the **VPoint** command. Notice how the tubes intersect around the center point of the torus.

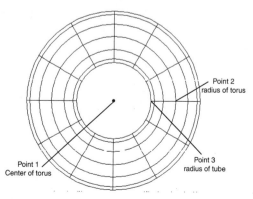

Figure 34.18 *Constructing a Torus*

You can also use a negative value for the torus radius. The tube radius, however, must be a positive number of greater value. For example, if the torus radius is −3, the tube radius must be greater than 3. Construct a torus with a torus radius of −3 and a tube radius of 4.

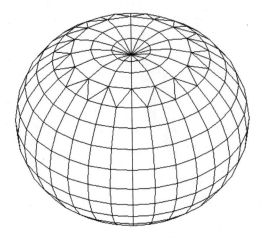

Figure 34.19 *Torus with Tube Radius Greater than Torus Radius*

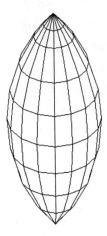

Figure 34.20 *Torus with Negative Torus Radius*

CONSTRUCTING A SOLID WEDGE

The Wedge command is used to create a solid wedge.

Construct the wedge by specifying the dimensions of the base and then the height of the end of the wedge. Alternatively, you can enter the length, width, and height in numeric values. Let's construct a wedge. Refer to Figure 34.22.

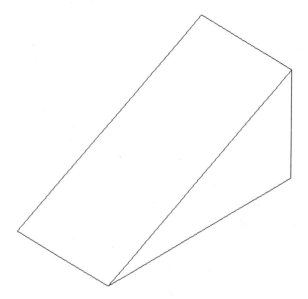

Figure 34.21 *Solid Wedge (with hidden line removed)*

Command: **wedge**

Specify first corner of wedge or [CEnter] <0,0,0>: *(Enter "Point 1".)*

Specify corner or [Cube/Length]: *(Enter "Point 2".)*

Notice how the base of the wedge is "rubber banded" from the first point. Let's continue in the command sequence.

Specify height: **2**

If you view the wedge in 3D, you will notice the height is applied to the end described by the first point entered, and the "point" of the wedge is placed at the end of the base described by the second point. The slope is always along the X axis.

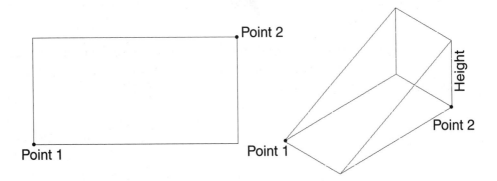

Figure 34.22 *Constructing a Wedge*

Length Option

If you wish to construct a wedge by entering the actual dimensions as numeric values, simply select the **Length** option. The following command sequence shows the prompts displayed when you select Length.

> Command: **wedge**
>
> Specify first corner of wedge or [CEnter] <0,0,0>: *(Enter a point.)*
>
> Specify corner or [Cube/Length]: **L**
>
> Specify length: *(Enter a numeric value.)*
>
> Specify width: *(Enter a numeric value.)*
>
> Specify height: *(Enter a numeric value.)*

Center and Cube Option

You can create a wedge centered on a point. As well, the wedge can have equal sides (the **Cube** option):

> Command: **wedge**
>
> Specify first corner of wedge or [CEnter] <0,0,0>:**CE**
>
> Specify center of wedge <0,0,0>: *(Pick.)*
>
> Specify opposite corner or [Cube/Length]: **C**
>
> Specify Length:

CHAPTER REVIEW

1. What is the primary difference between an object created with *3D surfaces* and one created as a *solid model?*

2. What is meant by a *building block approach?*

3. What is a solid *primitive?*

4. What are some solid primitive objects?

5. What are some ways to access AutoCAD's solid modeling commands?

6. What are the two types of solid cone bases that can be constructed?

7. What is a torus?

CHAPTER 35

Creating Custom Solids

In the previous chapter, you learned how to construct solid primitives. In this chapter, you practice techniques for creating solid objects of your own design. After completing this chapter, you will be able to

- Manipulate the **Extrude** and **Revolve** commands to create 3D solid objects from 2D objects

- Comprehend the rules and limitations of the commands

CREATING SOLIDS

You use AutoCAD commands to create many 3D solids objects from 2D patterns. We explore the following commands and their functions in this chapter:

Extrude

Creates solid extrusions from circles and polylines, including tapered extrusions.

Revolve

Creates solid revolutions by revolving an object around an axis.

Let's get going and create some interesting solids!

CREATING SOLID EXTRUSIONS

The **Extrude** command is used to create solid objects by extruding objects. The following objects can be extruded:

Closed Polylines	Closed Splines
Polygons	Doughnuts
Circles	Regions
Ellipses	

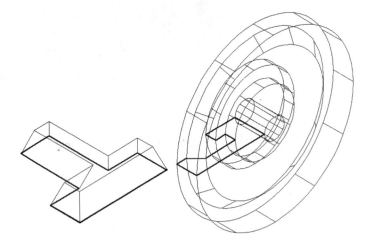

Figure 35.1 *Solid Revolution and Solid Tapered Extrusion*

The effect of the **Extrude** command is similar to extruding an object in 3D by specifying a thickness. One difference is that **Extrude** creates a solid extrusion as opposed to a wireframe 3D extrusion. Another difference is that the extrusion walls need not be parallel.

Solid Extrusion Rules and Limitations

To be extruded, a polyline must have at least three vertices and no more than 500 vertices. The polyline must be closed without crossing itself. If that (or any existing) segment crosses over another segment, the polyline cannot be extruded. If a polyline with a non-zero width is selected, it is extruded with a zero width at the centerline of the polyline. You can extrude several objects in one operation. If any selected objects cannot be extruded, they are simply ignored. Extrusions are created perpendicular to the *X,Y* plane of the object. You cannot extrude objects in a block.

Creating an Extrusion

1. Use the **PLine** command to draw an object similar to the one shown in Figure 35.2. Be sure to close the polyline with the **Close** option or with object snap ENDpoint.

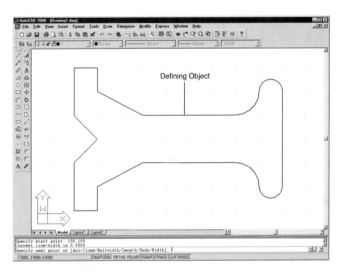

Figure 35.2 *Drawing the Object*

2. Now use the **Extrude** command to create an extrusion that is 3 units "thick." Select **Draw | Solids | Extrude** from the menu bar, and follow the command sequence:

   Command: **extrude**

   Current wire frame density: ISOLINES=4

   Select objects: *(Select the object to extrude.)*

   1 found Select objects: *(Press ENTER.)*

   Specify height of extrusion or [Path]: **3**

Specify angle of taper for extrusion <0>: *(Press Enter to accept 0.)*

3. Now use the **VPoint 1,1,1** command to view the object in 3D. When you are finished, use the **Plan** command to return to the plan view and **Zoom All**.

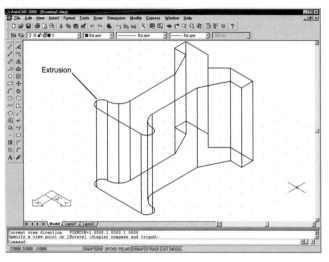

Figure 35.3 *Extruded Object in 3D*

Creating a Tapered Extrusion

Let's use **Extrude** to create a *tapered* extrusion. In a tapered extrusion, the "walls" are slanted inward.

1. Again, use the **PLine** command to draw an object similar to the one shown in Figure 35.4.

2. Now use the **Extrude** command to create an extrusion 5 units thick, with a taper of 30 degrees.

   Command: **extrude**

   Current wire frame density: ISOLINES=4

   Select objects: *(Select the object to extrude.)*

   1 found Select objects: *(Press* ENTER.*)*

   Specify height of extrusion or [Path]: **0.5**

   Specify angle of taper for extrusion <0>: **30**

3. Now use the **VPoint –1,–1,1** command to view the object as before. Notice how the sides of the polyline segments taper 30 degrees.

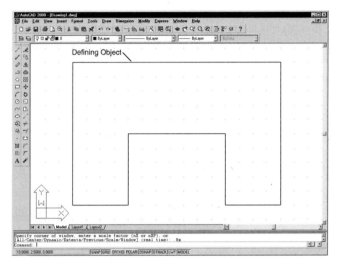

Figure 35.4 *Drawing the Object*

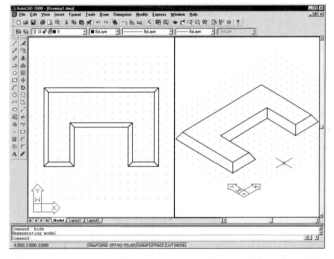

Figure 35.5 *Tapered Extrusion: Plan View (at left) and Isometric View (at right)*

The degree of taper specified must be greater than zero but less than 90 degrees. You cannot use negative values to create an "outward" taper. Extrusions can only taper inward from the original object. If the taper would intersect itself, the **Extrude** command fails.

CREATING A SOLID REVOLUTION

The **Revolve** command is used to create solid revolutions. This is similar to the **RevSurf** command. Revolving an object about a specified axis creates revolved solids. The following objects can be revolved:

Closed Polylines   Closed Splines

Polygons   Doughnuts

Circles   Regions

Ellipses

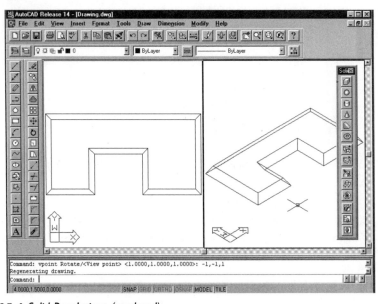

Figure 35.6 *Solid Revolutions (rendered)*

Revolved Solids Rules and Limitations

In order to be revolved, a polyline must have at least three vertices, but no more than 500 vertices. Polylines with non-zero width are changed to zero width. You can only revolve one object at a time. You cannot revolve a block. The closed polyline rules apply as with the **Extrude** command.

Creating a Revolved Solid

Let's use the Revolve command to create a revolved solid.

    1. Use the PLine command to draw an object similar to the one shown in Figure 35.7. Omit the notations shown.

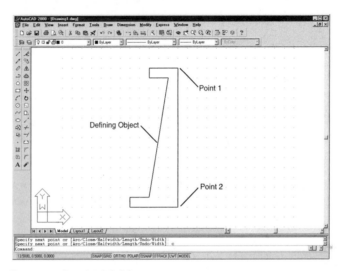

Figure 35.7 *Creating a Revolved Solid*

    2. From the menu bar, select **Draw | Solids | Revolve** and then follow this command sequence to create the revolved solid:

    Command: **revolve**

    Current wire frame density: ISOLINES=4

    Select objects: *(Select the defining polyline.)*

    1 found Select objects: *(Press* ENTER.*)*

    Specify start point for axis of revolution or define axis by [Object/X
        (axis)/Y (axis)]: **end**

    of *(Select "Point 1".)*

    Endpoint of axis: *(Select "point 2".)*

Specify angle of revolution <360>: *(Press Enter to accept a full circle revolution.)*

3. Again, use the **VPoint** command to view the object in 3D.

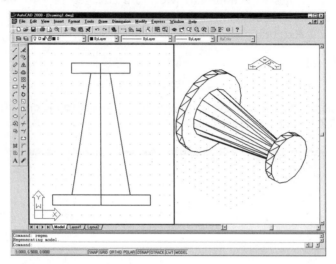

Figure 35.8 *Revolved Solid: Plan View (left view) and Isometric View (right view)*

4. After you have viewed the solid object, use **Undo** to return to the original polyline. Try a new revolved solid, using 180 degrees as the included angle.

Revolving Around an Object

You can use a separate object as the axis around which to revolve a polyline.

1. Draw a line in the position shown in Figure 35.9.

2. Now let's create a solid object with an open shaft along its central axis.

Command: **revolve**

Current wire frame density: ISOLINES=4

Select objects: *(Select the defining polyline.)*

1 found Select objects: *(Press* ENTER.*)*

Specify start point for axis of revolution or define axis by [Object/X (axis)/Y (axis)]: **OBJECT**

Select an object: *(Select the line.)*

Angle of revolution <full circle>: *(Press* ENTER *to accept a full circle revolution.)*

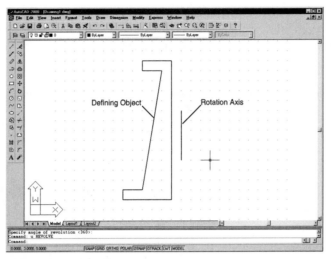

Figure 35.9 *Revolving Around an Entity*

3. View the object in 3D and notice the central shaft created by the location of the line from the polyline. If the axis line is at an angle and if that would cause the revolved object to intersect itself, AutoCAD complains, "Unable to revolve the selected object."

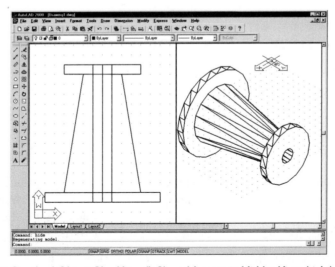

Figure 35.10 *Revolved Object: Plan View (left) and Isometric Hidden View (right)*

X and Y Rotation Options

The **X** and **Y** options of the **Revolve** command use either the X axis or Y axis of the current UCS as the axis of revolution. Figure 35.11 shows the effects of using each with a full circle revolution.

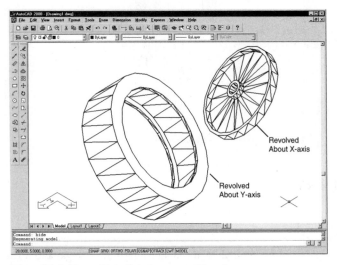

Figure 35.11 *Using X (left) and Y (right) Options of Revolve*

CHAPTER REVIEW

1. What is an *extrusion*?

2. What objects can be extruded?

3. What is created with the **Revolve** command?

4. What are some limitations to the Revolve command?

5. How can objects be revolved?

TUTORIAL

You have seen how AutoCAD creates 3D solid objects by extruding up from a closed object or by revolving a closed object around an axis. You can also extrude a closed object—such as a circle—along a path, such as a spline. This is the way to model tubes, handrails, and piping with bends. Let's see how to draw a tube:

1. Specify the path for the tube by drawing a spline curve:

   Command: **spline**

   Specify first point or [Object]: *(Pick a point.)*

Specify next point:: *(Pick a point.)*

Specify next point or [Close/Fit tolerance] <start tangent>: *(Pick a point.)*

Specify next point or [Close/Fit tolerance] <start tangent>: *(Pick a point.)*

Specify start tangent: *(Pick a point.)*

Specify end tangent: *(Pick a point.)*

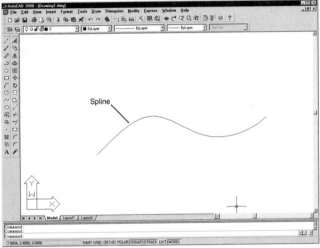

Figure 35.12 *Drawing the Spline Path*

2. Change the UCS so that you are looking at the end of the spline object. That lets you draw the object that defines the cross-sectional shape of the tube:

   Command: **UCS**

   Current ucs name: *WORLD*

   Enter an option
       [New/Move/orthoGraphic/Prev/Restore/Save/Del/Apply/?/
       World] <World>: **y**

   Specify rotation angle about Y axis <90>: **90**

3. Set the view to match the UCS with the **Plan** command:

   Command: **plan**

   Enter an option [Current ucs/Ucs/World] <Current>: *(Press ENTER.)*

   Regenerating drawing.

You may want to use the **Zoom Window** command to make the spline the right size.

4. Draw a circle to define the cross-sectional shape of the tube. (It doesn't need to be a circle; it can be any closed shape, such as an ellipse, a polygon, a rectangle, or closed polyline shape.) The circle can be drawn anywhere convenient; it is the circle that is extruded.

   Command: **circle**

   Specify center point for circle or [3P/2P/Ttr (tan tan radius)]: *(Pick point.)*

   Specify radius of circle or [Diameter]: *(Pick point.)*

   Don't draw the circle too big; otherwise the extrusion will fail.

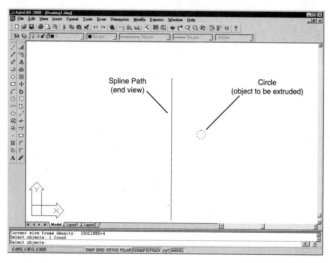

Figure 35.13 *Circle Defines the Cross Section of the Tube*

5. Finally, you are ready to apply the **Extrude** command:

   Command: **extrude**

   Select objects: *(Pick the circle.)*

   I found Select objects: *(Press* ENTER.*)*

   Specify height of extrusion or [Path]: **path**

   Select extrusion path: *(Pick spline curve.)*

   Path was moved to the center of the profile.

   Profile was oriented to be perpendicular to the path.

AutoCAD extrudes the circle along the path. Particularly complex paths and objects take a long time to extrude.

6. With isolines set to 4, the tube may not look like much. Use the **Isolines** system variable to make a more representative wireframe:

Command: **isolines**

Enter new value for ISOLINES <4>: **12**

Command: **regen**

Regenerating model.

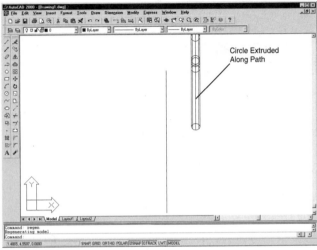

Figure 35.14 *Circle Extruded Along the Spline Path (12 isolines)*

7. Once AutoCAD is finished, change the viewpoint so that you can see the tube:

Command: **vpoint**

*** Switching to the WCS ***

Current view direction: VIEWDIR=1.0000,0.0000,0.0000

Specify a view point or [Rotate] <display compass and tripod>:
       -1,1,0.7

*** Returning to the UCS ***

Regenerating model.

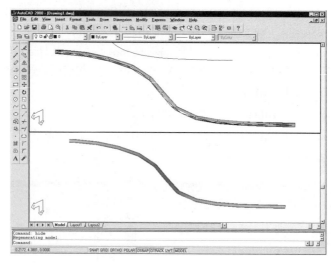

Figure 35.15 *Completed Tube*

8. Use the **Hide** or **Shade** command to see the tube with hidden lines removed:

Command: **hide**

Regenerating drawing.

Hiding lines 100% done.

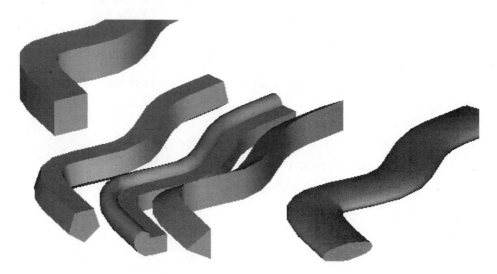

Figure 35.16 *Tubes Created with Other Cross Section Shapes: (left to right) Square, Pentagon, Closed Polyline, Triangle, Ellipse*

CHAPTER 36

Modifying Solid Objects

To create solid objects of any design, you must be able to modify basic solid shapes. After completing this chapter, you will be able to

- Perform Boolean operations on solid models

- Edit solid shapes

- Analyze solid models

INTRODUCTION

In this chapter, you learn about some of the commands for editing 3D solid models. The first three are known as "Boolean" operations:

Intersect

Returns the area common to two or more 3D solids.

Subtract

Subtracts part of a 3D solid model from another.

Union

Joins two solid models together.

The next two are already familiar to you from 2D editing:

Chamfer

Chamfers the edges of a solid model.

Fillet

Fillets the edges of the solid model.

The last few have trickier concepts to learn:

SolidEdit

Edits a solid object by extruding, moving, rotating, offsetting, tapering, copying, coloring, separating, shelling, cleaning, checking, and deleting faces and edges.

Slice

Slices a 3D solid model using a 2D plane.

Section

Creates a 2D region from a 3D solid.

Interference

Finds the interference of two (or more) 3D solids and creates a composite 3D solid from the volumes in common.

In addition, you can use many other editing commands on 3D solids, such as **Copy, Move, Stretch, Mirror, Color,** and **Explode.**

BOOLEAN OPERATIONS

Solid models can be constructed by the combination of custom and primitive shapes. You can intersect, combine, and subtract shapes to achieve the 3D model you desire. These operations are known as *Boolean operations*. Let's look at the commands used to modify solid shapes.

CREATING SOLID INTERSECTIONS

The **Intersect** command is used to create a solid from the intersection of two solids. The resulting solid is created from the common volume occupied by both solids. Figure 36.1 shows the result of using the **Intersect** command with a box and a sphere.

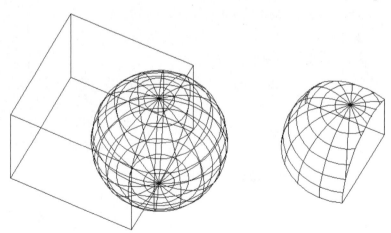

Figure 36.1 *Creating Solid Intersections*

When you select the **Intersect** command, you are prompted to choose the objects that are to be used for the operation.

Command: **intersect**

Select objects: *(Pick solid objects.)*

Select objects: *(Pick solid objects.)*

Select objects: *(Press* ENTER.*)*

AutoCAD removes everything not common to the selected objects.

SUBTRACTING SOLIDS

The **Subtract** command is used to subtract one solid shape from another, creating a new solid. Figure 36.2 shows a box with four cylinders. The **Subtract** routine has been used twice: (1) in the center, the box is subtracted from the cylinders, and (2) at right, the cylinders are subtracted from the box. This shows that the order of subtraction is important.

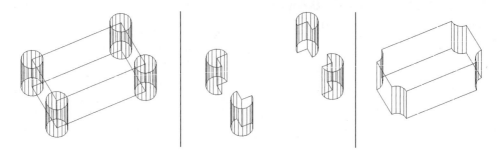

Figure 36.2 *Subtracting Solids*

Let's look at how the **Subtract** command prompts for the solids you will modify. **Subtract** first asks for the objects that will be subtracted *from*. The subtracted object(s) can be one or more objects. When you select the subtracted object(s), AutoCAD automatically performs a union of the selected objects.

Command: **subtract**

Select solids and regions to subtract from...

> Select objects: *(Select one or more source objects.)*
>
> Select objects: *(Press ENTER.)*

AutoCAD then prompts for the solid objects to be subtracted from the source object:

> Select solids and regions to subtract...
>
> Select objects: *(Select one or more subtraction objects.)*
>
> Select objects: *(Press ENTER.)*

AutoCAD then subtracts these objects from the source object, creating a new solid.

JOINING SOLID OBJECTS

The **Union** command joins two solid objects together as a single object. The new solid encloses both objects and their common spaces (if they overlap).

> Command: **union**
>
> Select objects: *(Select one or more objects.)*
>
> Select objects: *(Press ENTER.)*

AutoCAD joins the solids into a single object.

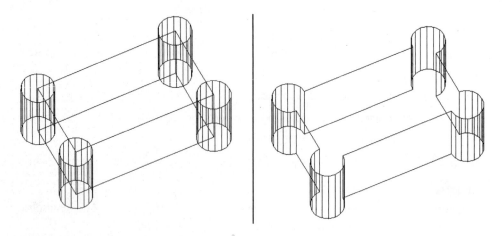

Figure 36.3 *Unioned Solids*

UNDO EDITING

You can use the **U**, **Undo**, and **Redo** commands to reverse the effects of your editing changes on 3D solid objects.

MODIFYING SOLID OBJECTS

You can use many of the editing commands familiar to you from 2D drafting and editing. Most of these operate exactly the way you expect them to (copy, move, array, and so on). Two of the commands—**Chamfer** and **Fillet**—operate somewhat differently than they do with 2D objects. A third editing command, **SolidEdit**, is unique to editing 3D solids.

CHAMFERING SOLIDS

The **Chamfer** command is used to create bevel edges on existing solid objects. You achieve the chamfer by first selecting a *base surface* and then selecting the edges of the base surface to be chamfered. Finally, the chamfer dimension is specified. Let's look at how the **Chamfer** command sequence is used.

> Command: **chamfer**
>
> (TRIM mode) Current chamfer Dist1 = 0.5000, Dist2 = 0.5000
>
> Select first line or [Polyline/Distance/Angle/Trim/Method]: *(Select an isoline on a solid.)*
>
> Base surface selection...
>
> Enter surface selection option [Next/OK (current)] <OK>: **next**

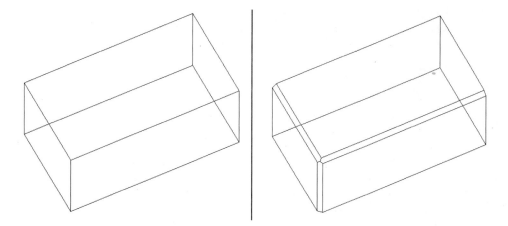

Figure 36.4 *Solid Object with Chamfers*

When you select a solid, you do so by selecting one of the solid's isolines. (The *iso-lines* define the surface of the solid shape. On a box or wedge, the isolines happen to coincide with the edges of the planes.) When you select an isoline, it is the intersection of two faces. AutoCAD highlights one face; if it is the wrong face, type **Next** to select the adjacent face.

> Enter surface selection option [Next/OK (current)] <OK>: *(Press Enter to accept face.)*

When the correct face is highlighted, press ENTER.

Next, you specify the chamfer distance. The default is 0.5 units.

> Specify base surface chamfer distance <0.5000>: *(Enter value or press ENTER.)*
>
> Specify other surface chamfer distance <0.5000>: *(Enter value or press ENTER.)*

If the chamfer distance is too large, AutoCAD complains, "Modeling Operation Error: Mitered vertex too complex to process. Failed to perform blend. Failure while chamfering." and does not perform the chamfer. Restart the **Chamfer** command and specify a smaller distance.

Finally, you choose the edges you want chamfered. Most faces have three or four edges. You can select individual edges, or all edges with the **Loop** option. Let's continue with the command sequence:

> Select an edge or [Loop]: *(Pick edges or enter L for all edges.)*

Select the edges to be chamfered as the response to this prompt. If the edge you select is not adjacent to the base surface, it will not be chamfered. You can select as many edges as you wish.

FILLETING SOLIDS

The **Fillet** command is used to create a filleted edge on a solid object.

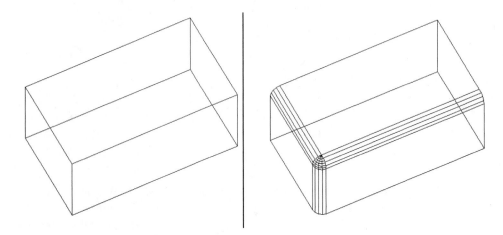

Figure 36.5 *Solid Fillet*

Fillet can create concave or convex fillets. You create the fillet by selecting the edges to be filleted.

> Command: **fillet**
>
> Current settings: Mode = TRIM, Radius = 0.5000
>
> Select first object or [Polyline/Radius/Trim]: *(Pick a solid.)*
>
> Enter fillet radius <0.5000>: *(Enter value or press* ENTER.*)*
>
> Select an edge or [Chain/Radius]: *(Pick edges.)*

AutoCAD fillets the edges.

EDITING SOLIDS

The **SolidEdit** command allows you to perform a number of operations on a 3D solid model. You can use it to extrude a face, taper a hole, offset faces, create a thin shell body, and change the solid's colors.

Command: **solidedit**

Solids editing automatic checking: SOLIDCHECK=1

Enter a solids editing option

[Face/Edge/Body/Undo/eXit] <eXit>:

The **SolidEdit** command allows you to change the face, edge, and body of the solid. A *face* is a "side" of the solid. The *edge* is the intersection of two faces. A *body* is the entire solid, no matter the shape.

The **SolidCheck** system variable validates the solid as an object that the ACIS solids modeler will accept.

Editing Faces

To edit one or more faces, enter **F**. AutoCAD displays these options:

Enter a face editing option
    [Extrude/Move/Rotate/Offset/Taper/Delete/Copy/coLor/Undo/
    eXit] <eXit>:

The options are as follows:

Extrude

The **Extrude** option extrudes selected planar faces to a height you specify, or along a path. This option operates like the **Extrude** command, except that this option operates on existing solids. This option is good for lengthening faces; you can specify a taper angle.

Move

The **Move** option moves a selected feature within a solid object to another location. For example, this option can be used to move a hole or stretch the object.

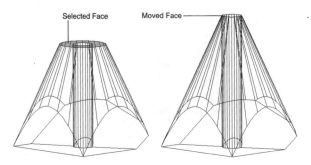

Figure 36.6 *Moving a Face*

Rotate

The **Rotate** option rotates a face or a feature.

Offset

The **Offset** option offsets faces equally by a distance or through a point. Use a positive value to increases the size of the solid; a negative value decreases the solid. Note that the opposite is true for holes: a negative value makes the hole larger, while a positive value makes the hole smaller.

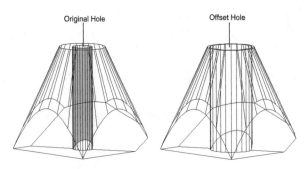

Figure 36.7 *Offseting a Feature*

Taper

The **Taper** option tapers faces with an angle. This is useful for turning a cylinder into a cone shape.

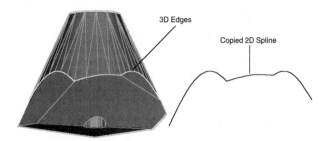

Figure 36.8 *Tapering a Feature*

Delete

The **Delete** option removes features, such as fillets, holes, etc.

Copy

The **Copy** option makes a copy of a feature, such as a hole.

Color

The **Color** option lets you change the color of a face. AutoCAD displays the **Select Color** dialog box.

Undo

The **Undo** option reverses the effect of the previous option.

Exit

Returns to the original **SolidEdit** prompt.

Editing Edges

To edit one or more edges, enter **E**. AutoCAD displays these options:

> Enter an edge editing option [Copy/coLor/Undo/eXit] <eXit>:

The options are:

Copy

The **Copy** option copies an edge from the solid as a 2D object— line, arc, spline, or ellipse. This is useful for creating 2D views of 3D objects.

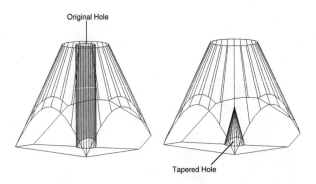

Figure 36.9 *Copying Edges From a 3D Solid*

Color

The **Color** option lets you change the color of an edge. AutoCAD displays the **Select Color** dialog box.

Undo

The **Undo** option reverses the effect of the previous option.

Exit:

Returns to the original **SolidEdit** prompt.

Editing the Body

To edit a solid body, enter **B**. AutoCAD displays these options:

> Enter a body editing option [Imprint/seParate
> solids/Shell/cLean/Check/Undo/eXit] <eXit>:

Imprint

The **Imprint** option "imprints" a 3D solid with another object—an arc, circle, line, 2D polyline, ellipse, spline, region, 3D polyline, body, or 3D solid. You can think of imprinting as a way of adding 2D objects to a solid object. In Figure 36.10, the circle has been imprinted on the cube. After imprinting, the resulting semi-circle is part of the 3D solid and is no longer a circle (or arc) object.

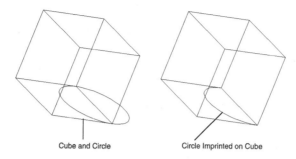

Cube and Circle   Circle Imprinted on Cube

Figure 36.10 *Imprinting a Circle on a 3D Solid*

Separate Solids

The **seParate** solids option separates a 3D solid with disjointed volumes into independent solid objects. This command does not operate on solids that have been created by Boolean operations, such as **Union** and **Subtract**.

Shell

The **Shell** option turns a 3D solid into a hollow, thin-walled object; you specify the wall thickness. Offsetting each existing face to the outside of its original position creates the new object. Note that a 3D solid can have one shell only.

At AutoCAD's "Enter the shell offset distance" prompt, specifying a positive value to create a shell that is offset to the inside of the solid; a negative value offsets the shell to the outside of the solid. After the **Shell** operation, the 3D solid may look identical, because it becomes "hollow" on the inside. Figure 36.11 shows a shaded, cut-away view of the shelled object.

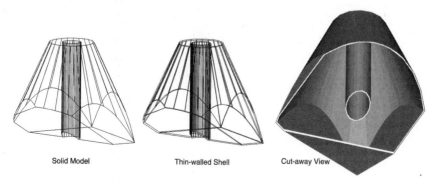

Solid Model   Thin-walled Shell   Cut-away View

Figure 36.11 *Creating a Shell*

Clean

The **cLean** option removes redundant edges and vertices, as well as imprinted objects and unused geometry.

Check

The **Check** option determines whether a 3D object is a valid ACIS solid. AutoCAD reports either "This object is a valid ACIS solid" or "A 3D solid must be selected."

Related to this option is the **SolidCheck** system variable, which toggles 3D solid validation on and off. By default, this option is turned on.

Undo

The **Undo** option reverses the effect of the previous option.

Exit

Returns to the original **SolidEdit** prompt.

SLICING 3D SOLIDS

The **Slice** command slices a 3D solid model in half. Actually, it doesn't need to be "in half." Anywhere you place a 2D plane that intersects the solid is where the **Slice** command splits the solid model apart. AutoCAD gives you the option to retain both halves or to erase one of the halves.

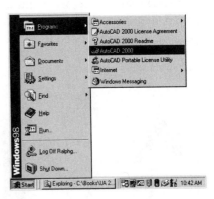

Command line: **slice**

Select objects: *(Pick one or more solid models.)*

Select objects: *(Press* ENTER.*)*

Specify first point on slicing plane by
[Object/Zaxis/View/XY/YZ/ZX/3points] <3points>: *(Pick slicing plane.)*

Specify a point on desired side of the plane or [keep Both sides]:
(Type B to retain both halves or pick the half you want to keep.)

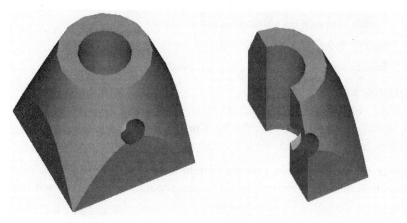

Figure 36.12 *Solid Model Rendered (left) and Sliced (right)*

Object

The **Object** option aligns the 2D cutting plane with an object. Legitimate 2D objects include circle, ellipse, arc, spline, and polyline. To use this option, you must have drawn one of those 2D objects *before* starting the **Slice** command.

Zaxis

This option defines the 2D-cutting plane with a point on the *Z* axis of the *XY* plane. When you enter **Z**, AutoCAD prompts:

Specify point on plane: *(Pick a point on the XY plane.)*

Specify point on Z-axis (normal) of the plane: *(Pick a point on the Z axis.)*

View

This option aligns the 2D cutting plane with viewing plane of the current viewport. AutoCAD prompts:

Specify point on view plane <0,0,0>: *(Pick a point.)*

XY or YZ or ZX

These three options align the 2D cutting plane with the plane of the current UCS. The plane is either in the *XY, YZ,* or *ZX* coordinate plane. You need to specify a point to define the location of the cutting plane in 3D space, as follows:

Point on XY plane <0,0,0>: *(Pick a point.)*

3points

This option defines the 3D cutting plane with three points. AutoCAD prompts:

Specify 1st point on plane: *(Pick first point.)*

Specify 2nd point on plane: *(Pick first point.)*

Specify 3rd point on plane: *(Pick first point.)*

SECTIONING SOLIDS

In many ways, the **Section** command is similar to the **Slice** command. Whereas the **Slice** command cuts a solid in half, the **Section** command gives you the slice. It uses the intersection of a plane and solids to create a region.

AutoCAD creates a region object on the current layer and inserts it at the location of the cross section. Selecting several solids creates separate regions for each solid.

Command: **section**

Select objects: *(Select object.)*

Specify first point on Section plane by

[Object/Zaxis/View/XY/YZ/ZX/3points] <3points>: *(Select an option.)*

The options are exactly the same as for the **Slice** command. After executing the **Section** command, you should see some new 2D objects where you performed the section. The new 2D objects are region objects. You can move the section with the **Move** command to see it more clearly or hatch the area.

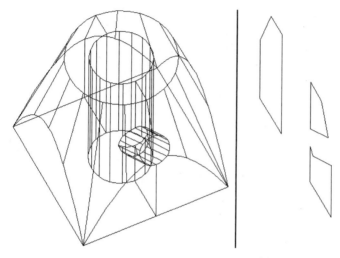

Figure 36.13 *Solid Model (left) and Section (right)*

FINDING AN INTERFERENCE

The **Interference** command is like a non-destructive **Intersection** command. **Interference** finds the volume common to two or more 3D solids, then optionally generates the resulting 3D solid.

Command: **interfere**

Select first set of solids:

Select objects: *(Pick solid.)*

1 found Select objects: *(Press* ENTER.*)*

Select second set of solids:

Select objects: *(Pick solid.)*

1 found Select objects: *(Press* ENTER.*)*

Comparing 1 solid against 1 solid.

Interfering solids	(first set)	: 1
	(second set)	: 1
Interfering pairs		: 1

Create interference solids ? <N>: **y**

Answering **Y** creates a third solid object, which is the result of the intersection of the two solids being checked for interference. To see the "interference" solid more clearly, use the **Move** command and select the **L** (last) object, moving it from out of the other solids.

When there are more than two objects that interfere, you get a chance to highlight the pairs of interfering solids. Answer **Y** to highlight and then choose **Next** to cycle through all possible pairs.

When objects do not interfere, AutoCAD reports, "Solids do not interfere."

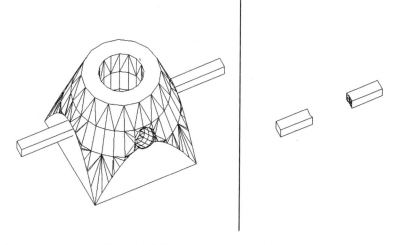

Figure 36.14 *Solid Model (left) and Interference Solid (right)*

CHAPTER REVIEW

1. Describe the function of the **Subtract** command.

2. Which command is used to create a new solid from the common area described by two overlapping solids?

3. How would you join two existing solids?

4. When chamfering a solid object, what is meant by the *base surface*?

5. How do the results of the **Slice** command differ from the **Section** command?

6. What is a *cutting plane*?

7. How does **Interference** differ from **Intersect**?

8. What is the **Interference** command good for?

TUTORIAL

Let's try using the **Subtract** and **Slice** commands now.

1. Start a new drawing.

2. Draw a cube. Use the **Box** command with origin at 0,0,0 and length of all three sizes is 10.

3. Draw a column. Use the **Cylinder** command with origin at 5,5,0, radius of 2, and a height of 10.

4. Use the **Subtract** command to subtract the cylinder from the box.

5. This leaves you with a cube with a hole cut out. To get a better view, set a viewpoint of 1,2,3.

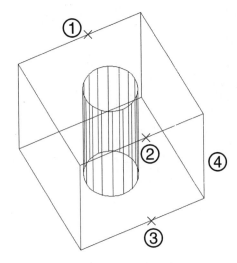

Figure 36.15 *Solid Model*

6. Let's get a view of the inside of the holed cube. You use the **Slice** command with the **3points** option:

Command: **slice**

Select objects: *(Pick solid object.)*

I found Select objects: *(Press* ENTER.*)*

Specify first point on Section plane by
     [Object/Zaxis/View/XY/YZ/ZX/3points] <3points>: **3**

Specify 1st point on plane: **mid**

of *(Pick point "1".)*

Specify 2nd point on plane: **mid**

of *(Pick point "2".)*

Specify 3rd point on plane: **mid**

of *(Pick point "3".)*

Specify a point on desired side of the plane or [keep Both sides]:
    (Pick point "4".)

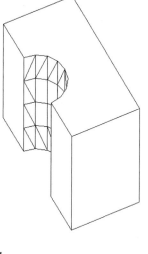

Figure 36.16 *Sliced Solid Model*

AutoCAD slices the solid model in half, erasing the half that you didn't select.

7. If you want, use the **Hide** or **Shade** commands to get a clearer view of the sliced solid.

8. Save the drawing as "Slice.Dwg."

CHAPTER 37

Creating a Composite Solid Model

To effectively use AutoCAD's solid modeler, you must practice using the commands in actual modeling conditions. After completing this chapter, you will be able to

- Demonstrate solid modeling commands through experience with an actual solid model

- Draw and edit a solid model

- Generate engineering views

INTRODUCTION

The primary focus of this chapter is a tutorial that lets you practice using some of the solids modeling commands you have learned in the previous chapters.

In addition, this chapter teaches you about a command for setting up engineering views of solid models—in effect, converting the 3D model back to traditional 2D views. The command:

SolView

Sets up standard engineering viewports.

DRAWING A COMPOSITE SOLID MODEL

Let's start a solid model drawing. You have the opportunity to construct the model, edit it, and display it as both a hidden line and solid object. Figure 37.1 shows the finished model.

Figure 37.1 *Completed Solid Model Drawing*

Beginning the Model

Start by beginning a new drawing. Use the **New** command to start a new drawing. This tutorial uses the default template drawing settings. If the **Acad.dwt** file has been changed, the actual results or display may vary from those shown.

Figure 37.2 is a dimensioned drawing of the object you will draw. You may want to refer to it as you construct the base model from solid primitives.

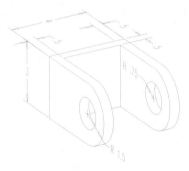

Figure 37.2 *Model I Dimension*

Begin the drawing by using these commands to construct the basic model shape. Figure 37.3 shows the building block components you will use to "assemble" the model.

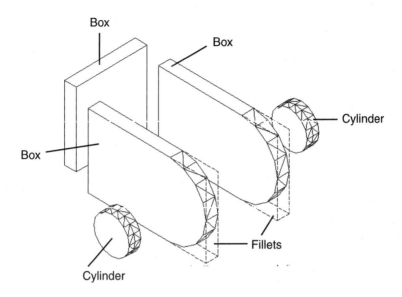

Figure 37.3 *Building Block Component*

Before starting any drawing, you should analyze the object to determine the best primitives to use. You are going to use both solid commands and some 3D principles to construct the model.

    1. Let's get started. Set Snap to 0.5.

Creating the First Box

    2. Select the **Box** command from the **Draw | Solids** menu, and enter the points indicated in the following command sequence. I'll discuss each step as you perform it.

        Command: **box**

Select the first corner of the box.

        Specify corner of box or [CEnter] <0,0,0>: **3,2**

Next, define the opposite corner of the box.

        Specify corner or [Cube/Length]: **@5,0.5**

Now define the height of the box.

        Specify height: **3**

    3. Type the **Box** command again. Refer to Figure 37.4 for the points to enter.

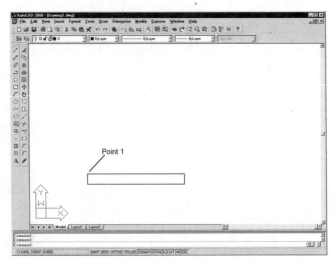

Figure 37.4 *Drawing the First Box*

Command: **box**

Specify corner of box or [CEnter] <0,0,0>: *(Select point "I".)*

Specify corner or [Cube/Length]: **@0.5,3**

Specify height: **3**

Your drawing should look like the one in Figure 37.5.

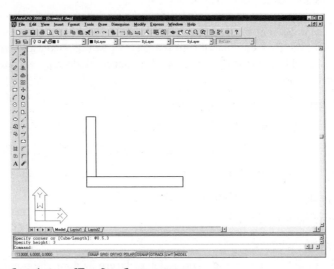

Figure 37.5 *Completion of Two Box Components*

Rotating the View

4. Use the **VPoint** command to rotate the view so you can complete your work.

   Command: **vpoint**

   Current view direction: VIEWDIR=0.0000,0.0000,1.0000

   Specify a view point or [Rotate] <display compass and tripod>: **R**

   Enter angle in XY plane from X axis <0>: **290**

   Enter angle from XY plane <0>: **23**

5. Use a zoom factor to "zoom out" the display of the drawing.

   Command: **zoom**

   Specify corner of window, enter a scale factor (nX or nXP), or
      [All/Center/Dynamic/Extents/Previous/Scale/Window] <real
      time>: **0.7X**

Your drawing should look similar to Figure 37.6.

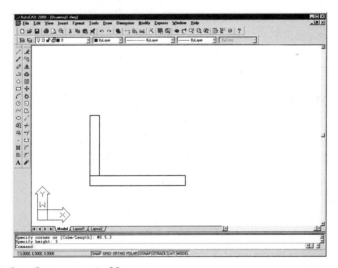

Figure 37.6 *Box Components in 3D*

Adding the Drill Holes

6. You want to add a drill hole to the side of the object. Begin by setting the UCS icon so you can see the origin of the UCS you will be working in.

   Command: **ucsicon**

Enter an option [ON/OFF/All/Noorigin/ORigin] <ON>: **or**

7. Now change the UCS. Refer to the following command sequence and Figure 37.7.

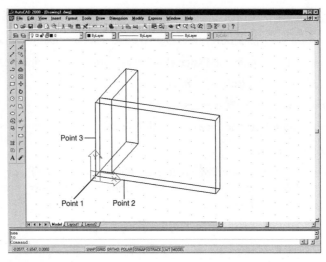

Figure 37.7 *Setting the UCS*

Command: **ucs**

Current ucs name: ***WORLD***

Enter an option
   [New/Move/orthoGraphic/Prev/Restore/Save/Del/Apply/?/
   World] <World>: **n**

Specify origin of new UCS or
   [ZAxis/3point/OBject/Face/View/X/Y/Z] <0,0,0>: **3**

Specify new origin point <0,0,0>: *(Select point "1".)*

Specify point on positive portion of X-axis
   <1.0000,0.0000,0.0000>:**nea**

of *(Select point "2".)*

Specify point on positive-Y portion of the UCS XY plane
   <0.0000,1.0000,0.0000>:**nea**

of *(Select point "3".)*

8. It is now time to add the drill hole. You construct this with the **Cylinder** command. I'll walk through each step of this command. Refer to Figure 37.8.

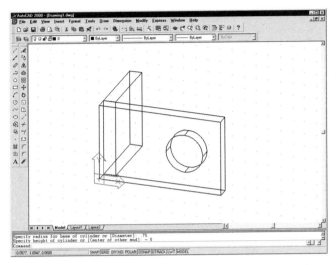

Figure 37.8 *Drawing the Drill Hole*

Command: **cylinder**

Current wire frame density: ISOLINES=4

Next, specify the center point of the cylinder. Note that the absolute coordinates you use are relative to the origin of the new UCS.

Specify center point for base of cylinder or [Elliptical] <0,0,0>:
3.5,1.5

Now specify the radius of the cylinder.

Specify radius for base of cylinder or [Diameter]: **0.75**

Finally, you designate the extrusion height of the cylinder. Since you want the extrusion to extend in a negative direction, the value is negative.

Specify height of cylinder or [Center of other end]: **–0.5**

Adding the Second Leg

9. Copy the first leg (with the cylinder) to create the second leg. Set the UCS back to world. Start the **Copy** command, select the first box and its cylinder, and specify a displacement of @3.5<90.

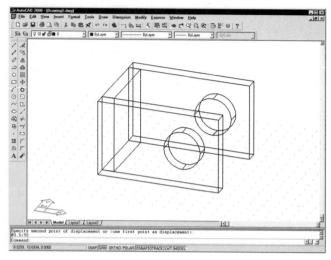

Figure 37.9 *Drawing the Second Drill Hole*

Converting to a Composite Solid

10. Now create a composite solid. To do this, you must change the three boxes into a single solid object and then subtract the cylinders that make the drill holes. You first use the **Union** command to combine the boxes. Use the following command sequence.

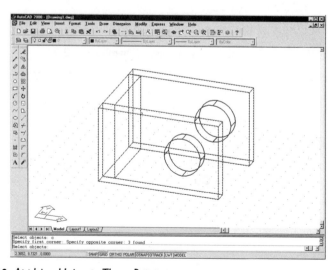

Figure 37.10 *Applying Union to Three Boxes*

Command: **union**

Select objects: *(Select each of the three boxes and press Enter.)*

AutoCAD goes to work and performs the necessary calculations to combine the boxes into a single solid object. You will notice that, when the drawing is redisplayed, the edge lines between the boxes are no longer a part of the object.

Subtracting the Cylinders

11. Continue by subtracting the cylinders from the object. You use the **Subtract** command.

    Command: **subtract**

    Select solids and regions to subtract from...

    Select objects: *(Select the unioned boxes.)*

    Select objects: *(Press* ENTER.*)*

    1 solid selected Select solids and regions to subtract...

    Select objects: *(Select each of the cylinders.)*

    Select objects: *(Press* ENTER.*)*

AutoCAD now evaluates the object and removes the cylinders from the solid object created by the union of the three boxes. Note that you will not see a visual difference after this step. Your drawing should look similar to Figure 37.11.

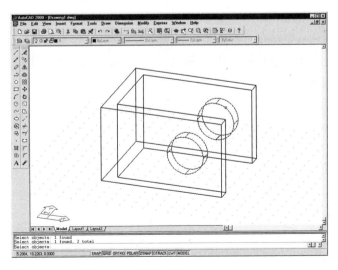

Figure 37.11 *Subtracting Cylinders to Make Holes*

Filleting the Corners

12. Next, you use the **Fillet** command to fillet the corners. Refer to the following command sequence and Figure 37.12. It may be easier to select the edges with snap turned off, and to zoom in closer.

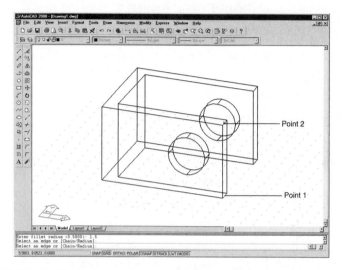

Figure 37.12 *Filleting the Box*

Command: **fillet**

Current settings: Mode = TRIM, Radius = 0.5000

Select first object or [Polyline/Radius/Trim]: *(Select point "1".)*

Enter fillet radius <0.5000>: **1.5**

Select an edge or [Chain/Radius]: *(Select point "2.")*

Select an edge or [Chain/Radius]: *(Press* ENTER.*)*

Your drawing should look like Figure 37.13. You may need to use the **Regen** command to clean up the view.

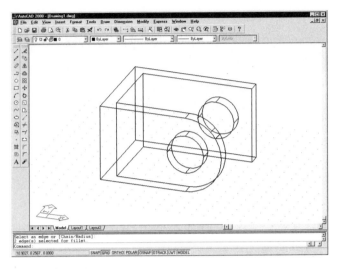

Figure 37.13 *Filleted Box*

13. Now use the **Fillet** command again to fillet the back leg of the object in the same manner. When completed, your drawing should look like Figure 37.14.

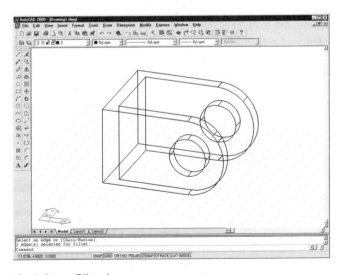

Figure 37.14 *Both Boxes Filleted*

Displaying Your Model

14. You can produce hidden-line and shaded models of your model.

Command: **hide**

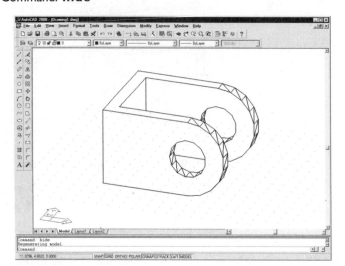

Figure 37.15 *Hidden Line View*

You may want to use the **Shade** command to produce a shaded model.

Command: **shade**

15. Save your work with the **Save** command, calling it "Model1.Dwg."

Figure 37.16 *Completed "Model1" Drawing*

GENERATING ENGINEERING VIEWS

Once you have completed drawing your solid model in 3D, you may need to create a plot of it using standard 2D engineering views, such as top, left, front, and isometric. The **SolView** command creates the viewports and specially named layers in layout mode.

THE SOLVIEW COMMAND

SolView is short for "solids viewports." It creates several viewports in paper space using orthographic projection. In each viewport, **SolView** sets up the viewing parameters for sectional views of 3D solid models.

> Command: **solview**
>
> Regenerating layout.
>
> Regenerating model.
>
> Enter an option [Ucs/Ortho/Auxiliary/Section]:

When you type the **SolView** command, AutoCAD creates a layout view and switches to **Layout1**. You select from one of these options:

Ucs

The **UCS** option creates a profile view relative to a UCS (user coordinate system). You can use a previously saved UCS name or specify a new UCS. This is the only one of **SolView**'s four options that does not require at least one viewport to already exist in the drawing.

Ortho

The **Ortho** option creates a folded orthographic view.

Auxiliary

The **Auxiliary** option creates an auxiliary view from an existing view. The auxiliary view is projected onto a plane perpendicular to one orthographic view and angled in the adjacent view.

Section

The **Section** option creates a sectional (sliced) view of the 3D object. Solid areas are hatched.

 Note: SolView creates layers to place hidden-lines views, hatching, and other information. Since AutoCAD uses these layers for its own purposes, it is important that you not purge these layers or draw anything on them. The layers are

Layer	Meaning
viewname-DIM	Dimensions
viewname-HAT	Hatch patterns for sectional views
viewname-HID	Hidden lines
viewname-VIS	Visible lines
VPORTS	Viewports

In order for lines to be drawn with the Hidden linetype, you must first use the **Linetype** command to load Hidden. **SolView** does not load the linetypes it needs.

TUTORIAL

Let's use the **SolView** command to create a standard engineering drawing of Model1.Dwg. When finished, the drawing will display the front, side, top, and one sectional view.

1. Open Model1.Dwg in AutoCAD. If necessary, switch the view to plan view, using the **Plan** command.

2. Select the Layout1 tab. Press ESC to cancel the **Plot** dialog box, which appears automatically.

3. The viewport takes up the entire sheet. You need to reduce its size. Select the viewport. Notice that its blue grips appear.

4. Grab the upper right grip, and drag the viewport inward.

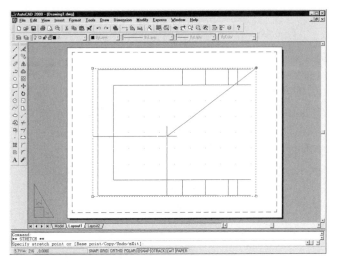

Figure 37.17 *Reducing the Size of the Viewport*

5. The model is probably too large for the viewport. You need to make the model smaller. Select the **PAPER** button on the status bar; notice the button changes to **MODEL**.

6. The easiest way to size the model to the viewport is to use these commands: (1) **Zoom Extents** and (2) **Zoom 0.9x**.

7. You now create the right side view in a new viewport. Start the **SolView** command by selecting **Draw | Solids | Setup | View** or type the **SolView** command, as follows:

   Command: **solview**

   Enter an option [Ucs/Ortho/Auxiliary/Section]: **o**

   Pick side of viewport to project: *(Pick right edge of viewport.)*

   In paper space, a viewport is just like a rectangle. Select one side of the viewport. For example, when you select the left side of the viewport, AutoCAD will later draw the left view of the 3D object. For this reason, it is good to be zoomed out a ways to give yourself room to work.

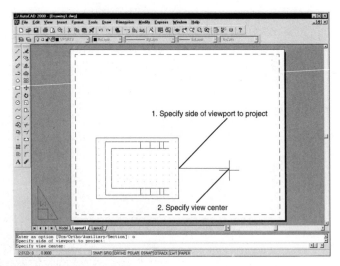

Figure 37.18 *Selecting Viewport Edge*

8. Here you locate the position of the new viewport.

   Specify view center: *(Pick.)*

   Specify view center: *(Press* ENTER.*)*

   AutoCAD turns on ortho mode. As you move the cursor to the right, AutoCAD draws a rubber-band line ortho to the side of the viewport. The point you enter is where AutoCAD places the center of the new view.

9. In this step, you specify the size of the viewport. Don't worry about getting the size exact; you can always come back later to resize and reposition the viewport using the **Stretch** and **Move** commands.

   Specify first corner of viewport: *(Pick one corner.)*

   Specify opposite corner of viewport: *(Pick other corner.)*

   Enter view name: **side**

   Give the view a name, such as "Side" for side view. AutoCAD redraws the right view of the 3D object.

   UCSVIEW = I   UCS will be saved with view

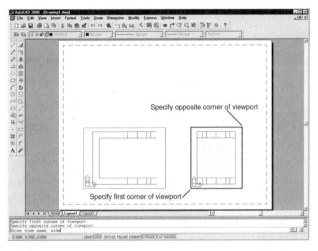

Figure 37.19 *Right View*

10. AutoCAD repeats the**SolView** options. Let's create the top view:

Enter an option [Ucs/Ortho/Auxiliary/Section]: **o**

Specify side of viewport to project: *(Pick top edge of original viewport.)*

Specify view center: *(Move cursor upward.)*

Specify view center: *(Press* ENTER.*)*

Specify first corner of viewport: *(Pick.)*

Specify opposite corner of viewport: *(Pick.)*

Enter view name: **top**

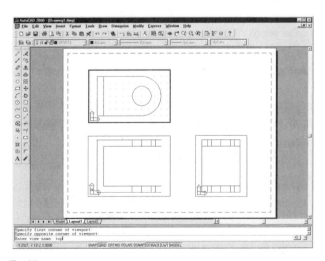

Figure 37.20 *Top View*

11. This time, you create a section view through the 3D object:

UCSVIEW = 1   UCS will be saved with view

Enter an option [Ucs/Ortho/Auxiliary/Section]:**s**

12. AutoCAD needs to know where you want the section and prompts you for two points. In the top viewport, select two points that go through the holes in the 3D model, as shown in Figure 37.21:

Specify first point of cutting plane: *(Pick.)*

Specify opposite point of cutting plane: *(Pick.)*

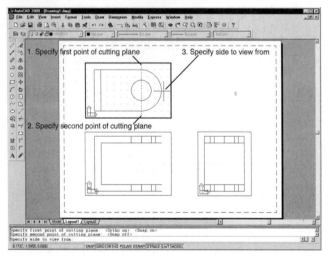

Figure 37.21 *Define Section View*

13. AutoCAD asks you on which side you want the section view. Choose a point to the right of the cutting plane:

Specify side to view from: *(Pick.)*

14. Since section views are often drawn at an enlarged scale, AutoCAD gives you the option to change the paper space scale factor (equivalent to the **Zoom XP** command). Don't worry about getting the view scale exactly correct at this point; you can always change the zoom factor later.

Enter view scale <1.0000>: *(Press* ENTER.*)*

15. The remainder of the prompts should be familiar to you. You need to turn off ortho mode to position the viewport at the upper right:

Specify view center: *(Click ORTHO on status line to turn off ortho mode.)*

<Ortho off> (Choose view's center point.)

Specify view center: *(Press* ENTER.*)*

Specify first corner of viewport: *(Pick.)*

Specify opposite corner of viewport: *(Pick.)*

Enter view name: **section**

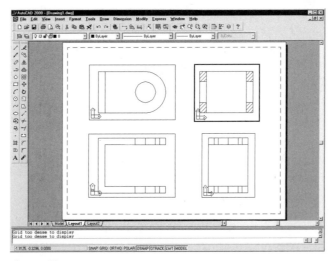

Figure 37.22 *Section View*

16. If you do not see lines drawn with the Hidden linetype, use the **Linetype** command to load the **Hidden** linetypes and then assign them to the *-HID layers.

17. Save your drawing as "Model2.Dwg."

18. You have now created four 2D views from the 3D model! But don't worry: your 3D drawing has not disappeared. Recall that layout mode lets you see a single model from several "port holes" at varying zoom levels, differing 3D view points, and various frozen layers. You can always get back your 3D model by returning to model view (such as for Chapter 38 on rendering). To do so, select the **Model** tab.

CHAPTER 38

Making a Realistic Rendering

To create a realistic rendering of a 3D model, you must know how to place materials and lights. After completing this chapter, you will be able to

- Create renderings of 3D objects
- Place lights and create scenes
- Apply material definitions to objects
- Save the rendering to a file on disk

INTRODUCTION

This chapter does not attempt to describe every detail of rendering. Instead, it gives you an overview that allows you to create a pleasing, realistic looking rendering of your 3D models. Commands covered in this chapter:

Render

Performs the rendering.

Light

Places lights in the drawing.

Scene

Collects lights and views in named scenes.

RMat

Attaches a material to an object.

MatLib

Loads material definitions into the drawing.

Background

Places a color or image in the background of the rendering.

Fog

Generates a fog-like effect.

SaveImg

Saves the rendering to file on disk.

THE RENDER COMMAND

The **Render** command is a more flexible and advanced version of the **Shade** and **3dOrbit** commands (Chapter 31 "Viewing 3D Drawings"). The **Render** command creates more realistic images. For example, **Render** smoothes all the facets you see on a 3D sphere, creating a round ball.

As with the **Shade** command, you can create a rendering by simply typing the **Render** command. Or you can set many options to create a complex rendering: you can set up many lights, apply materials to objects, place a background image, and save the rendering in one of several common file formats.

Render requires that your graphics board be capable of displaying at least 256 colors. For the most realistic renderings, your computer system should be set up with a graphics board that displays 16.7 million colors (also known as "24-bit color" or "true color"). To display that many colors, your graphics board needs 4 MB of RAM, a feature that is common and inexpensive today.

Before continuing, you should use the Windows Control Panel to change the display to 16.7 million colors, if possible.

YOUR FIRST RENDERING

Let's try using the **Render** command on the **Model1.Dwg** solid model created in the previous chapter.

1. Start AutoCAD and open the **Model1.Dwg** file. You may need to adjust the view to show the part in a manner similar to Figure 38.1.

2. From the **View** menu, select **Render | Render**. If this is the first time that the **Render** command has been used with this computer, AutoCAD takes a few seconds to load the rendering code.

3. When AutoCAD displays the **Render** dialog box, ignore all the options for now, and choose the **Render** button. In a second or so (depending on the speed of your computer), you see a beautifully rendered image of the **Model1** drawing appear.

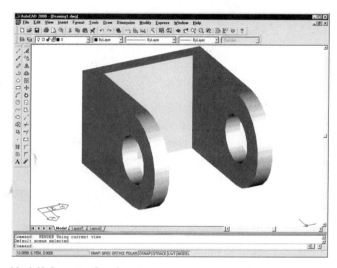

Figure 38.1 *Model1 Drawing Rendered*

ADVANCED RENDERINGS

For most renderings, you probably won't need to know anything more about the **Render** command. For total control, though, it is useful to know all about the **Render** command's options. Let's start the **Render** command again, and examine its **Render** dialog box.

Rendering Type

Unless you have installed another rendering program that runs inside AutoCAD, the three options are AutoCAD **Render**, **Photo Real**, and **Photo Raytrace**. Each option is progressively more realistic but takes longer to render.

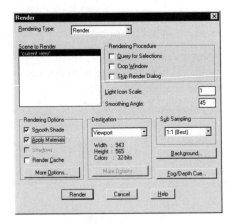

Figure 38.2 *Render Dialog Box*

Scene to Render

With your first rendering, there is only one scene defined: the current view. Later, we will use the **Scene** command to create named scenes whose names appear in this part of the **Render** dialog box.

Rendering Options

This is the core of the **Render** command. For the most part, you leave the options as they are set by default. If you want to change any, here is a summary of the effect they have on a rendering:

Smooth Shade

When on, AutoCAD creates a smooth transition through faceted areas of the 3D model. This makes the model look more realistic; the cutoff angle for smoothing facets is specified by the **Smooth Angle** option (default = 45 degrees). You would only turn off the **Smooth Shading** option for faster renderings.

Apply Materials

When on, AutoCAD renders the objects using the materials defined by the **RMat** command. This helps the object look more realistic. You would turn off this option for faster renderings.

Shadows

This option is only available with the **Photo Real** and **Photo Raytrace** rendering types. When on, shadows are generated by direct, spot, and point lights. Turning on this option slows down the rendering process but makes the image more realistic.

Render Cache

When on, this option speeds up the renderings after the first one by only rendering those parts that have changed from the previous rendering. Turning on this option makes the first rendering slower, but it speeds up subsequent ones.

Figure 38.3 shows the effect of turning off smooth shading: the facets of the curved areas are noticeable.

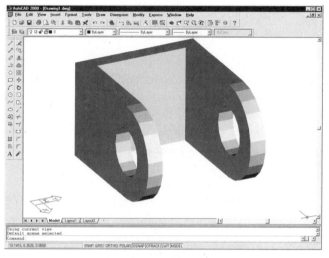

Figure 38.3 *A Faceted Rendering with No Material Applied*

Rendering Procedure

AutoCAD's **Render** modules give you three choices when performing each rendering. **Query for Selections** means that AutoCAD prompts you to select specific objects for rendering; this is useful for creating a faster rendering, since not every object in the drawing is rendered. **Crop Window** means that AutoCAD prompts you for a windowed portion of the drawing to render. Once again, this speeds up rendering by rendering a smaller portion of the overall drawing. **Skip Render Dialog** is a useful option when you have all rendering parameters already set up and don't want to see this dialog box every time you type **Render**.

Light Icon Scale

The lights that you place in a rendering scene are actually AutoCAD blocks. This option lets you specify a different size from the default scale factor of 1.0.

Destination

AutoCAD gives you three destinations to which you can have the rendered image sent: (1) the current viewport, (2) an independent rendering window, and (3) a file on disk. When you select **File**, there are additional options you need to decide on, such as the file format, resolution, interlacing, etc.

Sub Sampling

This option determines the number of pixels that are rendered. The default, **1:1**, renders all pixels in the scene. The "worst" setting, **8:1**, renders one pixel out of eight; the result is a very fast but a very chunky looking render.

Background

This button displays the **Background** dialog box, which lets you select the kind of background you want behind the rendered image. For example the background could be a raster image for the background of the rendering. The background image could consist of a grassy field, clouds, or cityscape. You find out more about the background with the **Background** command, discussed later in this chapter.

Fog/Depth Cue

This displays the **Fog/Depth** dialog box, discussed in greater detail later in this chapter.

Figure 38.4 *AutoCAD Render Options Dialog Box*

More Options

Choose this button to display the **Render Options** dialog box. When **Rendering Type** is **AutoCAD Render**, the following dialog box is displayed:

Render Quality

Gouraud and **Phong** are the names of the algorithms used by **Render** to create its images. Each is named after the computer scientist who came up with the rendering algorithms. **Phong** is more realistic when your model has lights, but it takes somewhat longer to render than does **Gouraud**.

Face Controls

AutoCAD uses the concept of face normals to determine which parts of the model are the front and which are the back. A face is the 3D polygon that makes up the surface of the model; a normal is a vector that points at right angles from the face. AutoCAD determines the direction of the normal by applying the right-hand rule to the order in which the face vertices were drawn. A positive normal points toward you; a negative normal points away from you. To save time in rendering, negative face normals are ignored, since they won't be seen in the rendering.

Turn on **Discard back faces** for faster rendering; turn off the option when mistakes appear in the rendering.

When the **Rendering Type** is **Photo Real**, the following dialog box is displayed:

Figure 38.5 *Photo Real Render Options Dialog Box*

Anti-Aliasing

This is a software technique for reducing the stairstep-like edges of a rendered image to make the edge look smoother. The stair-step effect—or "jaggies"—are due to the monitor's limited resolution of about 72dpi (dots per inch). That is a much coarser number than a typical 600-dpi laser printer. Thus, the Render software adds subtle pixels of gray and other colors to make the edges look smoother. Technically, AutoCAD uses the following styles of anti-aliasing:

Anti-aliasing	Meaning
Minimal	Analytical horizontal anti-aliasing
Low	Use 4 shading samples per pixel
Medium	Use 9 shading samples per pixel
High	Use 16 shading samples per pixel

The drawback to anti-aliasing is that the rendering process takes longer and the edges can look fuzzy. The higher the anti-aliasing setting, the smoother the look but the longer the rendering takes.

Figure 38.6 *Smooth Rendering (Left) and Anti-aliased Rendering (Right)*

Face Controls
Same as for the AutoCAD Render options.

Depth Map Shadow Controls
Helps prevent erroneous shadows, such as a shadow that casts its own shadow or a shadow that doesn't connect with its object (like Peter Pan's shadow). Autodesk recommends that the minimum bias should range between 2 and 20, while the **Maximum Bias** should range between 4 and 10 greater than the minimum. (Note that additional shadow controls are available when you define lights.)

Texture Map Sampling
A texture map is (usually) a raster image that is applied to a 3D object to make it look more realistic. For example, you could apply a picture of bricks onto a fireplace or a picture of a square of carpet onto the floor. This controls how a texture map is handled when it is applied to an object smaller than the texture map:

Sampling	Meaning
Point	The nearest pixel in the bitmap
Linear	Averages 4 pixel neighbors
MIP Map	Pyramidal average of a square sample

When the **Rendering Type** is **Photo Raytrace**, the dialog box in Figure 38.7 is displayed. This dialog box adds two more options.

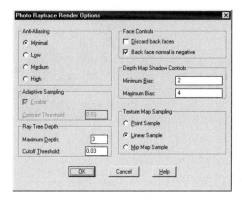

Figure 38.7 *Photo Raytrace Render Options Dialog Box*

Adaptive Sampling

This option is only available when you turn on **Low, Medium,** or **High** Anti-aliasing. Adaptive sampling allows for a faster anti-aliased rendering by ignoring some pixels. The **Enable** check box toggles this option. The **Contrast Threshold** adjusts the sensitivity between 0.0 and 1.0. For values closer to 0.0, AutoCAD takes more samples; for values closer to 1.0, AutoCAD takes fewer samples for faster rendering speed but possibly lower quality rendering.

Ray Tree Depth

In ray tracing, AutoCAD follows each beam of light as it reflects (bounces off opaque objects) and refracts (transmits through transparent objects) among the objects in the scene. This control allows you to speed up rendering by limiting the amount of reflecting and refracting that takes place. The **Maximum Depth** is the largest number of "tree branches" AutoCAD keeps track of as the light bounces and transmits through the scene; the range is 0 to 10. The **Cutoff Threshold** determines the percent that a ray trace must affect a pixel before ray tracing stops; range is 0.0 to 1.0.

CREATING LIGHTS

Whereas the **Shade** and **3dOrbit** commands are limited to a single light source, the **Render** command has four different kinds of light sources: distant, point, spot, and ambient. Except for ambient light, you can place as many of each of these lights as you want. All lights can emit any color at any level of brightness. The names of the lights have special meaning:

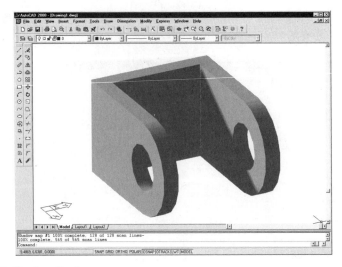

Figure 38.8 *Shadow Casting*

Point light

Emits light in all directions with varying intensity. The best example of a point light is the light bulb on the ceiling of your room.

Spotlight

Emits light beams in the shape of a cone. The best example of a spotlight is a high-intensity desk lamp or a vehicle-mounted spotlight. When you place spotlights in a drawing, you specify the hotspot of the light (where the light is brightest) and the falloff, where the light diminishes in intensity.

Distant light

Emits parallel light beams of constant intensity. The best examples of distant lights are the sun and moon. Typically, you want to place a single distant light to simulate the sun. To simulate a setting sun, you would change the color of the light to an orange-red color.

Ambient light

An omnipresent light source that ensures that every object in the scene has illumination. There is a single ambient light in every rendering; you would turn off the ambient light to simulate a nighttime scene.

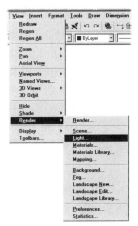

1. The **Light** command places lights in the drawing. Let's see how to place a sun in the **Model1** drawing:

2. Select **Render | Lights** from the **View** menu. AutoCAD displays the **Lights** dialog box.

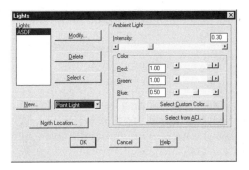

Figure 38.9 *Lights Dialog Box*

Initially, no lights are defined other than a single default light source located at your eye.

3. To place a new light, you need to decide on the type of light: spotlight, point light, or distant light. For now, select **Distant** light from the list box because a sun-like light is the easiest to work with (spotlights are the hardest lights to work with).

4. Choose the **New** button to give the light a name and to specify its parameters.

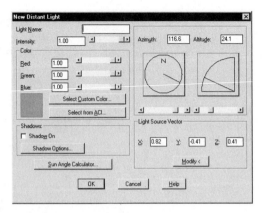

Figure 38.10 *New Distant Light Dialog Box*

5. The dialog box for each of the three light types is roughly similar, depending on their characteristics. Common to all three:

Name

6. Give the light a convenient name, such as "Sun."

Intensity

7. The brighter the light, the higher the intensity value. An intensity of zero turns off the light.

Color

8. You can select any color for each light. Choosing **Select Custom Color** displays the Windows **Color** dialog box. Select from ACI (the AutoCAD color index), the easiest option to use. Select the color yellow for the color of the light and choose **OK**.

9. The **New Distant Light** dialog box also lets you define the Sun's azimuth (how far around in the day) and altitude (height in the sky). Since it doesn't matter how far the Sun is from the object, AutoCAD places the distant light in the drawing for you.

10. **Shadows** lets you specify the type of shadow. **Volumes** is faster with sharp borders; **Map** is slower but has soft borders.

11. **Sun Angle Calculator** lets you select the sun's position by date, time and location.

12. Choose **OK**. Notice that an image of a light appears in the drawing.

Figure 38.11 *Shadow Options Dialog Box*

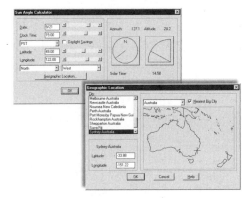

Figure 38.12 *Sun Angle Calculator Dialog Box*

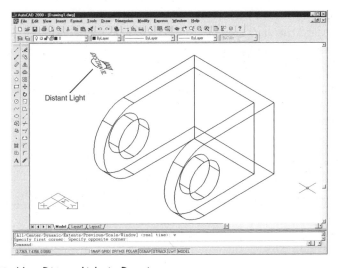

Figure 38.13 *New Distant Light in Drawing*

When you place lights in the drawing, AutoCAD actually places a block on a special layer named "AutoShade" (named after an earlier, now-obsolete rendering program no longer sold by Autodesk). Except for ambient light, each type of light has a unique block shape, as shown in Figure 38.14.

Direct.Dwg   Overhead.Dwg   Sh_Spot.Dwg

Figure 38.14 *The Three Light Blocks*

You have defined a single distant light called the Sun.

By default, AutoCAD uses all lights that you place in the drawing. If you wish to limit the rendering to some of the lights, use the **Scene** command.

CREATING SCENES

The **Scene** command does two things: (1) it lets you decide which lights should be used in a rendering, and (2) it lets you specify the name of the view for that rendering. The lights and the view are collected in a named scene. Let's see how the **Scene** command works.

1. Select **Render | Scenes** from the **View** menu. AutoCAD displays the **Scenes** dialog box.

Figure 38.15 *Scenes Dialog Box*

2. As the word *NONE* indicates, this drawing has no scenes defined yet. Choose the **New** button. AutoCAD displays the **New Scene** dialog box.

Figure 38.16 *New Scene Dialog Box*

3. Type a name for this scene, such as "First."

4. Then select the **Sun light**. (If you had used the **View Save** command earlier to define named views, these would be listed under **Views**.)

5. Choose the **OK** button twice to dismiss both dialog boxes.

6. Let's try rendering with the distant light. This time, the **Render** command's dialog box displays the name of the scene you just defined: FIRST. Select the FIRST scene, and then choose the **Render** button. After a moment, AutoCAD displays the rendering tinted with the yellow color you selected for the sunlight.

APPLYING MATERIALS AND BACKGROUNDS

To enhance the realism of the rendering, AutoCAD lets you apply material definitions to objects (with the **RMat** command) and place a background image with the **Background** command (formerly the **Replay** command).

The material definition uses four parameters to define the surface characteristics of objects:

Color

Reflection

Roughness

Ambient reflection

AutoCAD comes with many predefined material definitions stored in the **Render.mli** file. Let's use the **RMat** command to load and apply a material definition to the **Model1** drawing.

1. From the **View** menu, select the **Render | Materials** command. AutoCAD displays the **Materials** dialog box, which lists no materials.

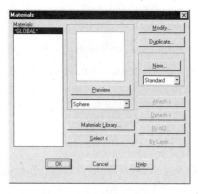

Figure 38.17 Materials Dialog Box

2. Before you can attach a material, you have to load its definition into AutoCAD, much like loading a linetype. Choose the **Materials Library** button. AutoCAD displays the **Materials Library** dialog box (you can get directly to this dialog box with the **MatLib** command).

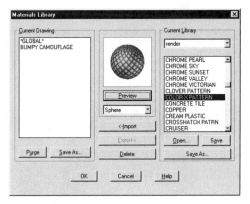

Figure 38.18 *Materials Library Dialog Box*

3. The dialog box lists the materials defined in the Render.mli file. Choose **Copper**, and then choose the **Import** button to add to the materials list.

4. For a preview of how a material looks on a sphere, choose the **Preview** button. AutoCAD quickly renders a sphere using the selected material. (This does not work when more than one material is selected.)

5. Choose the **OK** button to return to the **Materials** dialog box. Copper is listed in the **Materials** list box.

6. To attach the copper material to the Model1 object, choose the **Attach** button.

7. AutoCAD clears the dialog box and prompts you:

   Select objects to attach "COPPER" to: *(Pick the 3D model.)*

   I found Select objects: *(Press Enter.)*

   Updating drawing...done

   The dialog box returns. In addition to attaching the material to individual objects, you can attach a material definition to all objects of a specific color (choose the **By ACI** button) or to all objects on a layer (choose the **By Layer** button).

8. Choose the **OK** button, and use the **Render** command to re-render the model. Ensure that the **Apply Materials** option is turned on, and that **Photo**

Real or **Photo Raytrace** render is used. The newly rendered model looks more brown than before due to the copper material definition.

9   Try applying other material definitions to see the difference they make in color and shininess. Note that **Glass** makes the object transparent and takes a longer time to render.

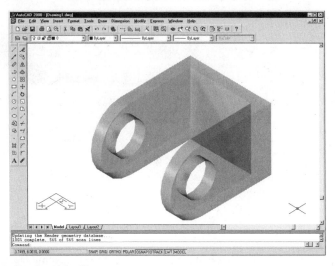

Figure 38.19 *Rendered As Glass*

10. Now, let's place an image behind the object with the **Background** command. This is useful for enhancing the rendering. For example, if you have designed a 3D house in AutoCAD, you could place a landscape image behind the rendering.

11. In order for a background image to be placed, the image must be in raster format. AutoCAD comes with many sample background images. Look in subdirectory \acad 2000\textures.

To load the background image, select **Render | Background** from the **View** menu.

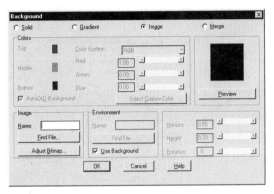

Figure 38.20 *Background Dialog Box*

12. AutoCAD displays the **Background** dialog box. Select a raster file. The samples provided with AutoCAD are all in TGA (Targa) format

13. AutoCAD displays the **Background** dialog box. Along the top, notice how AutoCAD gives you a choice of four different kinds of backgrounds:

Solid

A solid background means that AutoCAD replaces the default black (or white) background of the drawing screen with another color. Choose the **Preview** button to select the color. You chose the color from the **Colors** section of the dialog box.

Gradient

A gradient means the color changes from one end of the screen to the other, such as from red at the bottom to light blue at the top (to simulate a sunset). AutoCAD gives you three controls over creating a linear gradient—look carefully: they are tucked into the lower right corner.

> **Horizon** specifies where the lower color ends; a value closer to 0 moves the lower color lower down.
>
> When **Height** is set to 0, you get a two-color gradient; any other values gives you a three-color gradient.
>
> **Rotation** rotates the gradient.

Image

You select a raster image for the background. The image can be in BMP (bitmap), GIF, TGA (Targa), TIFF, JPEG (JPG), or PCX format. As mentioned earlier, the **\acadr 2000\textures** subdirectory contains TGA files, some of which are suitable as background images. To adjust the positioning, choose the **Adjust Bitmap** button; AutoCAD displays the **Adjust Background Bitmap Placement** dialog box.

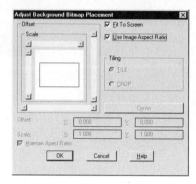

Figure 38.21 *Adjust Background Bitmap Placement Dialog Box*

The **Scale** section shows the relative placement of the background image in relation to the viewport. The red rectangle represents the extents of the objects that will be rendered; the magenta rectangle represents the background image.

Several check boxes and radio buttons let you automatically position the image, (such as stretching it to fit the entire viewport), retain the image's aspect ration (for no distortion), and determine whether to tile (repeat the pattern).

Back to the **Background** dialog box:

Merge

The current image is used as the background.

At any time, you can choose the **Preview** button to see the currently selected color, gradient, or raster image.

> 14. Now re-render the **Model1** drawing with the **Render** command. AutoCAD renders the object on top of the background image.

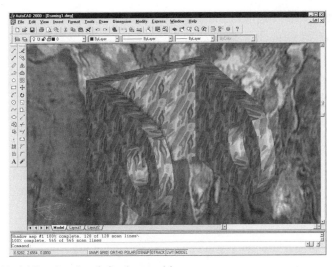

Figure 38.22 *Model Rendered with Background Image*

FOG AND DISTANCE EFFECTS

AutoCAD's fog effect simulates fog by applying an increasing amount of white with distance. The further away, the more dense the level of white.

The color need not be white. The subtle use of black, for example, can enhance the illusion of depth because objects further away tend to be darker in color. A limited application of yellow could create the illusion of a glowing lamp; the reckless use of green fog could simulate a Martian invasion.

You access the fog effect from the **Render** dialog box's **Fog/ Depth Cue** button or directly with the **Fog** command. Select **View | Render | Fog** from the menu bar.

The **Fog** command displays the **Fog/Depth Cue** dialog box.

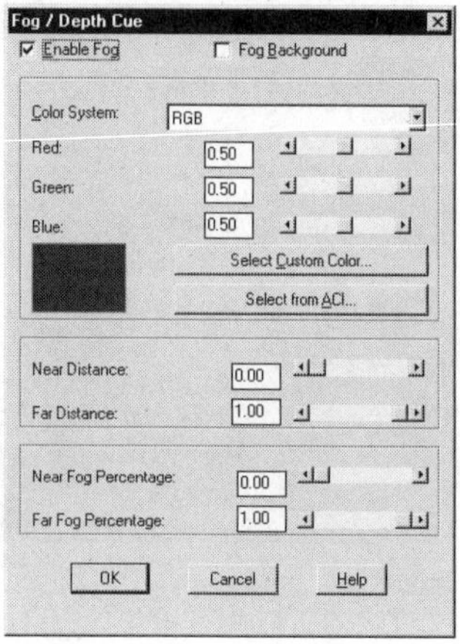

Figure 38.23 *Fog/Depth Cue Dialog Box*

Enable Fog

This toggle lets you turn the fog effect on and off without affecting any other parameters in this dialog box.

Fog Background

This toggle determines whether the fog affects the background. For example, if the background color in your rendering is normally white but you choose black for the fog color, then the background becomes black.

Color

Select a color for the fog. White is the default. Move all sliders to 0.0 for black, or choose a color from the Windows **Color** and AutoCAD **Color** dialog boxes.

Near Distance

The slider lets you position where the fog begins. The range is from 0.0 (the default) to 1.0. This slider can be tricky to understand, since it represents a relative distance from the camera to the back-clipping plane. You may have to adjust the fog's **Near Distance** a number of times—each time you use the **Render** command—until the effect looks right. For the rendering shown in Figure 38.24, we used a value of 0.45.

Far Distance

The slider lets you position where the fog ends. Like the **Near Distance**, **Far Distance** (default = 1.0) represents the percentage distance from the camera to the back clipping plane. For the rendering in Figure 38.24, we used a value of 0.60.

Near Fog and Far Fog Percentages.

These two sliders determine the percentage of fog effect at the near and far distances. Normally, these are 0.0 and 1.0, respectively. You would increase the value of **Near Fog** for a stronger fog effect; you should reduce the value of **Far Fog** for less fog effect.

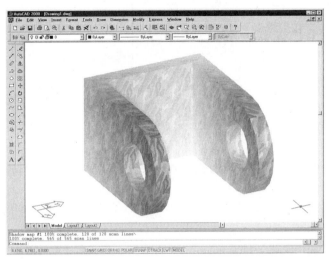

Figure 38.24 *Fog Effect*

SAVING RENDERINGS

There are three ways to save the result of the rendering:

- When the **Render** dialog box appears, select **File** from the **Destination** option. AutoCAD displays the **Rendering File** dialog box. Choose the **More Options** button, which allows you to choose from several raster file formats, including BMP, PCX, PostScript, Targ, and TIFF.

- Use the **SaveImg** command to save the rendering to a file. This command is used after the **Render** command has created the rendering in a viewport. Select **Display Image | Save** from the **Tools** menu. You have choices for the output file format: Targa, TIFF, and BMP. The **Options** button lets you select the type of image compression.

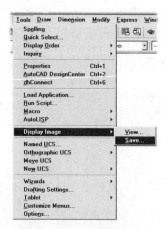

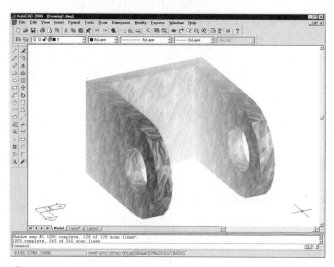

Figure 38.25 *Save Image Dialog Box*

- In the **Render** dialog box, select **Render Window** from the **Destination** option. Now choose **Render**, and AutoCAD outputs the rendering to an independent window. Select **Edit | Copy** from the menu bar. This captures the entire rendering window and copies it to the Windows Clipboard. In another application, such as Paint, press CTRL+V to paste the image, and then save it.

Figure 38.26 *Clipboard Viewer*

IMPROVING THE SPEED OF RENDERING

The time it takes for AutoCAD to create a rendering depends on many factors. These include

- Options selected for the **Render** command. Setting the following options results in a rendering that is at least twice as fast:

  Rendering type: **Render**

  Smooth shading: **Off**

  Apply materials: **Off**

  Render quality: **Gouraud**

  Discard back faces: **On**

  Destination: **small viewport or 320x200 file**

  Lights: **None**

  Render cache: **On**

  Shadows: **Off**

  Sub Sampling: **8:1**

  Enable Fog: **Off**

  Background: **Solid**

- Speed and memory of the computer. The faster the computer's CPU and the larger the computer's RAM, the faster the rendering.

- Complexity of the model. A complex model with many parts takes longer to render than a simple model with only a few parts.

- Amount of the model being rendered. Rendering the entire model takes longer than rendering a small portion.

- Size of the viewport being rendered. Rendering to a very small viewport greatly decreases the rendering time.

CHAPTER REVIEW

1. What is a *scene?*

2. What are the two ways of defining a viewpoint for a rendering?

3. Name three differences between the **Shade** and **Render** commands.

4. What are the four types of lights in AutoCAD Render?

5. List five ways to speed up a rendering.

6. Name the color of "fog":

   _____

   Color for increased depth illusion:

   _____

7. What is a *color gradient?*

8. How does *anti-aliasing* help a rendering? Hinder a rendering?

9. What type of light is the sun?

10. Name two types of shadow casting.

CHAPTER 39

Tailoring AutoCAD

At some point, all serious AutoCAD users want to learn to tailor their program. Increased drawing ease and performance will result from a properly tailored system. After completing this chapter, you will be able to

- Control the look and performance of the AutoCAD program with system variables
- Control the display and adjustment settings, such as the display of drawing blips and target box sizes
- Control display characteristics, such as the view resolution and regeneration rules
- Redefine existing commands
- Create shortcut aliases for commands

INTRODUCTION

In this chapter, you learn about ways of tailoring AutoCAD drawing with the following commands:

Setvar

Views and changes system variables.

Blipmode

Toggles the display of blips.

Aperture

Changes the size of the object snap target.

Regenauto

Controls when AutoCAD regenerates the screen.

ViewRes

Lets you specify the "roundness" of curved objects.

Undefine

Makes any command unavailable.

Redefine

Returns the availability of the undefined command.

SETTING SYSTEM VARIABLES

The **Setvar** (short for SET system VARiables) command allows you to change AutoCAD's system variables. All system variables are listed in Appendix C.

System variables control a variety of settings for AutoCAD, such as default aperture size, global linetype scale factor, and the current drawing name. Some variables are changed by commands, some are "read only," and most you can change to suit your needs.

To change a setting, type the system variable name, just as you would any other AutoCAD command name.

After entering the variable name, you can enter a new value. For example, you can change the name of the active layer by providing a different layer name for the **CLayer** system variable.

Command: **clayer**

Enter new value for CLAYER <"0">: *(Type valid layer name.)*

Note: Most variables can be executed directly at the command line, as shown above. A very few, however, require that you use the **Setvar** command first. Examples include **Aperture, Area, Blipmode,** and **Highlight**.

When you respond to the **Setvar** command with a "?", AutoCAD prompts:

Command: **setvar**

Enter variable name or ?: **?**

Enter variable(s) to list <*>:

If you respond by pressing ENTER, the name and current setting of all variables is displayed.

Some values are defined as "read only." This means you cannot change their value; only AutoCAD or the operating system is permitted to set the value. An example is the **Area** system variable:

Command: **setvar**

Variable name or ?: **area**

AREA= 0.0000 (Read only.)

USING SETVAR WHILE IN A COMMAND

There are times when you may want to change a system variable while in a command. AutoCAD allows you to do this by prefixing the Setvar command with the apostrophe ('). An example is

Command: **line**

From point: **'setvar**

\>\>Variable name or ?: **tracewid**

\>\>Enter new value for TRACEWID <default>: *(Enter new value.)*

(You can keep the default value by pressing ENTER in response to the prompt.) After you have entered the new value, AutoCAD returns you to the current command:

Resuming LINE command.

From point:

You now resume the previous command.

DISPLAYING BLIP MARKS

The marker that is left on the screen by your drawing activity is called a *blip*. Blips do not plot and may be thought of as "push pin" reference points to aid you.

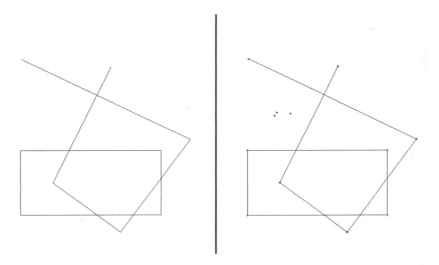

Figure 39.1 *Blipmode Off (Left) and On (Right)*

You turn the drawing of blips on and off by using the **Blipmode** command. To use the command, enter the following:

> Command: **Blipmode**
>
> Enter mode [ON/OFF] <ON>: **off**

When blipmode is on, you remove blips on the screen by issuing a **Zoom, Pan, Redraw,** or **Regen** command.

The initial setting of the **Blipmode** command is determined by the template drawing. You can change the setting at any time, and as often as you need. AutoCAD remembers the setting when you end the drawing and retains it as the initial setting when you reenter the drawing at a later time.

The value of the blipmode setting is stored in system variable **Blipmode**. When this variable is set to 0, blipmode is turned off; when set to 1, blipmode is turned on.

SETTING APERTURE SIZE

The **Aperture** command is used to adjust the size of the target box used in object snap modes, as in Figure 39.2.

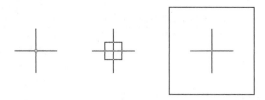

Figure 39.2 *Aperature = 1 (left), 10 (default, center), and 50 (right)*

The box size is described by the number of screen pixels specified. To set this number, enter:

> Command: **aperture**
>
> Object snap target height (1-50 pixels) <10>:

The last aperture size will be remembered when you reenter the drawing. The value is stored in system variable **Aperture**.

The aperture is used with object *snap* only. It is not the same as the pickbox, which looks slightly smaller and is used with object *selection*. (The size of the pickbox is changed through the **Pickbox** system variable).

CONTROLLING DRAWING REGENERATIONS

Some functions performed with AutoCAD cause an automatic regeneration. This regeneration reorganizes the data on the screen to ensure that all information is current. It is possible that you may need to perform several operations at one time that will force a regeneration each time. Since each regeneration takes some time to perform, AutoCAD provides a command to warn you before regenerations are performed by some operations.

> Command: **regenauto**
>
> Enter mode [ON/OFF] <ON>:

If **Regenauto** is **Off** and a regeneration is required, AutoCAD displays a warning dialog box: "About to regen—proceed?"

Choose **OK** to permit the drawing regeneration; choose **Cancel** to stop the regen from happening. You may not want a regen when the drawing is very large and the computer is very slow, because a regen then takes a long time.

The initial status of the **Regenauto** mode is determined by the template drawing. You can change it at any time and as many times as necessary. The last setting is remembered by AutoCAD and is restored when you reenter the drawing later. AutoCAD stores the setting in the **RegenMode** system variable.

SETTING THE VIEW RESOLUTION

The **ViewRes** command controls the fast zoom mode and resolution for circle and arc regenerations.

AutoCAD regenerates some zooms, some pans, and view restores, and when entering layout mode (paper space). This regeneration can, depending on the complexity of the drawing, take a long time. The Fast Zoom mode allows AutoCAD to simply redisplay the screen wherever possible. This redisplay is performed at the faster redraw speed. (Some extreme zooms still require a regeneration.)

AutoCAD calculates the number of segments required to make circles and arcs look "smooth" at the current zoom. (For display purposes, circles and arcs are actually made up of many short line segments.) The **ViewRes** command allows you to control the number of segments used. Using fewer segments speeds regeneration time but trades off in circle resolution. Although the displayed circles and arcs are not as smooth, plotter and printer plots are not affected.

To use the **ViewRes** command, enter:

> Command: **viewres**
>
> Do you want fast zooms? [Yes/No] <Y>: *(Choose Y or N.)*

Enter circle zoom percent (1-20000) <100>: **500**

Entering N at the first prompt causes all zooms, pans, and view restores to regenerate.

The default value for the circle zoom percent is 100. A value less than 100 diminishes the resolution of circles and arcs but results in faster regeneration times. I recommend you use a value of 500.

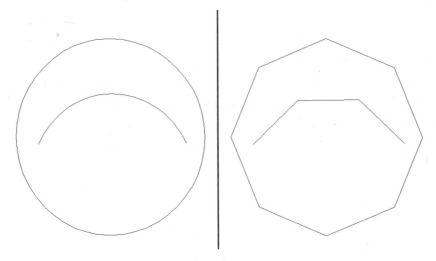

Figure 39.3 *Round (Left) and Octagonal (Right) Curves.*

A value greater than 100 results in a larger number of vectors than usual being displayed for circles and arcs. This is not important unless you zoom in a great amount.

For example, if you will be zooming in at a factor of 10, setting the circle zoom percent to 1,000 results in smooth circles and arcs at that zoom factor. If you want to maintain a smooth display and still achieve the optimum regeneration speed, set the percent equal to the maximum zoom ratio you intend to use.

Regardless of the setting, AutoCAD never displays a circle with fewer than eight sides. On the other hand, AutoCAD does not display any more circle or arc segments than it calculates to be necessary for the current zoom. If the circle uses less than two screen pixels at the maximum zoom magnification that does not require a regeneration, the circle is displayed as a single pixel.

If you wish to show the drawing to others at the maximum resolution, use a smaller percentage while drawing, then change the percentage and perform a regen. This allows you to perform the drawing at fast regeneration times and then redisplay the drawing at its optimum resolution for presentation.

REDEFINING COMMANDS

An existing AutoCAD command can be redefined to suit customized purposes. The **Undefine** command is used to facilitate this. **Undefine** is used extensively by AutoCAD third-party programmers to create new commands for their applications. You may want to use **Undefine** to customize some existing commands, making the use of AutoCAD more applicable to your particular tasks. See an example of using the **Undefine** command in the Exercise section below.

SHORTCUT ALIASES

If you prefer using the keyboard to enter commands, AutoCAD lets you reduce the number of keystrokes required to type a command name. Until now, you have been typing the full name, such as **Line**, **Circle**, and **Zoom**.

However, AutoCAD lets you type fewer characters for many frequently used commands. For example, **Line**, **Circle**, and **Zoom** can be entered as **L**, **C**, and **Z**, respectively. Try it now:

1. Open a new drawing.

2. At the command prompt, type **L**, as follows:

   Command: **L** *(Press* ENTER.*)*

   LINE Specify first point: *(Pick a point.)*

   Specify next point: *(Pick a point.)*

   Specify next point: *(Press* ENTER.*)*

To remind you which shortcut you are using, AutoCAD prints the full command name.

AutoCAD calls the keyboard shortcut L for **Line** an *alias*. Table 39–1, at the end of this chapter, lists all the aliases found in AutoCAD 2000.

You can create your own aliases—or change the aliases created by Autodesk—with the Notepad text editor and the **Acad.pgp** file, found in subdirectory **\acad 2000\support**. When you open the **Acad.pgp** (short for program parameters) file, you see lines of text that look like this:

        3F, *3DFACE

This is how AutoCAD defines aliases:

3F: The first item in each line is the alias. In this case, typing **3F** executes the **3dFace** command.

, * (comma, space, asterisk): A spacer that separates the alias from the command name.

3DFACE: The command name appears at the end of the line.

There are some rules in creating an alias that you should be aware of:

- You can create an alias for a keyboard command, such as **–BHatch**.

- You *cannot* create an alias for a command and one of its options. For example, you *cannot* use ZW for Zoom Window.

EXERCISE

1. Undefine the **Block** command, as follows:

   Command: **undefine**

   Enter command name: **block**

2. Now issue the Block command to see the message.

   Command: **block**

   Unknown command "BLOCK". Press F1 for help.

3. Let's now redefine the **Block** command. This returns it to its original status.

   Command: **redefine**

   Enter command name: **block**

The **Block** command is now redefined to its original form and the message will not be displayed if the command is subsequently issued.

CHAPTER REVIEW

1. What command can be used to set AutoCAD's system variables?

2. Under what condition can variables be entered directly from the command line?

3. How could you change a variable while actively in a command?

4. What is a *blip*?

5. How do you turn the blips on or off?

6. In what increment is the aperture setting measured?

7. Why would you sometimes want to set a low **ViewRes** value?

8. What difference in the displayed drawing would you notice if you set the **ViewRes** low?

9. Describe an example of redefining a command.

10. What is an alias?

11. What is the alias for **Line**? **Circle**? **Zoom**?

12. In which file do you create aliases?

Command	Alias(es)	
3DARRAY	3a	
3DFACE	3f	
3DORBIT	3do	orbit
3DPOLY	3p	
ADCENTER	adc	
ALIGN	al	
APPLOAD	ap	
ARC	a	
AREA	aa	
ARRAY	ar	
ATTDEF	att	ddattdef
-ATTDEF	-att	
ATTEDIT	ate	
-ATTEDIT	-ate	atte
ATTEXT	ddattext	
BHATCH	h	bh
BLOCK	b	
-BLOCK	-b	
BOUNDARY	bo	
-BOUNDARY	-bo	
BREAK	br	
CHAMFER	cha	
CHANGE	-ch	
CIRCLE	c	
COLOR	col	ddcolor

Table 39.1 *Aliases for Commonly Used AutoCAD Commands*

Command	Alias(es)	
COPY	co	cp
DBCONNECT	dbc	
DDEDIT	ed	
DDVPOINT	vp	
DIMALIGNED	dal	dimali
DIMANGULAR	dan	dimang
DIMBASELINE	dba	dimbase
DIMCENTER	dce	
DIMCONTINUE	dco	dimcont
DIMDIAMETER	ddi	dimdia
DIMEDIT	ded	dimed
DIMLINEAR	dli	dimlin
DIMORDINATE	dor	dimord
DIMOVERRIDE	dov	dimover
DIMRADIUS	dra	dimrad
DIMSTYLE	d	ddim
DIMTEDIT	dimted	
DIST	di	
DIVIDE	div	
DONUT	do	
DRAWORDER	dr	
DSETTINGS	ds	ddrmodes
DSVIEWER	av	
DVIEW	dv	
ELLIPSE	el	
ERASE	e	

Table 39.1 *Continued*

Command	Alias(es)	
EXPLODE	x	
EXPORT	exp	
EXTEND	ex	
EXTRUDE	ext	
FILLET	f	
FILTER	fi	
GROUP	g	
-GROUP	-g	
HATCH	-h	
HATCHEDIT	he	
HIDE	hi	
IMAGE	im	
-IMAGE	-im	
IMAGEADJUST	iad	
IMAGEATTACH	iat	
IMAGECLIP	icl	
IMPORT	imp	
INSERT	ddinsert	i
-INSERT	-i	
INTERFERE	inf	
INTERSECT	in	
INSERTOBJ	io	
LAYER	ddlmodes	la
-LAYER	-la	
-LAYOUT	lo	
LEADER	lead	

Table 39.1 *Continued*

Command	Alias(es)	
LENGTHEN	len	
LINE	l	
LINETYPE	lt	ltype
-LINETYPE	-lt	-ltype
LIST	li	ls
LTSCALE	lts	
LWEIGHT	lw	lineweight
MATCHPROP	ma	
MEASURE	me	
MIRROR	mi	
MLINE	ml	
MOVE	m	
MSPACE	ms	
MTEXT	t	mt
-MTEXT	-t	
MVIEW	mv	
OFFSET	o	
OPTIONS	ddgrips	gr
OSNAP	ddosnap	os
-OSNAP	-os	
PAN	p	
-PAN	-p	
PASTESPEC	pa	
PEDIT	pe	
PLINE	pl	
PLOT	print	

Table 39.1 *Continued*

Command	Alias(es)	
POINT	po	
POLYGON	pol	
PREVIEW	pre	
PROPERTIES	ch	ddchprop
PROPERTIESCLOSE	prclose	
PSPACE	ps	
PURGE	pu	
QLEADER	le	
QUIT	exit	
RECTANG	rec	
REDRAW	r	
REDRAWALL	ra	
REGEN	re	
REGENALL	rea	
REGION	reg	
RENAME	ren	
-RENAME	-ren	
RENDER	rr	
REVOLVE	rev	
RPREF	rpr	
ROTATE	ro	
SCALE	sc	
SCRIPT	scr	
SECTION	sec	
SETVAR	set	
SLICE	sl	

Table 39.1 *Continued*

Command	Alias(es)	
SNAP	sn	
SOLID	so	
SPELL	sp	
SPLINE	spl	
SPLINEDIT	spe	
STRETCH	s	
STYLE	st	
SUBTRACT	su	
TABLET	ta	
THICKNESS	th	
TILEMODE	ti	tm
TOLERANCE	tol	
TOOLBAR	to	
TORUS	tor	
TRIM	tr	
UNION	uni	
UNITS	un	ddunits
-UNITS	-un	
VIEW	ddview	v
-VIEW	-v	
VPOINT	-vp	
WBLOCK	w	
-WBLOCK	-w	
WEDGE	we	
XATTACH	xa	
XBIND	xb	

Table 39.1 *Continued*

Command	Alias(es)	
-XBIND	-xb	
XCLIP	xc	
XLINE	xl	
XREF	xr	
-XREF	-xr	
ZOOM	z	

Table 39.1 *Continued*

CHAPTER 40

Customizing the Menu

Using custom menu systems can multiply the productivity of any CAD system. You can buy expensive systems or you can become proficient at designing your own. After completing this chapter, you will be able to

- Utilize AutoCAD menu systems

- Understand the anatomy of a menu system

- Write the files necessary for the different sections of AutoCAD menus

- Construct special types of AutoCAD menus, such as tablet menus and custom pull-down menus

INTRODUCTION

This chapter provides a brief overview of customizing AutoCAD's menus. For a more in-depth treatment of the subject, refer to *Maximizing AutoCAD* and *The AutoCAD Programming and Customizing Quick Reference*, both from Autodesk Press. Here, you learn about the following commands:

Menu

Lets you load a new menu file.

Tablet

Configures AutoCAD for a new tablet overlay.

CUSTOM MENUS

Custom menus are an exciting part of AutoCAD. You can prepare a menu that is particular to your type of work or one that suits your style—or both!

A menu is nothing more than a text file. AutoCAD reads the item from the menu and executes it as though it were entered from the keyboard. Before you jump all the way in, let's learn two basic rules and look at a simple menu.

SIMPLE MENUS

Menu items are arranged one to a line. Each line contains one or more items that will be "typed" if chosen.

Menus are constructed with a word processor in non-document mode or in a text editor. The file must have a MNU extension to be loaded by AutoCAD. The following is a simple menu that could be written for AutoCAD:

LINE

ARC

CIRCLE

TRACE

REDRAW

ZOOM

COPY

MOVE

If you select any of these from the menu, it is executed as though you typed it at the keyboard.

Sometimes, it is desirable for the menu item to have a special listing on the menu. Consider the case of the **Quit** command. You may want the screen to display "Discard" instead of Quit.

You can include text that AutoCAD will not execute by enclosing it in square brackets ([ ]). The characters within the brackets are displayed on the menu and the items immediately following the closing bracket are executed. Consider the following menu items:

[Discard]QUIT

[Thick Line]TRACE

[Target]APERTURE

[Backwards]MIRROR

In the second line, **Thick Line** appears in the menu but the **Trace** command is executed. Be sure that you do not put a space between the closing bracket and the command, otherwise AutoCAD interprets the space as if you had pressed ENTER.

MULTIPLE MENUS

AutoCAD allows you to store, concurrently, several menus in one menu file. Each menu is compartmentalized in its own section. It is marked by a beginning label. The following table shows the section labels and the associated devices:

Menu Section	Meaning
***ACCELERATORS	Accelerator key definitions
***AUXn	Auxiliary device
***BUTTONSn	Pointing device buttons (where *n* is a number between 1 and 4)
***HELPSTRINGS	Text displayed on the status bar
***IMAGE	Image tile menu area (replaced Icon menu of earlier AutoCADs)
***MENUGROUP	Menu file group name
***POPn	Drop-down menu (where *n* is a number between 0 and 16)
***SCREEN	Menu area on screen
***TABLETn	Tablet menu area (where *n* is a number between 1 and 4)
***TOOLBARS	Toolbar definitions

Notice that each label starts with three asterisks. The asterisks, along with the label name, tell AutoCAD that this is the start of the menu items for that particular menu area or device.

The items that follow are contained in the associated section until another label is listed or the end of the file occurs. The following short menu shows two sections: POP1 (a drop-down menu), and AUX2 (an auxiliary menu for a device, such as a function key box):

```
***POP1
LINE
ARC
CIRCLE
[Backward]MIRROR
***AUX2
LINE
TRACE
ZOOM
```

Menus can contain *submenus,* which are useful for organizing alternate sets of macros.

LINKING MENUS

You have learned about section labels that separate menus in the same file, and sub-menus that are separate menu lists. Now you need to learn how to navigate them.

Many submenus are nice, but you need a way to move between them. When you select the **Circle** command from the screen menu, you then want a submenu that contains all the commands for working with circles.

AutoCAD provides a way to "jump" to a named submenu. The following format is used:

$section=submenu

The section refers to the section label. The sections can be referenced by the following letters:

Section Label	Meaning
$An=	AUXILIARY device *n* (n = 1 through 4)
$Bn=	BUTTONS menu *n* (n = 1 through 4)
$I=	IMAGE menu (replaces the ICON menu of earlier AutoCADs)
$P0=	Cursor menu
$Pn=	POPDOWN menu *n* (n = 1 through 16)
$S=	SCREEN menu
$Tn=	TABLET area *n* (n = 1 through 4)

If you want to access a submenu named ****Sublime** in the **Screen** section, the correct entry would be:

$S=Sublime

Notice that the double asterisk is *not* used when the submenu is referenced.

AutoCAD also provides a way to return to the last menu. The format for this is:

$S=

This returns you to the last screen menu item. The number of last (nested) menus allowed is eight.

The special code

=*

displays the current top level image, drop-down, or shortcut menu.

MULTIPLE COMMANDS IN MENUS

Commands can be linked together to perform several functions at once. This is called a macro. To do this, you must have a good understanding of AutoCAD's command sequences. The following special input items are used for this purpose:

Space

A space (or blank) is read as the ENTER key.

End of line

AutoCAD automatically inserts a blank at the end of each menu line. A blank is used interchangeably with the ENTER key in most commands.

Semicolon (;)

The semicolon can be used instead of a space to represent the ENTER key because it is easier to see when there are several in a row, as is the case in many macros.

Backslash (\)

A backslash is used where user input is desired. The command will pause and await your input.

Plus mark (+)

A plus mark is used to continue a long command string to the next line. If a plus mark is not present at the end of the line, AutoCAD will insert a blank.

Let's look at an example. Suppose you wanted to set up a new drawing by setting limits of 0,0 and 36,24, perform a zoom-all, and turn the grid on. The following menu sequence would perform this:

[Set Up Drawing]LIMITS 0,0 36,24 ZOOM A GRID ON

AutoCAD would read this as:

LIMITS 0,0 36,24

ZOOM A

GRID ON

Special functions can also be performed. Let's suppose that you want to insert a window in a solid wall made of a block. The window is a drawing called "WIN-1" and is stored in the C: drive. The window is 36 units in length, and the base point is at the right end of the window and four units down.

The following menu string can be used to insert the window in the wall:

[36" WIN]BREAK \@36,0 INSERT C:WIN-1 @ 1 1 0

AutoCAD will break the wall, leaving a 36-unit opening, and insert the window symbol in it.

Basepoint

Figure 40.1 *Window with 4-inch Offset*

LOADING MENUS

To load a menu, use the **Menu** command:

Command: **Menu**

The **Select Menu File** dialog box is displayed.

Figure 40.2 *Select Menu File Dialog Box*

A menu has a file extension of MNU up to the first time it is loaded. When it is loaded, AutoCAD compiles a menu (to an MNC file) for faster operation. A new copy of the menu is made with an MNS file extension. This is the menu used by AutoCAD.

If you edit the menu, you will need to edit the one with the MNU extension. You do not need to delete the menu of the same name with the .MNS extension. AutoCAD senses the change and automatically recompiles the menu.

TABLET MENUS

A digitizing tablet can be used with a tablet menu. You can specify up to four tablet menus and a drawing area. Figure 40.3 shows the tablet menu provided with AutoCAD 2000.

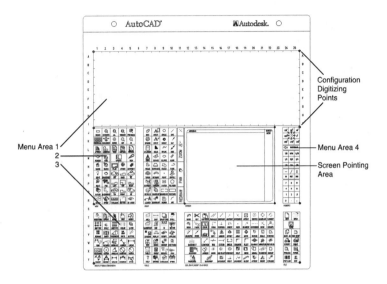

Figure 40.3 *Tablet Menu Template*

The four menu areas are designated TABLET1 through TABLET4. The menus are made up of smaller boxes of equal dimension in rows and columns. The menu boxes are labeled from left to right and from top to bottom. Each label corresponds to the item in the menu under the appropriate section label. The first command is labeled A1, and so forth. You can arrange the menus in any fashion you wish.

The **CFG** option of the **Tablet** command is used to configure the menus and drawing.

 Note: You can only use the **Tablet** command when your computer's pointing device is capable of being used as a tablet. If your computer has a mouse, the **Tablet** command retorts with this message, "Your pointing device cannot be used as a tablet."

Command: **Tablet**

Enter an option [ON/OFF/CAL/CFG]: **CFG**

Enter number of tablet menus desired (0-4) <default>:

If you have already used the tablet menu, AutoCAD will prompt:

Do you want to realign tablet menu areas? <N>: **y**

AutoCAD proceeds with these prompts:

Digitize upper left corner of menu area 1: *(Pick upper left corner.)*

Digitize lower left corner of menu area 1: *(Pick lower left corner.)*

Digitize lower right corner of menu area 1: *(Pick upper right corner.)*

After you enter the descriptive points, AutoCAD prompts:

Enter the number of columns for menu area 1: *(Enter a positive number.)*

Enter the number of rows for menu area 1: *(Enter a positive number.)*

You can now define the drawing area. You are prompted:

Do you want to respecify the screen pointing area? <N>: **y**

If you reply Y, AutoCAD prompts:

Digitize lower left corner of screen pointing area: *(Pick lower left corner.)*

Digitize upper right corner of screen pointing area: *(Pick upper right corner.)*

DROP-DOWN MENUS

Custom drop-down menus can be written that are accessed from the menu bar near the top of the window. Drop-down menus are designated as "POP1" through "POP16." They are written in the same fashion as screen and tablet menus, with some exceptions. POP0 is reserved for use as the cursor menu.

Each menu is handled as a separate menu, using, for example, *****POP1** as the menu area for the first drop-down menu. The equivalent to the **$S=** command in the drop-down menu is the **$P***n***=** command, where *n* is the number of the drop-down menu to access. In addition, the special command **$P***n***=*** line is used to automatically pull down the menu desired.

Let's look at an example of a drop-down menu (designated as **POP3**) that is to be automatically pulled down when accessed from a screen menu. You could use the following line in the screen menu area.

[PARTS]$p3=parts $p3=*

This would pull down the submenu named "PARTS" from the **POP3** menu (listed in the written menu under *****POP3**). The menu would then be forced down by the **$p3=*** command.

When you write the POP*n* menu sections, the first line under each area is used as the header bar title (listing in the top screen area). These listings will be displayed across the top of the screen. The menu under each will be as wide as the longest item in the corresponding menu. Keep in mind that lower resolution screens are capable of displaying only 80 columns in width. If the total width of the menus is longer, they will be truncated.

You can provide separation between items in the menu by placing a separator line between them. Placing a line as follows does this:

[—]

Recall that the square brackets [and] indicate a display item. The two hyphens automatically expand to the width of the menu and provide the separator line.

You can display a menu item label as "not available" by beginning it with a tilde (~). An unavailable item is typically shown by gray text and denotes an item that is not currently active. This could be used if a menu is in progress and will be completed at a later time. It could also be used for minor selections.

The > angle bracket creates a submenu (also known as a cascading menu). Use the other angle bracket, <, to indicate the last item of the submenu.

CHAPTER REVIEW

1. Why would you want to customize AutoCAD's menus?

2. How many different macro types can be served by an AutoCAD menu?

3. How do you designate a submenu within the menu file?

4. What is a macro?

5. What command is used to load a new menu?

6. What is a tablet menu?

7. What are the maximum menu areas on a tablet menu?

8. What is meant if a drop-down menu selection is gray?

CHAPTER 41

Programming Toolbar Macros

A toolbar macro is a simple, easy-to-create, "mini-program" that reduces the number of keystrokes and menu selections you need to make. After completing this chapter, you will be able to

- Understand the nature of macros

- Understand the syntax of macros

- Write your own toolbar macros

INTRODUCTION

Behind every icon button on every toolbar is a macro. A *macro* is a collection of one or more commands and options, such as "Zoom Window" or "QSave Zoom Extents Plot." By choosing an icon button, you ask AutoCAD to execute the macro. AutoCAD runs the commands in the macro many times faster than you could ever hope to type them or select them from the menus.

You should not confuse toolbar macros with menu macros; toolbar and menu macros share the same name, but they are programmed somewhat differently.

WHY USE MACROS?

Toolbar macros have a number of advantages and disadvantages over using AutoLISP in AutoCAD (see Chapters 42 and 43 for more about the AutoLISP programming language).

Advantages of Toolbar Macros

- Faster to write than AutoLISP programs; the programming environment is inside AutoCAD, not in an external text editor, as with AutoLISP.

- Always available when AutoCAD is running; AutoLISP programs must be specifically loaded or you must make special arrangements to have AutoLISP programs load automatically.

- Easier to start the macro than an AutoLISP routine: just choose the correct button.

Disadvantages of Toolbar Macros

- Macros are limited to 512 characters in length; AutoLISP programs can be thousands and thousands of characters in length.

- Macros are limited to what you can type at the keyboard; AutoLISP can be used to extend the functionality of macros.

- Both toolbar macros and AutoLISP are limited to running within AutoCAD. If you want to automate steps inside and outside AutoCAD, you must use another programming language, such as Visual Basic.

- Toolbar macros do not operate when dialog boxes are open.

You can think of toolbar macros as more user-friendly versions of script files. In short, you would use a toolbar macro when you want to execute one or more commands automatically. For anything more complex—such as performing calculations or parametric drafting—you need to use AutoLISP.

TOOLBAR MACRO BASICS

As mentioned, a macro is simply one or more AutoCAD commands in a row. For example, when you receive a drawing from a co-worker or client, you want to set up the four engineering views: top, front, side, and isometric.

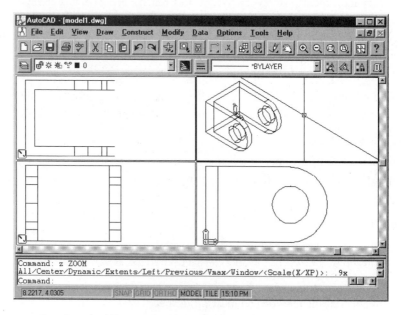

Figure 41.1 *Four Standard Engineering Viewpoints*

To set up the four views by entering the commands, you type:

> Command: **-vports 4**
>
> Command: *(Set side view.)* **vpoint 0,–1,0**
>
> Command: *(Pick lower left viewport for front view.)* **vpoint –1,0,0**
>
> Command: *(Pick upper left viewport for top view.)* **vpoint 0,0,0**
>
> Command: *(Pick upper right viewport for iso view.)* **vpoint 1,-1,1**

By my count, that's a total of 64 keystrokes and 3 picks.

Or, you can do the same commands by selecting items from the menus:

 Note: Whether you select Left of Front viewpoint depends on how the UCS is set up for the current drawing; similarly, you may need to select an isometric view other than SE.

> **View | Tiled Viewports | 4 Viewports**
>
> **View | 3D Viewpoint Presets | Left**
>
> *(Click on lower right viewport.)*
>
> **View | 3D Viewpoint Presets | Front**
>
> *(Click on upper right viewport.)*
>
> **View | 3D Viewpoint Presets | Top**
>
> *(Click on upper left viewport.)*
>
> **View | 3D Viewpoint Presets | SE Isometric**

As I count them, there's 18 menu and screen picks.

After doing this for a few drawings, you may start to find it tedious to set up each drawing this way. You decide to combine the many keystrokes and/or menu selections into a single toolbar macro. Here's how to do this:

1. Right-click any icon. AutoCAD displays a cursor menu.

2. Select **Customize**. AutoCAD displays the **Toolbars** dialog box.

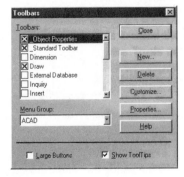

Figure 41.2 *Toolbars Dialog Box*

3. Choose the **Customize** button. AutoCAD displays the **Customize Toolbars** dialog box.

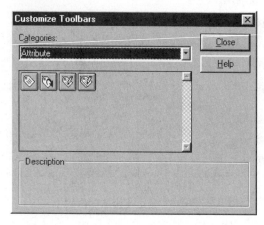

Figure 41.3 *Customize Toolbars Dialog Box*

4. Select **Standard** from the **Categories** list box.

5. Scroll through the many icons until you see the **Tile Model Space** icon—you'll find it in the third last row.

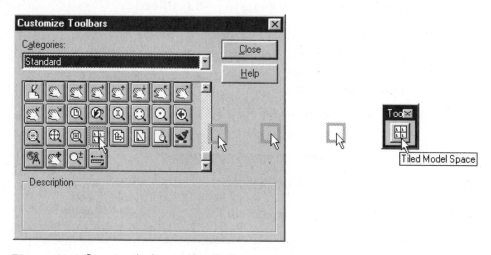

Figure 41.4 *Dragging the Icon to New Toolbar*

6. Drag the icon out of the dialog box and onto the AutoCAD drawing area.

7. Right-click the icon. The **Button Properties** dialog box appears. Notice that the dialog box has four areas of interest: **Name**, **Help**, **Macro**, and **Button Icon**.

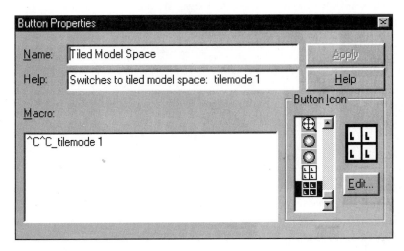

Figure 41.5 *Button Properties Dialog Box*

8. **Name:** This is the name that appears in the Tooltip when the cursor lingers over the button. Currently, the name reads "Tiled Model Space." Change the name text to read:

   Name: **Engineering Views**

9. **Help:** This is the text that appears on AutoCAD's status bar when you select the button. Currently, the help line reads "Switches to tiled model space." Change the help text to read:

   Help: **Creates four viewports with the standard engineering views.**

10. **Macro:** These are the commands that are executed when you choose the button. Currently, the macro reads, "^C^C_tilemode 1". You probably recognize the Tilemode command, but what about the other characters? Here's what they mean:

    ^C^C

    The ^ (caret) symbol is equivalent to holding down the CTRL key; the ^C is the same as pressing CTRL+C, which is the old AutoCAD method of canceling a command (C is short for Cancel). Two ^C^C in a row cancel out of nested commands, like **PEdit**.

    _tilemode

The _ (underscore) symbol internationalizes the command. AutoCAD is available in many languages other than English: German, Spanish, French, French-Canadian, and so on. In these international versions of AutoCAD, the commands have been localized. However, to ensure that any language version of AutoCAD understands the same macro, the underscore allows English language command names to work.

1

Finally, the number 1 is the value of **Tilemode** to create tiled viewports; 0 is the value for overlapping viewports.

Change the macro to read:

^C^C_vports 4 _cvport 2 _vpoint 0,-1,0 _cvport 3 _vpoint -1,0,0
_cvport 4 _vpoint 0,0,0 _cvport 5 _vpoint 1,-1,1

Much of this macro is similar to the commands we typed at the beginning of the chapter to create the four viewports and their viewpoints. Some characters you may not be familiar with are:

_z

This is the alias for the **Zoom** command. By using the alias, we reduce the number of characters in the macro. Recall that the macro can have a maximum of 255 characters. Recall, too, that the *underscore* prefix internationalizes the **Zoom** command.

_cvport

This is the system variable for switching viewports.

The dialog box should now look like the one in Figure 41.6.

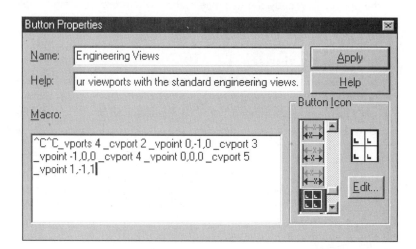

Figure 41.6 *Changes Made to Dialog Box*

11. **Button Icon:** This is the icon customization area. Here you can select an icon stored on disk or create a new one. Currently, the icon shows four window panes. Let's modify it to better represent the four engineering views. Choose the **Edit** button. AutoCAD displays the **Button Editor** dialog box.

Figure 41.7 *Button Editor Dialog Box*

12. Use the drawing tools to modify the icon. It is generally easier to use the paint tool for everything: to erase, paint over in another color.

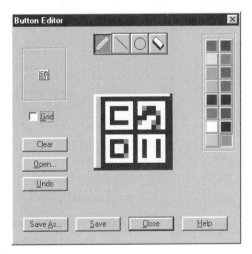

Figure 41.8 *Edited Button*

13. When done, choose the **Save** button.

14. Choose the **Close** button to dismiss the **Button Editor**.

15. Choose the **Apply** button. The icon is modified. AutoCAD gives the toolbar the generic title, "Toolbar 1."

Figure 41.9 *Modified Toolbar*

16. Dismiss any remaining dialog boxes; AutoCAD updates its menu files to store your changes.

17. To test the macro, open the **Model1.dwg** drawing file from the CD-ROM. It will probably appear in the 3D view you left it in.

18. Select the icon. The four engineering views should appear. If AutoCAD reports an error, check the wording of the macro carefully. A command might be misspelled; a comma or other character may be missing.

Toolbar macros are a quick and easy way to automate a number of command steps in AutoCAD.

SPECIAL CHARACTERS

Toolbar macros allow special characters. The complete list includes:

Character	Meaning
^B	Toggle snap mode between on and off, like CTRL+B
^C	Cancel current command (does not execute the **CopyClip** command), like CTRL+C
^D	Change coordinate display mode, like CTRL+D
^E	Switch to next isometric plane, like CTRL+E
^G	Toggle grid display, like CTRL+G
^H	Backspace, like CTRL+H
^I	TAB
^M	ENTER
^O	Toggle ortho mode, like CTRL+O
^V	Switch to next viewport, like CTRL+V
\n	Starts a new line
\t	TAB; has same effect as space
\nnn	Use ASCII character *nnn*
\\	Allows use of the \ character

Table 41.1 *Special Characters*

CHAPTER REVIEW

1. What is a macro?

2. When is it better to use a toolbar macro? An AutoLISP routine?

3. What does the ^ (caret) symbol mean?

4. Write a toolbar macro for preparing a drawing before plotting: Zoom to extents, save the drawing, and then start **Plot**.

CHAPTER 42

Introduction to AutoLISP

AutoLISP is a powerful programming tool that is a part of AutoCAD. After completing this chapter, you will be able to

- Understand the nature and purpose of the AutoLISP programming language

- Load and use existing AutoLISP routines

INTRODUCTION

In this chapter you learn three ways to load AutoLISP programs, by using these techniques:

(load)

AutoLISP's own function for loading programs.

Drag'n Drop

Use the Windows drag and drop technique.

AppLoad

Loads AutoLISP and other applications into AutoCAD.

USING AUTOLISP

AutoCAD is a powerful drawing and design program. While many people use this program for many purposes, most implement AutoCAD for a specific discipline. While the "generic" form of AutoCAD contains all the command routines needed to construct any type of drawing, most users benefit from the custom programming capabilities within AutoCAD.

In Chapter 40 "Customizing Menus and Icons," you learned how to customize menus and write macros to combine several commands into a single step. An even *more powerful* method of customization is the LISP programming language embedded within the AutoCAD package. This feature, referred to as AutoLISP, allows you to set and recall points, mix mathematical routines within a list of instructions, and automate many other duties.

WHY USE AUTOLISP?

AutoLISP allows the AutoCAD user to customize new commands that perform one or many functions. For example, programmers have used AutoLISP to create packages that automatically create a 3D contour map from site data, create "unfolded" patterns from three-dimensional objects, and construct a drawing from a list of dimensions that describes an object (parametric drawing construction). Over the years, users have employed AutoLISP to automatically generate stairs in architectural plans, gears for mechanical plans, and plans from survey data.

Using AutoLISP to customize routines for your work creates a more efficient drawing system. Whether you continue to increase your knowledge of LISP programming and write extensive routines, or just gain a general understanding and write simple timesaving routines, you will find that AutoLISP enhances your AutoCAD work!

USING AN AUTOLISP PROGRAM

You don't have to be a programmer to use AutoLISP routines. AutoCAD contains a number of AutoLISP routines for your use; other routines are freely available from the AutoCAD User Group International (AUGI) and from Web sites devoted to AutoCAD users.

Let's look at one AutoLISP program and then learn how to load and use it.

AXROT.LSP is an AutoLISP program used to rotate an object around any of its axes. Before starting, make sure this program is in the directory in which you have placed your AutoCAD files. Table 42.1 is a listing of the program.

Drawing the Box

Let's start a new drawing and try out this program. Start a new drawing. Use AutoCAD's 3D capabilities to draw a cube similar to the one shown in Figure 42.1. You can do this by using the **Ai_box** command. As you draw the cube, you will notice that it appears as a square in plan.

> Command: **ai_box**
>
> Specify corner point of box: *(Enter a point on the screen.)*
>
> Specify length of box: *(Move the crosshairs to show a length and pick.)*
>
> Specify width of box or [Cube]: **C**
>
> Specify rotation angle of box about the Z axis or [Reference]: **0**

Figure 42.1 *Solid Cube in Plan*

Loading the Axrot.Lsp Program

Now let's use the **Axrot LISP** program. The first thing you have to do is load the program. You load a LISP program in different ways. Enter the following exactly as listed, including the quotes and parentheses.

        Command: *(load "axrot")*

If you entered the line correctly, you will see a short message that the program was loaded.

Using the Axrot.Lsp Program

Now that it is loaded, you can use the program as a command. Let's do this now.

        Command: **axrot**

        Select objects: *(Select the cube.)*

        I selected, I found Select objects: *(Press* ENTER.*)*

        Axis of rotation X/Y/Z: **X**

        Degrees of rotation <0>: **30**

        Base point <0,0,0>: *(Select a point about the center of the cube.)*

The box has rotated 30 degrees about the X axis. Let's perform one more rotation. Press ENTER to repeat the command.

        Command: **axrot**

        Select objects: *(Select the cube.)*

        I selected, I found Select objects: *(Press* ENTER.*)*

        Axis of rotation X/Y/Z: **Y**

        Degrees of rotation <0>: **30**

        Base point <0,0,0>: *(Select a point about the center of the cube.)*

Your cube should now look similar to the one in Figure 42.2. Notice how you can see the rotations performed around the *X* and *Y* axes.

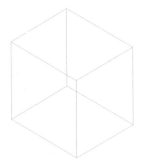

Figure 42.2 *3D Cube After Use of AXROT.LSP*

Now use the **Hide** command to see the cube more clearly.

Notice that the AutoLISP program acts just like any other AutoCAD command. Its prompts, such as "Select objects," are already familiar to you from commands built into AutoCAD.

OTHER WAYS TO LOAD AUTOLISP

In addition to using the **(load)** function, there are two other ways to load AutoLISP routines into AutoCAD: (1) drag and drop and (2) the **AppLoad** command.

DRAG AND DROP

The easiest way to load an AutoLISP routine is to drag the file name from Explorer (or File Manager) into AutoCAD, as shown in Figure 42.3.

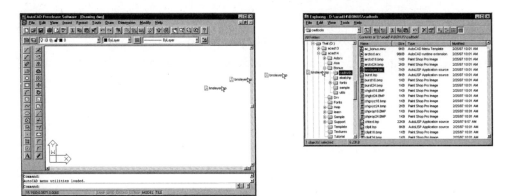

Figure 42.3 *Dragging an AutoLISP File into AutoCAD from Explorer*

For this to work, both AutoCAD and Explorer (or File Manager) must be running. In Explorer, go to the subdirectory where the AutoLISP routines are help. In AutoCAD 2000, there are a couple of folders:

> acad 2000\support

> \acad 2000\express

Find the AutoLISP program you are interested in loading. Drag the file from Explorer (or File Manager) into AutoCAD. To drag means to hold down the left mouse button on the file name as you move the cursor to AutoCAD. Once in AutoCAD, let go of the mouse button.

AutoCAD automatically invokes the **(load)** function and reports that the routine is loaded:

> Command: (LOAD "D:\\acad 2000\express/filename.lsp".)

THE APPLOAD COMMAND

It's not always convenient to hunt through subdirectories looking for an AutoLISP routine. For this reason, AutoCAD includes the **AppLoad** command.

Type the **AppLoad** command or select **Tools | Load Application** from the menu bar.

AutoCAD displays the **Load /Unload Applications** dialog box. When you first see the dialog box, it has quite a full list. That's because many AutoCAD commands are actually external programs that are loaded as needed.

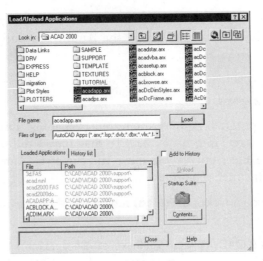

Figure 42.4 *The Load/Unload Applications Dialog Box*

The upper half of the dialog box acts just like any other file dialog in AutoCAD. It allows you to select a specific file to load. Select a file and then choose **Load**. AutoCAD immediately loads ObjectARX, VBA, and DBX applications. Note that LSP, VLX, and FAS applications are not loaded until you choose **Close**.

The lower half of the dialog box allows you to manage the applications that are loaded. The **Loaded Applications** list shows you the names of external programs already loaded by AutoCAD.

If you want AutoCAD to automatically load an AutoLISP routine when AutoCAD first starts up, choose the **Contents** button of **Startup Suite**. In the **Startup Suite** dialog box that appears, you can add (and delete) a list of applications.

CHAPTER REVIEW

1. Why would you program in AutoCAD with AutoLISP?

2. What are some uses for AutoLISP?

3. How would you load a LISP file named 3DARRAY.LSP into an AutoCAD drawing?

4. Describe an AutoLISP program that would be useful for you.

5. Name three ways to load an AutoLISP program into AutoCAD.

6. Can the **AppLoad** command automatically load LSP files?

7. What conditions must be met before drag and drop works?

```
;***************************************************
;                    AXROT.LSP
;
; By Jan S. Yoder                    May 11, 1988
;
; A routine to do 3 axis rotation of a selection set
;
;***************************************************


; Internal error handler

(defun axerr (s)                 ; If an error (such as CTRL-C)
        occurs
                                 ; while this command is active...
    (if (/= s "Function cancelled")
        (princ (strcat "\nError: " s))
    )
    (setq *error* olderr)        ; restore old *error* handler
    (setvar "gridmode" ogm)          ; restore saved modes
    (setvar "highlight" ohl)
    (setvar "ucsfollow" oucsf)
    (command "undo" "e")         ; complete undo group
    (setvar "cmdecho" oce)
    (princ)
)


; Main program

(defun c:axrot (/ olderr obpt ogm ohl oucsf ssel kwd dr bpt)
    (setq olderr *error*
        *error* axerr)
    (setq oce  (getvar "cmdecho")
        ogm  (getvar "gridmode")
```

```
        ohl   (getvar "highlight")
        oucsf (getvar "ucsfollow"))

(setvar "cmdecho" 0)
(command "undo" "group")
(setvar "gridmode" 0)
(setvar "ucsfollow" 0)

(setq ssel (ssget))

(setvar "highlight" 0)

(initget 1 "X Y Z")
(setq kwd (getkword "\nAxis of rotation X/Y/Z: "))
(setq dr (getreal "\nDegrees of rotation <0>: "))
(if (null dr)
   (setq dr 0)
)
(setq bpt (getpoint "\nBase point <0,0,0>: "))
(if (null bpt)
   (setq bpt (list 0 0 0))
)
(setq bpt (trans bpt 1 0))
(cond
    ((= kwd "X") (command "ucs" "Y" "90"))
    ((= kwd "Y") (command "ucs" "X" "-90"))
    ((= kwd "Z") (command "ucs" "Z" "0"))
)
(setq bpt (trans bpt 0 1))
(command "rotate" ssel "" bpt dr)

(command "ucs" "p")               ;restore previous ucs
(setvar "gridmode" ogm)           ;restore saved modes
```

```
(setvar "highlight" ohl)
(setvar "ucsfollow" oucsf)
(command "'redrawall")
(command "undo" "e")            ;complete undo group
(setvar "cmdecho" oce)
(princ)
)
```

CHAPTER 43

Programming in AutoLISP

To use AutoLISP, you must learn how to use the functions found in AutoLISP. You must also learn the basics of programming in AutoLISP. After completing this chapter, you will be able to

- Perform the basics of programming in AutoLISP
- Analyze an AutoLISP program file
- Use an AutoLISP file inside AutoCAD

AUTOLISP BASICS

AutoLISP programming can range from very simple to very complex. An AutoLISP program is nothing more than a list of instructions. The AutoLISP programming language embedded in AutoCAD contains many functions that are used to perform the tasks you need. Let's take a look at some of these.

ARITHMETIC FUNCTIONS

You can use AutoLISP to perform math functions. Start AutoCAD. We are going to enter some AutoLISP routines directly at the command line.

At the command line, enter the following exactly. Be sure to include the parentheses and spaces.

Command: **(+ 5 4)**

9

Notice how AutoCAD returns the value of 9. This simple routine adds the values of 5 and 4, then prints the sum to the screen. Continue and enter the following and notice the results of addition, subtraction, multiplication, and division.

Command: **(- 100 40)**

Command: **(* 5 6)**

Command: **(/ 100 20)**

Notice how the AutoLISP routine is typed at the command line with *parentheses*. The opening and closing parentheses to tell AutoCAD this is an AutoLISP expression. When AutoCAD detects an open parenthesis, the text that follows is passed to AutoLISP.

The closing parenthesis denotes the end of the expression. If you get a "1>" prompt from AutoCAD, enter a closing parenthesis, and then press Enter. This prompt means that there is not a closing parenthesis to match an opening parenthesis. There must always be an equal number of opening and closing parentheses. If you want to test this prompt, re-enter one of the previous expressions, leaving off the closing parenthesis.

AUTOLISP AND AUTOCAD COMMANDS

An AutoCAD command can be used within an AutoLISP routine. To place a command with the routine, use the **(command)** function together with the AutoCAD command name in quotation marks, followed by the proper *arguments*. The arguments are the responses to the command's prompts, such as typing 3 when prompted "Specify radius of circle or [Diameter]" by the **Circle** command.

If the command requires user input, the word "pause" is placed within the routine. For example, if you want to draw a circle with the center at the absolute coordinate of 5,5 and then have the user drag the circle to the desired radius, the following AutoLISP code is used.

```
(command "circle" "5,5" pause)
```

Notice that the coordinate of 5,5 is placed in quotes. This is because it is the actual value and not a variable.

Type the line at the "Command" prompt. Notice that the **Circle** command's prompts are echoed (displayed) to the command line, as shown below:

Command: **(command "circle" "5,5" pause)**

circle Specify center point for circle or [3P/2P/Ttr (tan tan radius)]:
    5,5

Specify radius of circle or [Diameter]: *(Pick a point.)*

Command: nil

There may be times that you wish these prompts not to display. For example, if the routine contains all the inputs for the prompts, it is useless to show them. Suppress the prompts by turning off the **CmdEcho** system variable within the expression. The following line can be included within an AutoLISP routine to do this.

```
(setvar "cmdecho" 0)
```

This line sets the AutoCAD system variable **CmdEcho** to zero, which suppresses command prompts. Set it to 1 to turn it back on. Appendix C contains a listing of all AutoCAD system variables.

THE SETQ FUNCTION

The **(setq)** function is used to assign a value to a variable. A variable is like a memory in a calculator. When you want to store the result of a calculation, you press the M+ key on the calculator. In AutoLISP, using the **(setq)** function is like pressing the M+ key. Unlike most calculators (which have a single memory), AutoLISP has as many memories as you need to use. As well, instead of calling the memory M+, you can name "memories" (variables) in AutoLISP just about any name you want. Here, for example, you assign values to variables A and B:

Command: **(setq A 10)**

The variable (that is, memory) called A now holds the value of ten. Continue and assign a value for B:

Command: **(setq B 25)**

B is now equal to 25. Now let's use an arithmetic expression with A and B, just as if you were in algebra class:

Command: **(+ A B)**

35

Notice how AutoCAD returned the value 35 for the sum of A and B.

GETTING AND STORING POINTS IN AUTOLISP

You can use AutoLISP to store X,Y coordinates, such as a point that you enter or a point selected on the drawing screen. For example, you may want to store a point with the **(setq)** function for later use in the routine. You do this with the **(getpoint)** function. You use this function later in an example.

You can also get and store a *distance*. The **(getdist)** function is used to obtain a distance. You can either enter the distance as a numeric value or show AutoCAD two points on the drawing screen.

PLACING PROMPTS WITHIN AUTOLISP

When you write your own AutoLISP routines, you may want to include *prompts*. Prompts are used to tell the user the kind of data you are expecting. AutoLISP allows you to include prompts with functions like **(getpoint)** and **(getdist)**. The text of the prompt is placed within quotes (" ") inside the AutoLISP routine. For example, the following line prompts "Reference point:" at the command line.

```
(setq DistA (getdist "Enter distance:"))
```

USING COMMENTS WITHIN A ROUTINE

As you write AutoLISP routines, you may want to include *comments* (notes). Comments are useful for documenting your programming work. Text that starts with a semicolon (;) is not executed as a part of the routine. The user does not see your comments.

It is sometimes helpful to include a heading that describes the routine you are writing. You include a description when you write an AutoLISP routine later in this chapter.

There are many more functions available in AutoLISP, but we will stop here and use the ones we have learned so far. If you want to learn all the available functions, the online *AutoLISP Reference* contains a complete listing. From the AutoCAD menu bar, select **Help | AutoCAD Help**. In the **Contents** tab, double-click on **Visual LISP and AutoLISP**.

WRITING AN AUTOLISP ROUTINE

It's time to write, store, and use a simple AutoLISP routine. Let's assume that you are a machine parts designer. You frequently draw parts that have drill holes of different diameters at a certain position on a plate. The position is always relative to a corner of the plate. You want to write a custom AutoCAD command to streamline your drawing process. The following AutoLISP routine could be used. Let's look at the routine and analyze the functions.

```
;————————————————————————————————
;DPCIR.LSP
;————————————————————————————————
;DESCRIPTION
;
;Constructs a circle of a specified radius at a designated
;point from a reference position.
;————————————————————————————————
;
(defun C:DPCIR ()
     (setvar "CMDECHO" 0)
     (setvar "lastpoint" (getpoint "Select corner of plate:"))
     (setq p1 (getpoint "\nEnter X,Y distance (using @): "))
     (setq p2 (getdist "\nEnter hole radius: "))
     (command "CIRCLE" p1 p2)
)
```

This routine is a simple ASCII text file; this means you can type it into the Notepad text editor provided with all copies of Windows.

The file is named "DPCIR.LSP," short for Designated Point Circle. Note that all AutoLISP files have an LSP file extension. Type the lines of text and save with the name DPCIR.LSP.

The file is a collection of lines of AutoLISP instructions. Let's discuss each section of the routine.

A semicolon prefixes the first nine lines. This means they are not processed as part of the routine. You use these lines to provide a description of the file. This might be useful if you looked through a library of AutoLISP routines later and could not remember the purpose of this routine.

You use the **(defun)** function to define a new command named DPCIR. The C: prefix allows you to run the routine as you would a regular AutoCAD command. If the C: were left out, you would have to type (DPCIR), like any other AutoLISP function.

The next line sets the variable **CmdEcho** to zero. You did this because you do not want the prompts for the **Circle** command, which occur later, to be displayed.

Next, you use the system variable **Lastpoint** (see Appendix C for explanations of this and all system variables) to set a reference point to locate the circle. If you remember, this reference point is to be the corner of the plate. You use the **(getpoint)** function to obtain this point, and include the prompt "Select corner of plate:".

You use the **(setq)** function to store a value of a point in a variable that you named "p1." The point for **p1** was obtained with the **(getpoint)** function; you included a prompt for a relative coordinate.

You used **(setq)** again to store a value for the point named **p2**. This time you used the **(getdist)** function to obtain this distance. You again included a prompt, this time for the radius distance.

Finally, you employed the **Circle** command. All the previous lines of code were for obtaining the values to provide for the **Circle** command. The variables **p1** and **p2** are the arguments for the prompts that the **Circle** command issues: center point and radius.

If you want to enhance the routine, you could use the **(setvar)** function to set the object snap to INTersection when you capture the plate corner; and then set the object snap back to NONE. The great part of AutoLISP programming is that you can write the command to suit yourself!

USING THE AUTOLISP ROUTINE

Let's try the AutoLISP routine we just looked at. The first step is to create a file named DPCIR.LSP. To do this, use a text editor (such as Notepad) or word processor in "non-document" or "text only" mode (that means that the file must be all ASCII text). Name the file DPCIR.LSP. Type each line exactly as it is printed, and then save the file to the folder where you have stored the AutoCAD program.

Next, start AutoCAD and open a drawing. Let's now load the AutoLISP program.

> Command: **(load "dpcir")**

If you get an error message, go back and check the contents of your file to verify that you copied it exactly as written. The most likely error is a missing parenthesis or quotation mark.

If the load was successful, AutoCAD displays "C:DPCIR" on the command line.

Now let's use the program. Draw a square that is 5 units on each side. Now place the drill hole using your **Dpcir** routine as a regular command:

> Command: **dpcir**
>
> Select corner of plate: *(Select the lower left corner.)*
>
> Enter X,Y-distance (using @): **@2,2**
>
> Enter radius: **.5**

Writing routines in AutoLISP takes time and practice to learn. The effective use of the AutoLISP programming techniques can be further explored in many excellent books dedicated to this purpose.

CHAPTER REVIEW

1. Write an AutoLISP routine that multiplies 2 times 4.

2. What does "n>" mean when n is a number?

3. How can an AutoCAD command be used within an AutoLISP routine?

4. How would you suppress the command prompts from an AutoCAD command contained within an AutoLISP file?

5. What does the **(setq)** function do?

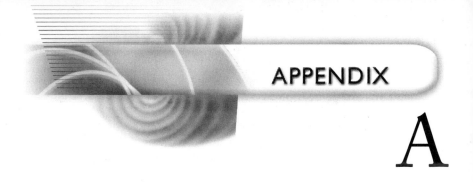

Professional CAD Techniques

Studying and learning the commands in this book are only a start toward becoming a professional CAD operator. Using proper techniques is just as important. The following is a list of rules that should be followed if you want to obtain the most from your expertise.

Learn every command.

Most CAD operators tend to learn just enough to get by. If this is your intention, stick with your pencil. You will not realize the full benefits of time savings and results that are more professional. Force yourself to use each command. You can do this best by introducing yourself to each new command and using it until you are comfortable with it. Some of the parts of AutoCAD that appear the hardest to learn become easy with just a little practice!

Practice, Practice, Practice!

Efficiency on a CAD system comes with practice. (Doesn't it also take practice to perform good board drafting?) Never pass up the opportunity to spend time with AutoCAD. After you are comfortable with the commands you are using, try new ones. Before long, you will be constructing drawings with ease!

Plan your approach.

Never just start drawing. There are dozens of approaches to every drawing when you are using CAD. The time saving you realize will come from your technique, not your speed. Drawing faster doesn't mean working harder or faster; it means working smarter. AutoCAD has some amazing capabilities; let it work for you!

Use layers.

If you are drawing a set of house plans, don't draw the floor plan more than once. Use the layer system to use the plan for each drawing that requires it. The electrical, plumbing, HVAC, lighting, framing, dimensions, furniture, and other plans exist on independent layers.

Use blocks.

Never draw anything twice! If you have drawn it once, write-block it. Before long, you will have a nice library to draw from. If you build a large enough library of the drawings of your trade, you will soon be assembling more than you will be drawing.

Customize toolbars.

Set up a toolbar that contains all the commands that you normally use in your work. Be sure to include any blocks that are normally used. This allows you to choose the commands and drawing parts quickly.

Buy proper equipment.

Don't try to save money on brain surgery or CAD equipment. You will purchase a CAD station to save time (that is, money). If the equipment doesn't suit the job, you won't obtain the savings and might end up not using it at all. It is cheaper to buy the proper equipment than it is to replace it later!

Set up a proper workstation.

As in any other type of job, the proper work setup is essential. If you have purchased the right equipment, plan a proper setting for it. A little planning will make your work-days (or those of your employees) more efficient.

Be patient.

Your first few months with CAD won't save you the time that you had anticipated. It will come! After a while, you will notice that your speed has increased dramatically. This time period will vary with each individual and each task. It is normal to experience a learning curve, but don't get discouraged; it shouldn't last too long.

Designate a CAD manager.

If you are using a multi-station CAD setup, assign one person the task of coordinating all operations. This person will emerge soon after you install the system. Some people just seem to gravitate to the position. This person will become a valuable employee.

Keep pushing!

Don't ever stop experimenting with ways to streamline your graphics procedures. Every task is different, with many ways to approach it. One of the advantages of a CAD system is the ability to customize it to your specific needs. With the capabilities of AutoCAD, your real challenge is to discover the ways it can serve you best!

Use Real-time Pan and Zoom.

Zoom in to a drawing to perform detail work. Don't attempt to draw complicated parts when the display shows the work in small scale.

Run check plots.

Run periodic check plots to check your work. You will also obtain a feel for the scale of the finished work. Use the laser printer for inexpensive plots.

Use both CAD and traditional techniques.

You will soon learn which work CAD is best for and which work traditional methods are best for. Sometimes it is best to lay out your drawing in rough form by hand, then digitize the base drawing in, and finish with CAD.

Use Copy and Array.

Use the Copy and Array commands for repeated work. Never, never draw the same thing twice!

Standardize.

Standardize your procedures and those of your office for CAD use. Create standards for base drawing layers, text sizes, fonts, linetypes, and other operations. Not only will your work be consistent, but your drawing operations will be easier to perform.

Use Associativity.

AutoCAD can do many of the editing tasks for you—if you let it. In particular, AutoCAD automatically updates dimensions and hatch patterns. Make sure you place associative dimensions and hatch patterns. It is far easier to include the dimension when you stretch a part, and have the dimension change automatically, than it is to erase and redraw the dimension.

Always Draw Full Size.

When you draw, always draw full size (1:1 scale). Not only is it easier (you save on the math involved in calculating scaled distances—which could cause an error), but you can use the CAD drawing to measure objects and distances to get accurate sizes.

Draw Accurately.

Use object snaps modes. Use direct distance entry. Use all the drawing tools that AutoCAD provides to make it easier to draw it once, correctly.

Command Summary

The following list of commands are documented by Autodesk as available in AutoCAD 2000. Commands prefixed with ' (apostrophe) are transparent commands, which can be executed during another command.

A complete reference of all AutoCAD 2000 commands, and how to access them, and how to use them can be found in *The Illustrated AutoCAD 2000 Quick Reference* by editor Ralph Grabowski, available through Autodesk Press.

A

'About Displays AutoCAD information dialog box that includes version and serial numbers.

AcisIn Imports an ASCII-format ACIS file into the drawing and creates a 3D solid, 2D region, or body object.

AcisOut Exports an AutoCAD 3D solid, 2D region, or body as an ASCI-format ACIS file (file extension .SAT).

 **AdcClose** Closes the AutoCAD DesignCenter window.

AdCenter (or CTRL+2) Opens AutoCAD DesignCenter; manages AutoCAD content.

AdcNavigate Directs the Desktop in AutoCAD DesignCenter to the file name, folder, or network path you specify.

Ai_Box, etc Collection of commands to draw 3D primitives from surface meshes: Ai_Box, Ai_Cone, Ai_Dish, Ai_Dome, Ai_mesh, Ai_Pyramid, Ai_Torus, and Ai_Wedge.

Align Uses three pairs of 3D points to move and rotate (align) 3D objects.

AmeConvert	Converts drawings made with AME v2.0 and v2.1 into ACIS solid models.
'Aperture	Adjusts the size of the target box used with object snap.
'AppLoad	Displays a dialog box that lets you list AutoLISP, ADS, and ARX program names for easy loading into AutoCAD.
Arc	Permits drawing of arcs using different parameters.
Area	Computes area and perimeter of a polygon.
Array	Makes multiple copies of an object or group of objects.
Arx	Loads and unloads ARX programs. Also displays the names of ARX program command names. The Unload option replaces the AseUnload and RenderUnload commands.
AttDef	Creates an attribute definition.
'AttDisp	Controls whether the attributes are displayed.
AttEdit	Used for editing attributes.
AttExt	Allows attribute objects to be extracted from a drawing and written to file on disk for use with another program.
AttReDef	Lets you assign existing attributes to a new block, and new attributes to an existing block.
Audit	A diagnostic command used to examine and/or correct errors in a drawing file.

B

Background	Sets up a background for a rendered scene. The background can be a solid color, gradient shades, an image (in BMP, PCX, GIF, Targa, JPEG, or TIFF format), or the current AutoCAD image.
'Base	Specifies point of origin for insertion into another drawing.

BHatch	Fills an automatically defined boundary with a hatch pattern; the use of a dialog box allows preview and adjustment without starting over.
'Blipmode	Controls whether a marker blip is displayed on the screen when picking a point.
Block	Forms a complex object from a group of separate objects in a drawing.
BlockIcon	Generates preview images for blocks created with Release 14 or earlier
BmpOut	Exports selected objects from the current viewport to a raster BMP file.
Boundary	Draws a closed boundary polyline.
Box	An ACIS command that creates a 3D solid box or cube.
Break	Erases part of an object or breaks it into two objects
Browser	Launches your Internet browser (as defined by the Windows system registry) from within AutoCAD with the URL you specify. A "URL" is a uniform resource locator and looks like http://www.autodesk.com

C

'Cal	The geometry calculator that evaluates integer, real, and vector expressions.
Camera	Sets the camera and target location.
Chamfer	Trims two intersecting lines and connects the two trimmed lines with a chamfer line.
Change	Permits modification of an object's characteristics.
ChProp	Similar to the Change command, except only the properties (linetype, color, etc.) of the object are changed.
Circle	Draws any size circle.
Close	Closes the current drawing.

'Color (or 'Colour)	Sets a new color for all subsequently drawn objects.
Compile	Compiles shape and font files.
Cone	An ACIS command that creates a 3D solid cone.
Convert	Converts 2D polylines and associative hatches in pre-Release 14 drawings to the new "lightweight" format, which saves on memory and disk space.
Copy	Copies selected objects.
CopyBase	Copies objects with a specified base point.
CopyClip (or CTRL+C)	Copies selected objects to the Windows Clipboard in several formats.
CopyHist	Copies Text window text to the Windows Clipboard.
CopyLink	Copies all objects in the current viewport to the Windows Clipboard in several formats.
CutClip (or CTRL+X)	Cuts selected objects from the drawing to the Windows Clipboard in several formats.
Cylinder	An ACIS command that creates a 3D solid cylinder.

D

DbcClose	Closes the dbConnect Manager.
DbConnect (or CTRL + 6)	Connects the AutoCAD drawing with external database tables.
DbList	Provides information about all objects in the drawing.
DdEdit	Edits text, paragraph text, and attribute text.
DDim	Opens the Dimension Manager dialog box.
'DdPType	Displays a dialog box that lets you choose the style and size of point.
DdVPoint	Displays a dialog box that lets you set a new 3D viewpoint.
'Delay	Allows for delay between operations in a script file.

Dim	Semi-automatic dimensioning capabilities.
Dim1	Executes a single Release 12-style dimension command.
DimAligned	Draws a linear dimension aligned to an object.
DimAngular	Draws an angular dimension.
DimBaseline	Draws a linear, angular, or ordinate dimension that continues from a baseline.
DimCenter	Draws the center mark on circles and arcs.
DimContinue	Draws a linear, angular, or ordinate dimension that continues from the last dimension.
DimDiameter	Draws diameter dimension on circles and arcs.
DimEdit	Edits the text and extension lines of associative dimensions.
DimLinear	Draws linear dimensions.
DimOrdinate	Draws ordinate dimensions in the X and Y directions.
DimOverride	Overrides current dimension variables to change the look of selected dimensions.
DimRadius	Draws radial dimensions for circles and arcs.
DimStyle	Creates, names, modifies, applies named dimension styles.
DimTEdit	Moves and rotates text in dimensions.
'Dist	Computes the distance between two points.
Divide	Divides an object into an equal number of parts and places either a specified block or a point object at the division points on the object.
Donut (or Doughnut)	Constructs solid filled circles and doughnuts.
'Dragmode	Permits dynamic dragging of an object to the desired position on the display.

	DrawOrder	Changes the order in which objects are displayed. The selected object or image is placed above or below other objects in the drawing. This is useful for ensuring text is not obscured, particularly in 3D and renderings.
2000	DSettings	Specifies drawing settings for snap, grid, polar, and object snap tracking.
	'DsViewer	Opens the Aerial View window.
	DView	Used to display a 3D view dynamically.
2000	DwgProps	Sets and displays the properties of the current drawing.
	DxbIn	Creates binary drawing interchange files.

E

	Edge	Changes the visibility of 3D face edges.
	EdgeSurf	Draws an edge-defined surface.
	'Elev	Sets current elevation and thickness.
	Ellipse	Used to construct ellipses.
	Erase	Removes entities from a drawing.
	Explode	Breaks down a block into the individual objects from which it was constructed; breaks down a polyline into lines and arcs.
	Export	Displays a dialog box to export the drawing in a variety of file formats.
2000	ExpressTools	Launches the AutoCAD Express Tools, if installed.
	Extend	Extends objects in a drawing to meet a boundary object.
	Extrude	Extrudes a 2D closed object into a 3D solid object.

F

	'Fill	Toggles the display of solid fills in the drawing.
	Fillet	Connects two lines with an arc.

'Filter	Creates a selection set of objects based on their properties.
2000 Find	Finds, replaces, selects, and zooms to specified text
Fog	Adds fog or depth effects to a rendering. The default fog effect is created by increasing white with distance; you can choose any color.

G

'GraphScr (or F2)	Switches to the drawing window from the Text window.
'Grid (or F7)	Displays grid of specified spacing.
Group	Creates a named selection set of objects.

H

Hatch	Performs hatching at the command prompt.
HatchEdit	Edits existing associative hatch patterns.
'Help (or ? or F1)	Displays a list of AutoCAD commands with detailed information available.
Hide	Removes hidden lines from the currently displayed view.
2000 HyperLink (or CTRL+K)	Attaches a hyperlink to a graphical object or modifies an existing hyperlink
2000 HyperLinkOptions	Controls the visibility of the hyperlink cursor and the display of hyperlink tooltips

I

'Id	Displays the position of a point in X,Y,Z coordinates.
Image	Controls the insertion of raster images with an Xref-like dialog box.
ImageAdjust	Controls the brightness, contrast, and fading of raster images.
ImageAttach	Attaches a raster image as an externally-referenced file to the current drawing.

ImageClip	Places a rectangular or irregular clipping boundary around an image object.
ImageFrame	Toggles the display of the image's frame.
ImageQuality	Controls the display quality of images.
Import	Displays a dialog box to import a variety of file formats into the drawing.
Insert	Inserts a block or another drawing into the current drawing.
InsertObj	Inserts an object generated by another Windows application.
Interfere	Determines the interference of two or more 3D solids.
Intersect	Creates a 3D solid (or 2D region) from the intersection of two or more 3D solids (or 2D regions).
'Isoplane (or F5)	Switches to another isoplane.

L

'Layer	Creates or switches drawing layers; toggles the state of layers; assigns linetypes, lineweights, plot styles, and colors to the layers.
Layout	Creates a new layout and renames, copies, saves, or deletes an existing layout.
LayoutWizard	Allows you to designate page and plot settings for a new layout.
Leader	Draws a leader dimension.
Lengthen	Lengthens or shortens open objects.
Light	Creates, names, places, and deletes "lights" used by the Render command.
'Limits	Sets the drawing boundaries.
Line	Draws a straight line.

'Linetype	Lists, creates, or modifies linetype definitions or loads them for use in a drawing.
List	Displays database information for a single entity in a drawing.
Load	Loads a shape file into a drawing.
LogFileOff	Closes the log file Acad.Log.
LogFileOn	Writes the text of the 'Command:' prompt area to the log file Acad.Log.
LsEdit	Edits a landscape object. A landscape object is a raster image (typically of a single object, such as a tree or book) with a mask. The "mask" lets objects behind the landscape show through.
LsLib	Accesses a library of landscape objects.
LsNew	Places a "landscape" item in the drawing.
'LtScale	Specifies a scale for all linetypes in a drawing.
[2000] LWeight	Sets the current lineweight, lineweight display options, and lineweight units.

M

MassProp	Calculates and displays the mass properties of 3D solids and 2D regions.
'MatchProp	Copies the properties from one object to one or more objects.
MatLib	Imports material-look definitions; used by the Render command.
Measure	Places point objects at a specified distance on an object.
Menu	Loads a menu of AutoCAD commands into the menu area.
MenuLoad	Loads a partial menu file.
MenuUnLoad	Unloads part of the menu file.

MInsert	Used to make an array of inserted blocks.
Mirror	Creates a mirror image of an object.
Mirror3D	Creates a mirror image of objects that can be rotated about a plane.
MlEdit	Displays a dialog box that lets you perform limited editing of multilines.
MLine	Draws multiple parallel lines (up to 16 parallel lines).
MlStyle	Displays a dialog box that lets you define named mline styles, including color, linetype, and endcapping.
2000 Model	Switches from a layout tab to the Model tab.
Move	Moves an object from one location to another.
MSlide	Creates a slide of the current display.
MSpace	Used to switch to model space.
MtEdit	An alias for the DdEdit command. Edits an mtext object.
MText	Places paragraph text that fits inside a rectangular boundary.
Multiple	When used before a command, causes the command to repeat after each use. For example: multiple circle.
MView	Used in paper space to create and manipulate viewports.
MvSetup	Sets up the specifications for a drawing.
N	
New (or CTRL+N)	Used to create a new drawing.
O	
Offset	Constructs a parallel copy to an object or constructs a larger or smaller image of the object through a point.
OleLinks	Controls the objects linked to the drawing.
2000 OleScale	Displays the OLE Properties dialog box.

Oops	Restores entities that were accidentally erased with the previous command.
Open (or CTRL+O**)**	Opens an existing drawing.
Options	Customizes the AutoCAD settings.
'Ortho (or F8**)**	Causes all lines to be drawn orthogonally with the set snap rotation angle.
'OSnap (or F3**)**	Allows geometric points of existing objects to be easily located.

P

PageSetup	Specifies the layout page, plotting device, paper size, and settings for each new layout.
'Pan	Moves the display window for viewing a different part of the drawing without changing the magnification in real-time.
PartiaLoad	Loads additional geometry into a partially opened drawing.
PartialOpen	Loads geometry from a selected view or layer into a drawing.
PasteBlock	Pastes a copied block into a drawing.
PasteClip (or CTRL+V**)**	Pastes an object from the Windows Clipboard into the upper left corner of the drawing.
PasteOrig	Pastes a copied object in a new drawing using the coordinates from the original drawing.
PasteSpec	Provides control over the format of the object pasted into the drawing.
PcInWizard	Imports PCP and PC2 configuration file plot settings into the current layout.
PEdit	Edits polylines.
PFace	Constructs a polygon mesh that is defined by the location of each vertex in the mesh.

Plan		Returns to the current UCS plan view.
PLine		(Polylines) Lines of specified width that can be manipulated.
Plot (or CTRL+P**)**		Plots a drawing to the printer or plotter.
2000	**PlotStyle**	Sets the current plot style for new objects, or the assigned plot style for selected objects.
2000	**PlotterManager**	Launch the Add-a-Plotter wizard and the Plotter Configuration Editor.
Point		Draws a specified point.
Polygon		Draws a regular polygon with a specified number of sides.
Preview		Windows-like plot preview.
2000	**Properties (or** CTRL+1**)**	Controls properties of existing objects
2000	**PropertiesClose**	Closes the Properties window
PsDrag		Controls the scale and position of an imported PostScript image that is being dragged into place.
2000	**PSetUpIn**	Imports a user-defined page setup into a new drawing layout
PsFill		Allows 2D polylines to be filled with PostScript fill patterns.
PsIn		Imports EPS (Encapsulated PostScript) files.
PsOut		Exports the current view of a drawing to an EPS (Encapsulated PostScript) file.
PSpace		Switches to paper space.
Purge		Selectively deletes unused blocks, layers, or linetypes.
Q		
2000	**QDim**	Quickly creates continuous dimensions.
2000	**QLeader**	Quickly creates a leader and leader annotation.

	QSave (CTRL+S)	Command that saves the drawing without requesting a file name.
2000	QSelect	Creates selection sets based on filtering criteria.
	QText	Redraws text objects as rectangles.
	Quit (or ALT+F4)	Exits AutoCAD.

R

	Ray	Draws a semi-infinite construction line.
	Recover	Attempts the recovery of corrupted or damaged files.
	Rectang	Draws a rectangle.
	Redefine	Restores AutoCAD's definition of a command.
	Redo (or CTRL+Y)	Restores operations deleted by the previous Undo command.
	'Redraw	Cleans up the display.
	'RedrawAll	Performs a redraw in all viewports.
2000	RefClose	Saves back or discards changes made during in-place editing of a reference (an xref or a block).
2000	RefEdit	Selects a reference for editing.
2000	RefSet	Adds or removes objects from a working set during in-place editing of a reference (an xref or a block).
	Regen	Causes the entire drawing to be regenerated and redraws the screen.
	RegenAll	Performs a regeneration in all viewports.
	'RegenAuto	Allows you to control when the drawing is automatically regenerated.
	Region	Creates a 2D region object from existing closed objects.
	Reinit	Reinitializes the I/O ports, digitizer, display, plotter, and PGP file.

Rename	Changes the name of blocks, linetypes, layers, text style, etc.
Render	Creates a rendering of 3D objects in the drawing.
RendScr	Redisplays the last rendering created with the Render command
Replay	Displays a dialog box that lets AutoCAD display a GIF, TGA, or TIFF raster file.
'Resume	Continues playing back a script file that had been interrupted by the + or keys.
Revolve	Creates a 3D solid by revolving a 2D closed object around an axis.
RevSurf	Draws a revolved surface.
RMat	Lets you define, load, create, attach, detach, and modify material-look definitions; used by the Render command.
Rotate	Rotates an entity or a group of objects around a specified center point.
Rotate3D	Rotates objects about a 3D axis.
RPref	Displays a dialog box that lets you set your preferences for renderings.
'RScript	Forces a script to be restarted from the beginning.
RuleSurf	Draws a ruled surface.

S

Save and SaveAs	Saves the current drawing by a different specified name.
SaveImg	Saves the current rendering in GIF, TGA, or TIFF format.
Scale	Changes the size of an object, equally in X, Y, and Z-directions.
Scene	Creates, modifies, and deletes named scenes; used by the Render command.

'Script	Allows user to invoke a script file while in AutoCAD.
Section	Creates a 2D region from a 3D solid by intersecting a plane through the solid.
Select	Used to preselect objects to be edited.
SetUV	Controls how a raster image is mapped onto an object; used for texture mapping in rendered scenes. "UV" refers to the x,y-direction of a 2D raster image.
'SetVar	Used to view and change AutoCAD's system variables.
ShadeMode	Shades the objects in the current viewport.
Shape	Places shapes from a shape file into a drawing.
Shell (or Sh)	Prompts you to run another program outside of AutoCAD.
ShowMat	Reports the material definition assigned to the selected object.
Sketch	Allows freehand sketching as a part of the drawing.
Slice	Slices a 3D solid with a plane.
'Snap (or F9)	Allows you to turn the snap on or off, change the snap resolution, set different spacing for the X- and Y-axis, or rotate the grid, and set isometric mode.
SolDraw	Creates profiles and sections of 3D solid models in viewports created with the SolView command.
Solid	Draws filled in polygons.
SolidEdit	Edits faces and edges of 3D solid objects.
SolProf	Creates profile images of 3D solid models.
SolView	Creates viewports in paperspace of orthogonal multi- and sectional view drawings of 3D solid models.
Spell	Checks the spelling of text in the drawing.
Sphere	Draws a 3D sphere.

	Spline	Draws a NURBS (spline) curve.
	SplinEdit	Edits a spline.
	Stats	Displays a dialog box that lists information about current state of rendering.
	'Status	Displays a status screen containing information about the current drawing.
	StlOut	Exports 3D solids to an SLT file, in ASCII or binary format.
	Stretch	Move selected objects while allowing their connections to other objects in the drawing to remain unchanged.
	'Style	Creates and modifies text styles.
2000	StylesManager	Displays the Plot Style Manager.
	Subtract	Creates a new 3D solid (or 2D region) by subtracting one object from a second object.
	SysWindows	Controls the size and position of the drawing window.

T

	Tablet (or F4)	Permits alignment of digitizer with existing drawing coordinates.
	TabSurf	Draws a tabulated surface.
	Text	Allows text to be entered into a drawing.
	'TextScr (or F2)	Displays the Text window.
	'Time	Keeps track of time functions for each drawing.
	Tolerance	Displays a dialog box that lets you select tolerance symbols.
	Toolbar	Controls the display of toolboxes.
	Torus	Draws a doughnut-shaped 3D solid.
	Trace	Draws lines of a specified width.

Transparency	Toggles the background of a bilevel image to transparent (see-through) or opaque.
'TreeStat	Displays drawing information on the current spatial index.
Trim	Trims objects in a drawing by defining other objects as cutting edges then specifying the part of the object to be cut from between them.

U

U (or CTRL+Z)	Used to undo the most recent command.
UCS	Used to manipulate the user-defined coordinate system.
UCSicon	Controls the on-screen display of the UCS icon indicator.
UcsMan	Manages user-defined coordinate systems.
Undefine	Disables a command.
Undo	Used to undo several command moves in a single operation.
Union	Creates a new 3D solid (or 2D region) from two solids (or regions).
'Units	Allows you to select display format and precision of that format.

V

VbaIDE (or ALT+F8)	Launches the Visual Basic Editor (short for Visual Basic for Applications Integrated Development Editor).
VbaLoad	Loads a VBA project into AutoCAD.
VbaMan	Loads, unloads, saves, creates, embeds, and extracts VBA projects.
VbaRun (or ALT+F11)	Runs a VBA macro.
VbaStmt	Executes a VBA statement at the AutoCAD command prompt.
VbaUnload	Unloads a global VBA project.

'View	Saves the display as a view or displays a named view.
ViewRes	Controls the fast zoom mode and resolution for circle and arc regenerations.
2000 VLisp	Launches the Visual LISP interactive development environment.
2000 VpClip	Clips viewport objects.
VpLayer	Controls the visibility of the individual viewport layers.
VPoint	Sets the viewpoint from which you view the drawing.
VPorts (or Viewports)	Sets the number and configuration of viewports displayed on the screen.
VSlide	Allows you to view a slide file.

W

WBlock	Writes objects to a new drawing file.
Wedge	Draws a 3D solid wedge.
2000 WhoHas	Displays ownership information for opened drawing files.
WmfIn	Imports a WMF file into the drawing as a block.
WmfOpts	Controls how a WMF file is imported.
WmfOut	Exports the drawing as a WMF file

X

XAttach	Attaches an externally-referenced DWG drawing file to the drawing.
Xbind	Binds externally referenced drawings, converting them to blocks in the master drawing.
XClip	Defines a clipping boundary; sets the front and back clipping planes.
XLine	Draws an infinite construction line.

Xplode	Breaks a compound object into its component objects, with user control.
Xref	Used to place an externally referenced drawing into a master drawing.

Z

'Zoom	Allows you to increase or decrease the size of the display for viewing purposes.

3

3D	Draws 3D surface objects out of polygon meshes: box, cone, dish, dome, mesh, pyramid, sphere, torus, and wedge.
3dArray	Creates a 3D array.
2000 3dClip	Switches to interactive 3D view and opens the Adjust Clipping Planes window.
2000 3dCOrbit	Swtiches to interactive 3D view and enables you to set the objects in the 3D view into continuous motion.
2000 3dDistance	Switches to interactive 3D view and makes objects appear closer or farther away.
3Dface	Creates a solid face in a defined plane.
3Dmesh	Draws a 3D mesh.
2000 3dOrbit	Controls the interactive viewing of objects in 3D.
2000 3dPan	Invokes the interactive 3D view and enables you to drag the view horizontally and vertically.
3dPoly	Draws a 3D polyline.
3dsIn	Imports 3D Studio geometry and rendering data; cannot import procedural materials and smoothing groups.
3dsOut	Exports AutoCAD geometry and rendering data.
2000 3dSwivel	Switches to interactive 3D view and simulates the effect of turning the camera.

3dZoom   Switches to interactive 3D view so you can zoom in and out on the view.

System Variables

AutoCAD stores information about the current state of itself, the drawing and the operating system in 358 *system variables*. The variables help programmers—who often work with menu macros and AutoLISP—determine the state of the AutoCAD system.

The following pages list all documented system variables, plus several more not documented by Autodesk. The listing uses the following conventions:

Symbol	Meaning
Bold Italicized	Undocumented system variable.
~~*Italicized*~~	System variable removed from AutoCAD.
⌨	Must be accessed via the **SetVar** command.
⋊⋉	System variable is new to AutoCAD 2000.
Default	Default values as set in the Acad.Dwg prototype drawing.
R/O	Read-only; cannot be changed by the user or by programming.
Loc	Location where the values of the system variable is saved:

Location	Meaning
ACAD	AutoCAD.
DWG	Current drawing.
REG	Windows registry.
...	Not saved.

TIPS

■ The **SetVar** command lets you change the value of all variables, except those marked read-only.

■ You get a list of system variables at the 'Command' prompt with the **?** option of the **SetVar** command:

> Command: **setvar**
>
> Variable name or ?: **?**
>
> Variable(s) to list <*>: **ENTER**

■ When a system variable is stored in the Windows registration file, the variable affects all drawings.

■ When a system variable is stored in the drawing, the variable affects only the current drawing.

■ When a system variable is not stored, the variable is set when AutoCAD loads. The value of the variable is either: (1) read from the operating system; or (2) set to the default value.

Variable	Default	R/O	Loc	Meaning
_LInfo	""	R/O	ACAD	Hardware lock's serial number; unlocked software returns "". Must be used with AutoLISP: (getvar "_linfo").
_PkSer	varies	R/O	ACAD	Software package serial number, such as "117-69999999".
_Server	0	R/O	REG	Network authorization code.
_VerNum	varies	R/O	REG	Internal program build number, such as "T.0.98".
A				
AcadLspAsDoc	0		REG	↖ Acad.Lsp is loaded into: 0 Just the first drawing. 1 Every drawing.
AcadPrefix	varies	R/O	...	Path spec'd by ACAD environment variable, such as "d:\acad 2000\support; d:\acad 2000\fonts" .
AcadVer	"15.0"	R/O	...	AutoCAD version number.
AcisOutVer	40	R/O	...	ACIS version number; can be 15, 16, 17, 18, 20, 21, 30, and 40.
AFlags	0	...	...	Attribute display code: 0 No mode specified. 1 Invisible.

Variable	Default	R/O	Loc	Meaning
				2 Constant.
				4 Verify.
				8 Preset.
AngBase	0.0000	...	DWG	Direction of zero degrees relative to UCS
AngDir	0	...	DWG	Rotation of angles:
				0 Clockwise
				1 Counterclockwise
ApBox	0	...	REG	AutoSnap aperture box cursor:
				0 Off.
				1 On.
Aperture	10	...	REG	Object snap aperture in pixels:
				1 Minimum size
				10 Default size
				50 Maximum size
Area	0.0000	R/O	...	Area measured by **Area**, **List**, or **Dblist**.
AttDia	0	...	DWG	Attribute entry interface:
				0 Command-line prompts.
				1 Dialog box.
AttMode	1	...	DWG	Display of attributes:
				0 Off.
				1 Normal.
				2 On.
AttReq	1	...	REG	Attribute values during insertion are:
				0 Default values.
				1 Prompt for values.
AuditCtl	0	...	REG	Determines creation of ADT audit log file:
				0 File not created.
				1 ADT file created.
AUnits	0	...	DWG	Mode of angular units:
				0 Decimal degrees.
				1 Degrees-minutes-seconds.
				2 Grads.
				3 Radians.
				4 Surveyor's units.
AUPrec	0	...	DWG	Decimals places displayed by angles.
AutoSnap	55	...	REG	Controls AutoSnap display:
				0 Turns off all AutoSnap features.
				1 Turns on marker.
				2 Turns on SnapTip.
				4 Turns on magnetic cursor.

Variable	Default	R/O	Loc	Meaning
				8 Turns on polar tracking ⟨⟩ .
				16 Turns on object snap tracking ⟨⟩ .
				32 Turns on tooltips for polar tracking and object snap tracking ⟨⟩ .
AuxStat	*0*	...	DWG	*-32768 Minimum value.*
				32767 Maximum value.
AxisMode	*0*	...	DWG	*Obsolete system variable.*
AxisUnit	*0.0000*	...	DWG	*Obsolete system variable.*

B

Variable	Default	R/O	Loc	Meaning
BackZ	0.0000	R/O	DWG	Back clipping plane offset.
BindType	0	...	...	⟨⟩ When binding an xref or editing an xref, xref names are:
				0 Converted from **xref|name** to **xref0name**.
				1 Converted from **xref|name** to **name**.
BlipMode	0	...	DWG	⌨ Display of blip marks:
				0 Off.
				1 On.

C

Variable	Default	R/O	Loc	Meaning
CDate	*varies*	R/O	...	Current date and time; format: YyyyMmDd.HhMmSsDd
CeColor	"BYLAYER"	...	DWG	Current color.
CeLtScale	1.0000	...	DWG	Current linetype scale.
CeLType	"BYLAYER"	...	DWG	Current linetype.
CeLWeight	-1	...	DWG	⟨⟩ Current lineweight in millimeters; valid values are limited to: 0, 5, 9, 13, 15, 18, 20, 25, 30, 35, 40, 50, 53, 60, 70, 80, 90, 100, 106, 120, 140, 158, 200, and 211.
				-1 Sets the lineweight to BYLAYER.
				-2 Sets the lineweight to BYBLOCK.
				-3 Sets the lineweight to DEFAULT, which is defined by **LwDdefault** system variable.
ChamferA	0.5000	...	DWG	First chamfer distance.
ChamferB	0.5000	...	DWG	Second chamfer distance.
ChamferC	1.0000	...	DWG	Chamfer length.
ChamferD	0.0000	...	DWG	Chamfer angle.
ChamMode	0	...	...	Chamfer input mode:
				0 Chamfer by two lengths.
				1 Chamfer by length and angle.

Variable	Default	R/O	Loc	Meaning
CircleRad	0.0000	...	...	Most-recent circle radius.
CLayer	"0"	...	DWG	Current layer name.
CmdActive	1	R/O	...	Type of current command: 1 Regular command. 2 Transparent command. 4 Script file. 8 Dialog box. 16 AutoLISPactive .
CmdDia	*1*	...	*REG*	*Formerly determined whether the **Plot** command displayed at the command line prompt or via dialog box; no longer has an effect in AutoCAD 2000; replaced by **PlQuiet**.*
CmdEcho	1	...	...	AutoLISP command display: 0 No command echoing. 1 Command echoing.
CmdNames	*varies*	R/O	...	Current command, such as "SETVAR".
CMLJust	0	...	DWG	Multiline justification mode: 0 Top. 1 Middle. 2 Bottom.
CMLScale	1.0000	...	DWG	Scales width of multiline: -*n* Flips offsets of multiline. 0 Collapses to single line. 1 Default. 2 Doubles multiline width.
CMLStyle	"STANDARD"	...	DWG	Current multiline style name.
Compass	0	...	...	↖ Toggles the display of the 3D compass: 0 Off. 1 On.
Coords	1	...	DWG	Coordinate display style: 0 Updated by screen picks. 1 Continuous display. 2 Polar display upon request.
CPlotStyle	"BYLAYER"	...	DWG	↖ Current plot style; values defined by AutoCAD are: "ByLayer" "ByBlock" "Normal" "User Defined"

Variable	Default	R/O	Loc	Meaning
CProfile	"<<Unnamed Profile>>"	R/O	REG	↖ Current profile.
CTab	"Model"	R/O	DWG	↖ Current tab.
~~CurrentProfile~~	"<<Unnamed Profile>>	...	...	*Removed from AutoCAD 2000; replaced by CProfile.*
CursorSize	5	...	REG	Cursor size, in percent of viewport: 1 Minimum size. 5 Default size. 100 Full viewport.
CVPort	2	...	DWG	Current viewport number: 2 Minimum (*default*)

D

Variable	Default	R/O	Loc	Meaning
Date	*varies*	R/O	...	Current date in Julian format, such as 2448860.54043252
DBGListAll	*0*	...	*ACAD*	*Toggle.*
DBMod	0	R/O	...	Drawing modified in these areas: 0 No modification made since last save. 1 Object database. 2 Symbol table. 4 Database variable. 8 Window. 16 View.
DctCust	"d:\acad 2000\support\sample.cus"	...	REG	Name of custom spelling dictionary.
DctMain	"enu"	...	REG	Code for spelling dictionary: ca Catalan. cs Czech. da Danish. de German; sharp 's'. ded German; double 's'. ena English; Australian. ens English; British 'ise'. enu English; American. enz English; British 'ize'. es Spanish; unaccented capitals. esa Spanish; accented capitals. fi Finish fr French; unaccented capitals. fra French; accented capitals.

Variable	Default	R/O	Loc	Meaning
				it Italian
				nl Dutch; primary.
				nls Dutch; secondary.
				no Norwegian; Bokmal.
				non Norwegian; Nynorsk.
				pt Portuguese; Iberian.
				ptb Portuguese; Brazilian.
				ru Russian; infrequent 'io'.
				rui Russian; frequent 'io'.
				sv Swedish.
DefLPlStyle	""	R/O	REG	↖ Default plot stype for new layers.
DefPlStyle	"BYLAYER"	R/O	REG	↖ Default plot style for new objects.
DelObj	1	...	REG	Toggle source objects deletion:
				0 Objects deleted.
				1 Objects retained.
DemandLoad	3	...	REG	AutoCAD loads app when drawing contains proxy objects:
				0 Demand loading turned off.
				1 Load app when drawing opened.
				2 Load app at first command.
				3 Load app when drawing opened or at first command.
DiaStat	1	R/O	...	User exited dialog box by clicking on:
				0 **Cancel** button.
				1 **OK** button.

DIMENSION VARIABLES

Variable	Default	R/O	Loc	Meaning
DimADec	-1	...	DWG	Angular dimension precision:
				-1 Use **DimDec** setting (*default*).
				0 Zero decimal places (*minimum*).
				8 Eight decimal places (*maximum*).
DimAlt	Off	...	DWG	Alternate units selected.
DimAltD	2	...	DWG	Alternate unit decimal places.
DimAltF	25.4000	...	DWG	Alternate unit scale factor.
DimAltRnd	0.00	...	dwg	↖ Rounding factor of alternate units.
DimAltTD	2	...	DWG	Tolerance alternate unit decimal places.
DimAltTZ	0	...	DWG	Alternate tolerance units zeros:
				0 Zeros not suppressed.
				1 All zeros suppressed.

Variable	Default	R/O	Loc	Meaning
				2 Include zero feet, but suppress zero inches ⬈ .
				3 Include zero inches, but suppress zero feet ⬈ .
				4 Suppresses leading zeros ⬈ .
				8 Suppresses trailing zeros ⬈ .
DimAltU	2	...	DWG	Alternate units:
				1 Scientific.
				2 Decimal.
				3 Engineering.
				4 Architectural; stacked.
				5 Fractional; stacked.
				6 Architectural.
				7 Fractional.
				8 Windows desktop units setting.
DimAltZ	0		DWG	Zero suppression for alternate units:
				0 Suppress zero ft and zero in.
				1 Include zero ft and zero in.
				2 Include zero ft; suppress zero in.
				3 Suppress zero ft; include zero in.
				4 Suppress leading zero in dec dim.
				8 Suppress trailing zero in dec dim.
				12 Suppress leading and trailing zeroes.
DimAPost	""	...	DWG	Suffix for alternate text.
DimAso	On	...	DWG	Toggle associative dimensions:
				On Dimensions are created associative.
				Off Dimensions are not associative.
DimASz	0.1800	...	DWG	Arrowhead length.
DimAtFit	3	...	DWG	⬈ When there isn't enough space between extension lines, dimension text and arrows are fitted:
				0 Text and arrows outside extension lines.
				1 Arrows first outside, then text.
				2 Text first outside, then arrows.
				3 Either text or arrows, whichever fits better.
DimAUnit	0	...	DWG	Angular dimension format:
				0 Decimal degrees.
				1 Degrees.Minutes.Seconds.
				2 Grad.
				3 Radian.

Variable	Default	R/O	Loc	Meaning
				4 Surveyor units.
DimAZin	0	...	DWG	⋈ Supresses zeros in angular dimensions:
				0 Display all leading and trailing zeros.
				1 Suppress zero in front of decimal.
				2 Suppress trailing zeros behind decimal.
				3 Suppress zeros in front and behind the decimal.
DimBlk	""	R/O	DWG	Arrowhead block name:
				Architectural tick: "Archtick"
				Box filled: "Boxfilled"
				Box: "Boxblank"
				Closed blank: "Closedblank"
				Closed filled: "" (*default*)
				Closed: "Closed"
				Datum triangle filled: "Datumfilled"
				Datum triangle: "Datumblank"
				Dot blanked: "Dotblank"
				Dot small: "Dotsmall"
				Dot: "Dot"
				Integral: "Integral"
				None: "None"
				Oblique: "Oblique"
				Open 30: "Open30"
				Open: "Open"
				Origin indication: "Origin"
				Right-angle: "Open90"
DimBlk1	""	R/O	DWG	First arrowhead block name; uses same list of names as under **DimBlk**.
				. No arrowhead.
DimBlk2	""	R/O	DWG	Second arrowhead block name; uses same list of names as under **DimBlk**.
				. No arrowhead.
DimCen	0.0900	...	DWG	Center mark size:
				$-n$ Draws center lines.
				0 No center mark or lines drawn.
				$+n$ Draws center marks of length n.
DimClrD	0	...	DWG	Dimension line color:
				0 BYBLOCK (*default*)
				1 Red.
				...

Variable	Default	R/O	Loc	Meaning
				255 Dark gray.
				256 BYLAYER.
DimClrE	0	...	DWG	Extension line and leader color.
DimClrT	0	...	DWG	Dimension text color.
DimDec	4	...	DWG	Primary tolerance decimal places.
DimDLE	0.0000	...	DWG	Dimension line extension.
DimDLI	0.3800	...	DWG	Dimension line continuation increment.
DimDSep	.	...	DWG	↖ Decimal separator (must be a single character.)
DimExe	0.1800	...	DWG	Extension above dimension line.
DimExO	0.0625	...	DWG	Extension line origin offset.
DimFit	3	...	DWG	Obsolete: Autodesk recommends use of **DimATfit** and **DimTMove** instead; placement of text and arrowheads between extension lines:
				0 Both text and arrows, if possible.
				1 Text has priority over arrowheads.
				2 Whichever fits between ext lines.
				3 Whatever fits.
				4 Place text at end of leader line.
				5 Place text without leader line.
DimFrac	0	...	DWG	↖ Fraction format when **DimLUnit** is set to 4 or 5:
				0 Horizontal.
				1 Diagonal.
				2 Not stacked.
DimGap	0.0900	...	DWG	Gap from dimension line to text.
DimJust	0	...	DWG	Horizontal text positioning:
				0 Center justify.
				1 Next to first extension line.
				2 Next to second extension line.
				3 Above first extension line.
				4 Above second extension line.
DimLdrBlk	""	...	DWG	↖ Block name for leader arrowhead; uses same names as DimBlock.
				. Supresses display of arrowhead.
DimLFac	1.0000	...	DWG	Linear unit scale factor.
DimLim	Off	...	DWG	Generate dimension limits.
DimLUnit	2	...	DWG	↖ Dimension units (except angular); replaces **DimUnit**:

Variable	Default	R/O	Loc	Meaning
				1 Scientific
				2 Decimal
				3 Engineering
				4 Architectural
				5 Fractional
				6 Windows desktop
DimLwd	BYBLOCK	...	dwg	⋈ Dimension line lineweight; valid values include BYLAYER, BYBLOCK, or an integer representing 0.01mm.
DimLwe	BYBLOCK	...	dwg	⋈ Extension lineweight; valid values include BYLAYER, BYBLOCK, or an integer representing 0.01mm.
DimPost	""	...	DWG	Default suffix for dimension text (maximum 13 characters):
				"" No suffix.
				<>mm Millimeter suffix.
				<>≈ Angstrom suffix.
DimRnd	0.0000	...	DWG	Rounding value.
DimSAh	Off	...	DWG	Separate arrowhead blocks:
				Off Use arrowhead defined by **DimBlk**.
				On Use arrowheads defined by **DimBlk1** and **DimBlk2**.
DimScale	1.0000	...	DWG	Overall scale factor:
				0 Value is computed from the scale between current modelspace viewport and paperspace.
				>0 Scales text and arrowheads.
DimSD1	Off	...	DWG	Suppress first dimension line:
				On First dimension line is suppressed.
				Off No suppressed.
DimSD2	Off	...	DWG	Suppress second dimension line:
				On Second dimension line is suppressed.
				Off Not suppressed.
DimSE1	Off	...	DWG	Suppress the first extension line:
				On First extension line is suppressed.
				Off Not suppressed.
DimSE2	Off	...	DWG	Suppress the second extension line:
				On Second extension line is suppressed.
				Off Not suppressed.
DimSho	On	...	DWG	Update dimensions while dragging:
				On Dimensions are updated during drag.

Variable	Default	R/O	Loc	Meaning
				Off Dimensions are updated after drag.
DimSOXD	Off	...	DWG	Suppress dimension lines outside extension lines:
				On Dimension lines are drawn outside the extension lines.
				Off Not drawn outside extension lines.
DimStyle	"STANDARD"	R/O	DWG	⌨ Current dimension style.
DimTAD	0	...	DWG	Vertical position of dimension text:
				0 Centered between extension lines.
				1 Above dimension line, except when dimension line is not horizontal &DimTIH = 1.
				2 On side of dimension line farthest from the defpoints.
				3 Conform to JIS.
DimTDec	4	...	DWG	Primary tolerance decimal places.
DimTFac	1.0000	...	DWG	Tolerance text height scaling factor.
DimTIH	On	...	DWG	Text inside extensions is horizontal:
				Off Text aligned with dimension line.
				On Text is horizontal.
DimTIX	Off	...	DWG	Place text inside extensions:
				Off Text is placed inside the extension lines, if room.
				On Forces text between the extension lines.
DimTM	0.0000	...	DWG	Minus tolerance.
DimTMove	0	...	DWG	⟱ Determines how dimension text is moved:
				0 Dimension line moved with text.
				1 Adds a leader when text is moved.
				2 Text moves anywhere; no leader.
DimTOFL	Off	...	DWG	Force line inside extension lines:
				Off Dimension lines not drawn when arrowheads are outside.
				On Dimension lines drawn, even when arrowheads are outside.
DimTOH	On	...	DWG	Text outside extensions is horizontal:
				Off Text aligned with dimension line.
				On Text is horizontal.
DimTol	Off	...	DWG	Generate dimension tolerances:
				Off Tolerances not drawn.
				On Tolerances are drawn.
DimTolJ	1	...	DWG	Tolerance vertical justification:

Variable	Default	R/O	Loc	Meaning
				0 Bottom.
				1 Middle.
				2 Top.
DimTP	0.0000	...	DWG	Plus tolerance.
DimTSz	0.0000	...	DWG	Size of oblique tick strokes:
				0 Arrowheads.
				>0 Oblique strokes.
DimTVP	0.0000	...	DWG	Text vertical position when **DimTAD** =0:
				1 Turns **DimTAD** on.
				>-0.7 *or* <0.7 Dimension line is split for text.
DimTxSty	"STANDARD"	...	DWG	Dimension text style.
DimTxt	0.1800	...	DWG	Text height.
DimTZin	0	...	DWG	Tolerance zero suppression:
				0 Suppress zero ft and zero in.
				1 Include zero ft and zero in.
				2 Include zero ft; suppress zero in.
				3 Suppress zero ft; include zero in.
				4 Suppress leading zero in dec dim.
				8 Suppress trailing zero in dec dim.
				12 Suppress leading and trailing zeroes.
DimUnit	2	...	DWG	Obsolete; replaced by **DimLUnit** and **DimFrac**; dimension unit format:
				1 Scientific.
				2 Decimal.
				3 Engineering.
				4 Architectural; stacked.
				5 Fractional; stacked.
				6 Architectural.
				7 Fractional.
				8 Windows desktop units setting.
DimUPT	Off	...	DWG	User-positioned text:
				Off Cursor positions dimension line
				On Cursor also positions text
DimZIN	0	...	DWG	Suppression of zero in feet-inches units:
				0 Suppress zero ft and zero in.
				1 Include zero ft and zero in.
				2 Include zero ft; suppress zero in.
				3 Suppress zero ft; include zero in.
				4 Suppress leading zero in dec dim.
				8 Suppress trailing zero in dec dim.

Variable	Default	R/O	Loc	Meaning
				12 Suppress leading and trailing zeroes.
DispSilh	0	...	DWG	Silhouette display of 3D solids:
				0 Off.
				1 On.
Distance	0.0000	R/O	...	Distance measured by **Dist** command.
~~Dither~~				*Removed from Release 14.*
DonutId	0.5000	...	...	Inside radius of donut.
DonutOd	1.0000	...	...	Outside radius of donut.
DragMode	2	...	REG	⌦ Drag mode:
				0 No drag.
				1 On if requested.
				2 Automatic.
DragP1	10	...	REG	Regen drag display.
DragP2	25	...	REG	Fast drag display.
DwgCheck	0	...	REG	↖ Toggles checking if drawing was editing by software other than AutoCAD:
				0 Supresses dialog box.
				1 Displays warning dialog box.
DwgCodePage	varies	...	DWG	Drawing code page, such as "ANSI_1252"
DwgName	varies	R/O	...	Current drawing filename, such as "Drawing1.dwg"
DwgPrefix	varies	R/O	...	Drawing's drive and subdirectory, such as "d:\"
DwgTitled	0	R/O	...	Drawing has filename:
				0 "Drawing1.Dwg".
				1 User-assigned name.
~~DwgWrite~~				*Removed from Release 14.*
E				
EdgeMode	0	...	REG	Toggle edge mode for **Trim** and **Extend** commands:
				0 No extension.
				1 Extends cutting edge.
Elevation	0.0000	...	DWG	Current elevation, relative to current UCS.
EntExts	*1*	*R/O*	...	*Controls how the drawing extents are calculated:*
				0 Extents calculated every time; slows down AutoCAD but uses less memory.
				1 Extents of every object is cached as a two-byte value (default).
				2 Extents of every object is cached as a four-byte value (fastest but uses more memory).

Variable	Default	R/O	Loc	Meaning
EntMods	*0*	*R/O*	...	*Increments by one each time an object is modified to indicate that an object has been modified since the drawing was opened; value ranges from 0 to 4.29497E9.*
ErrNo	*0*	...	...	*Error number from AutoLISP, ADS, Arx*
~~*ExeDir*~~				*Removed from Release 14.*
Expert	0	...	...	Suppresses the displays of prompts: 0 Normal prompts 1 "About to regen, proceed?" and "Really want to turn the current layer off?" 2 "Block already defined. Redefine it?" and "A block with this name already exists. Overwrite it?" 3 **Linetype** command messages. 4 **UCS Save** and **VPorts Save**. 5 **DimStyle Save** and **DimOverride.**
ExplMode	1	...	...	Toggle whether **Explode** and **Xplode** commands explode non-uniformly scaled blocks: 0 Does not explode. 1 Does explode.
ExtMax	-1.0000E+20, -1.0000E+20, -1.0000E+20	R/O	DWG	Upper-right coordinate of drawing extents.
ExtMin	1.0000E+20, 1.0000E+20, 1.0000E+20	R/O	DWG	Lower-left coordinate of drawing extents.
ExtNames	1	...	DWG	Format of named objects: 0 Names are limited to 31 characters, and can include A - Z, 0 - 9, dollar ($), underscore (_), and hyphen (-). 1 Names are limited 255 characters, and can include A - Z, 0 - 9, spaces, and any characters not used by Microsoft Windows or AutoCAD for special purposes.
F				
FaceTRatio	0	...	...	Controls the aspect ratio of facets on cylinder and cone ACIS solids: 0 Creates an *n* by 1 mesh. 1 Creates an *n* by *m* mesh.
FaceTRres	0.5	...	DWG	Adjusts smoothness of shaded and hidden-line objects: 0.01 Minimum value. 0.5 Default value.

Variable	Default	R/O	Loc	Meaning
				10.0 Maximum value.
FfLimit	...	...	...	*Removed from Release 14.*
FileDia	1	...	REG	User interface:
				0 Command-line prompts.
				1 Dialog boxes, when available.
FilletRad	0.5000	...	DWG	Current fillet radius.
FillMode	1	...	DWG	Fill of solid objects:
				0 Off.
				1 On.
Flatland	*0*	R/O	...	*Obsolete system variable.*
FontAlt	"simplex.shx"	...	REG	Name for substituted font.
FontMap	"acad.fmp"	...	REG	Name of font mapping file.
Force_Paging	*0*	...	...	*0 Minimum (default).*
				4.29497E9 Maximum.
FrontZ	0.0000	...	DWG	Front clipping plane offset.
FullOpen	1	R/O	...	⋏ Drawing is:
				0 Partially loaded.
				1 Fully open.
G				
GlobCheck	*0*	...	...	*Reports statistics on dialog boxes:*
				-1 Turn off local language.
				0 Turn off.
				1 Warns if larger than 640x400.
				2 Also reports size in pixels.
				3 Additional info.
GridMode	0	...	DWG	Display of grid:
				0 Off.
				1 On.
GridUnit	0.5000,0.5000	...	DWG	X,y-spacing of grid.
GripBlock	0	...	REG	Display of grips in blocks:
				0 At insertion point.
				1 At all objects within block.
GripColor	5	...	REG	Color of unselected grips:
				1 Minimum color number; red.
				5 Default color; blue.
				255 Maximum color number.
GripHot	1	...	REG	Color of selected grips:
				1 Default color, red.
				255 Maximum color number.
Grips	1	...	REG	Display of grips:

Variable	Default	R/O	Loc	Meaning
				0 Off.
				1 On
GripSize	3	...	REG	Size of grip box, in pixels:
				1 Minimum size.
				3 Default size.
				255 Maximum size.
H				
Handles	1	R/O	...	⌨ Obsolete system variable.
HidePrecision	0	...	...	⋈ Controls precision of hide calculations:
				0 Single precision, less accurate, faster.
				1 Double precision, more accurate, slower.
Highlight	1	...	...	Object selection highlighting:
				0 Disabled.
				1 Enabled.
HPAng	0	...	...	Current hatch pattern angle.
HPBound	1	...	...	Object created by **BHatch** and **Boundary** commands:
				0 Polyline.
				1 Region.
HPDouble	0	...	...	Double hatching:
				0 Disabled.
				1 Enabled.
HPName	"ANSI31"	...	...	Current hatch pattern name
				"" No default.
				. (*Period*) Set no default.
HPScale	1.0000	...	...	Current hatch pattern scale factor; cannot be zero.
HPSpace	1.0000	...	...	Current spacing of user-defined hatching; cannot be zero.
HyperlinkBase	""	...	DWG	⋈ Path for relative hyperlinks.
I				
ImageHlt	0	...	REG	⋈ When a raster image is selected:
				0 Image frame is highlighted.
				1 Entire image is highlighted.
IndexCtl	0	...	DWG	Creates layer and spatial indices:
				0 No indices created.
				1 Layer index created.
				2 Spatial index created.

Variable	Default	R/O	Loc	Meaning
				3 Both indices created.
InetLocation	"home.htm" ...		REG	Default browser URL.
InsBase	0.0000,0.0000,0.0000			
		...	DWG	Insertion base point relative to the current UCS for **Insert** and **DdInsert**.
InsName	""	...	...	Current block name:
	.			.(*Period*) Set to no default.
InsUnits				⋉ Drawing units when a block is dragged into drawing from AutoCAD DesignCenter:
				0 Unitless.
				1 Inches.
				2 Feet.
				3 Miles.
				4 Millimeters.
				5 Centimeters.
				6 Meters.
				7 Kilometers.
				8 Microinches.
				9 Mils.
				10 Yards.
				11 Angstroms.
				12 Nanometers.
				13 Microns.
				14 Decimeters.
				15 Decameters.
				16 Hectometers.
				17 Gigameters.
				18 Astronomical Units.
				19 Light Years.
				20 Parsecs.
InsUnitsDefSource				
	0	...	REG	⋉ Source drawing units value; ranges from 0 to 20.
InsUnitsDefTarget				
	0	...	REG	⋉ Target drawing units value; ranges from 0 to 20.
ISaveBak	1	...	REG	Controls whether BAK file is created:
				0 No BAK file created.
				1 BAK backup file created.

Variable	Default	R/O	Loc	Meaning
ISavePercent	50	...	REG	Percentage of waste in DWG file before cleanup occurs: 0 Every save is a full save.
IsoLines	4	...	DWG	Isolines on 3D solids: 0 No isolines; minimum. 4 Default. 16 Good-looking. 2,047 Maximum.

L

Variable	Default	R/O	Loc	Meaning
LastAngle	0	R/O	...	Ending angle of last-drawn arc.
LastPoint	*variers*	...	...	Last-entered point, such as 0.0000,0.0000,0.0000
LastPrompt	""	R/O	...	Last string on the command line; includes user input.
LazyLoad	*0*	...	...	*Toggle: 0 or 1.*
LensLength	50.0000	R/O	DWG	Perspective view lens length, in mm.
LimCheck	0	...	DWG	Drawing limits checking: 0 Disabled. 1 Enabled.
LimMax	12.0000,9.0000	...	DWG	Upper right drawing limits.
LimMin	0.0000,0.0000	...	DWG	Lower left drawing limits.
LispInit	1	...	REG	AutoLISP functions and variables are: 0 Preserved from drawing to drawing. 1 Valid in current drawing only.
Locale	"enu"	R/O		ISO language code.
LogFileMode	0	...	REG	Text window written to log file: 0 No. 1 Yes.
LogFileName	"d:\acad 2000\Drawing1_1_1_0000.log"			
		R/O	DWG	Filename and path for log file.
LogFilePath	"d:\acad 2000\" ...		REG	⚐ Path for the log file.
LogInName	" "	R/O	...	User's login name; max = 30 chars.
LongFName				*Removed from Release 14.*
LTScale	1.0000	...	DWG	⚐ Current linetype scale factor; cannot be zero.
LUnits	2	...	DWG	Linear units mode: 1 Scientific. 2 Decimal. 3 Engineering.

Variable	Default	R/O	Loc	Meaning
				4 Architectural.
				5 Fractional.
LUPrec	4	...	DWG	Decimal places of linear units.
LwDefault	25	...	REG	☌ Default lineweight, in millimeters; must be one of the following values: 0, 5, 9, 13, 15, 18, 20, 25, 30, 35, 40, 50, 53, 60, 70, 80, 90, 100, 106, 120, 140, 158, 200, or 211.
LwDisplay	0	...	DWG	☌ Toggles whether lineweight is displayed; setting is saved separately for Model space and each layout tab.
				0 Not displayed
				1 Displayed
LwUnits	1	...	REG	☌ Determines units for lineweight:
				0 Inches
				1 Millimeters

M

Variable	Default	R/O	Loc	Meaning
MacroTrace	*0*	...	...	*Diesel debug mode:*
				0 Off.
				1 On.
MaxActVP	48	...	...	Maximum viewports to regenerate:
				2 Minimum.
				64 Default.
				64 Maximum (increased from 48 in R14).
MaxObjMem	*0*	...	...	*Maximum number of objects in memory; object pager is turned off when value = 0, <0, or 2,147,483,647.*
MaxSort	200	...	REG	Maximum names sorted alphabetically.
MButtonPan	1	...	REG	☌ Determines behavior of wheel on Intellimouse:
				0 As defined by AutoCAD menu file.
				1 Pans when dragging with wheel.
MeasureInit	0	...	REG	Drawing units:
				0 English.
				1 Metric.
Measurement	0	...	DWG	Drawing units (overrides **MeasureInit**):
				0 English.
				1 Metric.
MenuCtl	1	...	REG	Submenu display:
				0 Only with menu picks.

Variable	Default	R/O	Loc	Meaning
				1 Also with keyboard entry.
MenuEcho	0	...		Menu and prompt echoing:
				0 All prompts displayed.
				1 Suppress menu echoing.
				2 Suppress system prompts.
				4 Disable ^P toggle.
				8 Display all input-output strings.
MenuName	"acad"	R/O	REG	Current menu filename.
MirrText	1	...	DWG	Text handling during **Mirror** command:
				0 Mirror text.
				1 Retain text orientation.
ModeMacro	""	...	...	Invoke Diesel programming language.
MTextEd	"internal"	...	REG	Name of the **MText** editor:
				. (*Period*) Use default editor.
				string See list below; must be less than 256 characters long and use this syntax:
				:AutoLISPtextEditorFunction#TextEditor
				0 Cancel the editing operation.
				-1 Use the secondary editor.
				"blank" MTEXT internal editor.
				"Internal" MTEXT internal editor.
				"Notepad" Windows Notepad editor.
				":lisped" Built-in AutoLISP function.

N

Variable	Default	R/O	Loc	Meaning
NodeName	*"AC$"*	*R/O*	*REG*	*Name of network node; range is 1 to 3 chars.*
NoMutt	1	...	...	⚹ Suppresses the display of message (a.k.a. muttering):
				0 Displays prompt, as normal.
				1 Suppresses muttering.

O

Variable	Default	R/O	Loc	Meaning
OffsetDist	1.0000	...	...	Current offset distance:
				<0 Offsets through a specified point.
				>0 Default offset distance.
OffsetGapHid				
	0	...	REG	⚹ Determines how to reconnect polyline when individual segments are offset:
				0 Extend segments to fill gap.
				1 Fill gap with fillet (arc segment).
				2 Fill gap with chamfer (line segment).

Variable	Default	R/O	Loc	Meaning
OleHide	0	...	REG	Display and plotting of OLE objects:
				0 All OLE objects visible.
				1 Visible in paper space only.
				2 Visible in model space only.
				3 Not visible.
OleStartup	0	...	dwg	☞ Loading OLE source application improves plot quality:
				0 Do not load OLE source app.
				1 Load OLE source app when plotting.
OleQuality	1	...	REG	☞ Quality of display and plotting of embedded OLE objects:
				0 Line art quality.
				1 Text quality.
				2 Graphics quality.
				3 Photograph quality.
				4 High quality photograph.
OrthoMode	0	...	DWG	Orthographic mode:
				0 Off.
				1 On.
OSMode	4133	...	DWG	Current object snap mode:
				0 NONe.
				1 ENDpoint.
				2 MIDpoint.
				4 CENter.
				8 NODe.
				16 QUAdrant.
				32 INTersection.
				64 INSertion.
				128 PERpendicular.
				256 TANgent.
				512 NEARest.
				1024 QUIck.
				2048 APPint.
				4096 EXTension ☞ .
				8192 PARallel ☞ .
				16383 All modes on ☞ .
				16384 Object snap turned off via **OSNAP** on the status bar.
OSnapCoord	2	...	REG	Keyboard overrides object snap:
				0 Object snap override keyboard.

Variable	Default	R/O	Loc	Meaning
				1 Keyboard overrides object snap.
				2 Keyboard overrides object snap, except in script.
P				
PaperUpdate	0	...	REG	⬉ Determines how AutoCAD deals with the plotting of a layout with a paper size different from the plotter's default:
				0 Displays a warning dialog box.
				1 Changes paper size to that of the plotter configuration file.
PDMode	0	...	DWG	Point display mode:
				0 Dot.
				1 No display.
				2 +-symbol.
				3 x-symbol.
				4 Short line.
				32 Circle.
				64 Square.
PDSize	0.0000	...	DWG	Point display size, in pixels:
				>0 Absolute size.
				0 5% of drawing area height.
				<0 Percentage of viewport size.
PEllipse	*0*	...	*DWG*	*Toggle Ellipse creation:*
				0 True ellipse.
				1 Polyline arcs.
Perimeter	0.0000	R/O	...	Perimeter calculated by **Area**, **DbList**, and **List** commands.
PFaceVMax	4	R/O	...	Maximum vertices per 3D face.
PHandle	*0*	...	*ACAD*	*Ranges from 0 to 4.29497E9*
PickAdd	1	...	REG	Effect of **SHIFT** key on selection set:
				0 Adds to selection set.
				1 Removes from selection set.
PickAuto	1	...	REG	Selection set mode:
				0 Single pick mode.
				1 Automatic windowing and crossing.
PickBox	3	...	REG	Object selection pickbox size, in pixels:
				0 Minimum size.
				3 Default size.
				50 Maximum size.
PickDrag	0	...	REG	Selection window mode:

Variable	Default	R/O	Loc	Meaning
				0 Pick two corners.
				1 Pick 1 corner; drag to 2nd corner.
PickFirst	1	...	REG	Command-selection mode:
				0 Enter command first.
				1 Select objects first.
PickStyle	1	...	REG	Included groups and associative hatches in selection:
				0 Neither included.
				1 Include groups.
				2 Include associative hatches.
				3 Include both.
Platform	"Microsoft Windows Version 4.10 (x86)"			
		R/O	ACAD	AutoCAD platform (name of the operating system).
PLineGen	0	...	DWG	Polyline linetype generation:
				0 From vertex to vertex.
				1 From end to end.
PLineType	2	...	REG	Automatic conversion and creation of 2D polylines by **PLine**:
				0 Not converted; old-format polylines created.
				1 Not converted; optimized polylines created.
				2 Polylines in older drawings are converted on open; **PLine** creates optimized polylines with Lwpolyline object.
PLineWid	0.0000	...	DWG	Current polyline width.
PlotId	""	...	REG	Obsolete; has no effect in AutoCAD 2000.
PlotRotMode	1	...	DWG	Orientation of plots:
				0 Lower left = 0,0.
				1 Lower left plotter area = lower left of media.
				2 X, y-origin offsets calculated relative to the rotated origin position ☇.
Plotter	0	...	REG	Obsolete; has effect in AutoCAD 2000.
PlQuiet	0	...	REG	☇ Toggles display during batch plotting and scripts (replaces **CmdDia**):
				0 Plot dialog boxes and nonfatal errors are displayed.
				1 Nonfatal errors are logged; plot dialog boxes are not displayed.

Variable	Default	R/O	Loc	Meaning
PolarAddAng	""	...	REG	⌐ Contains a list of up to 10 user-defined polar angles; each angle can be up to 25 characters long, each separated with a semicolon (;). For example: 0;15;22.5;45.
PolarAng	90	...	REG	⌐ Specifies the increment of polar angle; contrary to Autodesk documentation, you may specify any angle.
PolarDist	0	...	REG	⌐ The polar snap increment when **SnapStyl** is set to 1 (isometric).
PolarMode	0	...	REG	⌐ Settings for polar and object snap tracking:
				0 Measure polar angles based on current UCS (absolute), track orthogonally; don't use additional polar tracking angles; and acquire object tracking points automatically.
				1 Measure polar angles from selected objects (relative).
				2 Use polar tracking settings in object snap tracking.
				4 Use additional polar tracking angles (via **PolarAng**).
				8 Press **Shift** to acquire object snap tracking points.
PolySides	4	...	...	Current number of polygon sides:
				3 Minimum sides.
				4 Default.
				1024 Maximum sides.
Popups	1	R/O	...	Display driver support of AUI:
				0 Not available.
				1 Available.
Product	"AutoCAD"	R/O	ACAD	Name of the software.
Program	"acad"	R/O	ACAD	Name of the software's executable file.
ProjectName	""	...	DWG	Project name of the current drawing.
ProjMode	1	...	REG	Projection mode for **Trim** and **Extend** commands:
				0 No projection.
				1 Project to x,y-plane of current UCS.
				2 Project to view plane.
ProxyGraphics	1	...	REG	Proxy image saved in the drawing:
				0 Not saved; displays bounding box.
				1 Image saved with drawing.

Variable	Default	R/O	Loc	Meaning
ProxyNotice	1	...	REG	Display warning message: 0 No. 1 Yes.
ProxyShow	1	...	REG	Display of proxy objects: 0 Not displayed. 1 All displayed. 2 Bounding box displayed.
PsLtScale	1	...	DWG	Paper space linetype scaling: 0 Use model space scale factor. 1 Use viewport scale factor.
PsProlog	""	...	REG	PostScript prologue filename
PsQuality	75	...	REG	Resolution of PostScript display, in pixels: <0 Display as outlines; no fill. 0 Not displayed. >0 Display filled.
PStyleMode	0	...	DWG	⟆ Toggles the plot color matching mode of the drawing: 0 Use named plot style tables. 1 Use color-dependent plot style tables.
PStylePolicy	1	...	REG	⟆ Determines whether the object color is associated with its plot style: 0 Color and plot style not associated. 1 Object's plot style is associated with its color.
PsVpScale	0	...	...	⟆ Sets the view scale factor (the ratio of units in paper space to the units in newly created model space viewports) for all newly-created viewports: 0 Scaled to fit.
PUcsBase	""	...	DWG	⟆ Name of UCS defining the origin and orientation of orthographic UCS settings in paper space only.

Q

Variable	Default	R/O	Loc	Meaning
QAFlags	*1*	*...*	*...*	*Quality assurance flags:* *1 The ^C metacharacters in a menu macro cancels grips, just as if user pressed Esc.* *2 Long text screen listings do not pause.* *4 Error and warning messages are displayed at the command line, instead of in dialog boxes.*

Variable	Default	R/O	Loc	Meaning
				128 Screen picks are accepted via the AutoLISP (command) function.
QTextMode	0	...	DWG	Quick text mode: 0 Off. 1 On.
R				
RasterPreview	1	...	REG	Preview image: 0 None saved. 1 Saved in BMP format.
RefEditName	""	...	...	⟰ The reference filename when it is in reference-editing mode.
RegenMode	1	...	DWG	Regeneration mode: 0 Regen with each new view. 1 Regen only when required.
Re-Init	0	...	...	Reinitialize I/O devices: 1 Digitizer port. 2 Plotter port. 4 Digitizer. 8 Plotter. 16 Reload PGP file.
~~*RIAspect*~~				*Removed from Release 14.*
~~*RIBackG*~~				*Removed from Release 14.*
~~*RIEdge*~~				*Removed from Release 14.*
~~*RIGamut*~~				*Removed from Release 14.*
~~*RIGrey*~~				*Removed from Release 14.*
~~*RIThresh*~~				*Removed from Release 14.*
RTDisplay	1	...	REG	Raster display during realtime zoom and pan: 0 Display the entire raster image. 1 Display raster outline only.
S				
SaveFile	"auto.sv$"	R/O	REG	Automatic save filename
SaveFilePath	"d:\temp\"	...	REG	⟰ Path for automatic save files.
SaveName	""	R/O	...	Drawing save-as filename
SaveTime	120	...	REG	Automatic save interval, in minutes: 0 Disable auto save.
ScreenBoxes	0	R/O	ACAD	Maximum number of menu items 0 Screen menu turned off.
ScreenMode	0	R/O	...	State of AutoCAD display screen: 0 Text screen. 1 Graphics screen.

Variable	Default	R/O	Loc	Meaning
				2 Dual-screen display.
ScreenSize	575.0000,423.0000	R/O	...	Current viewport size, in pixels.
SDI	0	...	REG	⟰ Toggles multiple-document interface (SDI is short for "single document interface"):
				0 Turns on MDI.
				1 Turns off MDI (only one drawing may be loaded into AutoCAD).
				2 MDI disabled for apps that cannot support MDI; read-only.
				3 MDI disabled for apps that cannot support MDI, even when **SDI** set to 1; R/O.
ShadEdge	3	...	DWG	**Shade** style:
				0 Shade faces; 256-color shading.
				1 Shade faces; edges background color.
				2 Hidden-line removal.
				3 16-color shading.
ShadeDif	70	...	DWG	Percent of diffuse to ambient light:
				0 Minimum.
				70 Default.
				100 Maximum
ShortCutMenu	11	...	REG	⟰ Toggles availability of shortcut menus:
				0 Disables all Default, Edit, and Command shortcut menus.
				1 Enables Default shortcut menus.
				2 Enables Edit shortcut menus.
				4 Enables Command shortcut menus whenever a command is active.
				8 Enables Command shortcut menus only when command options are available at the command line.
ShpName	""	...	...	Current shape name:
				. (*Period*) Set to no default.
SketchInc	0.1000	...	DWG	Sketch command's recording increment.
SKPoly	0	...	DWG	Sketch line mode:
				0 Record as lines.
				1 Record as polylines.
SnapAng	0	...	DWG	Current rotation angle for snap and grid.
SnapBase	0.0000,0.0000	...	DWG	Current origin for snap and grid.
SnapIsoPair	0	...	DWG	Current isometric drawing plane:

Variable	Default	R/O	Loc	Meaning
				0 Left isoplane.
				1 Top isoplane.
				2 Right isoplane.
SnapMode	0	...	DWG	Snap mode:
				0 Off.
				1 On.
SnapStyl	0	...	DWG	Snap style:
				0 Normal.
				1 Isometric.
SnapType	0	...	REG	⋈ Toggles between standard or polar snap for the current viewport:
				0 Standard snap.
				1 Polar snap.
SnapUnit	0.5000,0.5000	...	DWG	X,y-spacing for snap.
SolidCheck	0	...	...	⋈ Toggles solid validation:
				0 Off.
				1 On.
SortEnts	96	...	DWG	Object display sort order:
				0 Off.
				1 Object selection.
				2 Object snap.
				4 Redraw.
				8 Slide generation.
				16 Regeneration.
				32 Plot.
				64 PostScript output.
SplFrame	0	...	DWG	Polyline and mesh display:
				0 Polyline control frame not displayed; display polygon fit mesh; 3D faces invisible edges not displayed
				1 Polyline control frame displayed; display polygon defining mesh; 3D faces invisible edges displayed
SplineSegs	8	...	DWG	Number of line segments that define a splined polyline.
SplineType	6	...	DWG	Spline curve type:
				5 Quadratic Bezier spline.
				6 Cubic Bezier spline.
SurfTab1	6	...	DWG	Density of surfaces and meshes:
				2 Minimum.
				6 Default.

Variable	Default	R/O	Loc	Meaning
				32766 Maximum.
SurfTab2	6	...	DWG	Density of surfaces and meshes:
				2 Minimum.
				6 Default.
				32766 Maximum.
SurfType	6	...	DWG	Pedit surface smoothing:
				5 Quadratic Bezier spline.
				6 Cubic Bezier spline.
				8 Bezier surface.
SurfU	6	...	DWG	Surface density in m-direction:
				2 Minimum.
				6 Default.
				200 Maximum.
SurfV	6	...	DWG	Surface density in n-direction:
				2 Minimum.
				6 Default.
				200 Maximum.
SysCodePage	"ansi_1252"	R/O	...	System code page.
T				
TabMode	0	...	...	Tablet mode:
				0 Off.
				1 On.
Target	0.0000,0.0000,0.0000	R/O	DWG	Target in current viewport.
TDCreate	*varies*	R/O	DWG	Time and date drawing created, such as 2448860.54014699.
TDInDwg	*varies*	R/O	DWG	Duration drawing loaded, such as 0.00040625.
TDuCreate	*varies*	R/O	DWG	☞ The universal time and date the drawing was created, such as 2451318.67772165.
TDUpdate	*varies*	R/O	DWG	Time and date of last update, such as 2448860.54014699.
TDUsrTimer	*varies*	R/O	DWG	Time elapsed by user-timer, such as 0.00040694.
TDuUpdate	*varies*	R/O	DWG	The universal time and date of the last save, such as 2451318.67772165.
TempPrefix	"d:\temp"	R/O	...	Path for temporary files.
TextEval	0	...	...	Interpretation of text input:
				0 Literal text.
				1 Read (and ! as AutoLISP code.
TextFill	1	...	REG	Toggle fill of TrueType fonts:

Variable	Default	R/O	Loc	Meaning
				0 Outline text.
				1 Filled text.
TextQlty	50	...	DWG	Resolution of TrueType fonts:
				0 Minimum resolution.
				50 Default.
				100 Maximum resolution.
TextSize	0.2000	...	DWG	Current height of text.
TextStyle	"STANDARD"	...	DWG	Current name of text style.
Thickness	0.0000	...	DWG	Current object thickness.
TileMode	1	...	DWG	View mode:
				0 Display layout tab.
				1 Display model tab.
ToolTips	1	...	REG	Display tooltips:
				0 Off.
				1 On.
TraceWid	0.0500	...	DWG	Current width of traces.
TrackPath	0	...	REG	↖ Determines the display of polar and object snap tracking alignment paths:
				0 Displays object snap tracking path across the entire viewport.
				1 Displays object snap tracking path between the alignment point and "From point" to cursor location.
				2 Turns off polar tracking path.
				3 Turns off polar and object snap tracking paths.
TreeDepth	3020	...	DWG	Maximum branch depth in *xxyy* format:
				xx Model-space nodes.
				yy Paper-space nodes.
				>0 3D drawing.
				<0 2D drawing.
TreeMax	10000000	...	REG	Limits memory consumption during drawing regeneration.
TrimMode	1	...	REG	Trim toggle for **Chamfer** and **Fillet** commands:
				0 Leave selected edges in place.
				1 Trim selected edges.
TSpaceFac	1	...	...	↖ Mtext line spacing distance; measured as a factor of text height; valid values range from 0.25 to 4.0.
TSpaceType	1	...	...	↖ Type of mtext line spacing:

Variable	Default	R/O	Loc	Meaning
				1 At Least: adjusts line spacing based on the height of the tallest character in a line of mtext.
				2 Exactly: uses the specified line spacing; ignores character height.
TStackAlign	1	...	DWG	⚟ Vertical alignment of stacked text.
				0 Bottom aligned.
				1 Center aligned.
				2 Top aligned.
TStackSize	70	...	DWG	⚟ Sizes stacked text as a percentage of the selected text's height; valid values range from 1% to 127%.
U				
UcsAxisAng	90	...	REG	⚟ Default angle for rotating the UCS around an axes (via the **UCS** command's using the **X**, **Y**, or **Z** options; valid values are limited to: 5, 10, 15, 18, 22.5, 30, 45, 90, or 180.
UcsBase	"World"	...	DWG	⚟ Name of the UCS that defines the origin and orientation of orthographic UCS settings.
UcsFollow	0	...	DWG	New UCS views:
				0 No change.
				1 Automatic display of plan view.
UcsIcon	1	...	DWG	⌨ Display of UCS icon:
				0 Off.
				1 On.
				2 Display at UCS origin, if possible.
				3 On and displayed at origin ⚟ .
UcsName	""	R/O	DWG	Name of current UCS view:
				"" Current UCS is unnamed.
UcsOrg	0.0000,0.0000,0.0000	R/O	DWG	Origin of current UCS relative to WCS.
UcsOrtho	1	...	REG	⚟ Determines whether the related orthographic UCS setting is restored automatically:
				0 UCS setting remains unchanged when orthographic view is restored.

Variable	Default	R/O	Loc	Meaning
				1 Related orthographic UCS setting is restored automatically when an orthographic view is restored.
UcsView	1	...	REG	⟋ Determines whether the current UCS is saved with a named view:
				0 Not saved.
				1 Saved.
UcsVp	1	...	DWG	⟋ Determines whether the UCS in active viewports remains fixed (locked) or changes (unlocked) to match the UCS of the current viewport:
				0 Unlocked.
				1 Locked.
UcsXDir	1.0000,0.0000,0.0000			
		R/O	DWG	X-direction of current UCS relative to WCS.
UcsYDir	0.0000,1.0000,0.0000			
		R/O	DWG	Y-direction of current UCS relative to WCS.
UndoCtl	5	R/O	...	State of undo:
				0 Undo disabled.
				1 Undo enabled.
				2 Undo limited to one command.
				4 Auto-group mode.
				8 Group currently active.
UndoMarks	0	R/O	...	Current number of undo marks
UnitMode	0	...	DWG	Units display:
				0 As set by **Units** command.
				1 As entered by user.
UserI1 *thru* UserI5				
	0	...	...	Five user-definable integer variables
UserR1 *thru* UserR5				
	0.0000	...	...	Five user-definable real variables
UserS1 *thru* UserS5				
	""	...	...	Five user-definable string variables
V				
ViewCtr	*varies*	R/O	DWG	X,y,z-coordinate of center of current view, such as 6.2433,4.5000,0.0000.
ViewDir	*varies*	R/O	DWG	Current view direction relative to UCS, such as 0.0000,0.0000,1.0000.
ViewMode	0	R/O	DWG	Current view mode:
				0 Normal view.

Variable	Default	R/O	Loc	Meaning
				1 Perspective mode on.
				2 Front clipping on.
				4 Back clipping on.
				8 UCS-follow on.
				16 Front clip not at eye.
ViewSize	9.0000	R/O	DWG	Height of current view.
ViewTwist	0	R/O	DWG	Twist angle of current view.
VisRetain	1	...	DWG	Determines xref drawing's layer settings — on-off, freeze-thaw, color, and linetype:
				0 Xref layer settings in the current drawing takes precedence for xref-dependent layers.
				1 Settings for xref-dependent layers take precedence over the xref layer definition in the current drawing.
VSMax	varies	R/O	DWG	Upper-right corner of virtual screen, such as 37.4600,27.0000,0.0000.
VSMin	varies	R/O	DWG	Lower-left corner of virtual screen, such as -24.9734,-18.0000,0.0000.
W				
WhipArc	0	...	REG	↖ Display of circular objects:
				0 Displayed as connected vectors.
				1 Displayed as true circles and arcs.
WmfBkgrnd	1	...	...	↖ Controls background of WMF files:
				0 Background is transparent.
				1 Background has same as AutoCAD background color.
WorldUcs	1	R/O	...	Matching of WCS with UCS:
				0 Current UCS is not WCS.
				1 UCS is WCS.
WorldView	1	...	DWG	Display during **3dOrbit**, **DView**, and **VPoint** commands:
				0 Display UCS.
				1 Display WCS.
				2 UCS changes relative to the UCS specified by the **UcsBase** system variable ↖ .
WriteStat	1	R/O	...	Indicates whether the drawing file is read-only:
				0 Drawing cannot be written to.
				1 Drawing can be writen to.

Variable	Default	R/O	Loc	Meaning
X				
XClipFrame	0	...	DWG	Visibility of xref clipping boundary:
				0 Not visible.
				1 Visible.
XEdit	0	...	DWG	⋊ Toggles whether the drawing can be edited in-place when being referenced by another drawing:
				0 Cannot in-place refedit.
				1 May in-place refedit.
XFadeCtl	50	...	REG	⋊ Fades objects not being edited in-place:
				0 No fading; minimum value.
				90 90% fading; maximum value.
XLoadCtl	1	...	REG	Controls demand loading:
				0 Demand loading turned off; entire drawing is loaded.
				1 Demand loading turned on; xref file opened.
				2 Demand loading turned on; a *copy* of the xref file is opened.
XLoadPath	""	...	REG	Path for loading xref file.
XRefCtl	0	...	REG	Determines creation of XLG xref log files:
				0 File not written.
				1 XLG file written.
Z				
ZoomFactor	10	...	REG	⋊ Controls the zoom level via the Intellimouse wheel; valid values range between 3 and 100.

ACAD.DWT Template Drawing Settings

The setup for each new drawing is determined by the settings in a template draw-ing. (See Chapter 5 "Getting In and Around AutoCAD.") The default template is named "Acad.Dwt", which is used when you select **Start from Scratch** and **English** units in the **Start Up** dialog box. See Chapter 5 "Getting In and Around AutoCAD" for an explanation of how to create your own template drawings.

The following list shows the default settings for the Acad.Dwt file.

APERTURE	10 pixels
Attributes	Displayed normally
BASE	Insertion base point at 0.0, 0.0, 0.0
Blipmode	On
CHAMFER	Distance 0.5
COLOR	Object color "BYLAYER"
Coordinate display	Updated continuously

DIM variables

DIMAFIT	3	DIMLUNIT	2
DIMALT	Off	DIMPOST	"" (None)
DIMALTD	2	DIMRND	0.0000
DIMALTF	25.4000	DIMSAH	Off
DIMAPOST	"" (None)	DIMSCALE	1.0000
DIMASO	On	DIMSE1	Off
DIMASZ	0.1800	DIMSE2	Off
DIMAZIN	0	DIMSHO	On
DIMBLK	"" (None)	DIMSOXD	Off
DIMBLK	"" (None)	DIMSTYLE	"Standard"
DIMBLK2	"" (None)	DIMTAD	0
DIMCEN	0.0900	DIMTFAC	1.0000
DIMCLRD	0	DIMTIH	On
DIMCLRE	0	DIMTIX	Off
DIMCLRT	0	DIMTM	0.0000
DIMDLE	0.0000	DIMTOFL	Off
DIMDLI	0.3800	DIMTOH	On
DIMEXE	0.1800	DIMTOL	Off
DIMEXO	0.0625	DIMTP	0.0000
DIMGAP	0.0900	DIMTSZ	0.0000
DIMLFAC	1.0000	DIMTVP	0.0000
DIMLIM	Off	DIMTXT	0.18
DIMLIND	-2	DIMUNIT	2
DIMLINE	-2	DIMZIN	0

DRAGMODE	Auto
DWGCHECK	0
DWGCODEPAGE	ANSI_1252
ELEV	Elevation 0.0, thickness 0.0
EXPERT	0
FACETRES	0.5
FILL	On
FILLET	Radius 0.5
FONTALT	Simplex.Shx
GRID	Off, spacing at 0.5, 0.5
GRIPS	Enabled; warm: blue; hot: red; 5 pixels
Highlighting	Enabled
INETLOCATION	C:\Program Files\Acad2000\Home.Htm
ISOPLANE	Left
LAYER	Current/only layer is "0", On, with color 7 (white) and line-type "CONTINUOUS"
LIMITS	Off, drawing limits (0.0, 0.0) to (12.0, 9.0)
LINETYPE	Object linetype "BYLAYER", no loaded linetypes other than "CONTINUOUS"
LTSCALE	1.0
MAXSORT	200
MENU	"acad"
MIRROR	Text mirrored with other objects
Object selection	Pick box size 3 pixels
ORTHO	Off
OSNAP	Intersection
PLINE	Line-width 0.0
Polar Angle	15
POINT	Display mode 0, size 0

QTEXT	Off
REGENAUTO	On
SKETCH	Record increment 0.10, producing lines
SHADE	Rendering type 3, percent diffuse reflection 70
SNAP	Off, spacing at 0.5, 0.5; standard nap, base point at 0.00, 0.00; rotation 0.0°
SPACE	Model
Spline curves	Frame off, segments 8, spline type = cubic
STYLE	One defined text style ("STANDARD"), using font file "txt", with variable height, width factor 1.0, hor-izontal orientation.
Surfaces	6 tabulations in M and N directions, 6 segments for smoothing in U and V directions, smooth surface type = cubic B-spline
TABLET	Off
TEXT	Style "STANDARD", height 0.20, rotation 0.0°
TILEMODE	On
TIME	User elapsed timer on
TRACE	Width 0.05
UCS	Current UCS equivalent to World, auto plan view off, coordinate system icon on (at origin)
UNITS (angular)	Decimal degrees, 0 decimal places, angle 0 direction is to the right angles increase counterclockwise
UNITS (linear)	Decimal, 4 decimal places
Viewing modes	One active viewport, plan view, perspective off, tar-get point (0, 0, 0), front and back clipping off, lens length 50mm, twist angle 0.0, fast zoom on, circle zoom percent 100, worldview 0
ZOOM	To drawing limits

APPENDIX

E

Hatch Patterns

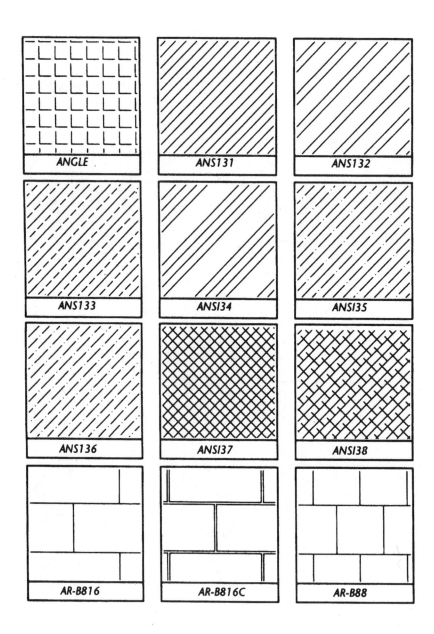

ANGLE   ANSI31   ANSI32

ANSI33   ANSI34   ANSI35

ANSI36   ANSI37   ANSI38

AR-B816   AR-B816C   AR-B88

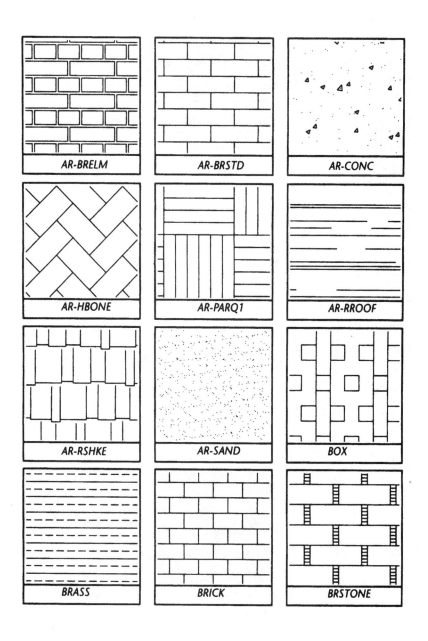

AR-BRELM	AR-BRSTD	AR-CONC
AR-HBONE	AR-PARQ1	AR-RROOF
AR-RSHKE	AR-SAND	BOX
BRASS	BRICK	BRSTONE

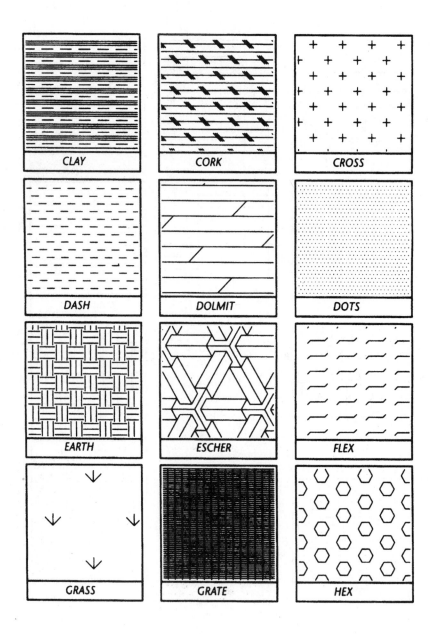

CLAY	CORK	CROSS
DASH	DOLMIT	DOTS
EARTH	ESCHER	FLEX
GRASS	GRATE	HEX

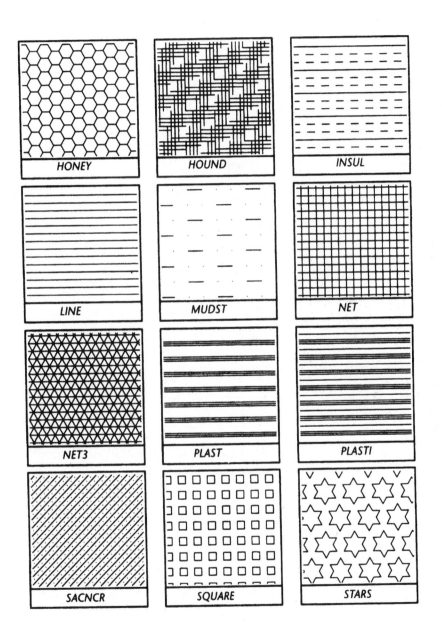

HONEY HOUND INSUL

LINE MUDST NET

NET3 PLAST PLASTI

SACNCR SQUARE STARS

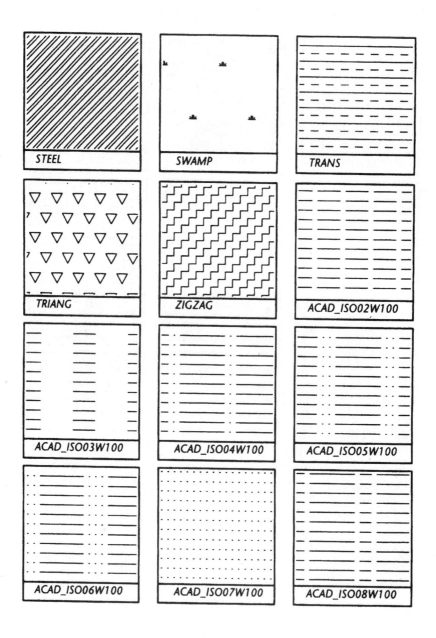

STEEL

SWAMP

TRANS

TRIANG

ZIGZAG

ACAD_ISO02W100

ACAD_ISO03W100

ACAD_ISO04W100

ACAD_ISO05W100

ACAD_ISO06W100

ACAD_ISO07W100

ACAD_ISO08W100

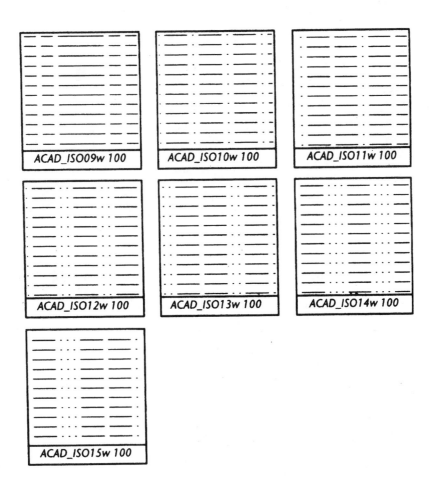

AutoCAD Linetypes

Standard Linetypes

AutoCAD provides a library of standard linetypes in the *acad.lin* file.

Linetype	
Border	—— —— · —— —— · —— —— · ——
Border2	— — — · — — — · — — — · —
BorderX2	———— ———— · ———— · ————
Center	—— — —— — —— — —— — ——
Center2	— — — — — — — — — — —
CenterX2	———— — ———— — ————
Dashdot	—— · —— · —— · —— · —— · ——
Dashdot2	—·—·—·—·—·—·—·—·—·—
DashdotX2	—— · —— · —— · —— · ——
Dashed	—— —— —— —— —— —— ——
Dashed2	— — — — — — — — — —
DashedX2	———— ———— ———— ————
Divide	—— · · —— · · —— · · —— · · ——
Divide2	—··—··—··—··—··—··
DivideX2	———— · · ———— · · ————
Dot	· ·
Dot2	···
DotX2	· · · · · · · · · · ·
Hidden	— — — — — — — — — — — —
Hidden2	--------------------------------
HiddenX2	—— —— —— —— —— ——
Phantom	———— —— —— ———— —— ——
Phantom2	—— — — —— — — —— — —
PhantomX2	———— —— —— ————
ACAD_ISO02W1000	—— —— —— —— —— —— —
ACAD_ISO03W1000	— — —— — —— — —
ACAD_ISO04W1000	———— · ———— · ———— ·
ACAD_ISO05W1000	———— · · ———— · · ————
ACAD_ISO06W1000	———— · · · ———— · · · ————
ACAD_ISO07W1000	· ·
ACAD_ISO08W1000	——— ——— — — ——— — —
ACAD_ISO09W1000	—— — — —— — — —— — —
ACAD_ISO10W1000	—— · —— · —— · —— · ——
ACAD_ISO11W1000	—— —— —— —— —— ——
ACAD_ISO12W1000	—— · · —— · · —— · · —— ·
ACAD_ISO13W1000	—— —— —— —— —— ——
ACAD_ISO14W1000	——— · ——— · ——— · ———
ACAD_ISO15W1000	—— —— · · —— —— · · ——

Complex Linetypes

AutoCAD provides some sample complex linetypes in the *ltypeshp.lin* file.

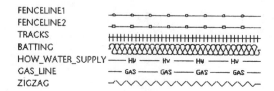

Linetype	
FENCELINE1	—o——o——o——o——o——o——o—
FENCELINE2	—□——□——□——□——□——□—
TRACKS	++++++++++++++++++++++++++++++++
BATTING	∞∞∞∞∞∞∞∞∞∞∞∞∞∞∞∞∞∞∞∞
HOW_WATER_SUPPLY	—— HW —— HW —— HW —— HW —
GAS_LINE	—— GAS —— GAS —— GAS —— GAS —
ZIGZAG	∿∿∿∿∿∿∿∿∿∿∿∿∿∿∿∿∿

G

Tables

INCHES TO MILLIMETRES

in.	mm	in.	mm	in.	mm	in.	mm
1	25.4	26	660.4	51	1295.4	76	1930.4
2	50.8	27	685.8	52	1320.8	77	1955.8
3	76.2	28	711.2	53	1346.2	78	1981.2
4	101.6	29	736.6	54	1371.6	79	2006.6
5	127.0	30	762.0	55	1397.0	80	2032.0
6	152.4	31	787.4	56	1422.4	81	2057.4
7	177.8	32	812.8	57	1447.8	82	2082.8
8	203.2	33	838.2	58	1473.2	83	2108.2
9	228.6	34	863.6	59	1498.6	84	2133.6
10	254.0	35	889.0	60	1524.0	85	2159.0
11	279.4	36	914.4	61	1549.4	86	2184.4
12	304.8	37	939.8	62	1574.8	87	2209.8
13	330.2	38	965.2	63	1600.2	88	2235.2
14	355.6	39	990.6	64	1625.6	89	2260.6
15	381.0	40	1016.0	65	1651.0	90	2286.0
16	406.4	41	1041.4	66	1676.4	91	2311.4
17	431.8	42	1066.8	67	1701.8	92	2336.8
18	457.2	43	1092.2	68	1727.2	93	2362.2
19	482.6	44	1117.6	69	1752.6	94	2387.6
20	508.0	45	1143.0	70	1778.0	95	2413.0
21	533.4	46	1168.4	71	1803.4	96	2438.4
22	558.8	47	1193.8	72	1828.8	97	2463.8
23	584.2	48	1219.2	73	1854.2	98	2489.2
24	609.6	49	1244.6	74	1879.6	99	2514.6
25	635.0	50	1270.0	75	1905.0	100	2540.0

The above table is exact on the basis: 1 in. = 25.4 mm

MILLIMETRES TO INCHES

mm	in.	mm	in.	mm	in.	mm	in.
1	0.039370	26	1.023622	51	2.007874	76	2.992126
2	0.078740	27	1.062992	52	2.047244	77	3.031496
3	0.118110	28	1.102362	53	2.086614	78	3.070866
4	0.157480	29	1.141732	54	2.125984	79	3.110236
5	0.196850	30	1.181102	55	2.165354	80	3.149606
6	0.236220	31	1.220472	56	2.204724	81	3.188976
7	0.275591	32	1.259843	57	2.244094	82	3.228346
8	0.314961	33	1.299213	58	2.283465	83	3.267717
9	0.354331	34	1.338583	59	2.322835	84	3.307087
10	0.393701	35	1.377953	60	2.362205	85	3.346457
11	0.433071	36	1.417323	61	2.401575	86	3.385827
12	0.472441	37	1.456693	62	2.440945	87	3.425197
13	0.511811	38	1.496063	63	2.480315	88	3.464567
14	0.551181	39	1.535433	64	2.519685	89	3.503937
15	0.590551	40	1.574803	65	2.559055	90	3.543307
16	0.629921	41	1.614173	66	2.598425	91	3.582677
17	0.669291	42	1.653543	67	2.637795	92	3.622047
18	0.708661	43	1.692913	68	2.677165	93	3.661417
19	0.748031	44	1.732283	69	2.716535	94	3.700787
20	0.787402	45	1.771654	70	2.755906	95	3.740157
21	0.826772	46	1.811024	71	2.795276	96	3.779528
22	0.866142	47	1.850394	72	2.834646	97	3.818898
23	0.905512	48	1.889764	73	2.874016	98	3.858268
24	0.944882	49	1.929134	74	2.913386	99	3.897638
25	0.984252	50	1.968504	75	2.952756	100	3.937008

The above table is approximate on the basis: 1 in. = 25.4 mm, 1/25.4 = 0.039370078740+

From Goetsch, Nelson, and Chalk, *Technical Drawing*, 3rd edition, copyright 1994 by Delmar Publishers Inc.

INCH/METRIC – EQUIVALENTS						
Fraction	**Decimal Equivalent**		**Fraction**	**Decimal Equivalent**		
	Customary (in.)	**Metric (mm)**		**Customary (in.)**	**Metric (mm)**	
1/64 —— .015625		0.3969	33/64 —— .515625		13.0969	
1/32 ———— .03125		0.7938	17/32 ———— .53125		13.4938	
3/64 —— .046875		1.1906	35/64 —— .546875		13.8906	
1/16 ———————— .0625		1.5875	9/16 ———————— .5625		14.2875	
5/64 —— .078125		1.9844	37/64 —— .578125		14.6844	
3/32 ———— .09375		2.3813	19/32 ———— .59375		15.0813	
7/64 —— .109375		2.7781	39/64 —— .609375		15.4781	
1/8 ———————————— .1250		3.1750	5/8 ———————————— .6250		15.8750	
9/64 —— .140625		3.5719	41/64 —— .640625		16.2719	
5/32 ———— .15625		3.9688	21/32 ———— .65625		16.6688	
11/64 —— .171875		4.3656	43/64 —— .671875		17.0656	
3/16 ———————— .1875		4.7625	11/16 ———————— .6875		17.4625	
13/64 —— .203125		5.1594	45/64 —— .703125		17.8594	
7/32 ———— .21875		5.5563	23/32 ———— .71875		18.2563	
15/64 —— .234375		5.9531	47/64 —— .734375		18.6531	
1/4 ———————————— .250		6.3500	3/4 ———————————— .750		19.0500	
17/64 —— .265625		6.7469	49/64 —— .765625		19.4469	
9/32 ———— .28125		7.1438	25/32 ———— .78125		19.8438	
19/64 —— .296875		7.5406	51/64 —— .796875		20.2406	
5/16 ———————— .3125		7.9375	13/16 ———————— .8125		20.6375	
21/64 —— .328125		8.3384	53/64 —— .828125		21.0344	
11/32 ———— .34375		8.7313	27/32 ———— .84375		21.4313	
23/64 —— .359375		9.1281	55/64 —— .859375		21.8281	
3/8 ———————————— .3750		9.5250	7/8 ———————————— .8750		22.2250	
25/64 —— .390625		9.9219	57/64 —— .890625		22.6219	
13/32 ———— .40625		10.3188	29/32 ———— .90625		23.0188	
27/64 —— .421875		10.7156	59/64 —— .921875		23.4156	
7/16 ———————— .4375		11.1125	15/16 ———————— .9375		23.8125	
29/64 —— .453125		11.5094	61/64 —— .953125		24.2094	
15/32 ———— .46875		11.9063	31/32 ———— .96875		24.6063	
31/64 —— .484375		12.3031	63/64 —— .984375		25.0031	
1/2 ———————————— .500		12.7000	1 ———————————— 1.000		25.4000	

From Nelson, *Drafting for Trades and Industry—Basic Skills.* Delmar Publishers Inc.

METRIC EQUIVALENTS

LENGTH

U.S. to Metric

1 inch = 2.540 centimetres
1 foot = .305 metre
1 yard = .914 metre
1 mile = 1.609 kilometres

Metric to U.S.

1 millimetre = .039 inch
1 centimetre = .394 inch
1 metre = 3.281 feet or 1.094 yards
1 kilometre = .621 mile

AREA

1 inch^2 = 6.451 centimetre^2
1 foot^2 = .093 metre^2
1 yard^2 = .836 metre^2
1 acre^2 = 4,046.873 metre^2

1 millimetre^2 = .00155 inch^2
1 centimetre^2 = .155 inch^2
1 metre^2 = 10.764 foot^2 or 1.196 yard^2
1 kilometre^2 = .386 mile^2 or 247.04 acre^2

VOLUME

1 inch^3 = 16.387 centimetre^3
1 foot^3 = .028 metre^3
1 yard^3 = .764 metre^3
1 quart = .946 litre
1 gallon = .003785 metre^3

1 centimetre^3 = 0.61 inch^3
1 metre^3 = 35.314 foot^3 or 1.308 yard^3
1 litre = .2642 gallons
1 litre = 1.057 quarts
1 metre^3 = 264.02 gallons

WEIGHT

1 ounce = 28.349 grams
1 pound = .454 kilogram
1 ton = .907 metric ton

1 gram = .035 ounce
1 kilogram = 2.205 pounds
1 metric ton = 1.102 tons

VELOCITY

1 foot/second = .305 metre/second
1 mile/hour = .447 metre/second

1 metre/second = 3.281 feet/second
1 kilometre/hour = .621 mile/second

ACCELERATION

1 inch/second^2 = .0254 metre/second^2
1 foot/second^2 = .305 metre/second^2

1 metre/second^2 = 3.278 feet/second^2

FORCE

N (newton) = basic unit of force, kg-m/s^2. A mass of one kilogram (1 kg) exerts a gravitational force of 9.8 N (theoretically 9.80665 N) at mean sea level.

From Goetsch, Nelson, and Chalk, *Technical Drawing*, 3rd edition, copyright 1994 by Delmar Publishers Inc.

MULTIPLIERS FOR DRAFTERS

Multiply	By	To Obtain
Acres	43,560	Square feet
Acres	4047	Square metres
Acres	1.562×10^{-3}	Square miles
Acres	4840	Square yards
Acre—feet	43,560	Cubic feet
Atmospheres	76.0	Cms. of mercury
Atmospheres	29.92	Inches of mercury
Atmospheres	33.90	Feet of water
Atmospheres	10,333	Kgs./sq. metre
Atmospheres	14.70	Lbs./sq. inch
Atmospheres	1.058	Tons/sq. ft.
Board feet	144 sq. in. × 1 in.	Cubic inches
British Thermal Units	0.2520	Kilogram—calories
British Thermal Units	777.5	Foot—lbs.
British Thermal Units	3.927×10^{-4}	Horsepower—hrs.
British Thermal Units	107.5	Kilogram—metres
British Thermal Units	2.928×10^{-4}	Kilowatt—hrs.
B.T.U./min.	12.96	Foot—lbs./sec.
B.T.U./min.	0.02356	Horsepower
B.T.U./min.	0.01757	Kilowatts
B.T.U./min.	17.57	Watts
Cubic centimetres	3.531×10^{-5}	Cubic feet
Cubic centimetres	6.102×10^{-2}	Cubic inches
Cubic centimetres	10^{-6}	Cubic metres
Cubic centimetres	1.308×10^{-6}	Cubic yards
Cubic centimetres	2.642×10^{-4}	Gallons
Cubic centimetres	10^{-3}	Litres
Cubic centimetres	2.113×10^{-3}	Pints (liq.)
Cubic centimetres	1.057×10^{-3}	Quarts (liq.)
Cubic feet	2.832×10^{4}	Cubic cms.
Cubic feet	1728	Cubic inches
Cubic feet	0.02832	Cubic metres
Cubic feet	0.03704	Cubic yards
Cubic feet	7.48052	Gallons
Cubic feet	28.32	Litres
Cubic feet	59.84	Pints (liq.)
Cubic feet	29.92	Quarts (liq.)
Cubic feet/min.	472.0	Cubic cms./sec.
Cubic feet/min.	0.1247	Gallons/sec.
Cubic feet/min.	0.4720	Litres/sec.
Cubic feet/min.	62.43	Pounds of water/min.
Cubic feet/sec.	0.646317	Millions gals./day
Cubic feet/sec.	448.831	Gallons/min.
Cubic inches	16.39	Cubic centimetres
Cubic inches	5.787×10^{-4}	Cubic feet
Cubic inches	1.639×10^{-5}	Cubic metres
Cubic inches	2.143×10^{-5}	Cubic yards
Cubic inches	4.329×10^{-3}	Gallons
Cubic inches	1.639×10^{-2}	Litres
Cubic inches	0.03463	Pints (liq.)
Cubic inches	0.01732	Quarts (liq.)
Cubic metres	10^{6}	Cubic centimetres
Cubic metres	35.31	Cubic feet
Cubic metres	61.023	Cubic inches
Cubic metres	1.308	Cubic yards
Cubic metres	264.2	Gallons
Cubic metres	10^{3}	Litres
Cubic metres	2113	Pints (liq.)
Cubic metres	1057	Quarts (liq.)
Degrees (angle)	60	Minutes
Degrees (angle)	0.01745	Radians
Degrees (angle)	3600	Seconds
Degrees/sec.	0.01745	Radians/sec.
Degrees/sec.	0.1667	Revolutions/min.

Multiply	By	To Obtain
Degrees/sec.	0.002778	Revolutions/sec.
Fathoms	6	Feet
Feet	30.48	Centimetres
Feet	12	Inches
Feet	0.3048	Metres
Foot—pounds	1.286×10^{-3}	British Thermal Units
Foot—pounds	5.050×10^{-7}	Horsepower—hrs.
Foot—pounds	3.241×10^{-4}	Kilogram—calories
Foot—pounds	0.1383	Kilogram—metres
Foot—pounds	3.766×10^{-7}	Kilowatt—hrs.
Foot—pounds/min.	1.286×10^{-3}	B.T.U./min.
Foot—pounds/min.	0.01667	Foot—pounds/sec.
Foot—pounds/min.	3.030×10^{-5}	Horsepower
Foot—pounds/min.	3.241×10^{-4}	Kg.—calories/min.
Foot—pounds/min.	2.260×10^{-5}	Kilowatts
Foot—pounds/sec.	7.717×10^{-2}	B.T.U./min.
Foot—pounds/sec.	1.818×10^{-3}	Horsepower
Foot—pounds/sec.	1.945×10^{-2}	Kg.—calories/min.
Foot—pounds/sec.	1.356×10^{-3}	Kilowatts
Gallons	3785	Cubic centimetres
Gallons	0.1337	Cubic feet
Gallons	231	Cubic inches
Gallons	3.785×10^{-3}	Cubic metres
Gallons	4.951×10^{-3}	Cubic yards
Gallons	3.785	Litres
Gallons	8	Pints (liq.)
Gallons	4	Quarts (liq.)
Gallons—Imperial	1.20095	U.S. gallons
Gallons—U.S.	0.83267	Imperial gallons
Gallons water	8.3453	Pounds of water
Horsepower	42.44	B.T.U./min.
Horsepower	33,000	Foot—lbs./min.
Horsepower	550	Foot—lbs./sec.
Horsepower	1.014	Horsepower (metric)
Horsepower	10.70	Kg.—calories/min.
Horsepower	0.7457	Kilowatts
Horsepower	745.7	Watts
Horsepower—hours	2547	B.T.U.
Horsepower—hours	1.98×10^{6}	Foot—lbs.
Horsepower—hours	641.7	Kilogram—calories
Horsepower—hours	2.737×10^{5}	Kilogram—metres
Horsepower—hours	0.7457	Kilowatt—hours
Kilometres	10^{5}	Centimetres
Kilometres	3281	Feet
Kilometres	10^{3}	Metres
Kilometres	0.6214	Miles
Kilometres	1094	Yards
Kilowatts	56.92	B.T.U./min.
Kilowatts	4.425×10^{4}	Foot—lbs./min.
Kilowatts	737.6	Foot—lbs./sec.
Kilowatts	1.341	Horsepower
Kilowatts	14.34	Kg.—calories/min.
Kilowatts	10^{3}	Watts
Kilowatt—hours	3415	B.T.U.
Kilowatt—hours	2.655×10^{6}	Foot—lbs.
Kilowatt—hours	1.341	Horsepower—hrs.
Kilowatt—hours	860.5	Kilogram—calories
Kilowatt—hours	3.671×10^{5}	Kilogram—metres
Lumber Width (in.) × $\dfrac{\text{Thickness (in.)}}{12}$	Length (ft.)	Board feet
Metres	100	Centimetres
Metres	3.281	Feet
Metres	39.37	Inches

From Goetsch, Nelson, and Chalk, *Technical Drawing*, 3rd edition, copyright 1994 by Delmar Publishers Inc.

MULTIPLIERS FOR DRAFTERS (cont'd)

Multiply	By	To Obtain	Multiply	By	To Obtain
Metres	10^{-3}	Kilometres	Pounds (troy)	373.24177	Grams
Metres	10^3	Millimetres	Pounds (troy)	0.822857	Pounds (avoir.)
Metres	1.094	Yards	Pounds (troy)	13.1657	Ounces (avoir.)
Metres/min.	1.667	Centimetres/sec.	Pounds (troy)	3.6735×10^{-4}	Tons (long)
Metres/min.	3.281	Feet/min.	Pounds (troy)	4.1143×10^{-4}	Tons (short)
Metres/min.	0.05468	Feet/sec.	Pounds (troy)	3.7324×10^{-4}	Tons (metric)
Metres/min.	0.06	Kilometres/hr.	Quadrants (angle)	90	Degrees
Metres/min.	0.03728	Miles/hr.	Quadrants (angle)	5400	Minutes
Metres/sec.	196.8	Feet/min.	Quadrants (angle)	1.571	Radians
Metres/sec.	3.281	Feet/sec.	Radians	57.30	Degrees
Metres/sec.	3.6	Kilometres/hr.	Radians	3438	Minutes
Metres/sec.	0.06	Kilometres/min.	Radians	0.637	Quadrants
Metres/sec.	2.237	Miles/hr.	Radians/sec.	57.30	Degrees/sec.
Metres/sec.	0.03728	Miles/min.	Radians/sec.	0.1592	Revolutions/sec.
Microns	10^{-6}	Metres	Radians/sec.	9.549	Revolutions/min.
Miles	5280	Feet	Radians/sec./sec.	573.0	Revs./min./min.
Miles	1.609	Kilometres	Radians/sec./sec.	0.1592	Revs./sec./sec.
Miles	1760	Yards	Reams	500	Sheets
Miles/hr.	1.609	Kilometres/hr.	Revolutions	360	Degrees
Miles/hr.	0.8684	Knots	Revolutions	4	Quadrants
Minutes (angle)	2.909×10^{-4}	Radians	Revolutions	6.283	Radians
Ounces	16	Drams	Revolutions/min.	6	Degrees/sec.
Ounces	437.5	Grains	Square yards	2.066×10^{-4}	Acres
Ounces	0.0625	Pounds	Square yards	9	Square feet
Ounces	28.349527	Grams	Square yards	0.8361	Square metres
Ounces	0.9115	Ounces (troy)	Square yards	3.228×10^{-7}	Square miles
Ounces	2.790×10^{-5}	Tons (long)	Temp. (°C.) + 273	1	Abs. temp. (°C.)
Ounces	2.835×10^{-5}	Tons (metric)	Temp. (°C.) + 17.78	1.8	Temp. (°F.)
Ounces (troy)	480	Grains	Temp. (°F.) + 460	1	Abs. temp. (°F.)
Ounces (troy)	20	Pennyweights (troy)	Temp. (°F.) − 32	5/9	Temp. (°C.)
Ounces (troy)	0.08333	Pounds (troy)	Watts	0.05692	B.T.U./min.
Ounces (troy)	31.103481	Grams	Watts	44.26	Foot—pounds/min.
Ounces (troy)	1.09714	Ounces (avoir.)	Watts	0.7376	Foot—pounds/sec.
Ounces (fluid)	1.805	Cubic inches	Watts	1.341×10^{-3}	Horsepower
Ounces (fluid)	0.02957	Litres	Watts	0.01434	Kg.—calories/min.
Ounces/sq. inch	0.0625	Lbs./sq. inch	Watts	10^{-3}	Kilowatts
Pounds	16	Ounces	Watt—hours	3.415	B.T.U.
Pounds	256	Drams	Watt—hours	2655	Foot—pounds
Pounds	7000	Grains	Watt—hours	1.341×10^{-3}	Horsepower—hrs.
Pounds	0.0005	Tons (short)	Watt—hours	0.8605	Kilogram—calories
Pounds	453.5924	Grams	Watt—hours	367.1	Kilogram—metres
Pounds	1.21528	Pounds (troy)	Watt—hours	10^{-3}	Kilowatt—hours
Pounds	14.5833	Ounces (troy)	Yards	91.44	Centimetres
Pounds (troy)	5760	Grains	Yards	3	Feet
Pounds (troy)	240	Pennyweights (troy)	Yards	36	Inches
Pounds (troy)	12	Ounces (troy)	Yards	0.9144	Metres

From Goetsch, Nelson, and Chalk, *Technical Drawing*, 3rd edition, copyright 1994 by Delmar Publishers Inc.

CIRCUMFERENCES AND AREAS OF CIRCLES
From 1/64 to 50, Diameter

Dia.	Circum.	Area	Dia.	Circum.	Area	Dia.	Circum.	Area	Dia.	Circum.	Area
1/64	.04909	.00019	8	25.1327	50.2655	17	53.4071	226.980	26	81.6814	530.929
1/32	.09818	.00077	8 1/8	25.5254	51.8485	17 1/8	53.7998	230.330	26 1/8	82.0741	536.047
1/16	.19635	.00307	8 1/4	25.9181	53.4562	17 1/4	54.1925	233.705	26 1/4	82.4668	541.188
1/8	.39270	.01227	8 3/8	26.3108	55.0883	17 3/8	54.5852	237.104	26 3/8	82.8595	546.355
3/16	.58905	.02761	8 1/2	26.7035	56.7450	17 1/2	54.9779	240.528	26 1/2	83.2522	551.546
1/4	.78540	.04909	8 5/8	27.0962	58.4262	17 5/8	55.3706	243.977	26 5/8	83.6449	556.761
5/16	.98175	.07670	8 3/4	27.4889	60.1321	17 3/4	55.7633	247.450	26 3/4	84.0376	562.002
3/8	1.1781	.11045	8 7/8	27.8816	61.8624	17 7/8	56.1560	250.947	26 7/8	84.4303	567.266
7/16	1.3744	.15033	9	28.2743	63.6173	18	56.5487	254.469	27	84.8230	572.555
1/2	1.5708	.19635	9 1/8	28.6670	65.3967	18 1/8	56.9414	258.016	27 1/8	85.2157	577.869
9/16	1.7671	.24850	9 1/4	29.0597	67.2007	18 1/4	57.3341	261.587	27 1/4	85.6084	583.207
5/8	1.9635	.30680	9 3/8	29.4524	69.0292	18 3/8	57.7268	265.182	27 3/8	86.0011	588.570
11/16	2.1598	.37122	9 1/2	29.8451	70.8822	18 1/2	58.1195	268.803	27 1/2	86.3938	593.957
3/4	2.3562	.44179	9 5/8	30.2378	72.7597	18 5/8	58.5122	272.447	27 5/8	86.7865	599.369
13/16	2.5525	.51849	9 3/4	30.6305	74.6619	18 3/4	58.9049	276.117	27 3/4	87.1792	604.806
7/8	2.7489	.60132	9 7/8	31.0232	76.5886	18 7/8	59.2976	279.810	27 7/8	87.5719	610.267
15/16	2.9452	.69029	10	31.4159	78.5398	19	59.6903	283.529	28	87.9646	615.752
1	3.1416	.78540	10 1/8	31.8086	80.5156	19 1/8	60.0830	287.272	28 1/8	88.3573	621.262
1 1/8	3.5343	.99402	10 1/4	32.2013	82.5159	19 1/4	60.4757	291.039	28 1/4	88.7500	626.797
1 1/4	3.9270	1.2272	10 3/8	32.5940	84.5408	19 3/8	60.8684	294.831	28 3/8	89.1427	632.356
1 3/8	4.3197	1.4849	10 1/2	32.9867	86.5902	19 1/2	61.2611	298.648	28 1/2	89.5354	637.940
1 1/2	4.7124	1.7671	10 5/8	33.3794	88.6641	19 5/8	61.6538	302.489	28 5/8	89.9281	643.548
1 5/8	5.1051	2.0739	10 3/4	33.7721	90.7626	19 3/4	62.0465	306.354	28 3/4	90.3208	649.181
1 3/4	5.4978	2.4053	10 7/8	34.1648	92.8856	19 7/8	62.4392	310.245	28 7/8	90.7135	654.838
1 7/8	5.8905	2.7612	11	34.5575	95.0332	20	62.8319	314.159	29	91.1062	660.520
2	6.2832	3.1416	11 1/8	34.9502	97.2053	20 1/8	63.2246	318.099	29 1/8	91.4989	666.226
2 1/8	6.6759	3.5466	11 1/4	35.3429	99.4020	20 1/4	63.6173	322.062	29 1/4	91.8916	671.957
2 1/4	7.0686	3.9761	11 3/8	35.7356	101.623	20 3/8	64.0100	326.051	29 3/8	92.2843	677.713
2 3/8	7.4613	4.4301	11 1/2	36.1283	103.869	20 1/2	64.4027	330.064	29 1/2	92.6770	683.493
2 1/2	7.8540	4.9087	11 5/8	36.5210	106.139	20 5/8	64.7954	334.101	29 5/8	93.0697	689.297
2 5/8	8.2467	5.4119	11 3/4	36.9137	108.434	20 3/4	65.1881	338.163	29 3/4	93.4624	695.127
2 3/4	8.6394	5.9396	11 7/8	37.3064	110.753	20 7/8	65.5808	342.250	29 7/8	93.8551	700.980
2 7/8	9.0321	6.4918	12	37.6991	113.097				30	94.2478	706.858
3	9.4248	7.0686	12 1/8	38.0918	115.466	21	65.9735	346.361	30 1/8	94.6405	712.761
3 1/8	9.8175	7.6699	12 1/4	38.4845	117.859	21 1/8	66.3662	350.496	30 1/4	95.0332	718.689
3 1/4	10.2102	8.2958	12 3/8	38.8772	120.276	21 1/4	66.7589	354.656	30 3/8	95.4259	724.640
3 3/8	10.6029	8.9462	12 1/2	39.2699	122.718	21 3/8	67.1516	358.841	30 1/2	95.8186	730.617
3 1/2	10.9956	9.6211	12 5/8	39.6626	125.185	21 1/2	67.5442	363.050	30 5/8	96.2113	736.618
3 5/8	11.3883	10.3206	12 3/4	40.0553	127.676	21 5/8	67.9369	367.284	30 3/4	96.6040	742.643
3 3/4	11.7810	11.0447	12 7/8	40.4480	130.191	21 3/4	68.3296	371.542	30 7/8	96.9967	748.693
3 7/8	12.1737	11.7932	13	40.8407	132.732	21 7/8	68.7223	375.825	31	97.3894	754.768
4	12.5664	12.5664	13 1/8	41.2334	135.297	22	69.1150	380.133	31 1/8	97.7821	760.867
4 1/8	12.9591	13.3640	13 1/4	41.6261	137.886	22 1/8	69.5077	384.465	31 1/4	98.1748	766.990
4 1/4	13.3518	14.1863	13 3/8	42.0188	140.500	22 1/4	69.9004	388.821	31 3/8	98.5675	773.139
4 3/8	13.7445	15.0330	13 1/2	42.4115	143.139	22 3/8	70.2931	393.203	31 1/2	98.9602	779.311
4 1/2	14.1372	15.9043	13 5/8	42.8042	145.802	22 1/2	70.6858	397.608	31 5/8	99.3529	785.509
4 5/8	14.5299	16.8002	13 3/4	43.1969	148.489	22 5/8	71.0785	402.038	31 3/4	99.7456	791.731
4 3/4	14.9226	17.7206	13 7/8	43.5896	151.201	22 3/4	71.4712	406.493	31 7/8	100.1383	797.977
4 7/8	15.3153	18.6655	14	43.9823	153.938	22 7/8	71.8639	410.972			
5	15.7080	19.6350	14 1/8	44.3750	156.699	23	72.2566	415.476	32	100.5310	804.248
5 1/8	16.1007	20.6290	14 1/4	44.7677	159.485	23 1/8	72.6493	420.004	32 1/8	100.9237	810.543
5 1/4	16.4934	21.6476	14 3/8	45.1604	162.295	23 1/4	73.0420	424.557	32 1/4	101.3164	816.863
5 3/8	16.8861	22.6906	14 1/2	45.5531	165.130	23 3/8	73.4347	429.134	32 3/8	101.7091	823.208
5 1/2	17.2788	23.7580	14 5/8	45.9458	167.989	23 1/2	73.8274	433.736	32 1/2	102.1018	829.577
5 5/8	17.6715	24.8505	14 3/4	46.3385	170.873	23 5/8	74.2201	438.363	32 5/8	102.4945	835.971
5 3/4	18.0642	25.9672	14 7/8	46.7312	173.782	23 3/4	74.6128	443.014	32 3/4	102.8872	842.389
5 7/8	18.4569	27.1085	15	47.1239	176.715	23 7/8	75.0055	447.689	32 7/8	103.2799	848.831
6	18.8496	28.2743	15 1/8	47.5166	179.672	24	75.3982	452.389	33	103.6726	855.299
6 1/8	19.2423	29.4647	15 1/4	47.9094	182.654	24 1/8	75.7909	457.114	33 1/8	104.0653	861.791
6 1/4	19.6350	30.6796	15 3/8	48.3020	185.661	24 1/4	76.1836	461.863	33 1/4	104.4580	868.307
6 3/8	20.0277	31.9191	15 1/2	48.6947	188.692	24 3/8	76.5763	466.637	33 3/8	104.8507	874.848
6 1/2	20.4204	33.1831	15 5/8	49.0874	191.748	24 1/2	76.9690	471.435	33 1/2	105.2434	881.413
6 5/8	20.8131	34.4716	15 3/4	49.4801	194.828	24 5/8	77.3617	476.258	33 5/8	105.6361	888.003
6 3/4	21.2058	35.7847	15 7/8	49.8728	197.933	24 3/4	77.7544	481.106	33 3/4	106.0288	894.618
6 7/8	21.5985	37.1223	16	50.2655	201.062	24 7/8	78.1471	485.977	33 7/8	106.4215	901.257
7	21.9912	38.4845	16 1/8	50.6582	204.216	25	78.5398	490.874	34	106.8142	907.920
7 1/8	22.3839	39.8712	16 1/4	51.0509	207.394	25 1/8	78.9325	495.795	34 1/8	107.2069	914.609
7 1/4	22.7765	41.2825	16 3/8	51.4436	210.597	25 1/4	79.3252	500.740	34 1/4	107.5996	921.321
7 3/8	23.1692	42.7183	16 1/2	51.8363	213.825	25 3/8	79.7179	505.711	34 3/8	107.9923	928.058
7 1/2	23.5619	44.1787	16 5/8	52.2290	217.077	25 1/2	80.1106	510.705	34 1/2	108.3850	934.820
7 5/8	23.9546	45.6636	16 3/4	52.6217	220.353	25 5/8	80.5033	515.724	34 5/8	108.7777	941.607
7 3/4	24.3473	47.1730	16 7/8	53.0144	223.654	25 3/4	80.8960	520.768	34 3/4	109.1704	948.417
7 7/8	24.7400	48.7069				25 7/8	81.2887	525.836	34 7/8	109.5631	955.253

From Goetsch, Nelson, and Chalk, *Technical Drawing*, 3rd edition, copyright 1994 by Delmar Publishers Inc.

ARCHITECTURAL

FINAL PLOT SCALE

	SHEET SIZE				
	A 11 x 8½	B 17 x 11	C 24 x 18	D 36 x 24	E 48 x 36
1/16	176', 136'	272', 176'	384', 288'	576', 384'	768', 576'
3/32	132', 102'	204', 132'	288', 216'	432', 288'	576', 432'
1/8	88', 68'	136', 88'	192', 144'	288', 192'	384', 288'
3/16	58'-8", 45'-4"	90'-8", 58'-8"	128', 96'	192', 128'	256', 192'
1/4	44', 34'	68', 44'	96', 72'.	144', 96'	192', 144'
3/8	29'-4", 22'-8"	45'-4", 29'-4"	64', 48'	96', 64'	128', 96'
1/2	22', 17'	34', 22'	48', 36'	72', 48'	96', 72'
3/4	14'-8", 11'-4"	22'-8", 14'-8"	32', 24'	48', 32'	64', 48'
1	11', 8'-6"	17', 11'	24', 18'	36', 24'	48', 36'
1½	7'-4", 5'-8"	11'-4", 7'-4"	16', 12'	24', 16'	32', 24'
3	3'-8", 2'-10"	5'-8", 3'-8"	8', 6"	12', 8'	16,' 12'

ENGINEERING

FINAL PLOT SCALE

	SHEET SIZE				
	A 11 x 8½	B 17 x 11	C 24 x 18	D 36 x 24	E 48 x 36
10	110, 85	170, 110	240, 180	360, 240	480, 360
20	220, 170	340, 220	480, 360	720, 480	960, 720
30	330, 255	510, 330	720, 540	1080, 720	1440, 1080
40	440, 340	680, 440	960, 720	1440, 960	1920, 1440
50	550, 425	850, 550	1200, 900	1800, 1200	2400, 1800
60	660, 510	1020, 660	1440, 1080	2160, 1440	2880, 2160
100	1100, 850	1700, 1100	2400, 1800	3600, 2400	4800, 3600
Full Size	11, 8.5	17, 11	24, 18	36, 24	48, 36

From Goetsch, Nelson, and Chalk, *Technical Drawing*, 3rd edition, copyright 1994 by Delmar Publishers Inc.

Glossary

Absolute coordinates	Points designated by a specific X, Y, and Z distance from a fixed origin.
Alphanumeric	Numbers, letters, and special characters.
ANSI	American National Standards Institute. ANSI is a professional organization which sets standards.
ASCII	American Standard Code for Information Interchange. ASCII is a standard computer data communications code.
Aspect ratio	The height-to-width ratio of an image on an output device.
Attribute	Information associated with a graphic object.
Baud	Data transmission rate of a computer. The baud rate is not the same as the number of bits sent per second.
Bill of materials	A listing of the parts required to assemble an object. Bills of materials may be extracted from a database.
Bit	The smallest unit of information of a binary system.
Buffer	Memory reserved for temporary storage of data.
Bug	A malfunction or design error in computer hardware or software.
Byte	Collection of eight bits that represent one letter.
CAD	Computer-aided design or drafting.
CAM	Computer-aided manufacturing.

Command	A specific word or other entry to provide an instruction.
Configuration	Providing setup for peripheral hardware by installing proper software drivers in a program.
Coordinate	A point located by X, Y, and Z directions and represented by real numbers.
CPU	Central processing unit. The part of a computer where the arithmetic and logic functions are performed.
Crosshair	Horizontal and vertical crossed lines on the display screen that designate the current working point of the drawing.
Crosshatching	Filling in of an area with a design.
CRT	Cathode ray tube. The display screen on which the computer display is shown. A CRT is similar to a television screen.
Cursor	Display device, such as a flashing bar or box, that represents the current working point on the display.
Database	The collection of information that is used to form a drawing or used to perform program functions.
Data extraction	Retrieval of data from the database.
Default	A predetermined value for a computer function.
Digitize	Electronically tracing a drawing.
Digitizer	An electronically sensitive pad used to translate movement of a hand-held device to cursor movement on a screen. May also be used to trace drawings.
Directory	A file listing of specific files or other data.
Disk	A circular piece of plastic or aluminum with a magnetic coating used to store computer data.
Display	Area on a CRT where graphic or text images are displayed.
Dot matrix	A dot grid that is used to display images by displaying specific patterns.

Dragging	Moving of objects relative to the displacement of an input device.
Drum or roll plotter	A pen plotter that moves both the paper and pen to achieve a plot.
Dual display	Use of two display devices simultaneously.
Entity	See Object.
Extension	The last three letters that follow a period in a file name, e.g., .DWG or .EXE.
Fillet	The corner fit or radius between two nonparallel lines.
Flatbed plotter	A pen plotter in which the paper remains stationary and the pen moves in both directions to create the plot.
Floppy disk	A thin plastic disk with a magnetic coating used to store computer data.
Fonts	The design of an alphabet.
Function key	Programmable keyboard key used for a specific purpose by a software program.
Gigabyte (GB)	One gigabyte (1 KB x 1 MB) is equal to 1,073,741,824 bytes.
Grid	A series of dots arranged in a designated X and Y spacing used for reference purposes when drawing.
Hard copy	The printed or plotted copy of a drawing or data.
Hardware	The physical computer equipment such as the computer, printer, mouse, etc.
Hatching	Filling an area with a pattern.
Input device	Any device used to enter data into a computer. Examples of input devices are keyboard, mouse, and digitizing pad.
Kilobyte (KB)	A unit of 1024 bytes.
Layer	An overlay used to store specific data on. A drawing may contain several overlays or layers with data on each.

Macro	Several keystroke operations executed by a single entry.
Mainframe	The largest and most powerful computer. Mainframes are used where large amounts of data are stored, such as by government and insurance companies.
Megabyte (MB)	Unit of storage equal to 1,048,576 bytes (1KB x 1KB).
Menu	A screen listing of options in a computer program.
Microcomputer	A small desktop computer. Usually referred to as a personal computer.
Minicomputer	A mid-sized computer of a size and capability between those of a microcomputer and a mainframe.
Mirroring	Reversal of a graphic image around a specified axis.
Modem	Modulator-demodulator. A modem is used to send computer data over telephone lines.
Mouse	A small, hand-held input device that is moved around on a surface. This movement is sensed by the computer and a relative displacement of the cursor or crosshair is made.
Network	The linking together of several computer systems.
Object	A single drawing element, such as a line, circle, or arc.
OLE	Object linking and embedding.
Operating system	Computer program that acts as a translator between the computer and applications programs.
Pan	Movement around a drawing while in a specified magnification.
Pen plotter	Output device used to record a CAD drawing by moving a pen over a drawing surface to obtain a hard-copy drawing.
Peripheral	Any device used in conjunction with a computer, e.g., printer, light pen, etc.
Pixel	Dots that make up the display. The arrangement of displayed pixels determines the image displayed.

Plotter	Device used to obtain a hard copy of a graphic display. Plotters may be pen, electrostatic, or dot matrix.
Polar	Coordinate system used to specify distance and angle, defined in reference to a specified starting origin.
Program	The set of instructions that the computer uses to perform tasks. The program is also referred to as software.
Prompt	A message or symbol displayed by the computer that tells the operator that the computer is ready for input.
Puck	The hand-held input device used with a digitizing tablet to move the crosshairs on the display.
RAM	Random access memory. Temporary memory storage area in a computer.
Relative coordinates	Points that are located relative to a specified point.
Resolution	The measure of precision and clarity of a display.
ROM	Read-only memory. Permanent memory storage area in a computer.
Rubberbanding	"Stretching" a line from a fixed point to the current location of the crosshairs.
Scrolling	Vertical movement of text lines.
Snap	Drawing aid that allows entered points to "snap" to the closest point of an imaginary grid.
Software	Set of instructions used by the computer to perform a task. Software is also referred to as the program.
Stylus	A pen-like input device used with a digitizing pad in the same manner as a puck.
Terrabyte	One terrabyte (TB) is 1KB∞1GB or 1,099,511,627,776 bytes.
Windowing	Enlargement or reduction of graphic screen images.

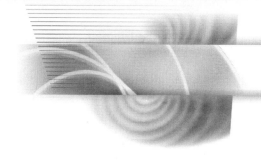